FRUIT PHYSIOLOGY AND PRODUCTION

FRUIT PHYSIOLOGY AND PRODUCTION

AMAR SINGH
M.Sc., D.Phil., FNA Sc.
Honorary Research Professor and
Honorary Senior Vice-Chairman
Siveast Consultant (U.S.A.)
Formerly Reader in Botany
University of Allahabad
ALLAHABAD (India)

KALYANI PUBLISHERS
LUDHIANA * NEW DELHI * NOIDA (U.P.)
HYDERABAD * CHENNAI * KOLKATA * CUTTACK

KALYANI PUBLISHERS

Head Office

B-I/1292, Rajinder Nagar,
Ludhiana-141 008

Administration Office :

4779, Ansari Road,
23, Daryaganj, **New Delhi**-110 002

Works

B-16, Sector 8, **Noida** (U.P.)

Branches

1, Mahalakshmi Street,
T. Nagar, **Chennai**-600 017

110/111, Bharatia Towers, Badambadi
Cuttack–9 (Orissa)

3-5-1108, Narayanaguda,
Hyderabad-500 029

10/2B, Ramanath Mazumdar Street,
Kolkata-700 009

O-27202 05

First Edition	1980
Fourth Revised Edition,	1995
Reprint	2000
Fifth Revised & Enlarged Edition	2003
Reprint	2009

Reprinted 2012

Reprinted, 2014

Reprinted, 2018

TYPESETTING & DIAGRAMS
(FILE NAME – FPPA)
At Gautam Graphic Printers,

ISBN 81-272-1178-8

PRINTED IN INDIA
At Plasticote (India), Shahdara Delhi-93

PREFACE TO THE FIFTH EDITION

The main purpose of the book is to meet fully the needs of all those interested in fruits – students, teachers, extension workers, nurserymen, orchardists, plant protection specialists and above all the common man growing fruit crops in his kitchen garden. With this aim in view the present edition has been revised, rearranged and enlarged to include additional information generated by intensive research in the field.

Newly evolved promising cultivars with their characteristics have been added for every fruit. In addition specific detailed account of latest recommended control measures for pests and diseases have been provided for all crops.

An outline of "Integrated Pest Management" has been given to lay emphasis on the basic principles of pest control which advocate limited use of chemicals so as not to destroy the beneficial flora and fauna, in the vicinity of the fruit tree. These may be profitably employed in biological method of control, often necessary, when the pests develop immunity to insecticides or other chemical treatments.

Endeavour has been made to enhance the utility of the book by addition of new topics *viz.* : (1) Propagation method of different fruit crops, (2) Pesticides and their use, (3) Use of herbicides in fruit crops, and (4) List of banned insecticides, in the appendix of the book.

I am indebted to large number of workers whose contributions are spread widely in a very large number of reports, journals, magazines, hand books, ICAR bulletins etc. Although it is not possible to make reference to each and every source of information the author expresses his gratitude to all the contributors past and present.

The author sincerely hopes that the book would help in the almost impossible task of keeping abreast with recent literature in the principles and practices of culture, protection and production of fruits.

2003 **AMAR SINGH**

PREFACE TO THE FIRST EDITION

FRUIT production based on complex of imperfect methods derived as a result of trial and error is unrewarding. It is the purpose of this book to bring out the application of plant physiology to the art and practice of fruit culture. All scientific discoveries and inventions however fundamental, get true recognition only when applied in day-to-day working for solving the problems of mankind. Some of the very basic facts revealed by the researches in the laboratory have been picked up and woven into a whole to provide to the fruit minded people their applications in the field of pomology. The biochemical or physiological processes involved have been referred to but the mechanism has not been discussed in detail as the book does not strive to answer the Hows or Whys ? Instead it pinpoints the Do's against the Dont's with their physiological support.

Fruits are invaluable and uncomparable in the calorific value, vitamin content, long storage life, easy all the year-round availability in a variety of shapes, sizes and tastes in temperature, subtropical and tropical belts throughout the world. In India, production of fruits and its utilization is much short of the world average. The situation can be improved only if commercial fruit growing is synchronised with the physiological and ecological demands of the plant. The potentialities of physiological laws to alleviate human problems of hunger and dread of diseases through the culture of fruit crops is amply reflected throughout.

Emphasis on the application of plant physiology in the control of the problems of fruit production is here for the benefit of the applied field scientist dealing with the life, health and output of plants. It should give a fruitful reading to all engaged in any facet of the science and art of fruit production because it seeks to narrow the gap between the two. It is well realized today that the classical boundaries of Agriculture and Botany are no longer operative as plant physiology has infused into culture and care of fruit plants on a large scale.

Practical aspects affecting propagation, culture, relation between vegetative and reproductive activities and rejuvenation have been generalised and presented. It is no wonder that scientific agriculture resting on the basic concepts of the life has the potentiality to wipe out food shortage by over production. Unless the fruit industry is cast on physiological fundamentals and the uncertainties in production eliminated not much progress is possible. Applied aspects of science can ill-afford to violate the principles of fundamental science and therefore an understanding into the physiological fundamentals of plants is a prelude to making fruit growing popular and economical.

Basic Aspects of Fruit Production have been dealt with in fourteen chapters, each independent and self-contained. Chapters dealing with cultural practises, elemental role and needs, regulation of form and function, protection against insect, pests and weeds, the plant, principles of propagation, stockscion relationships, incompatibility, technique of environmental control of plant growth, fruit physiology, picking, storage and preservation of fruits, and growth substances in fruit production from Part I of the book. This broad panoramic view of fruit physiology in different aspects is followed by detailed consideration of **'Science and Practice of Fruit Culture'** dealing with the origin, ecology, varieties, cultural aspects, pests and diseases etc. of 38 fruits in Part II. For the benefit of the reader each topic has been presented comprehensively. Inclusion of Selected References, Fruit Trees at a Glance, Glossary and Index further enhance the utility of the Book.

The treatise is intended for the fruit grower of today and tomorrow. It contributes to the understanding of the problems and solutions involved in fruit production. The book can be profitably used as a text by students preparing for honours course in Pomology. It is essentially an attempt to discuss the problems of fruit production in an easy style. References have been purposely omitted from the text to take away the stigma of being a turse reading.

I take this opportunity or recording my sincere thanks to Dr. Kanhaiya Lal who helped me in comparing the typescript with the handwritten manuscript.

AMAR SINGH

CONTENTS

Chapters *Pages*

PART–I
BASIC ASPECTS OF FRUIT PRODUCTION

PART–II
SCIENCE AND PRACTICE OF FRUIT CULTURE

PART–I
BASIC ASPECTS OF FRUIT PRODUCTION

1. THE APPROACH
2. ESTABLISHING AN ORCHARD : BASIC CONSIDERATIONS
3. SOIL MANAGEMENT
4. BASIC CULTURAL PRACTICES
5. ELEMENTAL ROLE AND NEEDS
6. REGULATION OF FORM AND FUNCTION
7. PROTECTION AGAINST INSECT PESTS AND WEEDS
8. THE PLANT
9. PROPAGATION : PRINCIPLES AND TECHNIQUES
10. STOCK-SCION RELATIONSHIP AND INCOMPATIBILITY
11. ENVIRONMENTAL CONTROL OF PLANT GROWTH
12. FRUIT PHYSIOLOGY
13. PICKING, STORAGE AND PRESERVATION OF FRUITS
14. GROWTH SUBSTANCES IN FRUIT PRODUCTION

CHAPTER

1

THE APPROACH

The practice of growing fruit plants is as old as antiquity yet the production level, in India, is far from satisfactory. Incidentally the population increase and the growth of the green plants both follow the cumulative-effect rule. The need of the day is massive consistent production. In U.S.A. one person is able to produce food for 40 others, almost five times as much as in 1920. This has been made possible due to the understanding and involvement of basic principles governing the performance of the plants and also proper use of tools and modern agriculture—water, power, fertilizers and pesticides—to the full extent. By the year 2,000 A.D. the population of the world has crossed six billions. Once it is admitted that man's welfare depends on his ability to control plant growth, careful application of the fundamentals of plant sciences have to be blended with cultural practices, the priority being given to rate-limiting step or factor. In India, 70% of the work force is engaged in farming with the per capita income equivalent to only 5% of that in U.S.A. where scientific agriculture has assumed prominence. It is no secret that unlike the rich countries India can ill-afford the use of automatically controlled work implements, self-propelled machines etc. on account of several considerations. India, as a representative of developing countries, uses only 10% of energy to produce unit quantity of protein as compared to U.S.A., a developed country. These and many other facts amply stress urgent need of working out our own technology instead of borrowing from other countries.

Fruits form a complete, wholesome food for the grown up, still growing and the invalids alike, leading to healthy body and mind. It is a ready source of energy with the unique capacity to guard against many deficiency diseases. In India, mango, cashewnut, bananas, citrus, guava, can be rated in descending order of preference if acreage under them may indicate so. Fruits can supply more than 1/3 of the requirements of calories, vitamins and minerals to mankind. With the active participation of plant scientists it should not be difficult to produce about 40% of our needs of fruits and fruit products by devoting to it only 10% of the area under cultivation. It is quite natural therefore that more stress is laid on the relationship of the physiology of the plant with cultural practices.

A plant has certain basic needs to thrive and prosper ; implementation of correct cultural practices including judicious use of biocides and fertilizers, timely and adequate irrigation and protection against adverse factors help to step up production because these meet the demands of the plant in elaborating more dry matter – the ultimate crude index of growth. In principle, no serious objection has ever been put forward against the successful adaptation of plant physiological researches to the growing of crops but in practice there are obstacles. Low appreciation of agricultural or horticultural science and less respect for its techniques is regretfully true. The outdated conventional methods are sought to be recast with a scientific approach detailed herein through intelligent meaningful adaptations so that the inherent production capacity of the plant is fully realized. Availability of fruits in the traditional cultivation area can be increased significantly by the use of improved methods of cultivation, better tillage, adequate nutrition, choice of better cultivars, meaningful use of growth regulators, early control of weeds, pests and pathogens, and above all timely picking and post-harvest care. It is in the appreciation and realization of this theme that the book is directed, this opening chapter just introduces the approach.

The ability of the orchardist to diagnose the factors which limit production, at any time during the life of the plant and correct the same is the touchstone to successful bumper harvest. It is well realized that the plant yields are circumscribed by the environment and the efficiency of cultural practices. The knowledge of the life activities of the plant is thus a must to make it yield more and of better quality. Production is an adjustment between heredity and environment and the latter includes cultural practices as well.

The achievement of 'Green revolution' could be possible not by traditional agriculture but through the two-prong venture of the science of inheritance and that of function. Plant production rests on a tripod, inherent potentiality to yield, availability of environmental optimum and the application of basic physiological truths.

Intensive and consistent production of fruit is rapidly becoming the domain of the specialists. The plant is a living dynamic entity with a host of metabolic reactions placed and manned in its body at different speed and directions according to its physiological make up. Truly speaking every plant, nay every occasion in the life of the plant has its own peculiarity and problem. It is the duty of the plant scientist to work out optimum conditions for efficient production and of the field worker to supply these.

Any change reflected, by the environment of the plant, in its physiological behaviour is of value to the plant grower. Let us then know these conditions with their promotive and inhibitive supplements to be able to mould the density and performance of the plant. The structural peculiarities of the plant organ determine the very basis of the functions and that the inner happenings in the plant body are indicated as 'visual symptoms' both in health and disease of the plant. A common example is the development of symptoms specific to even elements needed in traces.

Requirement of fruit plants, their response and adaptation to external factors vary with species, variety, rootstock, age, period of growth or of rest, chlorophyll level, nutrition supply, season etc. Interaction of the hereditary makeup of the plant with environment are complex and need an analysis and insight into the cultural practices necessary for high productivity.

Fruits may be either a garden or a field crop, it may be both under different conditions. Custard apple, banana, guava, papaya for example, are considered garden as well as field crops depending on the area these occupy in any particular locality. Nuts and pecans from trees growing wild may be classed as forest product while those obtained from planted trees are taken as horticultural product. Physiological considerations do not have separate dealings for these four situations of the same plant species unless there be a change in the nature of the produce desired to be had *e.g.*, leaves, stems, roots or any other storage organ in place of fruits. Plant species differ in their heritage of characters and requirements and one who could read the language of the plant can direct its activities for heavier, consistent and timely production of better quality. It is no secret that fruits have received very wide attention of the plant physiologists throughout the world.

Elements of mineral nutrition control the vegetative and fruiting performance of plants to a large extent. High nitrogen boosts vegetative growth, restricts flowering and fruiting. Role of micro-nutrients is also of immense consequence in keeping the plant healthy and active. Timely diagnosis of the ailment, with its cause and cure, results in economic production. Kind, mode, quantity and time of application of additives produce divergent effects on the plant so that the knowledge may be put to use in regulating the destiny and direction of plant.

Likewise irrigation practice, in all its aspects, rests on the amount of water available in the root zone, the relative atmospheric humidity, and the water needed by the plant. India can hardly afford to have 'drip-method' of irrigation (supplying water individually to fruit trees in regulated flow) which increases the efficiency of water, fertilizer and also labour productivity on account of economic considerations but irrigation on the basis of water requirement and critical period of water requirement should be made possible. Quite significantly every plant has certain criticial periods of water requirement and its knowledge and timely supply keeps the level of production high.

Some of the basic concepts of the cell and its working developed as a result of plant physiological studies are capable of rendering massive improvements in output. Even without much attention they have made themselves felt. The plant wilts during noon and recovers during night till the tissue moisture does not go below the critical level but as soon as the level drops beyond it, the wilting assumes a permanency which is irrevocable for all practical purposes. Proper irrigation facility is always useful ; on both the sides of optimum the plant lowers its productivity. The plant is a complete whole consisting of millions and millions of cells which are again independent units of structure and function. The coordination of the life activities of these basic units (cells) is all that one has to achieve for profitable ends. How to do it ? A knowledge of the solution and the capacity to manipulate determines the success. The scientific methods and techniques suggested to be used herein should benefit even the orchardist of meagre resources.

Stomatal response to stimuli (light, CO_2, temperature, water, relative humidity) and their reactions change the photosynthetic productivity and the transpirational efficiency. Use of plastics and antitranspirants to economise water losses to save the plants from drought has its roots in the structural physiology of the plant and the knowledge of the process of transpiration. Use of asphalt to increase better absorption of heat from sun's rays may facilitate growing of crops in marginal zones requiring slight yet meaningful adjustments in temperature. Pruning, girdling, grafting and similar horticultural practices are in no way independent of the existence and working of the maristematic and cambial tissues and the functional peculiarities of the structure. No fruit grower should legitimately be expected to know if the plant is a dicot or monocot but even without knowing that, if the propagation practices that entail active cambial participation are followed for monocots success may be only in very limited cases. Similarly success with a temperate plant in the tropics is scarcely possible on a large scale. No more examples need be cited to emphasize that the plant is dynamic responding to environment depending on its physiological condition.

Physiological diseases and the host-parasite relationship is of interest to the orchardist for keeping the plant in health. The biological control of insect pests and selective regulation of competition with the weeds should increase their efficiency. Exhaustion through weed is significantly more problematic in the tropics and subtropics as compared to the temperate regions. It brings down the expected returns through competitive action, the losses can be minimised through timely and effective control measures. Decision about the herbicide to be used and the mode of application rests on the nature of the action (pre-emergence or post-emergence) effectively besides their physiological selectivity.

Fruit ripening, taken as a manifestation of ageing in a living tissue, associated with physiological changes, controls the quality and prevention of wastage during storage. The discovery of climacteric and non-climacteric fruit kinds lays the foundation of their separate post-harvest care and technique to save losses. How far these and other aspects of fruit production need intelligent application has been portrayed in this treatise.

Physiological disorders, cracking of fruits, flower initiation, unfruitfulness, fruit-drop, alternate bearing and other problems can be mitigated or eliminated through the knowledge of the laws of plant growth and development. Many more problems and their scientific treatments have been unfolded herein to eliminate the gap in science and practice of fruit culture.

The fractorial study of plant metabolism reveals the role the internal and external factors play in stepping up production. It is possible to halt blooming, prevent sprouting, cause fruit set even without pollination, check fruit drop, arrest vegetative growth or reproductive phase and change the course and speed of several plant functions at will by the intelligent application of the principles of life. The grower needs guidance on the cultivar to grow, suitability of soil and site, planting details, care, ill-balance of factors of nutrition, all about the fruit and guarantee measures for better output. It should be of interest to note that the studies made on vegetables like tomato, eggplant, cucumber, melons etc. have contributed much to the understanding of the physiology of fruits. Many of the laws of economic production remain the same for fruits and vegetables alike but their interpretation in terms of basic needs and final returns have to be worked out for almost every occasion by the specialist so that there is complete synchronisation.

The increasing demand for cheap and more fruits of high quality justifies stress on increased production to strengthen the supply line by interweaving the physiological principles with the established practices. Existing methods and practices of fruit culture have been identified with the basic needs of the plant as a living mechanism and wherever necessary, alternatives have been suggested to suit the demand.

Several types of the fruits classed as pome (apple, pear, quince), drupe (peach, plum, apricot, cherry, almond), small fruits (grape, phalsa, berries), nuts (pecan, walnut, cashewnut) besides fruits like mulberry, persimmon among the decidous and mango, pineapple, date, banana, guava, papaya, jamun and citrus fruits as the examples of evergreens are grown in India. These and other fruits are considered here as temperate, sub-tropical and tropical fruits since this classification is of greater significance to the grower.

□□□

CHAPTER

2

ESTABLISHING AN ORCHARD : BASIC CONSIDERATIONS

Establishing an orchard is a prelude to production and as such certain basic considerations are discussed here.

Location and Site

Profit in fruit growing largely rests on location and site of the orchard. Even slightest error in this respect may make the fruit industry unprofitable. The fruit and the cultivar to be grown in a locality having been decided, the next step is to spot out the correct location. A thorough knowledge of the distribution of allied industries and the extent to which economic growing of fruits are possible in the locality are necessary for success.

Successful orcharding consists in not only yield, but economic production and efficient disposal. Among the factors determining the desirability of a piece of land for fruit growing are land values not from the quality of produce but from the point of view of other characters connected with the industry. Although the cost of land contributes towards cost of production, a cheap land may prove expensive in the long run, if the orchardist does not consider the probable expenses in making the land suitable for planting. It is mostly the economic side of the aspect which is important in deciding the land values.

Cultivars are well known for their specific ecological requirements *viz. Langra* and *Dashehri* cultivars of mango thrive well round about Varanasi and Lucknow respectively in Uttar Pradesh. The mango cultivars of South India fail to be of economic value in the North. Similarly commercial mango cultivars of North India *viz.* Langra and Dashehri do not flourish in South Indian conditions. It was formerly believed that papaya would not grow well in Allahabad but with the coming up of new cultivars these are now grown here successfully. Besides the climatic and soil requirements other factors that control fruit growing are land values, facility of labour, transportation and storage, fruit-product establishments, social conditions and the markets for disposal. Economic advantages of a location at times outweigh those of better soil or site.

Location denotes the situation in relation to a well known place. Site refers to the specific place with elevation, topography and the availability of the factors influencing the growth and development of the tree *viz.* climate, soil, labour, irrigation etc. Location *viz.* roadside, market, transport and storage facilities, cheap source of electricity, water supply, nearness to a processing plant etc. is more important from economic viewpoint. Where the fresh fruit market value is low, the availability of a cheap processing plant and storage facility is an asset.

Selection of proper site for an orchard is the pivot on which the industry flourishes. Fruit that can be marketed in dried form like the nuts, dates, grapes for raisins etc. may be grown in areas away from quick transport facilities or nearness to the market. It would be futile to assign value to any particular characteristic of site due to the fact that a large number of crops are involved which have differing demand for best performance. Size of the land is of importance since it must be an economic holding and should have space available for the development of the industry. While considering the site for orchard, consideration must be given to factors like soil, climate, temperature, market and transportation facilities.

Transportation is a necessity in most cases. Almost 30 to 50 per cent of the total cost of the fruit grower is expended on packing, transportation and other handling charges. The margin of profit can be increased only if such costs are reduced. Adequate selection of the site is most effective in eliminating such recurring type of expenses permanently.

If the produce has to be used fresh, good and cheap, transport facilities must be available to carry it to the market for quick and profitable disposal. Whereas site has more to do with the growth, bearing and health of trees, location mainly concerns the economic disposal of the produce. Factors which influence fruit production and go hand in hand with the locality also weigh while determining the location. Orchards away from forest have a preference since the menace of wild animals is rather remote.

In choosing the location it is of advantage to have a place where fruit growing is not new since the advantages of the experience of others, the knowledge of the profitable practices and availability of skilled labour are really useful. Location with a heavy output attract large buyers so that transport problem disappears and the profits are greatly enlarged. Medical, educational, social and religious facilities for those engaged in the profession are added attractions in favour of a location.

Topography

The contour of the land affects the system of planting of orchard. It is specially of importance from the standpoint of erosion, frost, freezing and desiccation. Slope of more than fifteen per cent makes the land uncultivable whereas one with four per cent slope may be kept clean without much loss. Uneven land involves heavy expenditure in levelling of the land and introduces other unfavourable factors and cultural difficulties.

Southern slopes are generally warmer and affect the plants in a manner different from the northern slopes because of the difference in the quantum of sunshine that the plants receive.

Air Drainage

In clear cool nights inversion of temperature takes place up to a certain altitude so that the valleys are cooler than the elevation due to the radiation of heat from earth during nights when cooler air which is heavier gets to the surface and the warmer air rises up. This is air drainage.

Air drainage involves the flow of air from areas of high elevation to areas of low elevation. The movement of cold air to low levels of adequate air drainage aids successful production of many fruits. Before orchards are planted, the frost hazard of proposed sites must be carefully evaluated. Low temperature accompanied by hail, frost etc. specially at blossom time is very harmful. Frosts cause injuries like black heart, crotch injury, crown or collar injury, splitting of bark and premature fall of newly expanded leaves. Injury to the pistil of the flowers as well as fruit drop is also the result of frosts. Hail injures the skin and deteriorates the quality of fruits. The evergreen fruit trees suffer more from frost than the deciduous fruit trees. Frost is more damaging to weak trees which suffer from mineral deficiency, lack proper water supply and have physiologically incompatible rootstock-scion union.

Protection from frost in plantation may be afforded by producing smoke with production of heat sufficient to raise the temperature of the orchard or by mechanically forced circulation of air. Hot air circulation method is also useful though costly.

Soils

Soils play a significant role in the success of fruit growing enterprises. The roots of fruit trees generally penetrate the soils to a much greater depth than field crops and as such the nature of the subsoil plays a more important role in the growth of the fruit plant. Physical properties of the soil *viz.* texture, structure etc. are more significant in fruit growing. Of course a soil rich in plant food material may prove better but fruit trees may be grown on poor soils also with adequate addition of nutrients and in most cases the practice responds. A loamy soil with good amount of organic matter content may be found to be ideal for many fruit trees although some flourish between in sandy loams. Presence of sand is a means to natural drainage condition with tendency to make it warm.

Soil colour may be used to identify soils as to its favourableness or otherwise.

Colour of the soil may vary depending on the amount of organic matter, the state of hydration or oxidation of iron compound and the moisture content. Soils are neither alike nor inert and static; these are dynamic and influence crops and cropping materially. Some weeds serve as soil indicators (Table 2.1) and the knowledge may be made use of in selecting soils for particular fruit crops or *vice versa*. The soil profile may be classified as best drained, well drained and imperfectly drained, the last condition being definitely harmful for an orchard. A fully drained soil with proper fertility is needed for fruit trees.

Table 2.1. Weeds as plant indicators

Soil characteristics	Weed flora
Chalk	Cichorium intybus, Silence vulgaris, Linaria vulgaris, Sinapis alba.
Heavy soil (clay)	Equisetum arvence, Plantago media, Primula vulgaris.
Damp soil	Juncus effusus, Lychnis Flos-cuculi.
Light soil (sandy)	Papaver rhoeas, Polygonum aviculare.
Acid soil	Chrysanthemum segetum, Digitalis purpurea.
Soil in need of lime	Pteridium aquilinum, Spergula arvensis.

Distribution of the plant root is a potent factor in the productiveness of the tree as well as its life span. Soils modify the distribution and nature of extension of roots. A plant having root distributed on the surface is easily subject to drought and may be easily damaged by wind besides having greater chances of early depletion of nutrients. Variants of the same species have generic differences in the rooting habits and whenever proper growth of the roots is not allowed (according to the character of the cultivar) the plantation becomes unprofitable. The practice of mulching encourages development of roots on the surface. In respect of soil reaction different plants have different requirements. The soil reaction may be modified by additives or manures or fertilizers, it has close relationship with calcium content of soils.

The pH range favourable for some of fruit crops is given in parentheses. Apple (5.5–8.0), blueberry (4.5–5.0), citrus (4.5–7.0), grape (5.2–6.7), pear (5.6–7.2), peach (5.1–6.7), pineapple (5.0–5.5), red raspberry (5.2–6.7) ; strawberry (5.0–6.0), plum (5.8–7.5) and cherry (5.8–7.0).

Fruit trees vary in their tolerance to salts present in the soil and only those fruit trees which are salt tolerant may be planted in such soils. Soil studies in this regard are therefore very meaningful. In salt tolerance the fruit trees arrange in descending order as grapes, olives, figs, almond, oranges, pears, peaches, *prunus*, apricots, lemons and mulberry. Usar soils which are not uncommon may be broadly put under three categories with high, medium and low salt content. Richards has classed date palm in the high salt tolerant group ; pomegranate, fig, oliva, grape in the medium salt tolerant group whereas pear, apple, orange, grape-fruit, plum, apricot, strawberry, lemon and avocado in the low salt tolerant group.

The general requirements of fruits in respect of the soil are portrayed in table 2.2.

Table 2.2. Soil Requirements of Some Fruits Trees

Fruits	Suitable Soil
Almond	Light loam.
Apple	Light loam or deep organic loam.
Banana	Deep, rich, alluvial loamy soils (Calcareous stony and undrained soils not suitable).
Ber	Deep, sandy loam, neutral or slightly alkaline prove best.
Chicku	Alluvial loams, lateritic soils, medium black soils and loamy soils are good.
Citrus sp.	Medium or light loam with slightly heavier sub-soil.
Custard apple	Red sandy soil is best. It can grow well in rocky soils also.
Datepalm	Light loam to heavy clay.
Fig	Heavy clay to light sands, heavy soils with fair degree of lime considered ideal.
Grapes	Well drained, shallow medium black loamy soil.
Guava	Light sandy loam.
Jackfruit	Rich, deep, well drained, alluvial.
Jamun	Loamy, deep and well drained soil.
Litchi	Clayey loams with good drainage.
Loquat	Sandy loam to heavier soil. Well drained light loam is ideal.
Mango	Rich alluvial loams (Sub soil must not be hard, rocky, limy or highly alkaline).
Papaya	Rich loams, medium black and alluvial soils (Low, marshy ground, and impeded drainage harmful).
Pineapple	New alluvial of plains, old alluvium of submountaneous regions and lateritic soil of hills preferred. Heavy clay must be avoided.
Pear	Deep, moist soil.
Pomegranate	Deep loam or alluvial soils.

Climate

Climate is a unit, it is not possible to express it by a figure or formula, it is necessary to know the seasonal rhythm of the factors that constitute the climate.

Ten principal climatic types are distinguished *i.e.* I. Equatorial, II. Tropical, III. Subtropical, IV. Mediterranean, V. Warm temperate, VI. Temperate, VII. Intemperate, VIII. Boreal, IX. Arctic and X. Maulan. Climate may be defined as the weather changing in a certain manner during the course of a year, taken as an average of several years.

Fruits are classified as temperate, tropical or subtropical on the basis of their climatic requirements although this classification is not very rigid and some fruits may be grown in more than one climatic setup. Mango is grown in tropics and subtropics and peaches do vary well in both temperate and subtropical regions. In the tropics freezing is unknown, a day-length does not vary much in the course of the year, but rainfall variations are large and common.

Weather is the overall effect of the constituent climatic elements such as temperature, rainfall, humidity, radiation etc. at any particular moment. Weather denotes a single occurrence of events in the series that constitute the climate, the average condition of atmosphere. Climate bears a very intimate relationship to the distribution of fruit crops. Weather may be favourable or otherwise for the growth of the plant and its yield.

The biological activities like flowering and fruiting are greatly influenced by the environmental conditions. The climate induces changes in the growth phases which may be of consequence to the orchardist. Plants may or may not blossom or the flowers may or may not develop into a fruit. The fruit may remain attached to the plant or experience a pre-harvest drop. Banana, a tropical fruit crop, suffers heavily if the temperature falls below 10°C. Coconut, cashew and pineapple are common examples of tropical fruits. The citrus, avocado, fig and datepalm generally assume commercial importance in the subtropical zone alone. Trees have a period of rest and growth which generally synchronises with the spells of unfavourable and favourable weather conditions respectively. Temperature conditions, for instance, are characterized by a cold period when the life activities are suspended temporarily and only those plants that have systems tuned to such conditions can be grown successfully. Similarly any venture to grow fruit plants unsuited to the typical conditions is bound to put the orchardist to losses of high magnitude.

Temperature

It is an important factor affecting the distribution of plant species. Temperature beyond the limits of tolerance of the plants kills them. Broad leaved evergreen trees may not be successful in regions where temperature falls below 5°C as their survival becomes doubtful. Deciduous plants which pass winters in dormant condition may thrive in temperate region. Extremes of temperature more often limits production ; the safe temperature limits for higher plants is considered to be between 5-27°C with some marginal adjustments.

In planting orchards the minimum temperature must be within the limit of tolerance of the fruit. The effect of temperature on a cloudy day is different from that on a clear day. The rest period of temperate fruits need be broken by low temperature and where it is not possible due to certain reasons the crop may be late or even lost.

High temperature may cause bolting and seed formation as in spinach and lettuce. In the case of tomatoes high summer temperature as well as low spring temperatures are unfavourable for fruit set. Melons grow best when the temperature is high. Temperature influences are great and well marked. Peaches tolerate more cold than citrus fruits whereas apple the best of these. The region selected must be suitable for the cultivar to be grown for most viewpoint of temperature requirements. The optimum temperature for most crop plants is 22° to 27°C whereas the maximum temperature limit may be 40° to 52°C at which most of the plants of temperate zone may not survive.

Large reservoirs or sources of water near the orchard moderate temperature-effects. When hot the water absorbs heat, lowers the air temperature whereas in the winter the water gives off heat and raises air temperature. Large bodies of water within and surrounding Indian Sub-continent make possible the growing of certain horticultural crops which could not have existed otherwise. The modifying influence of the Indian Ocean is partially responsible for the existence of the grape areas of south and modifying influence of great lakes make possible the growing of peaches and grapes in Kashmir.

Sunshine is of great value to fruit crops because of its influence on pollination and on the development and colouring of fruits. Interruption of sunshine (shade) makes great differences, the same temperature acts differently on the shaded buds as against those receiving sunshine. With low percentage of humidity sunshine is helpful in affording best opportunity for pollination and also for the growth of the pollen tube. Intensity and quality of light influence the formation of carbohydrates. Sunshine also influences the formation of flower bud.

Wind

Wind, as a factor, is also of importance in establishing an orchard. The cultivar of the plant to be grown is also controlled by the velocity of wind. High wind velocity does not suit leaved evergreen plants which experience economic loss due to mechanical shedding of fruits, flowers, leaves etc. Often the plants bend towards "Lee-ward" side under the stress of high winds. Dry winds, with both high as well as low temperatures after under-ripening of fruits besides disturbing other metabolic processes. Wind-breaks on the wind-ward side and close planting of the trees are practised to minimise the impact of occasional winds though disadvantages accompanying wind breaks like shade and competition for moisture and nutrients remain a disturbing feature. Adequate safeguards are needed to minimise such losses. In India trees like Babul (*Acacia arabica*), Shisham (*Dalbergia sisso*), Mulberry (*Morus sp.*), Jujube (*Ziziphus sp.*), Nim (*Azadirachta indica*), Bamboo (*Bambusa sp.*), Jamun (*Syzygium cuminii*) and Eucalyptus may be used as wind-break.

Rainfall

Water is the major constituent of the protoplasm. It is not possible to imagine a living cell devoid of water. Rainfall is the ultimate source of most of the water supply to all life and specially so to plants. Its spread and intensity cannot be overemphasised. Processes like absorption, ascent of sap, distribution of metabolites and photosynthates, syntheses of carbohydrate, protein and fats, respiration etc. are unthinkable in the absence of water. It controls both growth and development. Loss of water beyond a particular degree induces wilting of plants. As wilting sets in growth ceases and with the one set of permanent wilting plants fail to recover and ultimately perish. Plants have some critical periods of water requirements and adequate supply of water during these stages definitely raises the yield. Water shortage at the critical period of water requirement *i.e.* early growth, bud differentiation, blossoming, fruit set and development etc. lead to unprofitable effects. Plants receiving good supplies at the start fail to withstand drought later. High level of water supply and of fertility produces less extensive root growth and excessive water loss through transpiration leading to physiological disorders.

Rains, before harvest or at the time of bloom, have disastrous effects. Bananas become soft, cherries small and burnt while tomatoes crack. Late season rain may induce vegetative growth which is not good for the crop. Light soils with low organic matter content may accentuate the problem of short supply of water and uneven rainfall. Atmospheric humidity may result in slow drying and cause rapid development of diseases. Scab (apple), brown rot (peaches), black rot and downy mildew (grapes) etc. find favour in high atmospheric humidity. Humidity and cloudiness reduce photosynthetic activity of plants, day-length influences flowering.

Flood and Drought

Plenty or scarcity of rains may cause flood or drought and influence fruit growing. Drainage without loss of soil should be the aim to combat floods. Use of cover crops affords chances of minimising soil erosion due to flood and runoff. Rapid percolation in soils of good soil structure adds to the safety of the soil.

Drought conditions may be reversed by artificial supply of water though it may not be always economical. Maintenance under such artificial conditions may fall much below the expectations. Whenever water supply be limited the spacing between plants should be increased to promote large "feeding area" coupled with deep soil penetration by the roots. Water conservation practices including mulching, the use of certain elements and compounds and of antitranspirants may be beneficial . Only those plants be grown in such areas which have some adaptability to withstand drought at least for short periods.

Irrigation and Drainage Facilities

Successful orcharding needs adequate irrigation facilities. Although under the conditions prevailing in major parts of the subcontinent some mature trees may bear large crops even in the absence of irrigation owing to their deep and spreading root system, younger trees require ample watering for their economic survival. Availability of cheap source of irrigation and good drainage control economic bearing of orchards. For trees which thrive under conditions of water logging and in heavy soils it may sometimes be necessary to provide artificial drainage.

Labour and Manure

The site should be such that labourers may be cheap, skilled, and in plenty. As the fruit industry flourishes more skilled labour is needed. Manures should also be available cheaply in sufficient amounts. The application of manure is necessary for satisfactory growth of plants, even from very early stage when there is no production from orchard. An orchard can be maintained for longer periods in commercial bearing with adequate manuring. The nutrition of the tree is a matter of supreme importance, though maintaining fertility is likely to be one of the most expensive aspects of fruit growing, it repays.

CHAPTER

3

SOIL MANAGEMENT

Production is dependent to a great extent on the soil productivity which can be managed. Some of the principles of soil management are outlined here.

Properly managed soil can serve as inexhaustible source of nutrients for the plant provided, the debt and duty towards soil is realised by these dealing with soils. These must be maintained in productive state indefinitely to continue to yield profitably. The factors that count towards maintaining the nutrient status of the soil have been dealt with in detail elsewhere. Capacity to provide adequate nourishment to the vegetation without damaging the inherent capacity of the soil to produce is the crux of successful soil management.

The basic aim of soil management programme is to keep the oil fertile as well as productive. Fertility is maintained by continuous addition of elements necessary for growth and development of plants according to the needs of the soil either as fertilizers or manures. Once the soils are fertile *i.e.* they contain the necessary elements for the growth of the plants the prime need is to effect their availability. It is the availability of the adequate amounts of nutrients from the soil coupled with the conditions which keep the plant in health so as to be able to draw on the reserves that control production.

The value of choosing a good soil for an orchard has been stressed earlier. It is equally important that the soil be managed intelligently in respect of the physical conditions as well as nutrient and moisture status. The cultivation practices should be such as to maintain the productivity of soil. The role of commonly recognised orchard cultural practices has been dealt with elsewhere in this book. In this chapter reference is being made to soil management.

The Dynamic Soil

The earth nourishes every kind of life. Soil, the uppermost crust of lithosphere, is rightly called "the placenta of life". It may contain three billion organisms in every 100 gm or a handful in a hectare. Microbes live throughout the profile of the soil intermixed in response to the conditions of soil climate. Soil is living and dynamic in all respects. All types of organisms ranging from submicroscopic to the relatively gigantic forms are met within the soil. An idea of the size relationship between the visible forms found in the soil may be had from the fact that mite is about a millionth of the earthworm in size while the protozoa and virus are progressively reduced in size by a millionth each time.

The solid earth is not solid, it contains salts, water, air and microbes between the particles. The interspaces between the soil particles, offer a variety of habitat. Smaller pores and channels are filled with water whereas the larger with air alone. Clay may have more of the smaller pores whereas the sand predominates in large air pockets. Soil is never devoid of water, though the quantity, at times, may be such as to be unavailable to plants, extreme cases lead to drought conditions. With the addition of water to the soil the aerobic forms of life living therein become inert while the anaerobic forms flourish. The response of the microbes to light, temperature and pH are also of importance besides extreme sensitivity to soil moisture conditions. The abundance of organic raw material in soil encourages and sustains life. An orchard needs a deep soil of fine structure and good tilth of high water-holding capacity, free drainage with pH as near 7 as possible.

Conservation of Nutrients

Manuring is most responsive when it meets the specific need of the soil and the vegetation it supports with respect to the amount, time and form and application. Unrotten bulky organic matter when used arrests the available nitrogen due to increased activity of the microbes engaging nitrogen temporarily. Young plants should be supplied with well decomposed manures and worked in the soil. Fertilizer mixtures in proper ratio be applied to meet nitrogenous, phosphatic and potassic needs of the plant preferably with irrigation water but not with excessive rain. The mixtures must not be such as to adversely affect the pH, structure and texture of the soil. Whenever disorders due to the deficiency of minor elements occur it is advised to supply these are foliar spray for maximum profits.

Movement of Dissolved Minerals

With water move dissolved substances down the profile as well as in space. Movement in primarily influenced by texture, structure, nature and amount of cations present, base exchange reactions, and the size and amount of pore space.

Table 3.1. Elements contained in plant parts (quantities removed from soil)

Fruit tree	Yield/acre	Plant part	Kilograms/acre		
			Nitrogen	Phosphorus	Potash
Apple	400 bu	Fruit	10.0	3.5	15.5
		wood, leaves	5.0	1.5	2.5
Blackberry	4000 qt	Fruit	5.5	1.5	6.0
Cherry	800 lb	Fruit	8.0	2.0	10.0

Fruit tree	Yield/acre	Plant part	Kilograms/acre		
			Nitrogen	Phosphorus	Potash
Grape	4 tonnes	Fruit	5.0	3.5	10.0
		leaves, canes	7.5	2.0	7.5
Peach	500 bu	Fruit	15.0	7.5	27.5
Pear	300 bu	Fruit	3.7	1.5	7.5
Plum	200 bu	Fruit	7.7	2.7	10.7
Raspberry	4000 qt	Fruit	5.3	2.3	6.0
Strawberry	6200 qt	Fruit	6.7	5.1	13.5

Effect of soil conditions on the availability of water may be considered since reducing losses of water or increasing its availability to plants is essential for water economy and optimum growth of plants. Precipitation and irrigation form the chief source of water supply to the soil which loses it through percolation, runoff, evaporation and transpiration.

Infiltration

Infiltration of water in soil depends on the compactness and depth of the soil. Impacts of rain drops lessen infiltration considerably. Water-stable aggregates, organic matter content, macrospores and coarse surface texture tend to be directly related to infiltration. The rate of infiltration decreases with the saturation of soils. Once the soils are saturated further infiltration depends on the state of percolation. Such management practices which make the surface soil friable may be employed to increase infiltration of water if it is in the interest of plants. Tillage operations and coarse organic matter increase the movement of water down the profile. Soils with very fine texture, have very low infiltration rates (0-6 cm/24 hours). Sandy or silt loams or soils with blocky or prismatic structures have moderate infiltration rate (30-60 cm/24 hours) show infiltration rate of about 6-30 cm per 24 hours is found in soils which have good clay contents, low organic matter content and a plenty of massive structure. Soils with single grain or granular structure have rapid infiltration (more than 60 cm/24 hours). It may be commonly found in deep soils and well aggregated silt loam.

Permeability

Permeability of soil, that controls the movement of air and water is dependent on its pore size distribution. It increases with the number of the macrospores which in their turn are more in soils with larger aggregates. Compactness of soils reduces permeability while presence of organic matter increases it. The colloidal properties of the soil are also of importance in this respect, especially in the lower depth where organic matter content is often low. Soils with coarse structures have higher permeability. Irrigation water with sodium content decreases permeability unless the total salt content is high enough to prevent dispersion.

Permeability is also responsive to cultural practices and can be regulated within limit. Growing of legumes or deep rooted grasses or trees increases permeability while continued tillage operations reduce it. Permeability is high in wet soils and decreases with the drying of soils. The hourly permeability speed in soils may vary enormously, say from 10 mm to 15 cm or more.

Percolation

Water moves down the profile through column of soil due to percolation and is lost to the plant as ground water and carries with it dissolved nutrients. Percolation increases with precipitation/evaporation ratio and therefore exceeds with the intensity of rains specially in humid climate and affects the water table. Percolation checks water logging. Loose sandy soils show higher percolation rates than compact clayey soils. Growing of crops reduces percolation of water. Plant nutrients leach in the following order : Ca > Mg > S > K > N > P as the water containing dissolved substances move down.

The movement of water down the profile may be controlled by capillary tension of the soil in non-saturated conditions. Soils with excessive salts develop osmotic pressure of accountable magnitude. Another type of force which controls water movement is vapour pressure causing diffusion of water vapour from moist and warm soils to dry and cool ones.

Retention of Nutrients

Water is held on the surface of colloidal or other soil particles. Surface moisture tension is important for dry atmospheres. The force which the plant must overcome to take up water from soil is called soil moisture stress. the available water is between the range causing wilting and the field capacity, it almost equals capillary water. Water molecule is mainly held in capillary sized pores. Increased capacity of clay to hold water is responsible for higher available water content in loams, silt or clay in comparison to sandy soil.

Conservation of Soil

Soil and water conservation is a must for the survival of humanity. Regardless of soil type and kind of fruit grown, maintenance of adequate supply of available nutrients, control of runoff and of erosion to conserve both soil and moisture is needed. Of the natural forms of soil deterioration, erosion plays maximum havoc through the removal of soil, from one place to another, by such agencies as wind, water, snow, forces of gravity, etc. Erosion is influenced by factors like precipitation, wind velocity, topography, soil characteristics, condition of ground water and phenomena like landslides, earthquakes etc. To put in the form of function equation : Erosion $= f(cl, v, t, s, h)$ where f stands for function ; *cl* for climate ; *v* for vegetation ; *t* for topography ; *s* for soil, and *h* for human factors.

Man hastens erosion by clean cultivation, deforestation, overworking, overgrazing and shifting cultivation. Erosion robs soils of its fertility twenty times more than that removed by crops. The runoff and erosion losses may be minimised and offset by keeping the land under cover, use of green manuring, sod or pod mulch systems.

Irrigation is done to satisfy the needs of the plants in the event of shortage and uncertainty of natural precipitation. Proper irrigation practice is profitable only when it just fulfils the water needs of the plant, neither less nor more. It aids rapid growth of young trees even in the case of deep rooted trees that do not normally need watering. Irrigation water of low salt content should be supplied in such a way that loss of water through surface runoff, deep percolation, surface evaporation etc. be reduced to the minimum. Water should not be given in such heavy doses that promote deep percolation to avoid loss of water as well as nutrients. Reduction in soil and water losses as well as the maintenance and building up of the productivity of soil should form the chief basis of agricultural practices *viz.* crop rotation, strip cropping, mulching, planting of grasses and trees, contour tillage, contour bunding, terracing, control of gullies etc.

Growing of leguminous crops or grasses in the space between the fruit trees help in the turning over of the nutrients from lower depth to the surface. Green manuring of crops helps in keeping the nutrient cycle intact. Close growing crops should be grown to check erosion where rains are heavy. The value of mulching, grass canopy etc. is well recognised ; mulches minimise evaporation whereas grassy bunds, highways etc. protect badly eroded areas, control gully formation and establish sand dunes. Contour bunding, contour tillage, and terracing reduce runoff. Contour tillage specially used in undulating stretches of land consists of ploughing, sowing, harvesting and interculture operations in small bits at different levels created by passage of time without making any attempt to level the entire area in one and the same plane.

Maintenance of Organic Matter

Nutrients are supplied to plants in organic as well as inorganic forms, the former is spoken off as manure and the latter as fertilizer. Manure which is well decomposed and has a C : N ratio of 10 : 1 is used with profit. Manure keeps the soil healthy with favourable texture, conserves it in all its forms, maintains higher water holding capacity, good drainage, helps to retain its inherent capacity to feed the plant as inexhaustible source of balanced nutrients. Organic manures of the desired type do not disbalance nutrient availability. Manures are generally applied as basal dressing but these can also be useful when added *at least* two months before the onset of flowering. When flowering is about to commence or has started addition of manures lowers fruiting.

Microbes in the soil live by the organic matter present therein. Organic matter also reduces impact of rain on soil, surface runoff and erosion : it speeds up formation of aggregates and helps in the attainment of granular condition so that there is better aeration and permeability. The roots make use of the oxygen present in the pores of larger aggregates. The elements needed by the plants are added to the soil in the form of organic matter serving as a ready source of exchangeable and available bases like potassium, calcium, magnesium etc. Organic acids released and the carbondioxide produced by the activity of microbes are also helpful in affecting the availability of some elements. Ammonium form of nitrogen is held by humus and prevented from leaching. Microbial activity is enhanced in the presence of organic matter ; carbondioxide production indicates the extent of microbial activity on the soil. Organic matter also helps in the conditioning of the soil. The problem of replenishment of organic matter of the soil is greater necessity in the arid and semi-arid regions where high temperature tends to increase the losses of organic matter. Continuous cropping with cereals and legumes singly in regular succession or in association or green manuring aids retention of organic matter and nitrogen addition in the soil.

Nirogen Balance

Processes like ammonification, nitrification, denitrification, biological nitrogen fixation go a long way in maintaining the nitrogen status of the soil. Addition of undecomposed organic matter to standing crop is of little value to the plant since micro-organisms bind the nitrogen in their own self to the available only after their death and decay. Ammonium salts are converted by autotrophic bacteria, *Nitrosomans* and *Nitrobactor*, into nitrates specially in neutral soils.

Nitrification process may be accelerated by neutral pH, liming of acid soils, addition of organic matter to alkali soils, maintenance of soil moisture at 50% of the water holding capacity of the soil, and addition of nitrogenous and phosphatic fertilizers as well as drainage.

Denitrifying bacteria, *viz., Pseudomonas* and *Bacillus* help in denitrification. The rate of denitrification is governed by moisture, pH, soil structure and water logging, among other factors. Any pH below 6 is not conducive to denitrification. The structure of soil which does not provide free movement of air (oxygen) increase the rate of denitrification.

Biological fixation of nitrogen may be non-symbiotic through the activity of *Clostridium* and *Azatobacter* or symbiotic by the activity of *Rhizobia* working in association with legumes. Some non-legumes like Aldor shrubs and trees (*Alnus* sp.) also fix nitrogen symbiotically. It has been shown by the author that wheat also fixes nitrogen under the semi-arid conditions of Uttar Pradesh, India.

Nitrogen fixation in the tropics and sub-tropics may be encouraged by growing of adequate legume-cereal mixture in right proportion. The practice of mixed cropping *i.e.* growing of two or more crops in association, with proper seed rate proportionality and association must be encouraged since in such cases the addition of combined nitrogen to the soil is greater than growing of unprofitable associations in random ratios.

Problems of Saline and Alkali Soils

Saline soil (pH 7-8.5) known as white alkali, possess white crust on the surface due to the deposition of salts carried up by ascending water during dry summer days. Chlorides, sulphates and carbonates of calcium, magnesium and sodium present in such soils may be leached beyond the root zone by heavy participation or irrigation.

The alkali soils (black alkali, pH 8.8 to 10.0) have a high percentage of sodium which disperses both clay and organic matter reducing the pore size and restricting water and air in the soil. Soil which have both the characteristics of saline and alkali soils are termed saline-alkali. Saline soils though potentially productive with high wilting coefficient and low water contents may not support even a blade of grass. Impeded drainage, low aeration and reduced microbial activity, high pH reducing the solubility and availability of iron, manganese, copper and zinc make the alkali soils unfit for agriculture. Chlorosis may develop due to the accumulation of carbonate ions as shown by the author in several instances. Saline and alkaline soils and adequate drainage, surplus water supply to leach the salts and bunding for irrigation with good quality water. Sandy loam saline soils are reclaimed with greater ease than their clayey counter-parts. Addition of calcium salts replaces sodium ions and soils acquire good condition.

Fruit trees vary in their tolerance to salts and it is profitable to plant only those fruit trees which are able to tolerate the salts. In their capacity to tolerate salts, fruit trees arrange in descending order of grapes > olives > figs > almonds > oranges > pears > peaches > apricots > lemons > mulberry. According to Richards, dateplam may be placed in the high salt tolerant group ; and pear, apple, orange, grapefruit, plums, apricot, strawberry, lemon and avacado in the low salt tolerant group.

It transpires therefore that depending on the condition of the soil and other environmental factors the management practices must strive to provide adequate aeration, base exchange capacity, organic matter and other conditions for vigorous activity of the useful microbes as well as growth and well-being of the plant.

Each soil has specific need at any particular level of cropping and environment. The practices leading to restoration of soil fertility and maintenance of conditions beneficial for the growth of plants and also the activity of living organisms should be provided for optimum production.

CHAPTER

4

BASIC CULTURAL PRACTICES

Soils may be fertile or infertile, productive or unproductive. A fertile soil may not necessarily be productive although the reverse is true. Any soil containing the nutrients necessary for the growth of plants may be termed as fertile. Mere presence of nutrients is no guarantee of successful growth of plants since much depends on the rate, extent, and proportion of their availability. The crop producing power of the soil is usually linked to fertility and its physical status. Cultural practices should aim at maintaining the soil in good tilth and better productivity through better retention and supply of moisture and nutrients besides adequate aeration in the root zone. Cultural practices also help in the removal of weeds and maintenance of proper tilth of the soil. Though tillage is half manure deep tillage has no place in orchard cultivation where high frequency of cultural practices or clean cultivation is normally a wasteful process. Best period for soil cultural is between November-January when the roots are less active. Efficient ploughing followed by light hoeing keeps down the weeds considerably. Conservation of moisture in the root zone of the plant is required in most of the orchards.

Culture of orchard depends both on the kind of fruit and the existing conditions. In citrus, shallow ploughing may be done to keep down weeds. Green manuring, raising of cover or intercropping is useful only during prebearing stage of the fruit trees. As the trees grow only one tillage operation at the time of flowering may be profitable.

Soil management encompasses clean culture, clean culture with cover crop, artificial mulch, sod mulch, sod and intercropping. One or more of these practices may be found to be operative in a single orchard depending on the needs of the plant. Soil conservation and the need as well as the response of the trees determine the cultural practices which might be followed for profit.

Fruit trees respond to soil types more rapidly than garden or field crops, some fruit trees do not have an exacting demand on the cultural treatment while others have very specific needs. Older trees have the means and capacity to tolerate unfavourable growing condition to a greater extent than the younger trees. This is possible due to greater root development, more food reserves and sturdiness of the older trees. Thus the requirement of cultural practices may differ in nature, extent and periodicity with the age of the plants also. Seasonal changes and change in location of the orchard call for a change in the cultural operations. No particular pattern of cultural practice can therefore be listed for any fruit tree. This has to be selected on the basis of desired objective for any particular case.

The chief objectives of cultural practices are the maintenance of optimum moisture, nutrients and aeration, ensure sufficient organic matter to offset losses due to decomposition and runoff besides checking compaction of the soil. Predominantly two systems of orchard culturel are recognized *viz.*, tilling and sod culture. Soil management practices generally control the amount of water present in the top layer of soil but not the amount of seepage.

Sold Culture

Serious soil erosion is prevented by keeping the soil under sod crops. Establishment of vegetative cover by planting vines, legumes etc. is necessary to make soils free of erosion losses. Fruit crop enterprise suffers more from erosion than any thing else. Filling up of the gullies and the healing up of the surface soil is necessary in order to retain this sod in good condition. Check dams may be made before the planting of cover crops to hold the loose soil permanently. The banks of the temporary checks should have a slope of about 30°. Sod culture relates to a system of soil management wherein fruit plants are raised without any tillage or without mulching the soil with litter. This system is recommended for places where the topography of the land does not allow any other system of tillage. In the plains it has been of little use. In sod management or orchards the grass or vegetation may not be cut at all, may be cut and removed or may be cut and left on the ground. Where the grass is cut it is profitable to cut it twice or thrice and leave it to serve as a mulch.

Sod system in itself is not favourable for foliage colour, tree growth and fruit yield. This system in specially unfavourable for peaches and encourages nitrogen losses. Stone fruits and grapes also do not find it very favourable. Sod culture offers greater difficulty in control of fungal diseases and insect pests besides increasing fire hazard. Whenever a non-leguminous plant is grown in this scheme frequent fertilization becomes a necessity to maintain good vegetative growth as well as fruitfulness. Sod culture is more economical because fruits get ready a few days earlier, develop better colour and those that drop on the ground, before picking, are less damaged.

Any tree with about three metres deep root system is in a position to withstand the surface drying more than the shallow rooted plants, such a deep rooted tree can survive under a sod mulch method of control when a shallow rooted tree would suffer serious injury. Sod system may be followed where the trees are deep rooted or irrigation facilities exist. Whenever the surface soil moisture may be replenished and the plants be independent of the changes in the water of the surface soil the orchard may be kept under sod.

Shallow rooted crops are best for sod provided these make a low dense growth and also have a low water requirement. Alfalfa does not meet this requirement. *Crotolaria* and *Indigo* are satisfactory from this point of view as soil cover in the tropical fruit plantations.

Permanent Sod

Grasses like clovers, alfalfa, timothly etc. may be used in the permanent sod system for orchard either singly or in combination. The short sod-rotation method has been followed in those areas where the orchards decline in spite of the use of nitrogen along with sod. This short-sod-rotation consists in ploughing of grasses for some time and again seeding the same. Grass sod clearly resembles the cultivation and use of cover crops and is better adapted to spraying, and such other practices. Sod treatment of orchard soils usually involves a partial breaking of soil following rains or early spring. This causes the oxidation of organic matter accompanied by the liberation of nitrogen to the benefit of the fruit trees. The system checks erosion and runoff also.

Legume Sod

It is liked by fruit growers because of the capacity of biological nitrogen fixation that legumes possess. This type of orchard management is of special significance in the arid tropics and sub-tropics where the losses of nitrogen are alarmingly greater than in the temperate areas. This is profitable when irrigation facilities exist since in case of competition between the cover crop and the young fruit plantations the latter suffer a setback. Leguminous crops should be selected to suit the particular condition of growth *viz.*, areas with excessive water logging may do well with Dhaincha (*Sesbania* sp.), well drained soils with sannhemp (*Crotalaria juncea*), while cold regions prefer vetch (*Vicia* sp.) and under moderate conditions clover (*Trifolium* sp.) fares well. Soil, low in nitrogen, but with readily available irrigation facilities adds to its nitrogen content at much lower cost under legume sod.

Grass Mulch

Grass mulch or stubble mulch consists in the placing of or partially incorporating grass, straw, hairy stalks etc. on the soil surface between the trees. It is practised mostly for conservation of moisture in the top 30 cm layer of the soil. In no case mulching encourages the compaction of soil in the zone just below the layer where such operations are carried on as opposed to the cultivation system. Whenever addition of nutrients and conservation of moisture in the orchards is needed mulching is done. Mulching keeps the fertility status of soil high, reduces erosion losses, conserves soil moisture and increases organic matter. Soil mulching supplemented by the application of readily available nitrogenous fertilizer yields good fruit crops. The roots of the fruit crops have a tendency to abound the surface so that the nutrients added are well utilised. Mulching in fruit orchard is of several types from allowing cut grass to remain where it grew to the spreading of organic substances beneath the trees to a depth extending up to 30 cm. Although mulching has some disadvantages it proves definitely better than the sod system primarily because of the competition set in the sod system between the fruit trees and the grasses. As the litter decomposes it adds nutrients to the soils. Mulching adds organic matter in the top layer of the soil and has the greatest advantage that it reduces losses by restricting cultivation. Rain water is held and absorbed by the rough soil surface in large amounts even on moderately steep slopes. Mulching reduces evaporation of water from soil surface, prevents soil blowing, keeps down weed growth and improves soil texture.

Plant growth is limited in type, extent and quality by temperature variations. Extremes of temperature may be modified through certain cultural practices. One of the functions of mulches is the regulation of soil temperature though they conserve moisture by reducing evaporation and increasing the absorptive capacity of the surface soil. The temperature stabilizing effect of mulches is due to insolation, heat absorption and shading. It increases the availability of nutrients, improves root growth and thus the yield. Use of mulches are connected with problems like pests, weeds, disease encouraging micro-organisms, white ants and rodents. Highly inflammable materials should be avoided. It is not easy to produce litter for mulching of the orchard unless it is grown in the garden for the purpose. It is an expensive item too. The orchard in mulch system can be sprayed economically for the control of diseases.

Rodents are active in the sod system and still more in the mulch system yet the slopping or hilly areas where soil washing and erosion may be a problem mulching may be practised with ease and profit.

Clean Cultivation

Any cultural operation in undulating or sloping land is harmful. Provision of permanent cover of strips of grass between the narrow strips of clean litter of planting rows of fruits trees may be practised to minimise erosion losses. Sometimes it is found necessary to keep on working the soil, mainly to stimulate bacteria action, helping the formation of nitrates mainly from protein (insoluble nitrogenous compounds) present in the organic matter. Ploughing or disking may be done according to the needs of the orchard. This method (clean cultivation) has the obvious disadvantage of compacting the soil, depleting organic matter and increasing losses due to erosion besides loss of nitrates through leaching.

Shallow cultivation is more useful conservation of soil as against deep cultivation. The evaporation losses from cultivated soil is not much different from that of bare, uncultivated soil showing that cultivation is effective through its capacity to reduce the weeds in any given area. Increasing the frequency of cultivation increases the effectiveness of the method but to a certain limit only. For the arid or semi-arid areas where summer rains are unknown and the weed growth is limited intensive tillage shall not benefit.

Clean cultivation, in recent years, has not found favour with horticulturists since along with the weeds and vegetation nutrients are also removed. The tillage operations accompanying clean cultivation also proves injurious. With the supply of organic matter clean cultivation may give excellent vegetative growth and fruitfulness. Thus organic matter supply is one of the factors that determines the usefulness or otherwise of clean cultivation. The losses of organic matter and of nitrogen are greater in arid zones specially under the condition of clean cultivation. With the depletion of organic matter and humus, the working of the soil becomes more difficult so also the spread of the roots in space. In general therefore, clean cultivation, in the strict sense, should not be encouraged as a practice.

Tillage and Cover Crop System

This method, as the name implies, combines tillage operations as in clean cultivation and also the growing of cover crops at suitable time. Clean cultivation through cover crop usually to be ploughed in is a practice which has almost identical advantages as green manuring in fields. It helps in the conservation of moisture and nutrients, increase in organic matter and nitrogen, and in checking the losses due to erosion and runoff. This practice is of immense value in young orchards which need more nitrogen. The time of use of cover crops is to be adjusted to suit the climatic requirements and soil conditions of the orchard.

The soil is left uncovered for a period of about two months and this is helpful for the sterilisation of the soil, the destruction of insects and allows cultural operations with ease. Such growing of fruit crops lead to a competition of nutrients with the fruit plants but if the cover crop is a legume this aspect is minimised. Crops with shallow roots and thick cover prove better than deep rooted ones since it does not induce unhealthy competition in the immediate vicinity of the root zone of the fruit trees. The practice helps in the process of nitrification and makes available nitrates in greater abundance to the fruit trees. In orchards where the building up of organic matter and nitrogen supply along with thrifty water use are possible cultivation and cover crops pay.

Intercropping carrots, tomato, radish, cauliflower, onion, potato, chillies and short-term fruit crops like papaya, banana and pineapple is generally done for additional income from the land during the period when the chief fruit crop is still young to be remunerative. It also helps in the proper care of young trees and make them grow early to size. Orchardist must use such intercrops which have ready demand in the market and give good returns. Cover crops serve to minimise erosion losses as well as draw on excess of moisture during the monsoon period. These crops must not be allowed to be on the ground when erosion is not to be offset. Competition for the limited supply of soil moisture between the crop and the cover crop is an aspect needing attention. Cover crop must have quick growth and for that Sunnhemp (*Crotalaria* sp.) Cowpea (*Vigna* sp.), Senji (Clover, *Medicago, Lespedeza, Mililotus*) and Dhaincha (*Sesbania* sp.) are useful depending on the soil and climatic condition of the place and available water supply. Berseem (*Trifolium alexandrinum*) and Lucerne (*Medicago sativa*) in the north and *Pillipesara* (*Phaseolus trilobus*) in the south have been used as green manure or fodder crop in the orchards. Legumes prove better as cover crop than other grasses owing to added characteristics of nitrogen addition through symbiotic fixation. The cover crops suitable for winters are rye (*Secale cereale*), oats (*Avena sativa*), clover and alfalfa or lucerne (*Medicago sativa*) and for summers cowpea (*Vigna catjung*), soybean (*Glycine max*), sudan grass (*Sorghum* sp.) etc.

Use of cover crops is more harmful to the young orchard as against the old ones. Small grain crops need not be used as cover crops so far as possible because they drain heavily on soil moisture. This is more injurious specially when the fruit trees also have a heavy demand. Growing of tilled or hoed intercrops is not so bad. The cover crops generally remove excess of soil moisture and consequently the trees do not suffer due to water logging. Checking wind velocity in the immediate neighbourhood of the soil account for the advantages due to cover crops by lowering the temperature and checking evaporation.

Intercropping of Fruit Plantings

Intercrops are grown successfully in the young fruit plantings with a view to derive some income from the land before the fruit trees attain the bearing age. This is done profitably when care is taken to safeguard the long range need of the fruit trees. The chief precaution which need be taken to achieve the end in view is to grow such crops which do not make a heavy draft on the soil for moisture and nutrients. It is specially important since the fruit plants need heavy nutrition. Growing of wheat and oats are discouraged as against the growing of short-term leguminous crops. The crop should be such which does not encourage insect pests or diseases and should not stand in the way of normal pruning, spraying, fertilization or other practices that may be useful for the well being of the young fruit trees.

Intercropping should not be such as to require manuring, cultivation or stirring of the soil at a time which may force delayed tree growth so that it may not have time to mature properly before the onset of unfavourable season. Annual crops have to be preferred to perennials for obvious facilities in looking after the main crop. Intercrops should be taken in rows and at a distance of at least 3 metres from the fruit plantations on each side. This distance should be increased with the advancing age of the fruit plants to avoid any chance of damage to roots or branches. The space needed for the fruit trees should receive the culture and care which it needs. Better returns being the aim of intercropping, the selection of the intercrop must take into account its utility as a feed or food for home consumption or its value as an article of sale in the market. A crop which is in high demand and is likely to fetch good dividends should be preferred to one which does not ensure remunerative returns.

It may also be profitable to grow vegetables for home consumption. Most profitable returns are possible if the intercrop forms a valuable feed to the poultry or cattle at the farm itself. Pea, bean, tomato, brinjal, trailing gourd etc. may also be grown as intercrops. Growing of intercrops also forces cultivation of the soil which otherwise may not be cultivated at all only out of neglect.

Organic Matter

Organic matter has been used for long to promote porosity, availability of phosphorus, water, aeration, base holding power of the soil, among other things. It is specially useful in increasing the microbial activities of the soil. The soil is a dynamic body containing microbes that feed the soil constantly and effectively. The dead and decomposed micro-organisms also add to the organic matter content of the soil, the dead microbes release lot of nitrogen held up in their bodies as part of their microbiological activities.

The micro-organisms decompose organic material through enzymatic digestion, the release of carbon dioxide from the soil during the process being the best index of the activity of the micro-organisms. The water-soluble proteins are decomposed to aminoacids and then to available ammonium forms which could be converted to nitrates. The organic matter added to the soil as excreta of animals, through green manuring of farmyard manure and the decaying of roots or stubble form the chief source of organic matter. The substances that resist easy decomposition *viz.*, legnins, waxes, fats, proteins etc. form the humus which is non-crystalline and colloidal. Humus has better absorption capacity than many soils and is capable of affecting the structure and texture of soil which have direct bearing on the water-holding capacity and aeration of the soil. This store-house of nutrients is easily depleted and needs replenishment specially under tropical and subtropical climates. Legume or grass sod increases it permanently throughout the profile specially when deep rooted as well as shallow rooted crops are grown. Practices like green manuring do not have long-term effect. A temporary unavailability of the available nitrogen may be experienced when plant material with high C/N ratio are used as organic matter due to the fixation of nitrogen by the micro-organisms.

Effect of Cultural Practices on Temperature

Besides influencing the organic matter content of the soil cultural practices also affect the temperature of the surface. soil. The Sod-covered or mulched soil maintains a temperature different from the atmosphere whereas the base soil or one supportig a cover crop are more prone to the temperature changes of the atmosphere. Soil under sod system have lower temperature than either cover or base soil by about 3°C.

Effect of Moisture

Since there is a close relationship between soil water and the yield of crops the maintenance of proper moisture of the soil is of significance. Sod-mulched system contains more moisture than the base soil or that under cover crops alone. Sod method of management makes a demand on soil water to some extent and depletes the soil of the same to about 2.5 acre inches of water for every 0.5 ton of mulch produced. Surface evaporation leading to greater losses need be checked through such cultural practices that can obviate the losses. Moisture is conserved more with the increase in the frequency and depth of cultivation, within certain limits. Mulching by straw greatly increases the potassium availability of the surface soil which otherwise tends to be fixed.

The cultural practices which affect the increase and maintain the movement of soil capillary water should be encouraged. Optimum water content for the growth of plant is at about 50% of soil solution. Some water of the plant tissue is held in an absorbed or colloidal from and this determines its drought and frost resistant capacity. Soil texture controls the lateral movement of water. The method of application of irrigation water is very important in the maintenance of soil texture.

Root growth, a characteristic of the cultivar of species of plant is modified by moisture-air balance in the soil. Shallow soils are not considered adequate for fruit culture and even for the plants in which good proportion of the roots are limited to the top 37 cm of the soil. Tillage, use of cover crops, or intercrops may also affect depth of the roots in the orchard considerably. Peaches, apples and oranges may need more water than grapes and cherries. The cultivator may take recourse to such cultural practices that help in making proper use of the existing water supply and its conservation in the soil, by checking losses. The plants may be given such spacing which prove useful. Deep rooting rockstocks and better irrigation methods (basin, flooding, furrow or overhead) to suit the demands prove profitable.

Nothing is as damaging as lack of requisite water supply when the plants are in bloom or when the fruits are developing. Excessive irrigation is more harmful than scanty irrigation as most of excess water percolates and carries dissolved nutrients to the ground waters, beyond the easy reach of the plants. Growth rate of fruits when decreased due to shortage of water supply may be increased by irrigation. Incipient wilting during the noon or in the afternoon may be used as visual guideline for adjusting the frequency and level of irrigation. Lack of nutrients conditioned by inadequate moisture conditions often limit the growth of the plants and also the productiveness of the soil.

Existence of hard pan near the surface of the soil is definitely harmful to the fruit trees. Shallow soils that do not support good growth of the plants due to the existence of hard pan in the top 30 or 45 cm of the soil should be improved by breaking hard pans mechanically.

Effect of Cultural Practices

Fertile soils do not respond to the cultural practices to the extent as the non-fertile soils. Marginally fertile soils respond to cultural practices, improve their physical condition and become productive. No system of cultivation seems to be profitable unless it is economical. The cost of cultural practices including the price of land must be worked out to show a lucrative balance. Sod system is less costly if it is used with proper fertilization.

Cultural practices chiefly aim at larger production through better growth and development of plants. The system should strive to make available optimum quantities of nutrients and moisture to the crop especially at critical period of requirement. If it helps in the availability of more moisture or nutrients it may not prove useful. Excess of moisture and nutrients turn to be disadvantageous by stepping up vegetative growth and delaying the reproductive phase causing loss in yield.

CHAPTER

5

ELEMENTAL ROLE AND NEEDS

Fruit crops need more nutrition than field crops yet, in India, the fertilization of orchards is incomplete, faulty and uncommon. Fertilization must not be done at random since it benefits only when done in accordance with the set principles keeping in mind the specific needs of the plant as well as the soil. Experiments may be done on small scale to formulate the scheme of fertilization which shall prove profitable for a fruit orchard under the climatic conditions prevailing at the place.

Several types of manures are available, the one suiting the requirements should be used. Manures are termed as complete since these fulfil the requirements of a good majority of the elements required by a plant. A fertilizer may be called nitrogenous, phosphatic, potassic or calcareous in accordance with the element in which it is rich. Fertilizers may be used as a basal dressing or top dressing and also as foliar spray in very weak aqueous concentrations. The bulky manures like cowdung, farm-yard manure, compost and the like should be supplied as basal dressing alone. Green manuring may also be practised where the water supply is not a problem.

Plants may be likened to a machinery for manufacturing organic substances from the raw materials *viz.*, minerals, water, carbondioxide etc. available from the surrounding medium, the sun's energy serving as fuel and the chloroplast as combustion chamber. Enzymes and hormones control the tune and nature of the finished product. All the nutritional requirements controlling growth are as important replaceable by another. Any one or more of the essentialities needed by the plant may adversely influence the growth of plant when limiting. The subject of plant nutrition is fascinating as well as of interest since it forms the pivot on which the dynamics of the living world rests. The chief sources of the nutrients of plant are soil (Fig. 5.1). The medium of growth, and the atmosphere in which the plant grows and develops; water, carbondioxide, nitrogen, phosphorus, potassium and many other elements both in macro and micro quantities are needed by the living plant. Of the minerals considered essential for the nutrition of plants though not needed in very high amounts, sulphur, magnesium, iron, manganese, boron, zinc, copper and molybdenum are possibly of major significance and have been studies over long periods by plant physiologists, biochemists and agronomists.

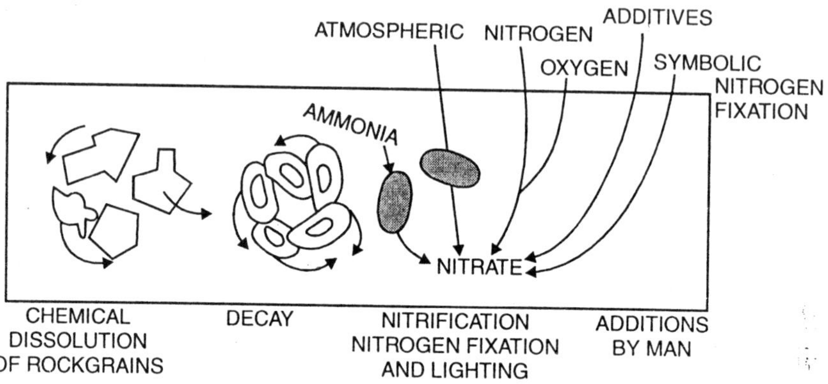

Fig. 5.1. Replenishment of soil

The elements needed by plants have been classified as (*i*) structural (*C, H, O, N, S, P, K*), (*ii*) biocatalyst (*Fe, Cu, Zn, Mn, Co*), (*iii*) indispensable (*K, Ca, B, Mo* and for some plants *Si, He, Al*), (*iv*) Radioactive (*Ur, Th, ha*) (*v*) stimulators (*Na, Cl, As, Ti, Co, Ni, I, Br,* etc.) by Shkolnik in 1950. The physiological role of the elements also portrays an insight into the elemental activities of the plant. How these elements affect the visual characters of the plant have been gathered and presented here to help the grower to identify and find out the corrective steps needed by the plant.

Carbondioxide

This gas is present in the atmosphere and is available to the plants without any chance of its deficiency being created. The importance of CO_2 as the chief source of carbon to plants and the form in which it is assimilated by these may be gauged from the fact that on dry weight basis plant contains about 50% of carbon. Atmospheric CO_2 makes its way into the tiny pores of the leaf known as stomata (Fig. 8.18). The gas then diffuses into the intercellular spaces and finally into the chlorophyll containing structure in dissolved state. The gas is utilized by the plant in the manufacture of carbohydrate and sugars which provide energy necessary for plant processes as these are broken down during the exergonic process of respiration. The plant is the central figure in the carbon- oxygen-water cycle found in nature (Fig. 5.2).

Plant respiration forms one of the chief processes that helps in the degradation of carbohydrates and other metabolites *viz.* protein, fat etc. elaborated in the plant for the release of energy. Living plant respires day and night irrespective of the light conditions while photosynthesis or carbon assimilation takes place in the chlorophyll bearing cells in the presence of light. During photosynthesis CO_2 of the atmosphere is utilized by the plant while during respiration it is returned to the atmosphere in lieu of the oxygen taken though not in the same proportion in every case.

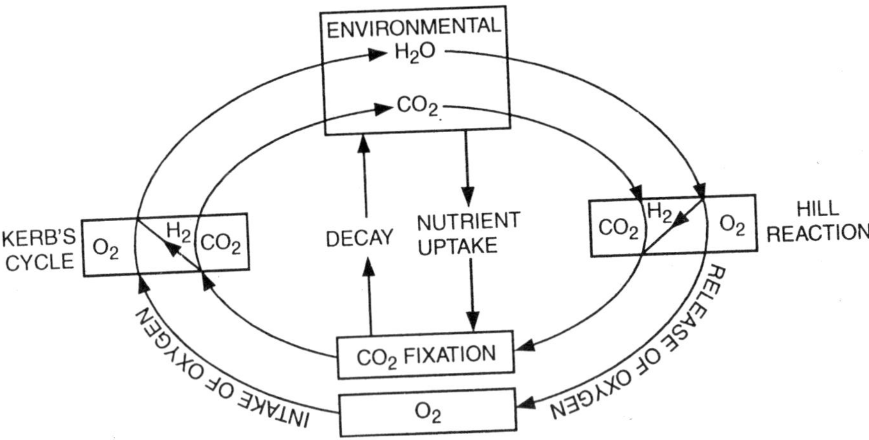

Fig. 5.2. Interlinking of the oxygen taken in by plants during respiration and its release in photosynthesis. Oxygen is incorporated into water and may be linked with the water or carbon cycle.

Water

Water, one of the initial requirements in the process of photosynthesis is needed by plants for the maintenance of life, as a constituent of protoplasm, the hydrogen therein being utilized in the formation of carbohydrates, amino acid, organic acids proteins, fats etc. It helps in translocation of raw food materials and also finished products from one place to another. Substances which have higher proportion of hydrogen give off more heat than those which have less of hydrogen *viz.* complete oxidation of fat gives off 9.3 cal while sugar yields only 4.1 cal of heat per gram of substance.

Nitrogen

Nitrogen is needed by plants in large amounts. It is an expensive element mostly used as combined nitrogen principally as nitrate and ammoniacal nitrogen. Nitrogen is important as an essential constituent of protein, it is also present in chlorophyll, alkaloids, aminoacids, amides etc. Supply of nitrogen to the plants gives a healthy green colour to the plants, the vegetative growth is stepped up and maturation delayed.

The supply of nitrogen controls, to a certain extent, the formation of new cells and therefore growth in which sugars are utilized as substrate of respiration. Plants conditioned, at optimum, for high rate of photosynthesis take in large quantities of nitrogen and the sugars formed are used up in the production of plant organs. If under similar conditions of photosynthesis nitrogen supply be limited the growth of plant organs shall be restricted and the utilization of sugar will be less rapid for vegetative growth. Sugars left over are utilized for the development of flower bud, fruit, seed etc. The carbohydrate/nitrogen ratio in the plants thus controls the type of growth. High C : N ratio favours reproductive development while the low favours vegetative growth.

Nitrogen deficiency leads to the yellowing of the middle vein of the monocot leaf while the entire dicot leaf turns yellowish green. The older tissues develop the symptoms early. Adequately liberal supply of nitrogen prompts increased vegetative growth leading to greater spread, more leaf area of dark green colour. The fruits which appear later and after some delay are usually, larger, healthier and often lack the colour of the variety. The cell walls of such plants are thin and the development of fibres in limited. The stems are soft and succulent. Nitrogen, which may be present in the free state in the atmosphere is not utilized by the plants. It is used only as combined nitrogen except in some forms which utilize atmospheric nitrogen through symbiotic fixation. Nitrogen therefore has to be transformed into such forms which are readily assimilated by the plants, *viz.* nitrate or ammonia. Nitrogen of the dead organisms and waste products appears as ammonia and this is converted to nitrite and finally to nitrate through bacterial action. The change leading to the formation of ammonia is called ammonification and is brought about by ammonifiers.

Further changeover is known as nitrification, *Nitrosomonas* being responsible for the formation of nitrite and *Nitrobacter* for nitrate. Nitrogen assimilated by the green plants in conjunction with the products of photosynthesis forms plant protein. If the nitrate is not utilized by the plant it may undergo denitrification which leads to the conversion of nitrates to free nitrogen. This is a kind of loss unless it is converted to nitric acid through lightening or fixed by algae, bacteria, and *Rhizobia* (Fig. 5.3).

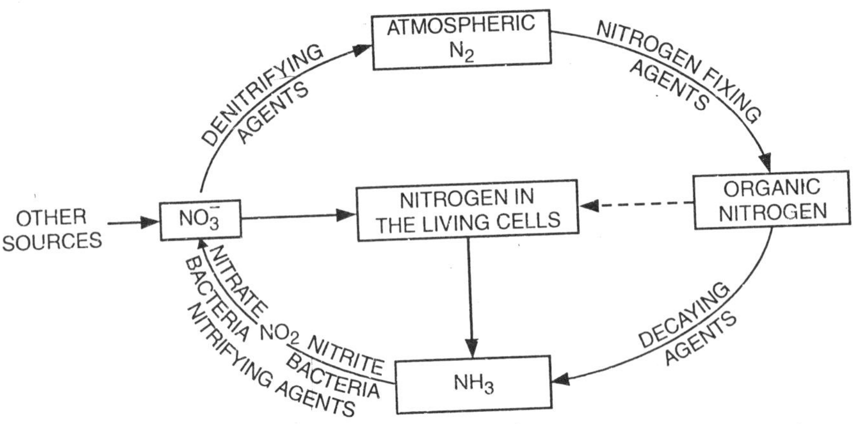

Fig. 5.3 The Nitrogen Cycle

The activity of the several organisms that take part in the nitrogen cycle is affected by environmental factor *viz.* temperature, moisture, pH and oxygen supply. The activity of the ammonifiers and nitrifiers is controlled by changes in temperature just in the same way as that of the plant. Major changes in soil temperature disturb the activity of the microbes considerably. Oxygen is essential for the ammonia to nitrate formers and these aerobes are also better suited to the decomposition or organic matter and making it available to the plants in dissolved state. The hydrogen ion concentration of the soil is a great controller of both the bacterial activity and the population in the soil. The optimum pH, for most soils, for nitrogen transformations has been worked out to be in the neighbourhood of 7.0.

It becomes necessary to resort to addition of nitrogenous fertilizers *viz.*, ammonium nitrate, amminophos, ammonium sulphate, urea, anhydrous ammonia, calcium nitrate, cynamide, sodium nitrate etc. containing 33, 11, 21, 46, 82, 16, 22 and 16% of nitrogen respectively to keep the plants healthy and the land productive. First four of the above nitrogenous fertilizers are classed as acid forming and the rest as base forming. Nitrogen in excess causes excessive vegetative growth, reduced flowering and fruiting.

Phosphorus

It is taken in by the plants as H_2PO_4 and is linked to organic combinations in highly oxidized form. It is an important constituent of nucleic acid and phospholipids. The nucleic acid may combine with proteins to form nucleoproteins. Phosphates are important as taking part in phosphorylation and as energy of phosphate bonds in metabolic reactions. Fruits and seeds contain a high percentage of plant phosphorus where it accumulates during storage. It is present in meristematic tissues. Phosphorus is essential for early maturity of seed and as a constituent of enzymes it plays a vital role in several plant processes.

The availability of phosphorus in the medium of growth is related to the absorption of nitrogen. Deficiency of phosphorus checks the synthesis of proteins. Thus the supply of phosphorus and nitrogen to the plants should be well balanced. Organic form of nitrogen favours better absorption of phosphorus. Phosphorus deficiency retards growth and induces late maturity. The young leaves are dark green, or greenish green. Sometimes phosphorus deficiency is reflected with purple colouration of leaves. The root as well as the stem are underdeveloped, the latter often turn woody. Only a few fruit buds develop, the blossoms drop off, the fruits that are unattractive and remain small.

Phosphorus may not be easily available as tricalcium phosphate or basic slag, the former existing as rock phosphate and bone meal. Both the forms are relatively insoluble and therefore of not much value to plants as readily available nutrient. Monocalcium phosphate $(CaH_4(PO_4)_2$ and dicalcium phosphate

$Ca_2H_2(PO_4)_2$ are easily soluble in water and in dilute citric acid respectively and therefore readily available to plants.

Excessive fertilization with phosphatic additive should be avoided since the vegetative growth is restricted and the maturity hastened leading to poor production. Phosphorus in excess causes very early maturity and also deficiency of potassium, iron and zinc.

Phosphorus, as phosphates, is redistributable in plants so that during deficiency it is translocated from older leaves to younger leaves, flowers and fruits.

Potassium

Certain salts of potassium, particularly KCl, K_2SO_4 and KNO_3 supply potassium. In general potassium is essential for the reduction of nitrate nitrogen in plants. Its role has been recognised in enzymes, citric acid cycle respiration and photosynthesis.

Symptoms of Potassium Deficiency

The deficiency symptoms of potassium vary with the plants. Potassium deficiency is reflected with the yellowing of leaf margin which proceeds along the veins while the margin turns dark brown, a condition known as scorch generally found in apple, pear, plum etc. In *Citrus* there is the drooping of branches and stitching of leaves. Young leaves of tomato, first to be affected, turn light yellow or yellowish green. The plant stops developing aerial portion and if there be any growth it is brittle and the margins curl downward in mild cases while in extreme cases the margins of the leaves turn brown. In cucumber the margins remain green under deficiency of potassium.

Though potassium is not definitely known to be built into essential organic compound in the plant, it is indispensable. The young growing regions are rich in potassium. The potassium content of seed and mature tissues is low while that of young leaves is high. The element is supposed to have a regulatory or catalytic role. Its supply augments carbon assimilation. Potassium deficiency is associated with higher percentage of aminoacids and amides in the tissues while it depresses the carbon assimilation per unit of leaf area and also rate of translocation of sugars from the leaves. Potassium is believed to be associated with the synthesis of proteins from aminoacids. Plants develop serious injury when supplied with ammonium nitrogen in the absence or deficiency of potassium possible due to accumulation of ammonia in the tissues. The reduced nitrogen compounds of sodium does not allow potassium deficiency to affect meristematic activity, tillering, leaf production, leaf size; the leaves, even with high water content, become pale yellowish-green (unaccompanied by reddening) and succumb. The plants maintain a high reducing sugar : sucrose ratio and less carbohydrates. Abundance of calcium and low sodium supply under conditions of potassium deficiency affects differently. The leaves become hard, dark green with a tendency to turn red and die early death. The meristematic activity is reduced, less carbohydrates are synthesized. Low utilization by the plant keeps the quantity of sugar high.

Calcium

Calcium is a constituent of cell wall and associated with permeability. Unlike phosphorus and potassium more calcium is generally present in older leaves. Calcium is found in the middle lamella in the pectic compounds. It may also be found as calcium oxalate in many plant species. Calcium is essential for continued growth of the apical meristem. Organic acids are synthesised as a result of the absorption of calcium and other cations. Absence of calcium renders some species unfit to absorb nitrates. Urea may serve as the source of nitrogen in the absence of calcium. This suggests that calcium is important in the reduction of nitrates in plants. Calcium does not move in plant tissues readily if the medium of growth is deficient in this element. Redistribution of calcium is rather a process in many cases. It plays a role in the formation of pectins.

In the absence of calcium, nitrate diversion becomes suppressed. Calcium deficiency leads to the accumulation of sugars and starch in plant tissues. The roots are stunted and discoloured. The leaves show brown spotting leading to death. The new leaves are mottled and chlorotic. Calcium possibly plays an important role in protein synthesis and the legumes are rich in the content. It seems to neutralise organic acids in plants since the mottled leaves or citrus with less calcium possess more acid. In excess of calcium, the deficiency of potassium, magnesium, zinc and born many arise.

Whenever deficiency symptoms for micro-elements appear foliar feeding is desirable. Two sprays a year is considered adequate to keep the plants free from deficiency : $CuSO_4$ (0.2%), $MnSO_4$ (0.2%), Slaked lime (0.4%), $ZnSO_4$ (0.4%) and Borax (0.5%) are some of the strengths and chemicals generally used. The timings of the year considered suitable for such sprays is February and September.

Sulphur

Associated with synthesis of chloropyll and proteins. A constituent of sinengerin, thiamine, biotine, cystine etc. it is fairly well distributed in the plants. Sulphur may be absorbed by the roots as sulphates and by the leaves as SO_2 when present in the atmosphere.

The chain of reactions which make protein sulphur present in organic matter, available to plants is depicted here (Fig. 5.4).

Sulphur is easily reduced to the sulfhydryl group in the formation of aminoacids and other sulphur containing organic compounds. Deficiency symptoms are more prominent in younger tissues. Sulphur deficiency causes the leaf veins to be of deeper colour than the surrounding leaf blade, a condition which is just akin to the symptom developed under the deficiency of manganese, iron and magnesium. "Leaf yellow" is formed in the case of sulphur deficiency. Sulphur is capable of increasing yield of crops rich in proteins. It is rarely in excess in the soil so that cases of its excess seldom arise.

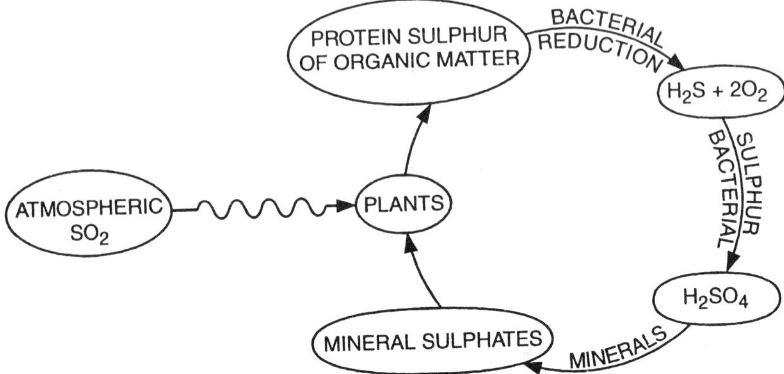

Fig. 5.4. Sulphur Transformation in Soil

Magnesium

The only mineral constituent of the chlorophyll molecule (Fig. 5.5), present in seed and considered essential for activities associated with the cambium.

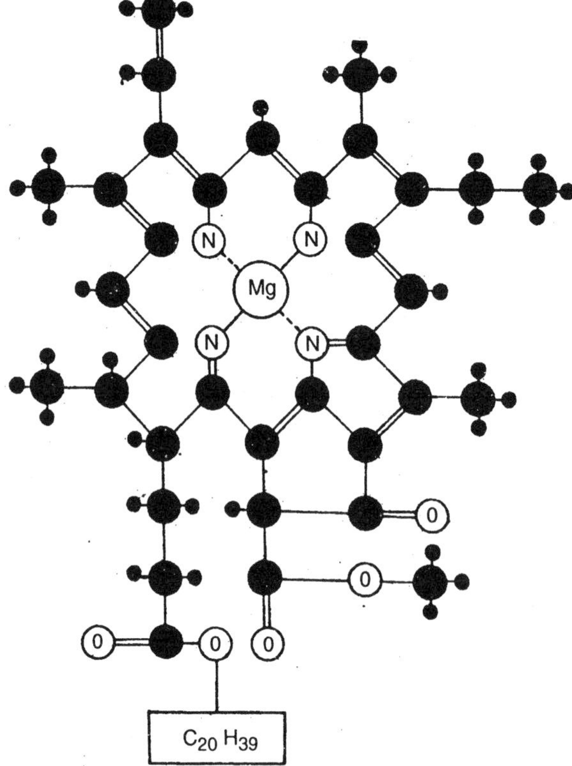

Fig. 5.5. Chlorophyll, a molecular structure

Magnesium play a vital role in the formation of chlorophyll, enzymes and photosynthesis. Magnesium is found in larger quantities in oil bearing seeds as compared to starchy seeds. It is considered essential for the formation of oils in plants.

Older leaves show signs of the deficiency before the younger ones. The deficiency of magnesium does not affect the colour of the leaf veins while the intervascular area gradually change to yellow, the margin being affected first. At times chlorophyll disintegrates, necrotic areas develop, the leaf margins turn brown and the tip of branches die. Cells of magnesium deficient plants are thin walled, strengthening tissue is ill developed and the carbohydrate content low. The flowering is delayed and flower colour is faint. The roots of magnesium deficient plants are ill developed. Excess of the element, toxic in culture solutions, may be offset by the presence of calcium in high doses. Magnesium toxicity in soil is not known except in the case of some alkali soils. Whenever magnesium is in excess it causes calcium deficiency.

Manganese

Manganese, a catalyst in chlorophyll synthesis, photosynthesis, activation of enzymes, is also essential in respiration and nitrate reduction. It is quite abundant in leaves and other physiologically active tissues. Its redistribution is rather slow, making the element more or less immobile. In the oxidation-reduction phenomena, specially of iron compound, manganese possibly plays a direct role. It activates enzymes like dehydrogenase and carboxylase. Manganese deficiency causes chlorosis which is of characteristic type. Common deficiency diseases are grey speck of oats, frenching of oats and yellow of sugarbeet. Manganese deficiency is not so severe as that of iron, the necrotic areas are similar and develop in the middle of the leaf and the veins remain green while the leaf gives a mottled appearance. Excess of manganese may, at times, give the symptoms of iron deficiency by converting the available iron into inactive ferric condition. "Yellow" disease of the pineapple on the black soils of Hawaii, rich in MnO_2, shows pronounced chlorosis of leaves. Root injury may be caused by toxic effects of manganese and immobility of iron. Excess of manganese causes die-back of leaves and shoots and blotchy leaves of peach, sometimes chlorosis may also take place.

Iron

Associated with enzymes cytochrome and formation of chlorophyll, iron is generally supplied in the form of iron sulphate or certain chelates. It is indispensable for chlorophyll synthesis. In association with enzymes like catalase, peroxidase, cytochrome oxidase and the cytochromes iron compounds are of great significance in cellular metabolism of plants. It is an immobile element as is seen by the transfer of a seedling from a complete nutrient solution to one deficient in iron when the new developing leaf exhibits chlorosis. Again if the seedling is transferred to a complete solution the succeeding leaves are healthy but the earlier ones continue to show chlorosis.

The element may not be available more commonly due to its insoluble form rather than its total absence. Excess of iron also causes similar effects. Iron deficiency is rarely a limiting factor in plant development. Characteristic chlorosis is developed in the deficiency of iron where intravascular area of young leaves turn light grey to yellow. At time the precipitation of iron as iron phosphate in the vascular tissue makes it immobile and it does not pass to the mesophyll causing chlorosis even though the quantity of iron present be considered more than adequate. Brown necrotic areas large than those caused by manganese deficiency develop in the efficiency of iron. When iron is in excess phosphorus and manganese deficiency are reported to set in.

Boron

Boron is considered to be associated with enzymes, synthesis glutamine, and necessary for the meristematic activities and also for the normal development of the conducting tissue. Crops like beans, cabbage, cauliflower, celery, citrus, grapes, lettuce, peas, turnips, tomatoes need boron for their growth. Boron is needed in traces, 1-5 ppm in general, and 20-25 ppm rarely. It becomes toxic with slight increase in dose and in most cases it is considered to be safe not to touch the limit of maximum requirement of the element. Excess of boron may cause browning of old leaves and gummosis on *Citrus*. It is considered more necessary for leguminous plants, and for the development of nodules and fixation of nitrogen. Boron deficiency symptoms are characterized by cessation of protein synthesis evidenced by the accumulation of carbohydrates, nitrogenous and ammonium compounds. Death of the apical meristems of the root and the stem is also caused under the conditions of boron deficiency. The younger leaves tend to role half circle from the tip to the base and they become thick. The common physiological disease caused by the deficiency of the element are 'Heart Rot' of sugarbeet, 'Brown-Heart' of turnip, 'Browning' of cauliflower and 'Leaf-fall' of potato. These diseases may be eliminated by foliar spray of boron which ensures quicker action. The affect of boron deficiency in the case of sugarbeet is generally more severe in alkaline soils. The physiological disorder known as "black tip disease" of mango is easily curable by the application of boron as borax @ 6-7 kg/ha as foliar spray in three divided instalments between the flush and the time the fruit is about 6 cm in diameter, the deficiency may also be caused when the rootstock is not in position to take-in sufficient boron to meet the requirement of the scion, or the intake amount is not in excess of the needs of the root stock itself.

Zinc

Role of zinc lies in formation of auxin and also association with enzymes. Zinc is considered necessary for the growth and development of cereals, vegetables etc. Essential in traces though highly toxic even in slight excess. Iron deficiency and partial leaf necrosis may develop in excess of zinc. It is generally supplied as sulphate or chloride. Zinc is considered to be essential for the formation of chlorophyll and thence useful towards photosynthetic activity. Zinc is a constitutent of some enzymes and possibly takes part in the synthesis of indoleacetic acid in plants.

Formation of rosettes of 'pecan', 'little leaf' or deciduous fruit trees takes place in the deficiency of zinc, other symptoms are the development of 'mottle-leaves' in citrus, bronzing of tung, 'white-bud' of maize and yellow leaves in walnut. There is dwarfing of vegetative growth and seeds are not formed. It has been reported that zinc deficiency causes structural defects in the root tips. *Khaira* disease in paddy crops is controllable by $ZnSO_4$ + slaked lime sprays in the proportion of 1 : 0.5.

When in excess, the leaves of affected plants, become chlorotic and develop small reddish brown interveinal patches. The symptoms resulting from an excess of zinc resemble those of manganese deficiency. In geneal, excess of zinc affects plants unfavourably.

Copper

Activation of enzymes, synthesis of ascorbic acid are considered to be the role of copper. Copper, a constituent of the oxidizing-reducing enzymes, is considered an essential element. It may be generally supplied to the plants as copper sulphate. Copper deficiency is responsible for the "Reclamation Disease" and the "Dieback" disease of citrus. The reclamation disease occurs on newly reclaimed peats or moorland soils. When the plants turn yellow, the top leaves become white and death accrues. Cereals, sugarbeet and leguminous crops are generally affected by this disease.

Molybdenum

Molybdenum changes nitrates to nitrites besides synthesis of tanin. Yet another element considered indispensable for the autotrophic plants. It has been shown to be essential for tomato. Myrobalan seedling of the plum and for cabbage. In the absence of molybdenum the seedlings develop slowly, possess short internodes, and remain dwarf while leaves have light brown areas of dead tissue at the tip and margins. The need of this element is perhaps the least *viz.* 0.1 ppm (tomato). Molybdenum helps in the reduction of nitrates. Molybdenum deficient soils may be corrected by the addition of 42 gms of molybdenum oxide (MoO_3) to an hectare of soil.

Excess of the element is toxic, produces tints on leaves, and such plants when fed to cattle cause ailments.

Sodium

Essentiality of the element to plant metabolism is not established. The element forms a constituent of plant ash almost in all cases. Flax, barley, carrot, cotton and tomato benefit by the application of sodium where potassium is lacking in the medium of growth. Sugarbeet, turnip, celery benefit by the addition of sodium when potassium is sufficient. The element as sodium carbonate causes increase in the nitrogen status of wheat grains in certain soils besides increase in yield. Sodium chloride is useful for plants under certain conditions but beyond a specific dose it becomes toxic. Its excess causes calcium deficiency and inhibits plant growth.

Cobalt

It affects plant development by activating the effect of cobalt ions on plant enzymes *viz.* carboxylases and peptidases. Cobalt is a constituent of vitamins B_{12}. Its deficiency causes diseases to cattle when fed on forage raised on cobalt deficient soils.

Lithium

It is toxic to plants, mottled leaf disease of *Citrus* is a well known example. Lithium, in minute quantities, enables wheat plants to withstand rust.

Iodine

Iodine has not been shown to be essential but when present in soil it is taken in by the plants and may be helpful in correcting the iodine deficiency symptoms in men and other animals. The tomato plant shows sign of iodine toxicity with a supply of 10 ppm of iodine as potassium iodide.

Flourine

Its apparent role in preventing dental carries in men have given it some importance recently. It is easily absorbed by plants in acid soils but 10 ppm of flourine as NaF proves toxic.

Aluminium

It is present in plant ash and is considered essential to plants. Beyond 1-5 ppm concentration it may produce toxicity, addition of lime or phosphate reduces the toxicity by lowering the solubility of aluminium salts present in the soils. It reduces considerably the absorption of ammonium ion by cut potato tuber tissues. Soils with a pH of 5 or less are good in availability of aluminium as against those of higher pH where the availability is reduced very considerably. The blue colour of the flowers of *Hydron macrophylla* is linked to the presence of more aluminium than that present in the pink ones. When in excess shoots and roots appear to be as in phosphorus deficiency, roots are ill developed.

Silicon

Silicon influenced the phosphate metabolism of plants possibly through increase in its availability also. It is abundant in bark of trees, in *Equisetum,* sunflower and in certain other plants to a considerable extend. But for the experiments of Lipman and Rale˙gh silicon was considered unessential to plants because trees could be grown to maturity in culture solutions devoid of silicon. For sunflower, barley and beet the element is considered essential.

Chlorine

Chlorine in excess causes browning and shrinking of leaves as in potassium deficient and current plants.

Selenium

The mineral is a strong antioxidant thought to reduce the risk of certain cancers, as well as heart diseases. Brazil nuts are the richest known food source of the mineral selenium.

CHAPTER

REGULATION OF FORM AND FUNCTION

It is infrequent to find cultivated fruit plants with scanty bearing and wild growth. For a long time the training and/or pruning of fruit trees have been practised to get increased fruit production. While training concerns form, pruning affects function, the two practices have been regarded as inseparable or complimentary to some extent. A plant may need some training which is done by providing support (over a trellis) or staking or fostering in certain ways. Training controls the shape of the plant while pruning aims to assist the working of the fruiting machinery at full efficiency, year after year. Both in the case of training as well as pruning the knowledge of growth and fruiting habits of the plant is essential. Peach trees bear fruits on shoots of the past season only from nodes at which no lateral branches arise. The walnut bears on the shoots produced by the subterminal buds of the previous season. The functioning of fruiting machinery influenced by age, vigour, food supply, environmental and other factors may be modified to advantage by adequate pruning and thinning. The exact nature of the fruiting habit of different species and method of their modifications should be known for the operations to be profitable.

Pruning aims to increase yield, reduce cost of production and improve the quality of the produce. Pruning done for the form is known as training. It aims at proper distribution of fruit bearing parts, control of insects and diseases, checking loss from breakage and development of proper colour of the fruit. The chief operations in training are heading-back, provision of reasonable number of well spaced scaffold branches, besides preventing the formation of weak crotches.

Training for Form

The growth habits of a plant are characteristic. These cannot be altered but can certainly be modified by inducing the top growth on a particular side. Such a control on the direction of growth is known as training. It develops a satisfactory framework and promotes early production. The shaping of plants, referred to as training, *e.g.*, peach trees to an open centre (Fig. 6.1B), apple trees to a modified leader (Fig. 6.1C), grape vines to a cordon pruned and double-guyot-pruning systems (Fig. 6.2) and the dwarf fruit trees to a cordon.

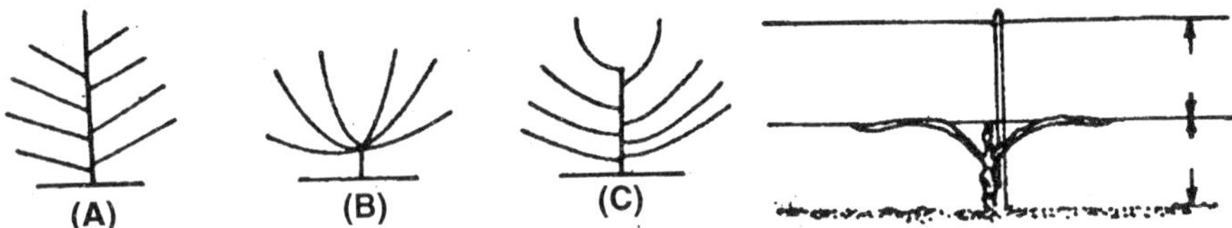

Fig. 6.1. Training system for fruit trees, A Central leader ; B. open centre system ; C. modified leader.

Fig. 6.2. Double guyot pruning system.

Training has little effect on fruit bearing and is considered to be a secondary importance in production. It nevertheless determines the grade of quality of the fruit. This effect is the outcome of the facility provided by the operations like spraying that control pest and disease and may be manipulated. The adjustment between shade and sun falling on the plant controls the colour of the fruit thereby affecting its market value.

Training controls the distribution of fruiting wood and the facility and ease of handling of orchard practices. It aims at development of good framework. The tree may be made low headed (four feet of unbranched trunk from the base) or high headed (trunks free from the base). The number of scaffold branches allowed to develop on a trunk may be above 6-8 with strong crotches, large number of branches may induce more formation of crotches, and may lead to the weakening of the trunk besides giving a bushy top growth. The distribution of the scaffold branches is very important ; again these must be well spaced and allowed to run parallel. Open centred trees have access to more height and give bettur colour to the fruits.

Mechanical Strength

One of the chief objects of training is to give a mechanically strong framework to the trees needed to support the heavy load of fruit and withstand the strain of wind. Breakage of brances at random under the strain of circumstances is a loss in many ways. The highlights of maintaining mechanical strength are (*i*) no tree should be allowed to develop more than one main branch, (*ii*) the primary lateral branches should be smaller than the main trunk, and (*iii*) the secondary branches should be subordinated to the primary branches. The branches which arise at wide angles are mechanically stronger than those which have narrow angle due to the fact

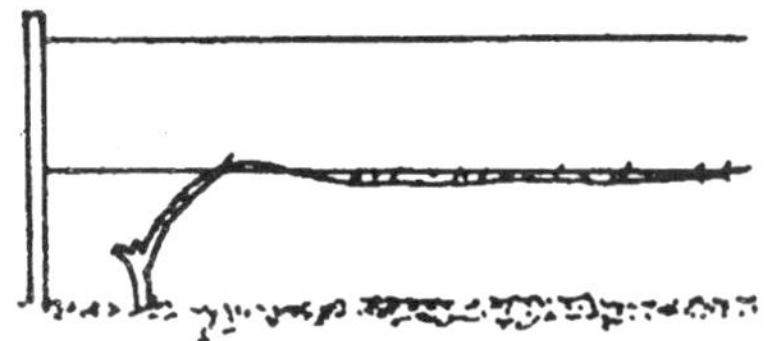

Fig. 6.3. Single guyot pruning system.

the crotch is weaker in the latter. Low trees with well spaced branches exposed to light and air allow better utilization of the solar energy, in larger synthesis of carbohydrate, provide convenience in spraying which is not possible with closely spaced branches specially in the interior regions of the plants. We have examples of tomato trained to stakes being better producers and disease-free than those crawling on the ground. Similar is the case with grape vines and several other plants.

Bearing Characteristics

Plants with many small closely packed branches do not bear well. These may be made to bear heavily by increasing the vigour of the plants, increasing the capacity of the branches by the supply of more nutrition to bear the strain of the fruits through pruning, *i.e.*, removal of surplus branches adequately in a scientific manner. The extent and time of pruning shift the carbohydrate/nitrogen balance of plants leading to change in quality and quantity of bearing. The training of plants is usually done by cutting or by guiding their direction of growth or even through a combination of the two. Training of horticultural plants depends on the kind of plant and the object in view.

Severing of Plant Parts

The operation is of different types and magnitudes. Heading-back is the art of severing of the terminal bud to encourage lateral branching down below the cut. The heading-back must be done within a distance of 15-25 cm from the point where first branches are desired. Clipping of the shoot tips, in the case of gram (*Cicer arietinum*) in India, is an age-long practice. It promotes branching and also encourages symbiotic nitrogen fixation. The twig to be subordinated should be pruned more heavily to reduce carbohydrate supply. Thinning out, removes entirely a shoot, spur, branch, limb etc. (Fig. 6.4). Whereas heading-back removes only a portion (the terminal head) leaving the rest from which new growth develops, in thinning the terminal head of the main branch is intact. Heading-back (Fig 6.4) also leads to dwarfing for reasons stated elsewhere. Thinning out is the process of removal of twig or part thereof according to needs. It is used to provide adequate space to the remaining parts for better flowering and fruiting. Old trees are kept within their limits by thinning. Vertical branches secure more nutrition, water, sunlight etc. and have greater speed of growth. These are therefore removed if the growth of primary branches is desired. Parallel branches should not be allowed to remain since one of these will be

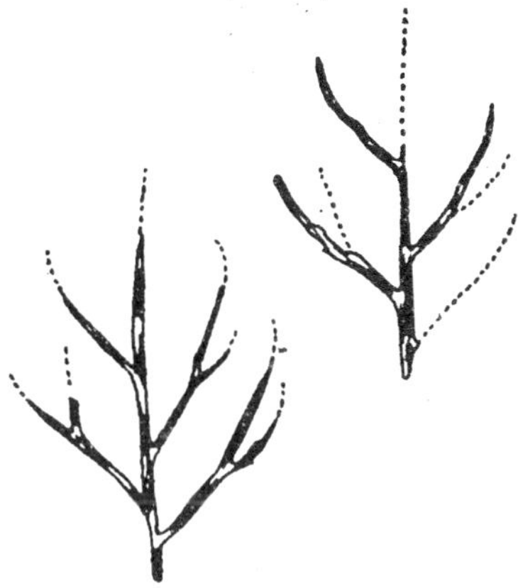

Fig. 6.4. Heading-back (left) and thinning out (right).

shaded to the detriment of the plant and therefore all but one of these must be removed. Proper placing of laterals is also of significance specially in case of vines.

System of Training

Various horticultural plants respond to different systems of training according to the growth habit of the plant. The system is different for vines, shrubs and trees. Vines are trained to acquire the shape and spacing among the branches according to the purpose for which these are grown. The support provided to the vine has close relationship with the method by which the vine holds the support *viz.* adhesive disk, aerial roots, leaf petioles. Taking the example of grape, the best system used depends on the fact that pruning diverts the flow of the sap to the fruiting zone, keeps the plants in manageable shape and increases fruit production.

Vines are trained in many ways according to the single stake system where the vine is fastened to an upright support of a tree called Pangara (*Erythrina indica*). In the kniffen system the vine is trained on a two or three wire horizontal trellis. Bower system consists in the climbing and spreading of vines in the trallised roof. Head system has the vines trained as dwarfish bush about 90-120 cm high. The cordon system has vines trained with long straight trunk which is bent and fastened horizontally on the vine trellis. Training is therefore done in a manner that the vine gets the shape obtained by the desired system.

In the case of shrubs training is mainly done to maintain either informal or formal shape. When the shrubs are trained to formal shapes according to the need of the occasion, it needs some pruning.

With trees the different types of training consist of the following forms : (*i*) normal, (*ii*) central leader, (*iii*) open centre, (*iv*) modified leader, (*v*) espalier. Trees are trained with the sole object of eliminating the development of weak crotches, too tall, too dense or too low trees and also for easy harvesting of fruits. Early bearing reduces size of produce and also the duration of profitable production. Training helps to rectify this defect also.

Normal

When plants are allowed to grow without any training many small branches develop making the tree compact. Very dense foliage when poorly placed encourage low grade fruits of small size. Operations like thinning, picking, spraying are difficult and costly.

Central Leader

The main trunk is allowed to develop without interruption. Primary lateral branches arise from this main trunk at various intervals (Fig. 6.1). Training mainly aims at the selection, spacing and arrangement of the primary laterals. Adequate subordination is maintained in these laterals. The trees are not trained for keeping the height under check.

Open Centre System

It is also known as the open-head or vase form wherein diverging branches of approximately equal diameter at the base, are developed. Generally the growth the uppermost branch is prevented and endeavour is made to have almost equal leaf surface on the lateral forming the main scaffold branch. This system leads to a spreading structure, though with somewhat weaker crotches, facilitating thinning, spraying, picking, removal of diseased portions and other operations. In this type the main trunk is severed to a lateral branch and 3-6 nearly equally developed primary lateral branches are allowed to develop from the trunk (Fig. 6.1). In this system, as the primary lateral develops, mechanically weak structure is developed which needs support. It exposes the tree to the maximum extent and this has a uniform distribution of fruits on the branches. The fruits are highly coloured. Since the tree is of short stature, the cost of spraying and harvesting are minimised. It is generally practised in peaches, apples, cherries and some types of plum.

Modified Leader

A newly planted plant often shows a branch more prominent than others so as to be known as 'leader'. This leader may be modified to support other scaffold limbs properly. This is known as modified leader system which should provide, when properly formed, stronger crotches, better spacing between laterals and more height than open-head tree. The dominant lateral to be selected as leader must be a fast growing one near the trunk terminal. The desired result *i.e.*, disproportionate development of other laterals and also the leader is achieved by cutting the laterals at unequal lengths to induce unequal photosynthetic surface and diameter of the branches. Whenever the object is not achieved another leader is selected and the operation repeated. The time of the operations to develop the modified leader has advantages not to allow more than a couple of secondary laterals to develop from a primary arising on the main axis as crowding causes shading which has its own disadvantages besides that arising due to structural defects.

The central leader is allowed to grow to about 1.8 to 2.5 cm in height and then cut back. The tree assumes a shape as may be expected by a combination of the open centre and central leader type. Heading-back and disbudding is followed for obtaining the modified leader type of tree (Fig. 6.1). Pruning is done systematically for four years with the sole object of getting trees in which upright growing branches, parallel competing branches, large rubbing branches, and weak crotches are eliminated. The top must not overgrow the bottom scaffold. Too many

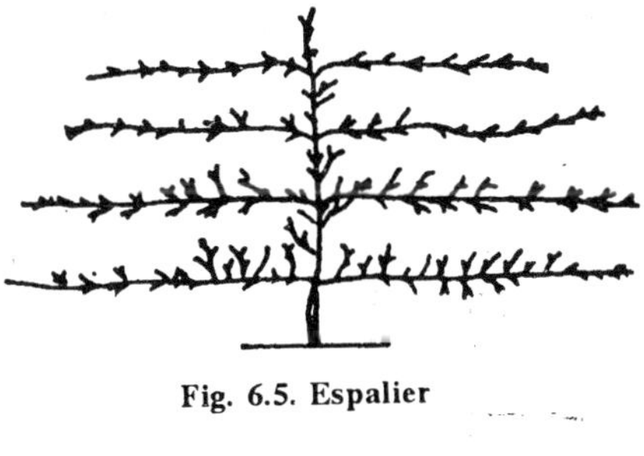

Fig. 6.5. Espalier

scaffold branches are got rid of through pruning. Disbudding may also be done to achieve the desired shape and object.

Espalier

It denotes training of fruit trees to walls, trellises and fences (Fig 6.5). It is common in case of pears.

PRUNING FOR PRODUCTION

Pruning is not an easy job as is apparent by the popular saying "everybody cuts but few prune". Pruning is a science involving the removal of a portion of the plant for the benefit it gives to the grower, in shape of better fruiting over longer periods. It improves the shape of the plant, influences the growth behaviour, flowering time and fruitfulness besides improving the quality of the fruits. Young trees are pruned to give proper shape or appearance and to develop strong wide angled crotches with the sole object of giving the plant capacity to support more fruit without breakage, at the fruiting time. It is also used to remove damaged, diseased or dead parts of the plant. Besides giving structural strength it has dwarfing effect thereby increasing the vigour, removal of water sprouts and suckers. It also reduces the number of flowers on shoots of the current season. It is possible to produce large flowers. Above all it maintains good appearance of trees.

Pruning is an operation based on scientific principles to improve the overall performance of fruit trees. The growth and flowering habits of specific cultivar must be taken into account for deciding the nature and extent of pruning. It is not enough to have pruning schedule based on recommendations on the kind of fruit alone since cultivars differ in their growing habits. A spreading cultivar may be converted into an upright one and *vice versa.* Cultivars vary in the proportion of small branches on them and the pruning style has to be set accordingly. Location of fruit buds and the length of the internodes are other points which need attention for deciding the nature, extent and time of pruning. No pruning is required if the tree is open and the growth, flowering and fruiting are satisfactory.

This operation reduces the tops and also affects roots detrimentally and proportionately. Root growth is detrimentally affected due to the lowering of the supply of carbohydrates and phytohormones whose synthesis has dwindled down as a result of pruning. Heavy pruning is specially harmful, moderate pruning is always recommended as safe. It helps in increased availability of nutrients and water to the remaining portions. Bulk pruning encourages water sprouts. Bulk pruning consists in large cuts whereas details pruning strives to remove small branches only.

Pruning should not be done in young, yet to flower trees unless it is desired to delay flowering. The stimulation of more shoots and spurs considered profitable for bearing can be achieved through pruning. Larger proportion of flowers develops into fruit as a consequence of pruning wherein the remaining growing points get a better supply of nutrients. Although the potential bearing surface is reduced as a result of pruning, there is more production specially when the plants remain unmanured.

Heading-back, based on the principle of apical dominance consists in cutting back of a shoot usually to a lateral so that the growth in particular direction is prevented and the latent buds quickly become active and grow in other direction. Thinning out, another process is generally carried out to remove small portions of the shoot branches etc., from positions lateral to the main stem or axis. The terminal portion of the main stem remains intact in thinning out. Heading-back checks height or expanse of the plant but thinning out only seeks to remove small branches so that shading is minimised, the former operation dwarfs the tree more than the latter. Occasionally heading back encourages growth of water sprouts near the cut.

In effect, pruning improves the size and colour of fruits unless the leaf to fruit ratio is lowered to any great extent since the operation essentially controls the photosynthetic surface and through it the physiological activities of the plant connected directly or indirectly with the leaf. Another important purpose of pruning is the development of strong crotch which needs some experience with the variety for success.

A bud needs proper moisture, humidity and nutritive condition with an accumulation of carbohydrates for differentiation of the floral parts leading to the development of fruit. Some buds grow into a leafy structure and no fruits can be formed on the shoot. New fruit bud develops at some new part on the newly formed shoot. The position of the formation of flower are formed either terminally, laterally or adventitiously. The terminal buds may be formed on long or short (spur) growth, laterally on present or past season's growth and adventitously from any point on the trunk or root. Plants having terminal fruit buds (raspberry or blackberry) do not possess enormous spreading habit and the trees are rather compact. Similarly plants possessing flower buds on the spurs (terminally as in apple and laterally as in sweet cherry) are more compact than those, where these are borne on long shoots (graphs and peaches). Depending on the position of the flower bud developing into a fruit the orchardist has to use short, medium or long pruning system. Essentially three different types of flower bearing shoots are found.

1. Only flower parts are present (single flower as in peach or flowers in clusters, *e.g.*, cherry). No leaves.
2. Mixed bud, developing a short or long leafy shoot terminating in an inflorescence (apple, pear).
3. Mixed bud develop a short or long leafy shoot bearing flowers or flower clusters in some of the leafy axillary ones (Persimmon). These three classes are generally subdivided.

Fruit trees may be classified according to their bearing habits. We have six main types of fruiting habits and consequently training and pruning must be done accordingly to promote the existence of fruit bearing buds.

I. Fruit buds borne terminally containing flower parts and giving rise to inflorescence without leaves, *e.g.*, loquat, mango.

II. Fruit buds borne terminally unfolding to produce leafy shoots that terminate in flower clusters as in most of the pome fruits and a few-others (apples, pear, quince, pecan).

III. Fruit buds borne terminally unfolding to produce leafy shoots with flowers or flower clusters in the leaf axils as in guava.

IV. Fruit buds borne laterally containing flower parts only and giving rise to inflorescence without leaves or if leaves are present they are reduced in size *viz.*, date, peach, plum, apricot, cherry, almond, curant, citrus fruits, walnut (staminate flowers only).

V. Fruit buds borne laterally unfolding to produce leafy shoots that terminate in flower clusters as in blackberry, raspberry, grapes, cashew nut.

VI. Fruit buds borne laterally unfolding to produce leafy shoots with flower clusters in the leaf axils (mulberry, fig, chestnut, jujube, avocado).

Pruning is essentialy a dwarfing practice since total growth of the unpruned tree is always more than the pruned tree under similar conditions of growth. It is so due to the loss of carbohydrates and also the lowering of the photosynthetic processes due to removal of the leaves, the centre of photosynthetic activity. Root growth is also reduced alongside the top growth by pruning method. Heavy clipping (a form of disbudding) is detrimental to the growth of both root as well as shoot growth and also nitrogen fixation as in *Cicer* and Indigo. Pruning is done to maintain plant's vigour, quantum of growth besides improving the quality of yield and factilitate horticultural operations. Pruning, its extent, nature and type depends on the kind of plant, the bearing habit, the age of the plant and purpose for which it is done. Same degree of pruning is not justified for mature and young plants.

In young trees which have not yet come to the flowering stage, pruning delays the differentiation of floral primordia in proportion to the severity of pruning. It stimulates vegetative growth at the expense of materials which are needed for flower initiation. Pruning delays bearing but increases the size of individual fruits as well as of harvest. The delay is due to the fact that pruning stimulates and prolongs vegetative growth, reduces the carbohydrate balance leading to a check in the formation of flower and fruit buds. The varieties that bear flower on spurs, naturally, do not flower till spurs are not allowed to develop. The 'heading-back' delays flowering to a much greater extent than thinning out.

Pruning of young fruit trees at flowering does not inhibit flower formation much since the flowering process once initiated is not suppressed easily. Pruning encourages more flow of nutrients and water to the remaining shoots which flower and that is how the percentage of flowers that develop to form fruits is increased invariably. So long a balance between the loss of potentital bearing surface and the nutrient supply is kept at optimum pruning always encourages better bearing. The extent of pruning alone or the amount of thinning and pruning to be practised is to be decided by the grower from experience to get more yield at minimum cost and for larger number of years. Pruning may be of the heading-back, thinning out bulk or scanty type. The type of pruning determines the yield of tree. Pruning reduces the number of fruit spurs.

Pruning alone at the right moment and to the adequate extent improves the size, colour and quality of fruits. The colour is improved by making more sunshine fall on the leaves, fruits and on a larger portion of the plant. It reduces the transpiration areas and thus the need of the plant at any particular time for water is met with from resources at hand. Pruning, when transpiration rate is high, increases growth.

The old saying that "Pruning may be done any time the knife is sharp" or that "summer pruning for fruit and winter for wood" are no longer held to be scientifically tenable. Pruning when carried out for a number of years systematically gives best results. Light annual pruning is more desirable than heavy pruning done at an interval of two, three or even four years. Annual pruning is also not the rule for all fruit trees, *i.e.*, it may not suit as in apple. Best time for routine pruning is the season when the plants are dormant.

□□□

CHAPTER

PROTECTION AGAINST INSECT PESTS AND WEEDS

Insect-pests and weeds rob plants of nutrients, vigour and reduce their life span besides lowering production level. Although all fruit trees do not have the same problems in respect of these natural enemies, knowledge of the methods that may be of use in combating the menace and minimising the hazard in fruit orchard is essential.

Pest Control

All insects do not always behave as pests. San Jose scale (*Aspidiatus perniciosus*) of considerable consequence in USA was introduced from Southern Asia where it was not considered to be a pest. Similarly the Japanese bettle, *Popillia japonica* (Newman) quite a harmless insect in Japan when introduced into the USA became a serious pest. The European corn borer (*Ostrinia nubilalis* Hbn.) infected maize only after being introduced in America. Any newly introduced pest in a country either assumed importance soon thereafter or after it had adapted itself to the multiphagous mode of life. Competition or high density of population also tends to reduce the rate of increase of a species through physical, physiological or genetical control.

Insects are specific in their demand for temperature, humidity and biotic environment at certain critical stages in their life to be of serious importance as pest. Insect pests prove a menace to plants, animals and their products in storage whenever the environmental conditions so permit. Many insect pests affect the fruit plants, their flowers and fruits to the level of importance. These have been largely mentioned in the description of the individual fruits, if these have assumed significance, in the book. Without going into the details of the types of insects and the nature of damage, the ways to combat them may be outlined here. Cultural, legal, physical, mechanical, biological and chemical methods of control of pests coupled with the use of resistant varieties etc. in addition to the use of **male-sterilization** technique, the pheromones, hormones, anti-feedants, genetic sterility etc. have received varying degrees of success. Artificial measures adopted to control the pests have not been able to eliminate them possibly due to the killing of the predators and natural parasites disbalancing nature's drive to keep these under control.

Cultural Control

Cultural methods of control are generally applied against the tissue-borers which evade the insecticides, natural predators and parasites. Crop rotations, tillage, variations in the method and time of planting and harvesting, clean cultivation, thinning, pruning etc. have been tried successfully in the control of plant pests and diseases.

(*i*) *Stirring the soil :* Tillage operations disturb the soil inhabiting pests to their disadvantage ; packing of the soil helps in controlling cutworms, ploughing exposes the hibernating forms to the natural enemies and renders majority of eggs useless by the blocking of their holes. Frequent hoeing and interculture operations are disadvantageous to the soil-dwelling insects.

(*ii*) *Fumigation :* Protection of new orchards from San Jose scale, *Aspidiotus perniciosus* Coms., infestation may be done with by fumigating of nursery plants before plantation. Citrus plants should be similarly treated, so that the spread of red citrus scale (*Aoniella aurantii* Mask) is checked.

(*iii*) *Regulating Irrigation :* Flooding of the fields kills large number of insects present in the soil and some are driven out and exposed to their natural enemies. It may also adversely affect the stages of development of the insect so that the life cycle is cut short.

(*iv*) *Resistant Cultivars :* Use of resistant strains which are not damaged by insects starves and eliminates the insect population.

(*v*) *Vigorous and early growth of plants :* Growing healthy plants from healthy disease and pest-free seeds also helps in their survival as compared to weak plants. Cucurbits sown early in well-manured fields, have a fast rate of growth early in the spring and withstand the attack of red pumpkin beetle (*Raphidopalpa foveicollis* Lucas) which otherwise completely destroys the plants.

(*vi*) *Clean cultivation :* It helps by the removal of undesirable plants and plant debris, in orchards, that give protection and provide food to the newly emerged nymphs of the mango mealy bug (*Drosicha mangiferae,* Green), for instance, in January and February. For about one month pests feed on th weeds and therefore clean cultivation starves them.

(*vii*) *Cropping scheme and the trap crops.* Elimination of alternative host through specialized farming or crop rotations is also useful though in the case of fruit crops it has large limitations. During intercropping care must be taken to avoid crops which may serve as alternative host to the pests of fruit crops grown in orchard.

(*viii*) *Pruning and thinning :* Proper pruning of the affected portions of plants is useful for keeping under check the citrus leaf miner (*Phylloenistis citrella*

Stainton), citrus psylla (*Diaphorina citri* Kuway), the citrus red scale (*Aonidiella aurantii* Mask), San Jose scale (*Aspidiotus perniciosus* Coms.), the woolly apple aphid (*Eriosoma lanigerum* Haus) and peach leaf curl aphid (*Anuraphis helichrysi* Kaltenback).

(*ix*) *Destruction of crop residues* : Infested fruits of guava, melon, peach, plum, apricot etc. which contain maggots of the fruitflies and the fruits of walnut which contain grubs of the walnut weevil should be destroyed to say the new crop from the emerging adults.

Mechanical and Physical Control

The time consuming and laborious mechanical and physical methods of control consists in the use of devices that physically prevent insects from reaching the plants. Mechanical control comprises (1) hand destruction, (2) mechanical screening, (3) use of traps, (4) crushing, grinding, (5) humidity control, (6) temperature manipulation, (7) electric shocks, (8) light and other radiations, (9) sound waves. Wrapping individual fruits of pomegranate and citrus with butter paper envelopes to save them from the attack of the anaar butterfly (*Virachola isocrates* F.) and the fruit sucking moths (*Ophideres* spp.) respectively. Superheating of empty godowns to a temperature of 50°C or more for 10-12 hours kills the hibernating storage pests. Entrapping, burning or killing of insects is found to be useful. Refrigeration to a temperature of 5-10°C of eatables and woollen clothes kills the insects.

Bio-Ecological Methods of Insect Control

Protection and encouragement of insectivorous wild birds and other animals not harmful to plants, use of domesticated fowls and mammals, introduction, artificial increase and colonization of parasitic and predacious insect, protozoa, mites, worms, fishes etc. Infestation, spread of diseases on pests are examples of methods of biological control. In the biological control of insect pests, the parasites and predators are collected and dispersed wherever these are found useful. These may be reared in large numbers and released, even imported from other countries or parts of the globe for the purpose. The natural enemies may be conserved and augmented and used when needed. The bio-ecological methods of pest control avoid use of pesticides and check pollution of environment, these are grouped under four categories.

(*i*) *The Biological control* : Biological control consists in the use of one living organism to control another living organism. It has been possible by the help of insects, mites, nematodes, viruses, bacteria, fungi, predacious snails, fishes, amphibians, birds and mammals for the control of insects, mites, snails, rabbits and weeds. The method has become a more powerful tool and also popular since the pests develop resistance to chemical pesticides, the chemical pesticides are also not 100% safe for the eatables and fruits etc. on which these are sprayed. The pesticide is also not welcome due to the annihilation effect on the existing natural enemy population causing severe upset in the balance of nature. It is also desirable to reduce pollution of environment caused by the use of pesticides. Elimination of pests through the use of parasites, predators and pathogens, known as biological control is self perpetuating, once established does not need any finances. The woolly apple aphid, *Eriosoma lanigerum* (Hausm.) has been controlled in the western Himalayas by *Aphelinus mali* (Hald.) subsequently American strain of *Prospaltella perniciosi* Tower succeeded in the control of the San Jose scale, *Aspidiotus perniciosus* (Coms).

Biological methods of control thus consist in the introduction, conservation and preservation of exotic species of parasites or predators. The parasites are periodically multiplied and released, in the areas where required, under suitable environmental conditions. The fungal pathogen should be spread by their spores on the pest population while the virus strains need a high host density. Competitive displacement, *i.e.*, displacement of the more harmful ecological homologue with lesser evil over extensive areas have been practised. The predators and parasites of pests generally comprise of arthropoda, micro-organisms, protozoa and vertebrates.

Arthropoda :The spiders, mites, and insects are important in biological control. *Ascyltus pterygodes* is known to have controlled the outbreak of coconut moths in Fiji. Other well known species are *Chilomenes sexmaculatus* F., *Brumus suturalis* F. and *Coccinella septempuncata* L. The last mentioned being very commonly used against cabbage aphid.

Microorganisms : The fungi, bacteria and viruses are also used in biological control. The spore forming bacteria have been particularly useful in controlling certain insects. Use of insect virus in the control of insect pests is gaining importance. There are 5 types of control through insect viruses. (1) cytoplasmic polyhedrosis, (2) granulosis, (3) non-inclusion, (4) nuclear polyhedrosis and (5) polymorphic inclusion. Micro-organisms have been used widely by the farmers in Thailand against cabbage looper.

Protozoa : *Glugea pyraustae* (Pail) has given encouraging results against the European corn-borer, *Ostrinia nubilalis* (Hbn.) and *Glugea lageri* (Pail) against the cabbage butterfly, *Pieris brassicae* (L).

Invertebrates : Predacious snails have been used to control phytophagous snails. Examples of these are the check of giant African snail. *Achatina fulica* have been checked by snails belonging to the genus *Gonoxis*. The giant African snail is an intermediate host of *Angiostrongylus cantoneusis* which causes *Eosinophilic meningoen cephalitis* of man.

Verterbrates : Fishes, tadpoles, toads and frogs, destroy a large number of mosquito larvae and other harmful insects. On land also frogs, toads, salamanders and lizards are active all day eating insects many of which are pests.

(*ii*) *The male-sterilization* : This method is practicable in isolated populations only. The technique involves rearing, sterilizing and releasing of insects in such numbers that overpower the natural population. The species to be used to control must be such that could be reared economically in large numbers and must not cause economic losses through direct damage to the plants. The males and the females should mix and the sterilized insects should not be affected in their vigour, longevity, behaviour and mating competitiveness.

(*iii*) *Pheromones, repellents, antifeedants and hormones :* Pheromones (chemical attractants produced by insects), chemical repellents (repel the insects from their food), antifeedants (hinder the feeding by insects), and hormones (hormone-mimicking compounds) may be used in pest control. The brain hormones, the ecdysone (moulting hormone), the juvenile hormones and hormone-mimcking compounds when applied to insects prevent normal development. Chemical compounds inhibit the utilization of body metabolites by their antagonistic action.

(*iv*) *Resistance in plants and animals :* Wild plants are comparatively more resistant to the attack of insects than cultivated cultivars, susceptibility of the latter being partly due to higher doses of fertilizer and greater number of irrigations given. Recently there has been an improvement in our knowledge of insect behaviour, their physiology and the methods of determining chemical substances contributing to resistance needed in plants in very minute quantities. The winter Majetin variety of apple and resistant stocks (Merton 779, 997) are resistant to the wooly aphid (*Eriosoma lanigerum* Hausm). Grape phylloxera (*Phylloxera vitifoliae*) menace was combated by use of resistant rootstocks. Efforts have been made to determinate the chemical, physical or other factors that contribute to resistance. Antibiosis may be due to biochemical constituents (*i.e.*, toxin, growth or reproduction deterrent) in the plants. The effect of a resistant strain on the population of the insect resisted is specific, persistant and cumulative. The basis of resistance should be known for success in evolving a new resistant cultivar.

Besides these, there are methods of the natural control *viz.*, (1) climatic factors (rainfall, sunshine, cold, heat and wind). (2) topographic features (river, lakes, mountains etc.) acting as natural barriers. (3) natural predators and parasites in any particular environment (4) diseases caused by entomophagous fungi, bacteria and other micro-organisms present in some regions.

Integrated Pest Management (IPM)

Pesticides have provided good control of many major pests but at the same time they have adversly affected the enviornment. Many crops have even failed in some years in spite of heavy application of pesticides. Hence, to sustain the high yield better pest control is needed in an integrated approach as the indiscriminate use of pesticides has changed the pest complex and led to many problems like resistance in insects, development of secondary pests and adverse effects on parasites, predators, pollinators etc. Therefore, to sustain the crop production there is no choice but to follow the integrated pest management (IPM)

Definition

1. According to **Kogan** (1988), "A decision support system for the selection and use of pest control tactics, singly or harmoniously coordinated into a management strategy, based on cost/benefit analyses that take into account the interest of and impact on procedures, society and environment."

2. IPM is holistic approach which involves various techniques for the management of pest problems.

IPM means not to go for a 100 **percent** killing of insect-pests in pest management rather the pest is allowed to a level which can be tolerated *i.e.* economic threshold. This also provides opportunity for parasites and predators to survive feeding upon the insect pests kepping them at a certain equilibrium level.

Economic threshold : The economic threshold (ETL) indicates the pest population that produces incremental damage equal to the cost of preventing that damage. (Heady 1972).

ETL is a fluctuating value which depends on the following factors–

(*i*) Market value of the crop,

(*ii*) The price of pesticides and equipment

(*iii*) The grower's concern about the danger of pollution,

(*iv*) Expected benefit of growing another crop instead, and

(*v*) Many other factors.

Biotic Agents

The principal biotic agents are as follows–

1. Parasites. Parasites spend their life on host temporarily or permanantly. Each parasite requires a host, which it kills for its development into a free-living adult. Important parasites are trichogrammatid, ichneumonidi, scelionid, braconid, encyrtid, eulophid, chalcid and tachinid flies.

2. Predators. Predators are free-living and larger in size than their prey. They require several preys to complete their life cycle while parasites require only one host to complete their life cycle. Examples of predators are dragon flies, damsel flies, ground beetles, ladybird beetles, syrphid flies, spiders, mites, shrews, monkeys, toads, frogs and birds.

3. Pathogens. Pathogens kill their host of debilitate the future generation. Dead insect pests are often found hanging on the plants or a fungal growth on the body of dried insect is easily seen. Pathogens are disease causing organisms which include viruses, fungi, bacteria and protozoa.

(*a*) **Viruses.** Viruses are occluded in a crystalline protein matrix that contain a molecule of circular super coiled double stranded DNA. Viruses cause disease in insect pest belonging to order Lepidoptera, Diptera, Hymenoptera, Coleoptera and Trichoptera. The genera of bacculo viruses *viz.* nucleopolyhedrosis viruses (NPVs) and granulo virsus (GVs) are most promising to cause diseases in insect-pests. In India 24 and 14 species of insect pests are found to be infected with NPV and GV respectively on various crops.

(*b*) **Fungi.** Fungi causes diseases in insects. Body of infected insect become hard. About 9 endomogenous fungi have been recorded from India attacking more than 23 different species of insects. Genera that have been more investigated for mycocides are *Metarrhizium, Beauveria, Nomuraea* and *Verticillium* in India. The fungi *Metarrhizium anisopliae* and *Beauveria brogniort* attack the soil inhibiting insect pest. *Beauveria bassiana* is pathogenic to several insect-pests. *Nomuraea rileyi* attacks the caterpillars and *Verticillium lecanii* recorded from scale insect and aphids.

(*c*) **Bacteria.** Seven species of entomopathogenic bacteria *viz. Bacillus thuringienesis, B. cereus, B. magaterium, B. subtilis, Pseudomonas aeruginosa, Serratia marcescens* and *Streptococcus faecaulis* have been recoded in India attacking about 115 insect species. Among them *B. thuringiensis* is most effective particularly noctuids.

(*d*) **Protozoa.** Sporozoa and Cnidospora of Protozoa cause death of the parasitized hosts. The parasitism is caused by infective food and water within 72 hours of ingestion.

(*e*) **Namatodes.** Nematodes of Steiner nematidae and Rhabotitidae group attack the insect-pests. High humidity is a pre-requisite for the nematodes to find their host. It has been recorded that about 1500 speceis of insects are attacked by nematodes. The use of commercial product DD-136 in humid ecosystem is highly encouraging.

Biological Control of Plant Diseases

High capital investment, non-remunerative, temporary effect, selectivity, development of resistance, pollution, health hazard, effect on physico-chemical and biological factors are some of the limitations which have been growing awareness towards biological control.

According to Darwin, "in the struggle for life everything is born to eat and get eaten". This maintains natural biological balance. Biological control is the reduction of inoculum density or disease producing activities of a pathogen or parasite in its active or dormant state, by one or more organisms, accomplished naturally or through manipulation of the environment, host or antagonist or by mass introduction of one or more antagonists.

Biological Agents

All classes and groups of organisms existing in an ecosystem are to be included as biological agent but maximum emphasis has been given to fungal and bacterial biocontrol agents.

Important Genera of Fungi

Trichoderma, Gliocladium, Aspergillus, Chaetomium, Penicillium, Neurospora, Fusarium (saprophytic), *Rhizoctonia,* spp., *Dactylella, Arthrobotrys, Catenaria, Paecilomyces, Glomus* etc.

Among above, species of *Trichoderma* and *Gliocladium* have been explored and recommended for use.

Important Bacteria

Agrobacterium, Actinoplanes Bacillus, Enterobacter, Erurinia, Pseudomonas, Streptomyces etc. Among these, currently Rhizobacteria-fluorescent *Pseudomonas* are under intensive research.

Mode of Application of Biocontrol Agent

The biocontrol agent is being used in following manner :

1. Seed treatment

1.1. Seed coating. The biocontrol agent is dissolved in water and suspension is made which is applied to the seed. Pelgel, Carboxymethyl cellulose (CMC), Xanthan gum and Talc are some of stickers used in suspension solution of biocontrol agent.

1.2. Seed biopriming. In this, first the seeds are ccated with the suspension solution of biocontrol agent and then incubated under warm, moist conditions until just prior to radical emergence. The seeds may be planted immediately or dried for storage purposes.

1.3. Seedling dip. Tubers, bulbs, rhizomes cutting and seed roots of seedlings are dipped in the suspension of biocontrol agent before transplanting.

2. Soil application

Biocontrol agents are either applied in furrows, at the base of the plants or spread over the entire field preferably along with FYM.

3. Foliar spray

Suspension solution of biocontrol agents are sprayed on aerial part of the plant in evening so that the extreme of temperature and dust during the day may be avoided.

Different products of *Trichoderma virde, T. harzianum, Aspergillus niger, Pseudomonas fluoroscens* and *Bacillus subtillis* are available in India which are used as seed treatment, root dip, soil application and as foliar spray against soil borne pathogens.

Realistically, biocontrol may not totally replace chemicals but the judicious use of biocontrol agents can significantly better our lives, the environment and agricultural productivity.

Pest Control with Chemicals

A good insecticide should be very potent against insects in general or preferably against a specific group of insects (Table 7.1). Cythion is an insecticide with varied uses which may be recommended as a first-aid measure specially when the attack is severe or the availability of the specific insecticide is remote.

Table 7.1. Recommendations for effective control of insect pest of some fruit crops through high volume spray of Cythion* 50% EC.

Fruit crops	Insect pests	Recommended quantity of cythion per (l water)
Apple	Wooly aphid, San Jose scale Defoliating and fruit eating beetle	1 ml 2 ml
Banana	Aphid, Lace wing bug	2 ml
Ber	Fruit fly Lêaf curling aphid	Bait : 1 ml Cythion + 5 gm Jaggary 1 ml
Citrus	Whitefly, Leaf miner, Psylla Green scale Citrus butterfly Fruit sucking moth	1.5 ml 2.0 ml 1.0 ml Bait : 1 ml Cythion + 100 gm Jaggary
Grape	Thrips, Leaf roller, Jassid, Flea beette	1 ml/vine
Guava	Fruit fly Scab (Helopeltis)	Bait : 1 ml Cythion + 5 gm Jaggary 2.0 ml
Litchi	Leaft curl mite, Thrips, Jassid	1.0 ml
Loquat	Thrips, Jassid	1.0 ml
Mango	Mango hopper Mealy scale Fruit fly Mite, Red ant	2.0 ml 2.0 ml Bait : 1 ml Cythion + 100 gm Jaggary 2.0 ml
Papaya	Aphid, Whitefly	1.0 ml
Peach	Whitefly Peach black aphid	Bait : 1.0 ml Cythion + 5 gm Jaggary 1.0 ml
Pomegranate	Fruit Bug	2.0 ml

***Cythion** (0, 0- dimethyl phosphorodithiote of dimethyl marcaptosuccinate) an organophosphorus insecticide which controls a wide range of pests infecting various crops.

Insecticides

Chemically the insecticides may be (*i*) elements (sulphur, phosphrous, thallium, mercury), (*ii*) inorganic compounds such as lead arsenate or other arsenicals, sodium fluosilicate, zinc phosphide, copper acetoarsenite (Paris green), (*iii*) organic compounds *viz.* DDT, BHC, poisonous gases as hydrogen cyanide, ethylene dichloride, carbon tertrachloride, methyl bromide, phosphine, etc. (*iv*) plant extracts from neem (*Azadirakhata indica)* nicotine sulphate, pyrethrum, rotenone. Several types of insecticides are known (See Appendix).

(A) *Stomach poisons.* These are generally applied against insects with chewing, sponging, siphoning or lopping mouth parts. For the stomach poisons to be acceptable in common use, they should have pre-requisites such as the concentration which is applied to the plant for insect control, should not be injurious to the foliage or those using the foliage or any other portion of the plant. The insecticide should have the physical properties of adhering to the plant surface on application and yet after a certain period should disintegrate or be washed away by rain etc. so that it does not persist on the plants for long periods. The stomach poisons are applied as sprays, dusts, dips or baits.

(B) *Systemic poisons.* The systemic insecticides (*i.e.* dimecron, metasystox, menazone, phorate etc.) should be absorbed and translocated to various parts of the plant in amounts lethal to insect pests without being harmful to the parasites, predators and other natural enemies of the pests. Sevin is contact insecticide but has a mild systemic action in some plants. It is good for all major apple and pear insects, bollworms of cotton, leaf hoppers and leaf rollers of grapes.

(C) *Contact poisons.* Contact sprays kill insects by clogging in spiracles and respiratory system or by acting as general tissue poisons due to the high penetrating capacity in the cuticle covering of insects. These insecticides should not be phytotoxic ; BHC and DDT are toxic to cucurbits and citrus flowers. Toxaphene, chlordane, heptachlor, methoxychlor, eldrin, dieldrin, endrin, thiodan (endosulfan) are other common contact sprays in the market. Besides these, parathion, diazinon (Basudin), malathion and other compounds are being widely used. Sulphur, sodium fluoride, arsenicals, mineral oils including diesel oil, crude oil and the animal fats including fish oil, soaps are prepared from them. Nicotine sulphate, anabasine, rotenone, pyrethrum, **neem** extract, **dharek** extract etc. are also examples of contact poisons. Nicotine, an alkaloid, is effective contact and stomach poison with a marked fumigant action. It is not phytotoxic at reasonable concentration.

(D) *Miscellaneous chemicals.* Compounds that cause changes in the physiology and behaviour of insects may be used to kill them these include antifeedants, chemosterilants, repellents etc. (*i*) Antifeedants make the food distasteful so that the insects do not feed on them and are starved to death.

Table 7.2. Insecticides, Acaricides and Fumigants

Name of insecticide	Range/Nature of action	Toxicity	Residual effect	Remark
1	2	3	4	5
1. Nicotine sulphate	Successful against soft bodied insects with sucking type of mouth parts. Controls alphids, capsids, midges, mealy bugs etc.	High, warm blooded animals	Quite stable	Addition of soap liberates nicotine quicker and makes it more stable and effective.
2. Pyrethrum	Contact insecticide, with quick knockdown action. Toxicity may be increased by addition of piperonyl butoxide. Toxic to most insects. Safe for edible crops and as household spray. Controls aphids, house flies, mosquitoes, bed bugs and pests of livestock.	Non phytotoxic, not harmful to mammals	Disintegrates quickly	Extracted from flower of *Chrysanthemum cinarariifolium* grown commercially in Japan and Africa.
3. Rotenone	It is found mostly in leguminous plants *i.e., Derris* and *Lonchochocarpus*. A selective insecticide acting as stomach poison and as contact spray.	Low mammalian toxicity	Deteriorates in storage	

1	2	3	4	5
4. Neem extract	Extract from the fruits or seeds or leaves of neem plant. It is repellant and insecticidal. It contains azadirachtin. Shade-dried leaves are highly effective against the wooly bear (*Anthrenus vovax* Waterhouse), clothes moth (*Tineola bisselliela* blummel).	Not harmful to mammals		
5. Methoxychlor	Also known as dimethoxy-DT, DMDT, dianisyl trichloroethane. It is non-systemic contact and stomach insecticide. Its activity is as wide as that of DDT but it is not stored in body fat.	–	–	Used for contro
6. Endosulfan	In plants it is metabolized to sulphate, does not accumulate in milk and fat but has high toxicity.	Very toxic to fishes but safe for bees.	–	

Organophosphorus Insecticides				
1	2	3	4	5
7. Parathion and methyl-parathion.	It is a potent contact and stomach insecticide with some fumigant action but is non-systemic. Highly effective against scale insects, aphids and mites. Crops in bloosoms.	Non-phytotoxic, highly toxic to mammals but non-cumulative. Methyl parathion has lower toxicity in mammals	Less stable	Not be used in houses of livestock.
8. Malathion	Liquid, non-systemic insecticide with moderate persistence. Effective against sap sucking insects and mites. A contact and a stomach poison besides acting as a fumigant. Controls aphids, scale insects etc.	Generally non-phytotoxic but cucumbers affected.	Moderate persistence	Safe on ornamentals, crops, storage.
9. Schardan	Systemic compound, effective against sap-sucking insects and mites but has little effect as a contact poison. Less destructive to parasites, predators and insect pollinators.	Non-phytotoxic at insecticidal concentrations but high mammalian toxicity	–	Its use on edible crops is also restricted.

1	2	3	4	5
10. Tetraethyl pyrophosphate	Hazardous to insect pollinators	High mammalian toxicity.	–	40% pyrophosphate is generally marketed.
11. Diazinon	Very effective against aphids, thrips and flies, good acaricide. Non-systemic, translocated by capillary action if mixed in irrigation water. Kills the borers (ring stem borer of paddy).	Low dermal toxicity, non-phototoxic.	–	Safer than parathion.
12. Trichlorphon	Mainly for the control of DDT-resistant houseflies. Other trade names are Neguvon, Tugon and Dylox. Effective in crop protection and against coparastie. Non-systemic contact and stomach poison. Also non-hazardous to bees.	Low mammalian toxicity.	–	–
13. Mevinphos	Systemic acaricide. Compound is readily soluble in water.	High mammalian toxicity.	Short residual activity	Useful aphicide
14. Azinphosmethyl	Non-systemic Sinsecticide and acaricide.	–	–	Formulated as dust, wettable powder and emulsified concentrate.

1	2	3	4	5
15. Thiometon	Systemic insecticide and acaricide. Very effective against sucking insects and mites in the orchards. It is metaboilized into water-soluble compounds.	Dermal toxicity is moderate	Quite stable	Its effect on the plant persists for 2-3 weeks.
16. Phorate	Systemic insecticide, seedlings and plants are protected from the attack of sap-sucking insects. The bees are not affected. The use of granules reduces the hazards to some extent.	Very high mammalian toxicity.	Short residual action	Used for seed or soil treatment at the time of flowering.
17. Demeron-methyl	Systemic and contact insecticide and acaricide.	–	–	Mixture of dematon-O-methyl and demetion-S-methyl.
18. Carbofenothion	Non-systemic acaricide and insecticide.	Less toxic to honeybees than DDT and less toxic to natural enemies than parathion.	Long residual action	Used in deciduous fruits.
19. Dichlorvos	It has contact, stomach and fumigant action. It is effective against sucking and leaf-biting insects. Also used as fumigant.	Low dermal toxicity but is toxic to bees. Non-phytotoxic.	Non-persistent	Used against mosquitoes and other Deptera inside the building.

1	2	3	4	5
20. Dimethoate (Rogor)	Systemic insecticide. It controls the fruit flies. It is hazardous to bees visiting flowers. Used against sap feeding insects and mites	Low mammalian toxicity.	Short life in plant tissues.	Soluble in water, can be used systematically for the control of cattle grubs and fruit flies.
21. Menazon	Systemic aphicide. Useful for tuber and seed treatment.	Non-phytotoxic, toxic to bees with low mammalian toxicity.	—	Also used on foliage should not be applied, when the bees are forging.
22. Formation	Suitable for application on all crops.	Toxicity to chrysanthemum, low mammalian toxicity.	—	It is particularly useful for edible crops.
		Acaricides		
23. Chlorobenzilate	No action on insects. Effective against phytophagous mites.	No phytotoxicity, moderately toxic to mammals.	Prolonged	Paper strips carrying its deposits are prepared and used for the fumigation of live bees inside a hive. Strips are marketed as Falbex.
24. Dicofol	Non-systemic. Effective against all stages of phytophagous mites on a wide range of crops. Strongly acaricidal and poorly insecticidal. High initial kill.	Low mammalian toxicity.	Low residual activity against plant mites.	'H' of the bridge in DDT is replaced by 'OH' in this. Residue decrease rapidly in soil but traces may appear to persists even for a year

1	2	3	4	5
25. Carbaryl	Contact insecticide with slight systemic action. Used against insect pests of fruits and vegetables. Good for those pests which have developed resistance to chlorinated hydrocarbons and organophosphatic insecticides. Comparactively safe for insect pollinators, parasites and predators.	No phytotoxicity at the insecticidal concentration.	More safe, long residual effect on crops.	—
26. Methyl bromide	It has high insecticidal and acaricidal properties. Very useful for the fumigation of grain godowns, fresh fruits, vegetables, ornamental flowers, nursery stock, flour-mills and ships. Also good soil fumigant against nematodes, fungi and woods.	Highly toxic to mammals.	—	Packed in metal cylinders or glass ampules. Does not inhibit the germination capacity of dry seed.
27. Ethylene dichloride	3:1 ratio of ethylene dichloride and carbon tetrachloride is used for stored grain pest but not on plants. Non-injurious to seeds.	—	—	EDCT is mixture used against stored grain insects but it cannot be used on growing plants.

1	2	3	4	5
28. Ethylene bromide	It is effective against pests of stored products and against insects and nematodes in the soil. In gardens and in soil it should be applied at least one week before planting. 60 cm penetration depth, colourless, odourless, non-inflammable.	To avoid phytotoxicity to plants the soil treatment should precede by one week of actual sowing of the crop.	—	Available in small ampules. The fumigant does not go beyond 60 cm and should be distributed at different depths.
29. Dichloropropane-di chloropropene	It has been in use as an effective nematicide in cooler climates (4-27°C). Should be stored away from food.	Phytotoxic, high mammalian toxicity.	—	Its contact with the skin or the inhailing of its fumes should be avoided. Dark liquid at ordinary temperature. Very inflammable.
30. Fensulfothion	It is a nematicide and soil insecticide. Systemic. Effective in the control of root-knot and golden nematodes into potato roots.	—	Long persistent action.	

Note : More details are provided under the caption "Pesticides and Their Use" in the Appendix.

(*ii*) Chemosterilants fed to the insects cause sterility leading to ultimate elimination of the pests.

(E) Fumigants. The poisonous gases are used as fumigants to kill insect pests of stored grains, soil and other products. Fumigants include nicotine, hydrogen cyanide, carbon bisulphide, sulphur dioxide, paradichlorobenzene, naphthalene, ethylene oxide, ethylene dichloride and carbon tetrachloride mixture, methyl bromide, formate, phosphine, nemoagon, phostoxin, chloropicrin, formalin, vapam etc.

Materials which are not toxic or poisonous to insects may also be needed to help the insecticides in doing their job better. These substances are known as (*i*) Carriers (Water in spray, talk or gypsum in dust,.or bran in a poison bait), (*ii*) Emulsifiers (soaps, flowers, calcium and potassium arsenate), (*iii*) Strechers (lime, rosin, glues), (*iv*) Wetting agents and detergents (several sulphur compounds), (*v*) Stabilisers (caustic soda, glue, gelatin), (*vi*) Attractants (methyl eugirol), (*vii*) Repellents (Bordeaux mixture, creosote, naphthalene, tars, cinchona, alkaloids). Repellents make the food unattractive or offensive. Bordeaux mixture and lime sulphur repel leaf hoppers and chewing insects, (*viii*) trichlorobenzene is effective against crawling insects. Adhesive tapes carrying bichloride of mercury fixed at places check the movement of ants across. Heavy oils applied at the base of poultry roosters act as barrier against poultry lice, (*ix*) fumarin, rodafarin, warfarin when fed to rodents cause internal hameorrage and kill them in few days.

Fungicides

For the control of plant disease caused by fungus the following fungicides are widely used :

(*a*) *Bordeaux mixture.* It is a mixture of a solution of copper sulphate and lime suspension (2 kg each) in 250 litres of water. The freshly prepared mixture with better fungicidal properties, better adherence, persists for a longer time. It can be prepared with varying ratios of the ingredients. If the mixture is acidic in reaction it is phytotoxic in nature. It can be used alongwith any other insecticide but no parathion, demetòn, malathion, endosulfan, tetraethyl phyrophosphate (TEPP), dimethoate and carbaryl. The lime may be replaced with washing soda to get Burgundy mixture. Perenox, fungimar, cuprocide, containing copper oxide as active ingredient and bliox, blue copper, kirti copper containing copper oxychloride are in market.

(*b*) *Sulphur and allied compounds.* Compatible with most of the insecticides, Sulphur dust is widely used against powdery mildews. Tetramethyl thiuram disulphide (TMTD) as foliar application controls apple scab and late blight of potato. It is also effective against disease like tikka disease of groundnut, dry rot and seedling, blight of maize, the damping off of chillies and peas and the smut of onion.

(*c*) *Zinc compounds.* Ziram and cuman (zinc dimethyl dihiocärbamate) are useful against paddy pest, tikka disease of groundnut, late blight of potato etc.

Zineb or Dithane Z-78, with zinc ethylene bisdithiocarbamate as active ingredient are multipurpose fungicides safe on edible crops.

(*d*) *Organic mercury compounds.* Ethyl mercury chloride, ethyl mercury phosphate and phenyl mercury acetate have been used extensively against seed borne diseases *e.g.* cereasan, agrosan G.N. agallol, aretan which are very toxic to warm blooded animals.

(*e*) *Nirogen compounds.* Best examples are captan and Esso fungicide 406 which are useful for the control of diseases of grapes.

(*f*) *Antibiotics.* Blasticidin and Aureofungin are effective against paddy blast. Streptocycline controls the bacterial blight of paddy.

PROTECTION AGAINST WEEDS

Weeds have their role in influencing the life of plants useful to man. Protection against weeds is as important as against pests and diseases. Weeds may be (*i*) annual *e.g., Digera arvensis* (**Lahsua**); *Chenopodium album* (**Bathua**) etc. (*ii*) binnial *e.g., Ipomoea reptans* (**Karmi sag**), *Launaea asplieniifolia* (**Jungli gobbi**), and (*iii*) perennial *e.g., Sorghum halepense* (**Baru**), *Cyperus rotundus* (**Motha**) etc.

METHODS OF WEED CONTROL

Weeds are controlled by (*i*) Chemical and (*ii*) Non-chemical methods, the latter includes (*a*) biological, (*b*) mechanical and (*c*) cultural techniques.

Chemical Control

Only such chemicals are used for which the standing crop has a high degree of resistance or tolerance. Several types of chemicals may be used on the basis of their effect on weeds and crops.

1. *Contact herbicides.* These chemicals kill the plant tissues with which hey come into contact, but do not move in the plant to any extent. They are again distinguishable as (*a*) selective and (*b*) non-selective. A selective herbicide kills certain weed species with little or no injury to other plant species. For example, MCPB is toxic to few plans, but not to peas. A non-selective weedicide kills all the weeds covered by spraying *e.g.*, chlorates, dinitrohenol and pentachlorophenol.

2. *Translocated herbicides.* These herbicides are usually applied to the leaves or stem of the plant, sometimes to the roots. They are absorbed by the plant and then distributed all over, where they accumulate. Translocated herbicides are also widely known among growth regulators. These herbicides may be either (*a*) selective *viz.* 2, 4-D ; 2, 4, 5-T ; M CPA *or* (*b*) non-selective like sodium arsenate, chlorates and thiocynate.

3. *Soil applied or Residual herbicides.* These, when present in the soil, check the growth of green plants. These are applied to the soil and are primarily effective against germinating seeds and seedlings. Examples are CIPC, TCA, Monuron etc.

Time of application of herbicides. The time of application may largely determine their effectivity in various crops :

(*i*) *Pre-planting.* Any treatment made before the crop is planted or sown is known as pre-planting treatment. Eptam is normally incorporated into the soil before sowing of crops.

(*ii*) *Pre-emergence.* Any treatment made prior to emergence of specific crop's weed is known as pre-emergence treatment. For example, Simazine may be applied before the germination of the crop and weed.

(*iii*) *Emergence.* Any application after emergence of crop or weeds is known as post-emergence treatment.

Formulation of herbicides : Most herbicides are usually commerical formulations that contain the herbicides and can be (1) dissolved, emulsified or suspended in a liquid carrier, (2) distributed dry by a spreader or by hand, and (3) injected into soil for vaporization and fumigation.

Active chemical content : The containers for all commercial herbicides have label expressing the amount of active chemical contained in the particular product either in percenage of active ingredient (*a.i.*) or acid equivalent (*a.e.*).

Calculation for herbicide application : All recommendations are made on the basis of active ingredient (*a.i.*) or acid equivalent (*a.e.*). The following formula may be used to calculate the quantity of a commercial product required to give a specific dose of the active ingredient.

$$\text{Wt. of the commerical material required} = \frac{\text{Dose of } a.i. \text{ required} \times 100}{\% \; a.i. \text{ in commercial product}}$$

If a herbaicide with 50% active ingredient is desired to be applied @ of 1.0 kg *a.i.*/ ha then 1 × 100/50 = 2 kg of the commercial product is needed.

Precaution for safe use of herbicides : Some herbicides cause injury to susceptible plants growing nearby, therefore it is necessary to avoid spray drift of these. Sprayer used for herbicides should not be used for any other kind of spray on crop or ornamental plants. It is necessary to clean the sprayer with warm water every time after use. Do not store herbicides near seed, feed, fungicide, or insecticides. Spray at recommended rates and proper stage of the crop only. Avoid inhaling vapours, dust or spray mist. Wash hands and face thoroughly with soap and water after use of herbicides.

Phenoxyacetic acids : 2, 4-D ; 2, 4 5-T and MCPA are most important post-emergency selective herbicides used to control broad-leaved weeds in any situation. The toxic symptoms of this group are visible within a few hours of application in bending and twisting of shoots, dropping of leaves and yellowing. Phenoxy compounds are formulated as under.

Amine salts : Amine salts of 2, 4-D are commonly used as liquids. These are highly soluble in water and have a power of penetration more than sodium salt but less than ester.

Ester formulation : There are also available as liquids ; when properly formulated and mixed with water esters form emulsion ; these appear milky and are not clear. Ester formulations of 2, 4-D have a tremendous knock down effect if sprayed post-emergently.

Phenoxypropionic acids : The most important herbicide of this group of Silvex or Fenoprop 2-(2, 4, 5-trichlorphenoxy) is propionic acid which is more effective than 2, 4, 5-T. It is used for the control of woody bushes and certain aquatic weeds.

Phenoxybutyric acids : MCPB, 4-(2 methyl-4-chlorophenoxy) butyric acid and 2, 4-DB, 4(3, 4-dichloro-phenoxy) butyric acid belonging to this group have shown promise for the post-emergence control of broad leaved weeds in pea and groundnut.

Pehnoxyethyl (X) : Sesone (2, 4, dichlorophenoxyethyl hydrogen sulphate) is formulated as the sodium salt and is a white crystalline powder that is soluble in water. When applied to moist soil Sesone is converted into a herbicide with properties similar to 2, 4-D. It is effective as pre-emergence herbicide.

Phenylacetic acids : 2, 3, 6-TBA (2, 3, 6-Trichlorobenzoic acid) and Amiben (3, amino-2, 5-dichlorobenzoic acid) are examples of this group which are used as pre-emergence application for control of deep rooted noxious perennial weed as *Convolvulus*, tomatoes, sweet potato and groundnut. Action of trichlorobenzoic acid is generally similar to that of 2, 4-D and is much more persistent in the soil than phenoxyacetic acid.

Sodium salts : The sodium salt of 2, 4-D is only four per cent water soluble and hence is less likely to penetrate through either cuticle or stomates.

Halogenated aliphatic acids : These are more toxic to grasses than broad leaved weeds. TCA (Trichloroacetic acid) and Dalapon (2, 2-dichloropropionic acid) are examples of this group which are applied to the soil and the foliage depending upon their solubility in water. These are soil and the foliage depending upon their solubility in water. These are generally used in uncropped lands for the control of deep rooted weeds like *Kans*, *Baru* and *Doob*.

Amids : CDAA or Randox (2-chloro-N, N-diallylacetamide), 3, 4-DAP (3, 4 dichloropropionanilide) and MH (maleic hydrazide) are important chemicals of this group.

CDAA : It is available in emulsifiable concentrate formulation and is a pre-emergence herbicide against germinating annual grasses and broad leaved weed in maize, sorghum, beans and onion. It is not effective against established annual or perennial weeds.

3, 4-DPA : 'Stam F-34' and 'Rogue' are the two herbicides under this group. They are used as post-emergence herbicides for the control of *Echinochloa, Pancium* and other weeds in paddy and potato fields.

Dalapon : This herbicide is similar to TCA in properties. It has proved less erratic and more effective than TCA when applied as a foliage-spray for the control of most of the annual grasses. It is much more effective on *Doob, Baru* and other perennial grasses.

Trichloroacetate. The sodium salt of TCA is widely used. It has shown varying degrees of effectiveness in controlling *Doob, Baru* and other annual and perennial grasses. But results are obtained when it is applied in combination with tillage and cultural practices.

Substituted urea : Substituted urea are listed below :

(*i*) *Cotoran [N-3-trifluoro-methylphenyl)-N, N-dimethyl urea] :* It is slective herbicide with a long residual action, lasting over 2-5 months and specially recommended for the control of weeds in cotton, sugarcane, sorghum, wheat and pea.

(*ii*) *Diuron :* It has been used successfully in cotton and sugarcane @ 0.5-5.0 kg/ha. It is also being effectively used as soil sterilant and is preferred over monuron in areas of high rainfall or on light sandy soils.

(*iii*) *Fenuron :* It is widely used as a soil treatment to kill woody plants.

(*iv*) *Monuron :* It is used both as soil sterilant and for selective annual weed control in resistant crops. For selective weed control purpose, the rate varies from 0.5-5.0 kg./ha as pre-emergence ; as soil sterilent, the rate varies from 5-10 kg/ha.

(*v*) *Tenoron [N-4 (P-Caloro-phenoxy)-pheyl-N'. N-dimethyl urea] :* It is a selective pre emergence herbicide wettable powder for the control of annual broad-leaf weeds, annual grasses under moist soil conditions. It is recommended for use in carrots, beans, peas, groundnut, onion, garlic, tomatoes, chillies and fruit orchards such as a mango, citrus, apple etc.

Other formulations of this group are Telvar, Karmex and Lorax. Best results from urea herbicides are obtained if rainfall or irrigation occurs within two weeks of application.

Carbamates : These are effective through soil medium, are highly volatile and are therefore incorporated immediately after application. The following members are grouped as given below :

(*i*) *CDEC :* It has shown promise for the pre-emergence control of certain weeds in vegetable crops. It is more effective on grasses than on broad-leaved weeds.

(*ii*) *Eptam (S-ethyl N, N dipropyl thiolcarbamate):* It has been successfully used as a pre-sowing herbicide for control of grassy weeds as they germinate.

Although it is not easy to spell out any universal rule for the use of herbicides, certain guidelines may be provided on the basis of the response of the plants to a large number of herbicides. Plants have been classed into six categories depending on their response to herbicides. Among the monocotyledonous plants, there are following three classes :

I. *Annual* (Crops like cereals, small grains, maize, sorghum etc.) : Weeds like wild oats in these crop field may be controlled by the use of chlorophenoxys, substituted ureas, triazines and benzoics. The annual weeds such as wild oats, Bromus, foxtails etc. are susceptible to amitrole chlorinated aliphatic acid, substituted ureas and carbamates.

II. *Tree-like perennials* (palms ; banana, bamboo) : Herbicides like triazines, chlorophenoxys, paraquate may be used to kill the weed in these crops without harming them. Bamboos and palms coming under this group are susceptible to amitrole, aliphatic acids.

III. *Herbaceous perennial* including pineapple, sugarcane, asparagus, ornamental and edible bulbs. The herbicides which can be used in these crops to kill the weeds without injuring the crops are chlorophenoxys, substituted ureas, triazines. The perennial grasses like Johnson grass, canary grass. *Agropyron* etc. are susceptible to amitrole, and chlorinated aliphatic acids.

Among the dicotyledonous plants again three classes have been suggested on the basis of their response to herbicides :

I. *Herbaceous annuals :* Pulses, peanuts, tomatoes, potatoes, cucurbits, cole crops, bulbs, cotton are safe with carbamates, phenols, substituted ureas, toluidines. The weeds put under this category *viz.* pigweed, knotweed, nightshades, mustards, borages, mallows, dogbane, fennel etc. are susceptible to chlorophenoxys, benzoic acids and picloram.

II. *Herbaceous perennials i.e.,* mint, alfalfa, clover and strawberries : The herbicides used : substituted ureas, triazines, carbamates, phenols. Weed like bindweed, thistles (*Cirsium* sp.) Knapweeds, perennial nightshades, lettuce are susceptible to chlorophenoxys, benzoic acid, picloram. The weeds like gorse, poison-ivy, honeysuckle, blackberry, dogbane, brambles are also susceptible to chlorophenoxys, picloram besides those listed earlier.

III. *Woody perennials :* Fruit trees, forest trees, rubber, tea, coffee, fruiting and ornamental shrubs. Herbicides which may be used with safety are triazines, substituted ureas, phenols, amitrole (avoid foliage).

Use 2, 4-D, simazine, diuron, amitrole, paraquate and dalapon for controlling weeds round plantation tree crop like rubber, tea, coffee, cocoa etc. The trees may be killed with 2, 4, 5-T or arsenic to permit replanting. Berries and grapes have triazines, substituted ureas, and dinitros as safe herbicides for the control of weeds in these orchards. Plants are usually resistant when dormant or after fruit harvest—let that be the time of application of herbicides in this type of bushy fruits which have shallow root system. Avoid the time when the plants are in bloom, fruit formation stage or in active vegetative growth. Chemicals should be used around the base of trees where cultivation is undesirable. Citrus trees are very suspectible to dalapon but resistant to diuron. Oils fortified with DNBP (4, 6-dinitro-O, *sec* butyl phenol) may be used in citrus orchard. Strawberries may be treated with sesone and simazine as pre-emergence treatments and with 2,4-D as dormant sprays. The role of species, differences, in age, stage of growth, nutritional estimated in determining the precise selective herbicide which must be used. Herbicides act in different ways *viz.* DNBP inhibits ATP production. DCMU also called Diuron prevents the reduction of cytochrome of an electron carrier in the photosynthetic electron transport chain. Linuron, substituted urea similar to diuron inhibits Hill reaction.

Perennial weeds in apple orchards can be controlled by growth regulators but the chemicals must not reach the tree foliage. Blossoming period should be avoided in this treatment. Dalapon can be effecitve against weeds in well established pear and apple orchards. It is not safe under plums and cherries. Low rates of simazine (1-3 kg/ha of active ingredient) may be used to control annual weeds in established plantings of bush and cane fruits through the summer to clean ground in the spring. Soil should be moist to have full effect of simazine but it must be kept off the fruit plants as much as possible. Simazine is not harmful to maize and may be used to control all types of annual weeds in maize fields. Use of dalapon is done only when the trees are dormant. Useful against annual and perennial weeds, can be effectively used in citrus, grapes, lucern and sugarcane. Currant bushes may tolerate 10 kg/ha, gooseberries and raspberries tolerate much less (only 1 kg/ha). Tinoran and Kotoran are effective against weeds in sugarcane and cotton. Propanil is effective against weeds in rice fields.

Multiplicity of Effects of Herbicides

Table 7.3. Multiplicity of effects of herbicides

Effect	Herbicides	Remarks
1	2	3
1. Colour changes	Spray of (140/g/ha) 2, 4-D very dilute solution.	Increase in red colour of red potato.
	Dalapon applied in soil at the time of tuber formation.	Reduces the red colour of the red potato.
2. Decortication	Sodium arsenite 2, 4-D and 2,4,5-T	Loosening of bark by killing the cambium and the adjacent tissues. Valuable in pulpwood species for paper making and in the preparation of fence posts.

1	2	3
3. Defoliation	Contact herbicides	Desired injury to the leaf blade should occur but none to the abscission zone. This is a temperature sensitive and oxygen requiring process inhibited by respiratory poisons, 85°F or more temp. is needed. Important internal factors are correlative changes accompanying hormonal, nutritional or moisture relations among plant pests.
4. Desiccation	Contact herbicides may be used. Sodium arsenite, paraquat, pentachloro-phenol sodium chlorate and magnesium chlorate may be used @ twice the conc. used for defoliation.	Chemically accelerated through injury of cell membranes that permit rapid water loss. High temp. and low humidity favour. The cells are effectively injured and abscission is prevented.
5. Dwarfing	MH-interferes with cell division. Amo-1618, phosphon, cycocel, retard growth by inhibiting bud growth.	Limiting the size of grasses, hedges, suckering of tobacco, checking growth in height of wheat and increasing tillering. High humidity and rainfall are helpful in success.
6. Flowering	2, 4-D, (Pineapple cultivars) and MH.	Delays growth and flowering.
7. Inhibition of fruit drop	2, 4-D, 2, 4, 5-T and other auxins.	Low conc. between 10-100 ppm. Formation of the abscission layer is delayed and all the fruits can be picked up at the same time. Not effective in the case of young fruits.
8. Growth stimulation	Gibberellic acid	Increase in height, early production of flowers, better quality of fruits of larger size. Production of more amylase in barley seed.

1	2	3
9. Hastening maturing	2, 4-D	Unripe detatched fruits of banana, pears, apples ripen quickly. Preharvest sprays hasten maturity.
	2, 4, 5-T	Hastens maturity of developing fruits still attached to the plant.
10. Increasing latex	Appln. of 2, 4-D, 2, 4, 5-T to the bark	In case of rubber plant *Hevea brasiliensis*.
11. Prolonging storage	2, 4-D, 2, 4, 5-T, CPIC, MH	Longer storage of citrus fruits has been reported with 2, 4-D and 2, 4, 5-T. Sprouting of potato is prevented by treatment with CPIC, MH, and 2, 4, 5-T.
12. Stimulation of fruit drop	MA, NAA, IPC, 2,4-D, 2, 4, 5-T	When the fruits are very small or a few hours before pollination, damage to the pollen tube or the young embryo causes subsequent drop in fruits.

Note. For classified information on "Use of Herbicides in Fruit Crops", please refer to Appendix.

CHAPTER

8

THE PLANT

Classification

Plants have been classified for the sake of convenience of reference. The system of classification of plants in use are botanical, ecological, agricultural or horticultural to enable the botanist, ecologist, agriculturist and horticulturist to place the plants according to their viewpoint.

Botanists classify the plant kingdom into divisions, sub-divisions, sub-classes, orders, families, genera and species on the basis of simiarities within each group which indicate relationship. Plants that interbreed freely are known to belong to one species and are generally distinct from those of other species which interbreed freely among themselves but not with those of the first group. A genera may have one, two, or more species. The botanical name of the plant constitutes generic and species names *viz. Pyrus malus* or *Pyrus communis* belonging to the same genera (*Pyrus*) are two different species as indicated by the second word in each case *i.e., malus* (apple) and *communis (pear).*

One family may have many genera in it. Take the case of family *Rosaceae* which includes the genus *Pyrus, Prunus, Fragaria, Jugulans* etc. If the species has many forms and their recognition be difficult it is known as polymorphic (containing many forms). Several families may be included in one order, many orders in one subdivision and many subdivisions in a division. That is how the botanical classification conveys the idea of similarity or dissimilarity among plants.

Such a classification is useful since it helps to identify and name the plants, convey their relationship within the same class or with a number of other classes. These may also suggest their environmental requirement, climatic adaptations etc., at times.

Ecologists classify the plants as desert, prairie, bog, aquatic etc., according to the condition in which these flourish. In the agricultural classification we have cereals, pulses, millets sugar and starch crops, oilseeds, condiments, spices and narcotics, plantation crops etc. Besides these we also have forage crops and grasses. Horticultural classification is based on growth characteristics which broadly determine cultural pecularities and cultural requirements *viz.* annuals and perennials.

Annuals

Annuals complete their life cycle in one year or one season. Annuals taking more than 180 days for completing the life cycle may be termed as long season and those requiring less than that as short season plants. Some strains of cotton, sugarcane etc., take almost a year to complete their life cycle from seed to seed in contrast to cereals, cucurbits, tomato, pea and beans which need only three to four months. If the annuals could survive and grow indefinitely these are known as plurannuals *viz. Coleus,* some species of Chillies and Bitter gourd, *Pelargonium* etc.

Biennials

Plants which need at least two growing seasons to complete their life cycle are known as biennials. Under natural environment a biennial is a herbaceous plant that continues to grow more or less indefinitely under suitable conditions continuing to produce a succession of flowers, fruits and seeds. Biennials may be forced to behave like annuals by change in environmental conditions or special treatment and *vice versa*. The terms annual and biennial are therefore flexible. The tree biennials and perennials have the capacity to survive the stress of unfavourable conditions.

Perennials

Plants that normally live, or may live for more than 2 years are known as perennials. These are further subdivided into herbaceous and woody according to the nature of the growth.

Herbaceous Perennials

These plants are characterized by more or less succulent stems. In temperate climates the tops die after a season of vegetative growth. The tops are annual, but the crown and often roots remain alive and produce new stems and tops in another favourable reason.

Woody Perennials

There is hardly any difficulty in distinguishing between woody and herbaceous perennials. Though the stems of herbaceous plants may become somewhat woody, the difference in hardness and toughness of woody fibres is quite distinct. Woody perennials consist mostly of trees with typical, upright growth and stem or trunk, taking the form of a central axis, which is a dominant part of framework. The shrubs with no main central axis and vines with stems too weak to support the growth may also be a woody perennial.

Fruit plants may be grouped in several classes on the basis of their morphological features.

Morphological Classification of Fruit Trees

(*i*) Highly developed stem *viz.* walnut, pecan, sweet chestnut, sweet cherry, pear, mango.

(*ii*) Moderately developed stem *viz.* apple, persimmon, quince, apricot, guava, litchi.

(*iii*) Shrub-like trees *viz.* pomegranate, fig, peach, sour cherry.

(*iv*) Shrubs *viz.* gooseberry, currants forming a shrub with several stems.

(*v*) Semi-shrubs or brambles *viz.* raspberries.

(*vi*) Vines *viz.* grape vines.

(*vii*) Hebacious perennials *viz.* strawberry.

The longevity of the trees is very variable and plants may be classified on that basis as well :

1000 years	Sweet chestnut
100 – 300 years	Walnut, pecan
50 – 100 years	Persimmon, avocado
30 – 70 years	Apricot, fig
20 – 40 years	Peach, plum, pomegranate
25 – 30 years	Currant, gooseberry and raspberry
4 – 5 years	Strawberry

Plants exhibit bountiful diversity in size, shape, habit of growth and other characteristics all over the world. Roots, shoots, leaves, buds, flowers, fruits with identical structures, origin and early development show considerable variation in their anatomical make up from one genus to another. Similar organs of various plants differ in their structural orientation even though they possess identical basic structure and functions. Plant organs have important basic functions and utility to the plant and their structure has been modified to large extent to enable these to fulfil the needs of the plants. The structural peculiarities in relation to their functions and growth are considered in this chapter so that the horticulturist may harness the inherent potentialities of the structures for better production. The pattern of the tissues that makes up the root, for instance, differs from the stem. Between the structure of the stem of hydrophytes, xerophytes and mesophytes also there are anatomical differences. The monocot and dicot have dissimilar anatomical features, organ for organ.

The fruit trees has several parts important being root, trunk, stem, branches etc. (Fig. 8.1). Impact of environment and diversity in ability to adapt have put their seal on the root and shoot characters of fruit plants.

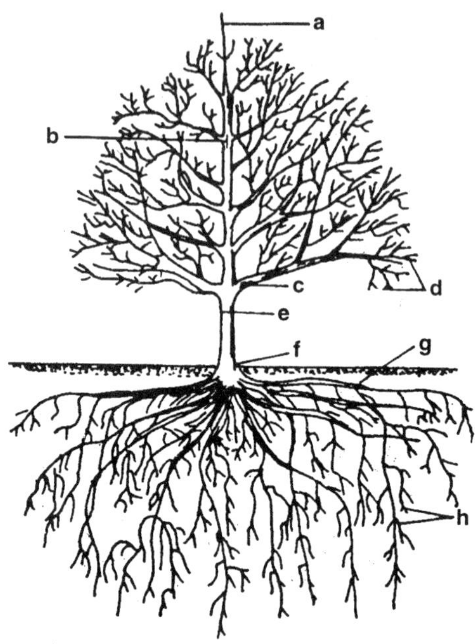

Fig. 8.1. Relative aerial and underground spread of a fruit-tree.
a. central leader ; b. stem ; c. main branches ; d. laterals ;
e. trunk ; f. root collar ; g. horizontal roots ; h. vertical roots.

About Cell, the Unit of Life

Living matter called protoplasm is a viscous transluscent mixture of complex materials. It may contain materials like fat or starch which are non-living. Cells are functional as well as the structural building block of organisms. Organism may be unicellular or multicellular. Plants have rigid cell walls composed chiefly of cellulose formed by the cytoplasm during its inception and growth. The protoplasm has a nucleus and cytoplasm is located in the cell. The nucleus is separated by a nuclear membrane, while the protoplasm has nucleoplasm with one or more nucleoli and a mass of chromatin. Mitochondria are scattered in the cytoplasm so also microsomes, the two being closely associated with endoplasmic reticulum forming a "skeleton" for the cell (Fig. 8.2). The number of vacuoles may be one or many. The vacuole within the cytoplasm are filled with non-protoplasmic watery

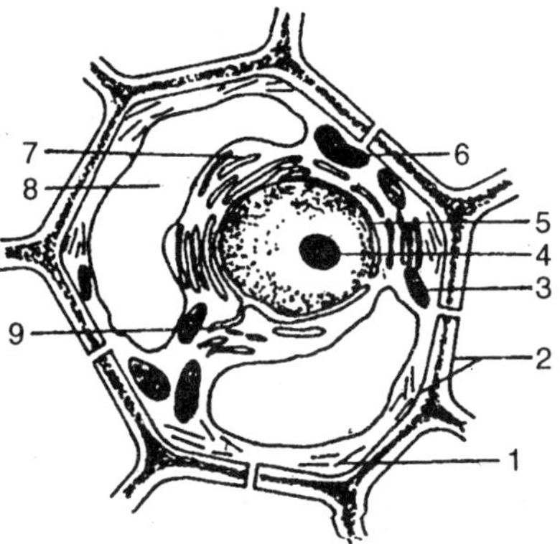

Fig. 8.2. A generalized plant cell.
1. microtubules ; 2. cellulose cell wall ;
3. mitochondrion ; 4. nucleolus ;
5. nucleus ; 6. chloroplast ;
7. ribosomes ; 8. vacuole ;
9. endoplasmic reticulum.

fluid. Centrioles are seen with proper staining. The centrioles are closely associated with cell division. Plastids are found in the plant cells and chlorophyll is the most common plastid found in green plants.

Cells are the ultimate structural as well as functional unit of a living organism of several types with specific arrangement for every organ of the plant. They vary in their number, form and arrangement depending on the type of the organ and the environmental conditions and are of great significance to the practical plant grower. The basic characteristics of the cell may be considered before we take up the study of important types of cells. Young cells present in any part of the plant tissue are similar. They change in shape, size etc., as they differentiate. The morphological parts of plant cell have bearing on the life activities and therefore are of importance.

The protoplasm has calcium, carbon, chlorine, hydrogen, iron, magnesium, nitrogen, oxygen, phosphorus, potassium, sodium and sulphur in good amounts while boron, copper, manganese, zinc, florine, molybdenum, cobalt etc., may be present only in traces. The protoplasm is composed of 75-85% of water which helps the solution of ions and molecules, for the suspension of molecular aggregates and in chemical reactions. The salts like sodium chloride and calcium carbonate that are present in the protoplasm, as ions, play a definite role of great importance in metabolism. Iron, copper and magnesium also take part in the synthesis of metabolites. Besides the indispensable inorganic salts present in the protoplasm it contains (*i*) carbohydrates which may be monosaccharide, disaccharide or trisaccharide or even polysaccharide. Cellulose, starch etc., are common examples of polysaccharides, (*ii*) proteins and their building units known as amino acids, (*iii*) fats, rich in potential energy and are generally composed of glycerol and fatty acid, (*iv*) enzymes, 'spark plugs' that make chemical reactions possible are proteins which catalyse many thousand reactions in no time, (*v*) nucleic acid composed of nucleotides which as carbon sugar (ribose or deoxyribose) as base besides a phosphate group.

The nucleotides may be either a ribonucleotide or a deoxyribonucleotide. The two important nucleic acids existing in the cell are ribonucleic acid (RNA) and deoxyribonucleic acid (DNA). The RNA molecule has four different ribonucleotides and the DNA molecule four different deoxyribonucleotides. The arrangement of the nucleotide to form nucleic acid within the protoplasm of an organism is specific. Besides these the protoplasm has vitamins which play an important role in the formation of co-enzymes (CoA), for example, being formed in the presence of pantothenic acid. The physical relationship which the various components of the protoplasm have to each other is the factor that makes it "living".

The protoplasm of plant cell, almost invariably enclosed by a more or less solid yet elastic membrane secreted by it, is an organized, living, simplified, granular material differentiated into nucleus, cytoplasm, plastid etc., which have not only difference in structural make up but also in function. Nucleus is the dense portion of protoplasm, roundish in shape, endowed with a distinct outline and surrounded by cytoplasm. The nucleus has in it chromosomes, the carrier of heredity. The cytoplasm is capable of making streaming movements which can be seen under the microscope. Cytoplasmic streaming may be seen in root cortex or in the parenchyma of the leaves. Most possibly the concentration gradient, leading to diffusion is responsible for cytoplasmic streaming. Different substances move to different directions at the same time in any cell under the directive influence of cytoplasmic streaming.

Plastids, small dense bodies of various sizes and shapes often containing starch, fats, proteins and pigments lie either embedded, suspended or floating in the cytoplasm. The colour shades and patterns of plants is due to pigments present in the plants. Chloroplasts have the capacity to trap solar energy during the process of photosynthesis. Cells also have one or more vacuoles in the cytoplasm that contain a solution of organic and inorganic substances known as cell-sap. As the cells mature smaller vacuoles coalesce to form a large central vacuole. We may now consider the different types of cells generally found in plant kingdom designed to carry out specific functions.

The process of division, elongation and differentiation is characteristic of meristematic tissues. The tips of growing plants and the endosperm of seeds may possess undifferentiated mass of parenchymatous tissue derived from numerous parechymatous cells of identical shape, size etc. as a result of the repeated cell division and enlargement of the cell generally found in the stem pith, root tip and storage organs.

Differentiation leads to the formation of different types of cells. The development of three types of cells in the rind of watermelon affords a good example of differentiation, beginning from the outermost layer we have epidermal or subepidermal layer consisting of cells with highly thickened wall, this is followed by smaller more compact, thin-walled cells bordering the coarser, thin-walled cells which lie in contact with edible portion. It must be realized that all plant tissues have generally a combination of different types of cells. The leaf consists of several kind of parenchyma, vessels and fibres. The differentiated anatomical structure of the root, stem and leaf are shown (Figs. 8.8, 8.9, 8.10, 8.17).

Parenchymatous Cell

These are the most widely distributed, thin walled slightly elongated cells possessing simple structure (Fig. 8.3), mostly concerned with synthesis or storage. These are located near the tips of the root and stems with capacity to divide and multiply under proper condition. Whenever there is a mechanical injury parenchymatous cells give rise to wound tissue. The parenchymatous cells possess the capacity of division even when these have become relatively more mature and may be modified to form the epidermis (Fig. 8.4). The epidermis persists as one cell thick external layer serving as a protective tissue with no chloroplast except in the guard cells. Epidermal cells withstand stretching to a considerable extent and consequently allow for rapid increase in size.

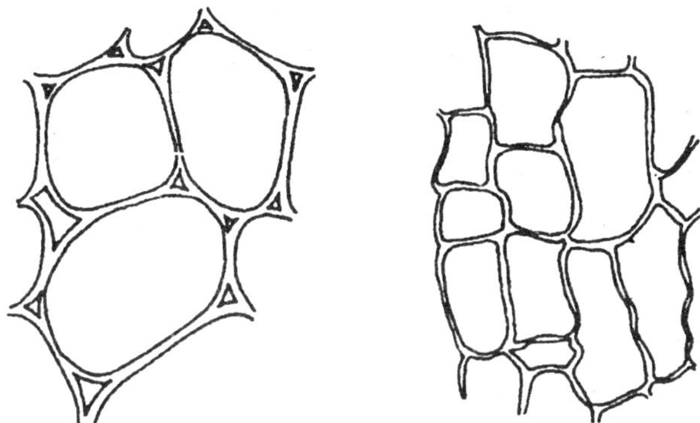

Fig. 8.3. Left, typical parenchyma tissue. Right, cork cells.

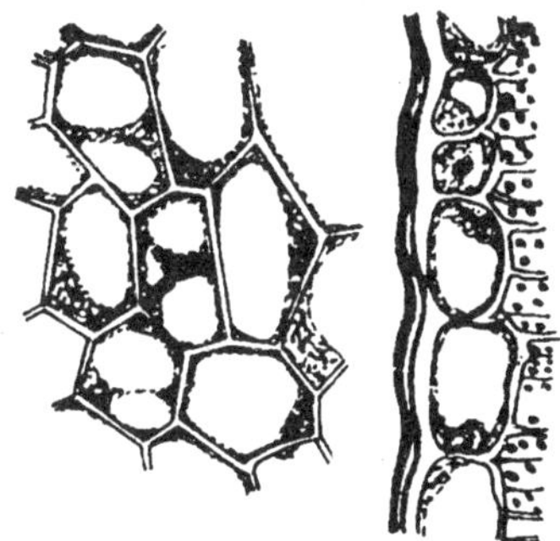

Fig. 8.4. Epidermal cells. Left, surface view. Right, cross section.

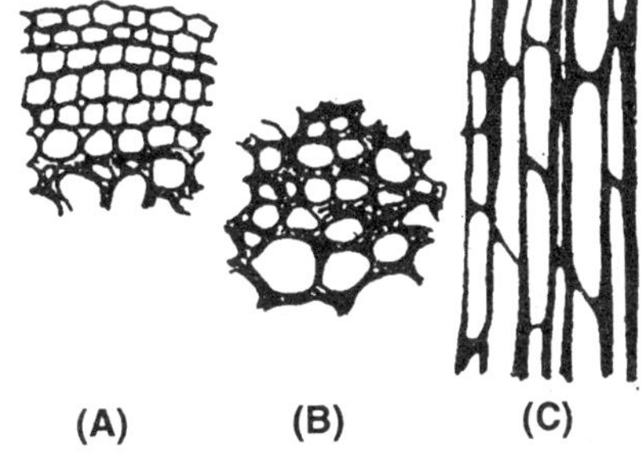

Fig. 8.5. Collenchyma cell as seen in T.S. (A and B), enlarged view (B) and L.S. (C)

Parenchyma containing chlorophyll, as in leaves and green stems, is known as *collenchyma* and consists of somewhat elongated cells with thickening strands along the sides and corners. The herbaceous stems, petioles and midrib of leaves frequently have these cells located just below the epidermis (Fig. 8.5). The collenchyma cells exhibit the properties of living cells *viz.* division, extension etc. even when mature unlike the schlerenchyma that cease to have living protoplasm at maturity. It is chiefly concentrated at sites where strength and support are required by the young plant.

The parenchymatous cells sometimes develop thick and hard walls giving rise to fibrous or stony cells. These cells are called the *schlerenchyma* (Fig. 8.6). Schlerenchyma may be present in different shapes to give rise to stone cells and fibres. The hard shell of nuts or the grit cells of pear are formed by the group of roundish schlerenchymatous cells. Elongated schlerenchyma cells may occur in groups to form fibres of economic importance and value. There is no protoplast in the cells so that extremely thin cell wall surrounding a small lumen results. The strength of the schlerenchymatous cells is due to lignin which is associated with the cellulose of young cell walls. Mature cells like the pith cells of the maize stem consist of the cell wall without the protoplasm when mature. Similar type of cells exist freely in plant kingdom and these provide for storage of food material, support for the young meristematic tissue, passage for flow of sap and framework.

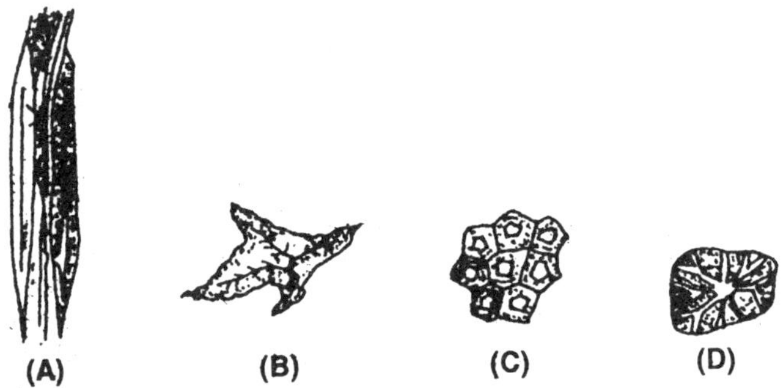

Fig. 8.6. Schlerenchyma (diagrammatic) : A, fibres in longitudinal view. B, Section through a whole selereid. C, fibres in cross section and D, section through a part of selereid.

The Root

Root systems differ in their origin, morphological, anatomical and physiological characteristics. Fruit plants have two distinct types of roots based on their origin *viz.* generative and vegetative. Root system of generative origin is characteristic of such fruit trees which are budded or grafted on the root system of seed grown stocks. Their root systems are deep in contrast to the others, own-rooted or grafted on vegetatively propagated types. Root system of vegetative origin is common with soft fruits (strawberry, currants and gooseberry), sub-tropical and certain other fruit trees. Of the subtropical fruits the olive, lemon, pomegranate, fig and those where root forms runners, or stem cuttings are also included. Among the tree fruits may be listed, apple grown on Paradisi or other clonal rootstocks and pears on quince whose root systems have developed from vegetatively propagated layers and cutting *i.e.*, clonal rootstock (not seedlings). Root systems of vegetative origin may also be developed from stem or suckers growing out of the roots of the parent plants as in sour cherry, plum, raspberries.

Roots differ in size and form. These may be main (scaffold), lateral, adventitious or fibrous. The scaffold roots are long (0.3-14 cm) and thick (10 cm or more) and include the primary as well as some secondary roots. The fibrous roots are short and thin (up to 3 mm) and include roots of third, fourth and higher order of branching. In their distribution Roots may be horizontal or vertical that grow deep in soil up to 6-12 metres. Both the types of spread are found in fruit trees. They may differ in the extent of the two types of spread.

Horizontal roots roaming parallel to the soil surface are in greater contact with high humus layer of the soil where micro-organisms are more active and nutrients are available in greater quantity. Radicles of the seed embryo give rise to primary roots as in the case of seedlings. Adventitious roots develop from stem cutting and layering. Four types of fibrous roots have been recognised (Fig. 8.7) : (*i*) growing or axial, (*ii*) absorbing or active, (*iii*) intermediate, and (*iv*) conducting.

Fig. 8.7. (A). Initial stages in the development of young seedling ; (B). one year old seedling showing primary and secondary root systems ; (C). The root system (enlarged) ; 1. Conducting roots ; 2. Absorbing roots ; 3. Region of elongation ; 4. Root cap.

The axial roots are fast growing primary structures possessing large meristematic zone. Though short-lived, rooting away or self thinning of these branches give rise to the absorbing or feeding roots. The latter are white like the former and account for about 90% of the root system. The absorbing roots may have mycorrhiza which is totally lacking in the axial roots. The intermediate roots are also of primary structure, light grey in colour. These may perform either absorbing or conducting functions. Intermediate roots are generally found in roots which are active. Conducting roots are of secondary structure unlike others and grow to form the

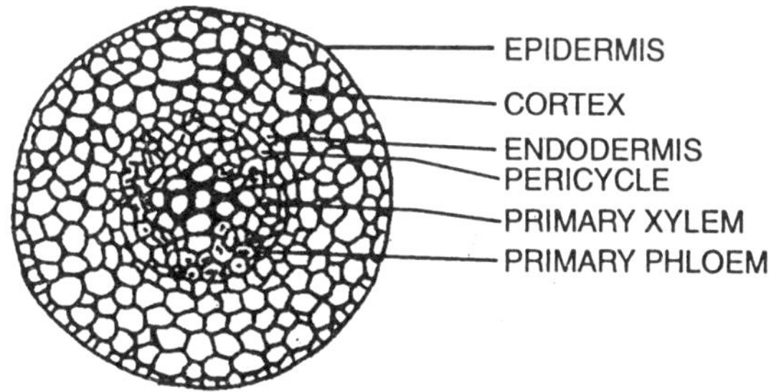

Fig. 8.8. The primary tissues in a cross section of a young root.

lateral or main root. They help in anchorage as well as in conduct of water and nutrients in both directions. The four types of roots may be found in plant at one and the same time. The general anatomical characteristics of the root (Fig. 8.8) is quite different from that of the stem (Figs. 8.9 and 8.10).

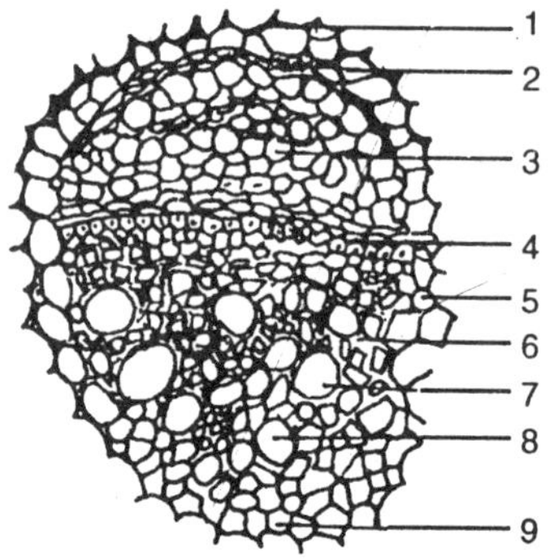

Fig. 8.9 (A). T.S. of young vascular bundle of a dicotyledonous stem 1. Parenchyma of pericyle , 2. Crushed phloem, 3. Sieve tube, 4. Fascicular cambium, 5. Medullary ray cells, 6. Trachieds, 7. Tracheal tubes, 8. Xylem (wood) parenchyma cells, 9. Pitch cells.

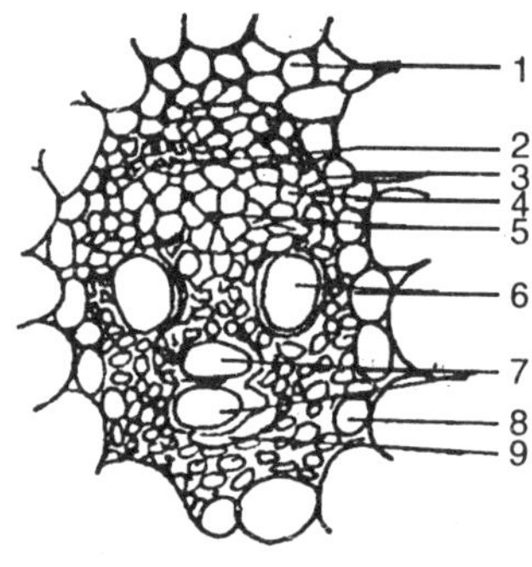

Fig. 8.9 (B). T.S. of young vascular bundle of monocotyledonous stem. 1. Pith cells, 2. Crushed protophloem, 3. Sieve tube, 4. Sieve plate, 5. Companion cells, 6. Pitted tracheal tube, 7. Annual or spiral tracheal tube, 8. Bundle sheath, 9. Intercellular passage.

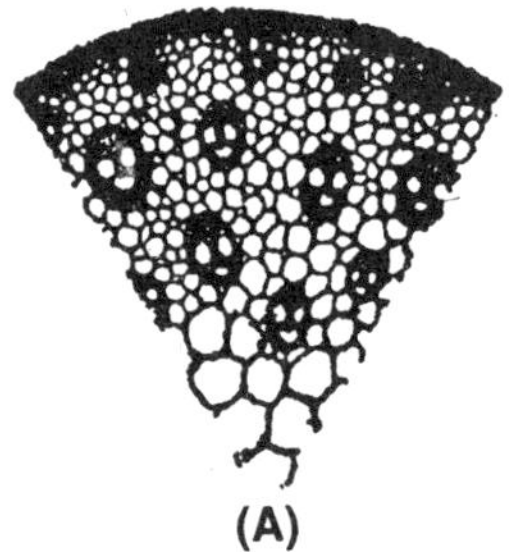

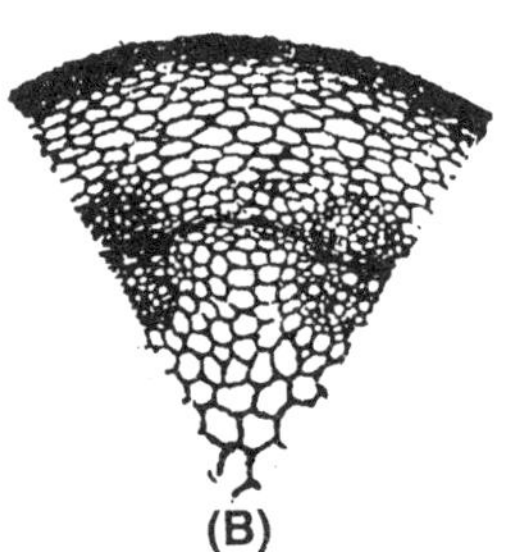

Fig. 8.10 T.S. of stem ; (A). monocotyledonous ; (B). dicotyledonous.

Functions

Roots have several functions : (*i*) anchorage of the plant in the soil, (*ii*) accumulation and storage of food resources, (*iii*) uptake of nutrients and water from the soil, (*iv*) formation of amides, amino-acids, proteins, lipids, nucleoproteins, hormones and other organic substances, (*v*) excretion of organic substances in the soil for the benefit of micro-organisms and the associate plants, (*vi*) supply of water and products of its own manufacture to the leaves in lieu of the utilization of energy and carbohydrates received from the latter, (*vii*) regulate the oxidation-reduction process, (*viii*) propagation of certain plants by root suckers *viz.*, cherry, raspberry, plum etc. Some of the more important functions of roots as reported by Kolesnikov (1966) are : (*i*) initial stages of synthesis of organic nitrogen, (*ii*) reduction of nitrates, (*iii*) protein synthesis in the active roots, (*iv*) source of supply of amino acids and asparagine to buds.

Mycorrhiza has symbiotic relationship with higher plants though exact nature of its usefulness remains to be identified. Nuts (walnuts, hazel, pecan, chestnut, almond), pome fruits (apple, quince, pear), stone fruits (sour cherry and plum) substropical fruits (persimmon), citrus fruits and all berried fruits have been reported to have mycorrhizal associations. These are generally present in developed roots under proper moisture conditions. Water stress destroys mycorrhiza. The root system is controlled by hereditary characteristics, an apple tree does not have a tap root but possesses a branched and lateral root system while the walnuts develop a deep tap root with less branching. Similarly the roots of carrot, radish, turnip, beet etc., are considerably thickened and shorter. The roots of lettuce are fibrous. In the case of sweet potato, radish, carrot and the like, the roots serve as storage organs. The spread of the root system is very much influenced by soil conditions in the vicinity of the roots. Inadequate aeration, high fertility and high water table tend to limit branching and development of deep root system

The different functions of roots are accompanied by some adaptations. Better the spread down the profile and in space of lateral roots, more effective is the mechanical support rendered to the plant. The spread is specially of importance in the case of perennial trees growing to large sizes. The extent of development of root hair controls, to a large measure, absorption of water and dissolved nutrients by the plant. Root hairs are the outgrowth of epidermal cells of the roots and are located in the roots to the maximum extent at the region showing greatest growth, adjacent to the root cap. Most fruit plants possess root hairs which are markedly different in their length and distribution. The length of root hairs varies from 50 microns in lemons to 1,200 microns in strawberry. Larger the number of root hairs more is the area in contact with fresh zone of soil mass which is capable of supplying more of nutrients. Good aeration, consequent upon adequate drainage, aids better development of the root hairs and also their efficiency to conduct dissolved substances from the zone of absorption into the plant body.

In certain cases (*Pinus sylvestis*) root hairs do not develop. Their function is amply discharged by highly extensive root system possessing thinner and more permeable walls of the epidermal cells specially in the younger stages. The part played by the absorptive surface of the root hairs is shared by the cells interior to the epidermis so that the problem of the movement of water and salts from the soil to the tops of the tallest trees is tackled without much apparent efforts on the part of the plant.

Growth of roots in the soil and the formation of new vascular or conducting tissue go hand in hand. This results in a very efficient transport system. Some land plants, for example, *Gleditschia tricanthos* do not absorb water freely after the death and decay of root hairs. The epidermal cells though a little more thick-walled allow absorption of water in small quantities for short periods. As pressure is exerted by the expansion of the differentiating tissues cracks take place in the epidermal layer.

The Stem

The stem of a plant or the trunk of a tree may be of many dimensions, shapes, habit or nature depending on the species, environmental factors and training. Whereas good forest tree has high unbranched trunk, the orchard tree is deemed to be good when branched at lower level of the stem. The stem of carrot, radish, onion and the like is extremely short and insignificant in stature, just sufficient to connect the leaves with the roots. The stem of pumpkin, bottlegourd, grape vine, bean, etc. may creep for considerable distances on ground or on any other support (Fig. 8.11). Irrespective of the nature of stem or its dimensions it performs certain functions in each and every case *viz.* serving as physcial connection that allow the transport of water and dissolved solutes from the root to the leaves and the reverse transport of manufactured food materials besides to the roots, serving as support for the leaves and reproductive organs of the plant and at times also as storage organs.

Stems have several modifications to suit the conditions of growth for greater benefit of the plant. The stems may be erect and self supporting (*e.g.*, sunflower, lily etc.), weak (creeping), reduced (carrot, radish) etc. The weak stems may be climbing (*Vitis, Passiflora. Smilax, Vanilla* etc.) or twining (*Ipomea, Thunbergia, Meyenia* etc.) or trailing (*Cynodon, Boerhaavia* etc.). The stems may also be found as runners, offset or suckers. The runner or stolon is the type found in strawberry while mint and rose form good examples of sucker. It may also be as bulbil as in some lilies. Similarly stem may be underground (rhizome, corm, bulb, stem tuber etc.). In asparagus (*Asparagus officinalis*) and Muehlenbeckia stems and branches are much flattened performing the ordinary functions of the leaves, with leaves reduced to inconspicous bracts, such stems are called cladophyll.

Pomologically, the aerial portion of the tree has stem trunk, central leader, extension shoot, main branches, crown and root crown or collar. Stem is defined as that portion which is upright and carries the crown. The portion between the collar and the lowest main branch forms the trunk. The portion of the stem above the trunk from the first branch to the base of the current season's terminal growth is distinguished as the central leader. The current season's terminal growth is called the leader or extension shoot (Fig. 8.1). The limbs comprise of the main branches forming the biggest branches of the tree making the framework. The crown in the upper part of tree which bears the branches and leaves while the collar is the point of junction of stem and root. Collar, the junction point of the stem and root has been used to denote diffcrent mcaning in the different context creating lots of confusion. In forestry, for example, it implies the position of the trunk where the bob (straight trunk portion) ends and branching begins.

Fig. 8.11. 1. Creeping stem of *Lagenaria* sp.; 2. Tendrial ; 3. Male flower ; 4. Female flower ; 5. Fruit.

The tissue of the stem is essentially alike that in the root though the proportion and position of the component structures differ. Woody stems have longer fibres and other supporting tissues. Similarly the protective tissue in a stem is more than in the root. Stems have nodes unlike the roots.

Young growing tips of stems consist of thin-walled meristematic cells while slightly older portion showing differentiation have relatively thick-walled epidermal layer which protects the stem from desiccation. Stem has vascular bundles just as in the root. The vascular bundles (Fig. 8.9) have a systematic distribution pattern in primary stem of dicots forming a ring though in the monocots these are scattered unsystematically throughout the cortex. Between the vascular bundles some differentiation takes place though much less in the bundles themselves or in the epidermal layer (Fig. 8.10). Similar to well developed roots the older stem also possesses a woody central cylinder consisting of thick walled fibres, conducting vessels, wood parenchyma. The woody cylinder is covered by bark, differentiated into an outer cortical layer and an inner layer of fibrous and conducting tissues, the phloem. Between bark and wood lies a thin layer of thin-walled undifferentiated tissue, the cambium (Fig. 8.12) which performs functions similar to its counterpart in the roots.

The plant elaborates organic substances like starch, sugars, proteins, fats, amino-acids, organic acid etc. in their metabolic cycle. These substances are stored in one form or the other in almost all the organs of the plants. The chief purpose of storage in roots is to provide food material, during unfavourable periods, for the growth of new tissues.

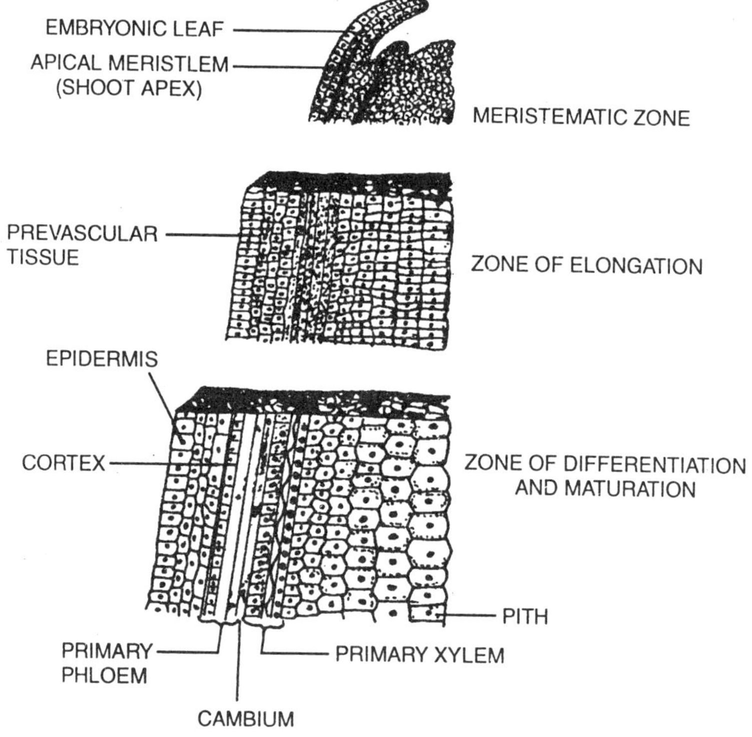

Fig. 8.12. Zones of growth. As seen in the terminal longitudinal half (section) of a stem under a microscope.

Storage in roots is neither continuous nor haphazard. It is cyclic beginning with the cessation of growth and also just before fall. This explains why pruning of fruit trees should be done when the roots have sufficient stored food material since this ensures quick recovery for profitable production. If pruning is done when the roots do not have enough stored food material dwarfing results. Adjustment of the time of pruning can also be of value in controlling the nature of the growth of the plants. Low reserves or organic materials at the time of pruning checks bushy growth.

Plants have definite growth period depending on the climatic conditions. With the end of the growing season new tissue formation ceases. The impact of the growing season on growth is reflected in the growth ring which are quite prominent with inner coarser and outer finer texture, one usually grading imperceptibly into the other. These growth or annual rings (Fig. 8.13) are helpful in assessing the age of the twigs or branches.

Annual rings are formed due to the differential activity of the cambium cells forming larger cells in early spring and smaller ones later in the season. This behaviour gives rise to formation of concentric rings. The number of rings that develop in a particular year may be one, two or even more depending on the number of flushes during the period. Each flush gives rise to false ring. The true or early flush may be distinguished from the false or secondary flush and the age may be determined on the basis of the rings developed during the span of one year. The terminal bud scales slaughed off every year produce a ring-like structure between the longitudinal growth of several seasons. The annual rings have also been made use of in determining the age of the plant. The cambial activity leads to the increase in growth.

With ageing of the stems and branches waste products are deposited, these assume dark colour, become physiologically inactive and are known as heartwood. Heartwood is easily distinguished from the younger, lighter coloured, actively functioning sapwood. Trees having larger amounts of heartwood are considered to be source of good quality wood. In orchards where better fruiting is desired slower encroachment of heartwood on sapwood is considered more profitable since it helps in the better transport of food materials leading to quick growth.

Fig. 8.13. A part of transverse section of a tree trunk showing growth rings of twelve years.

Linear growth mainly takes place near the tip. The growing point consists of young cells full of protoplasm and possesses thin walls. These divide actively, take up positions and form. Depending on the pressure, these enlarge and cause linear growth. A little away from the tip such growth rarely occurs.

As buds, ten or fifteen cm apart, send out branches that survive their centres remain similarly distant over considerably long periods.

Growth may be vigorous from the terminal bud (tobacco) or subterminal bud (apple, dahlia etc.). Where the basal bud has vigorous growth the plant assumes a bushy appearance upon injury, the exposed cambial cells or those beneath enlarge and divide to form daughter cells which divide again and this sequence is repeated over and over again till a mass of cambium like-tissue (Callus tissue) results covering the exposed portion.

Growth and differentiation take place in callus in the same way as in the stem. The process of callus formation is of special interest in the union of stock and scion, in the process of grafting from the exposed cambium tissues of the two units. Callus formation serves as a cover over the wounded surface made in grafting or budding. Success, in grafting, is dependent on the closeness or the fit for the two parts, the greater the fit, the better the results. The callus tissue of the two parts unite to form a single entity.

The Bud

Morphogenetically a bud may be regarded as a rudimentary structure or embryo-stem with a short central axis having nodes lying very close. The cells of the bud show little differentiation, most of them being meristematic. Its leaves consist of overlapping scales that protect the delicate stem tissues within (Fig. 8.14). Compare it with a fruit bud (Fig. 8.15). With the onset of conditions favourable for the growth of the buds, these unfold to form a branch.

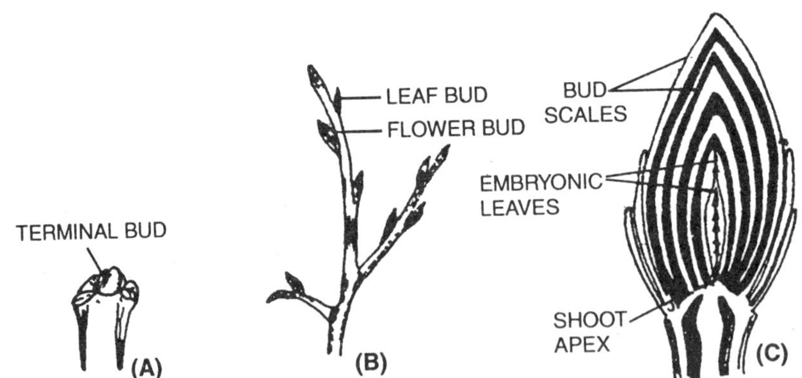

A. Terminal bud B. A twig with leaf and flower bud C. L.S. through terminal bud

Fig. 8.14. Buds, location and form.

Types of bud. Buds of several types have been recognized.

(*i*) *Terminal and axillary.* Terminal buds may be formed at the apex of a shoot of varying size, length and position *i.e.* on the side of a shoot and ordinarily in the axil of a leaf. Generally terminal buds (Fig. 8.14) are larger and more vigorous and may provide for the vegetative extension. When present on short shoots of species of the plant these give rise to flowers and fruits. Axillary buds are relatively much more numerous and may produce side branches, flowers and fruits or may fail to grow for an indefinite period of time. Such lateral buds that fail to develop in first year are termed dormant while those that are not forced into growth in the second year are spoken of as latent. Usually only one single bud is formed in the axil of a leaf though sometimes, as in peach, two or more are formed. These extra buds are known as supernumerary.

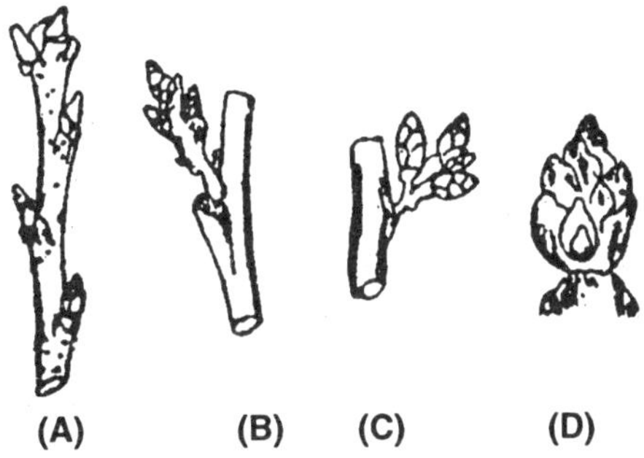

Fig. 8.15. Fruit buds : Sweet cherry A, Sour cherry B, Plum C, and Peach D.

(*ii*) *Adventitious or adventive buds.* Internodes generally do not have buds but if present these are termed adventive or adventitious buds. These may be present on the roots (blackberry) and even on the leaves (*Bryophyllum*) and can be used for propagation. Heavy pruning or fertilization induces new shoot development from both latent as well as adventitious buds.

(*iii*) *Simple or compound buds.* Buds may be (*a*) simple that give rise to vegetative growth only, (*b*) mixed, that develop a shoot with a flower or cluster of flowers, (*c*) compound, that produce both leaves and flowers. Simple bud is known as leaf bud and the mixed one as flower bud. The compound but at times, incorrectly called a flower bud should be correctly termed as a compound bud. Buds may develop in two ways *i.e.* one single bud at node as in apple or in a group as in sour cherry.

In the case of apple, compound bud gives rise to a group of leaves and blossom. In the cherry however buds of separate flowers appear close together. In addition to a large primary bud in grape, secondary and occasionally compound buds may also be formed. The secondary and tertiary buds ensure further growth and extension even after the loss of the first bud.

The Leaf

Leaves are usually bifacially flattened (Fig. 8.16), occasionally needle or scale like appendages of the stem that serve as 'kitchen' of the plant where carbon assimilation takes place by the help of dissolved nutrients absorbed by the roots from the soil and conducted to the leaves through the xylem. The fixation of carbon from the carbon dioxide of the atmosphere in the leaves is accomplished by the chloroplast cells in the presence of solar energy.

Besides carbon assimilation the leaves also regulate the transpiration stream. These vary in size, shape and structure considerably and consist mainly of layers of palisade cells containing chloroplasts and some spongy parenchyma cells, both being enclosed by layers of epidermal cells protected to varying degree, by a waxy layer called cuticle (Fig. 8.17). Leaves have vascular bundles which keep continuous connection with that of the root through the stem and branches. Leaves have small openings (stomata) through epidermis mostly on the lower side. These are bordered by two guard cells containing chloroplasts (Fig. 8.18).

Fig. 8.16. Leaf.

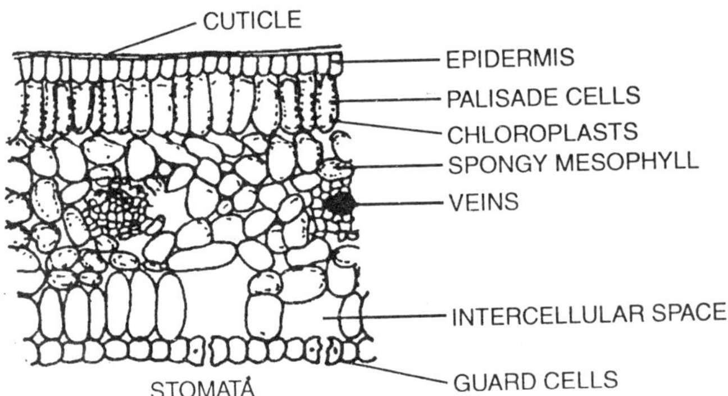

Fig. 8.17. Diagrammatic cross-section of a green leaf

Stomata (Fig. 8.18) serve for the gaseous exchange and foliar absorption. Ordinarily a leaf may consist of stipule, petiole and blade. The stipules are usually small, bract like structures found at the base of petiole which either remain attached to the stem or drop off with the unfolding of the leaf. Petiole holds the blades and also serves as conducting tissue for the transport of water, nutrients etc. from the stem to the leaf blade and the photosynthates from the leaf to other parts. Petioles may serve as storage organ *viz*. rhubarb and celery where they possess considerable parenchymatous tissue.

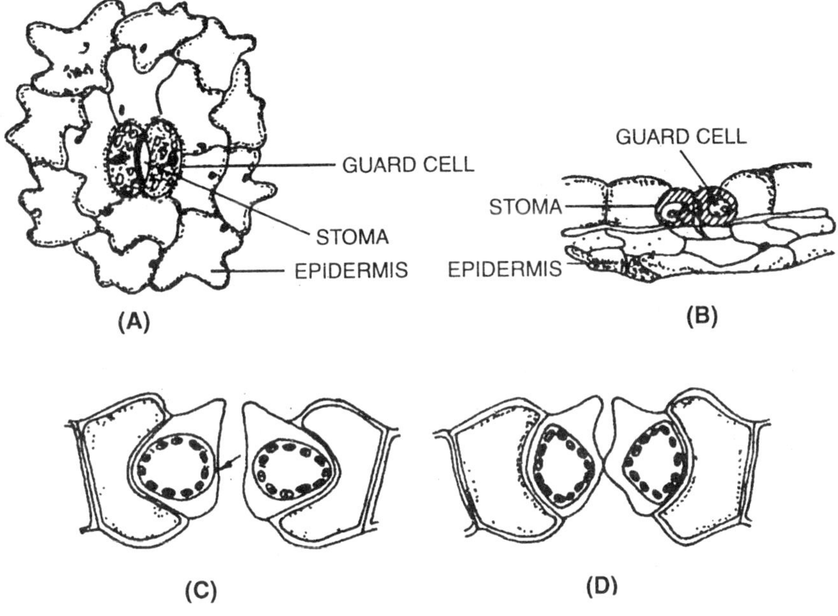

Fig. 8.18. Stoma. (a) Surface view ; (b) Cross- section ; (c) Open stoma ; (d) Closed stoma.

The Flower

A flower (Fig. 8.19) may be taken as a shoot bearing modified leaves arranged in whorls or parts on a broadened stem tip called the receptacle. A perfect flower is hermaphrodite and it must possess sepals constituting the calyx, the petals making the corolla, the male sexual organ (*androecium*) and the pistil (Fig. 8.19). If the flower contains only the male part it is called staminate while that containing only the female part, is called pistillate. The stamen usually consists of a filament and the another which produces pollen grains. The female portion of the flower consists of ovary, stigma and style. The flowers of apple, grape, strawberry, cranberry, citrus, cabbage, radish, carrot, tomato, bean etc. are hermaphrodite.

Some plants may contain pistillate and staminate flowers while others may have only one type. Plants of the former type are called monoecious and include among others the pecan, walnut, chestnut, cucumber, muskmelon, pumpkin, squash, watermelon, sweet corn. Those plants which have either pistillate or staminate flowers are spoken of as dioecious and include the papaya, holly, persimmon, certain varieties of grape, date, asparagus etc.

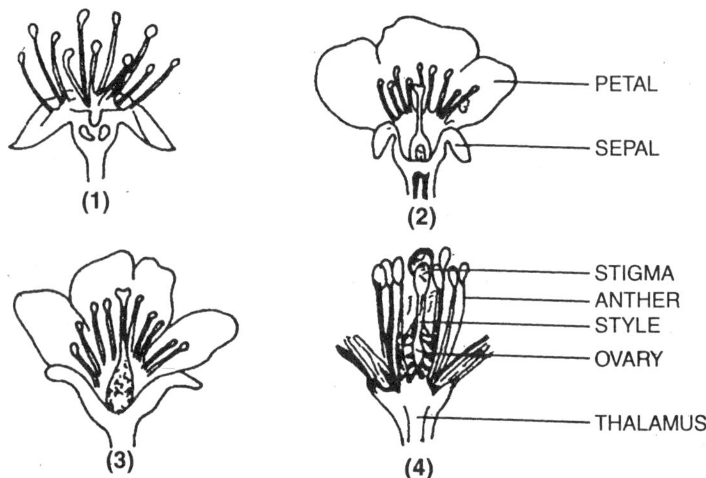

Fig. 8.19. Flowers of (1) apple, (2) plum, (3) almond, (4) lemon.

The development of fruit is influenced by the position of ovary and time of opening of flowers.

The Fruit

A fruit is the ripened ovary together with variable accessories. In production of fruits the development of ovary following fertilization is important. The fleshy receptacle is of value to the horticulturist and is the main portion consumed *viz.* in apple and peach. The development of fruit receptacle depends upon fertilization and seed development in almost all cases except in the Naval orange. A detailed consideration of fruits is given in Chapter 12.

Suckers and Water Sprouts

Suckers and sprouts are commonly found on trees and shrubs. Suckers are leafy shoots developing from adventitious buds at the crown of the tree or from the roots at some distance from the crown.

In own rooted plants, the growth from the suckers is the same as that of the top plant, with grafted material, however, the sucker growth resembles the stock in most cases. With some plants the tendency to form suckers is common *viz.,* raspberry, lilac, locust. Suckering is stimulated by injury to roots. In all grafted plants suckers serve no purpose and must be eliminated. Water sprouts are actively growing vegetative shoots arising from adventitious buds. These are characterized with upright growth, crowding and weak mechanical union and slow to flower and fruit. Sprouts may be developed to replace other branches or may be used for propagation. Water sprouts usually occur as a result of heavy fertilization or loss of tops.

Spurs and Specialized Stems

Besides suckers and water sprouts, some plants produce relatively short stubby growths, usually from older wood, called spurs. Similar to small branches, spurs differ from ordinary branches in that they make very short annual growth.

Fruits are borne, in most cases, largely or exclusively on spurs. A tuber is an underground stem functioning as a storage organ. The common potato possessing "eyes" or buds is an example of a tuber. The strawberry develops from runners which are slender trailing stems with widely spaced nodes running along the surface of the ground. Plants with horizontally growing branches in the soil strike root and send up shoots at intervals. Such plants spread rapidly and may be easily propagated. The common quack grass is one of the most persistent plants of this sort. The horizontally growing underground part of banana is called a rhizome.

Types of Growth

The main branches have different types of characteristic fruit formation. In the case of pome fruits, for example, following vegetative and fruiting growths are recognized.

(*i*) The shoot also known as leader or extension shoot is stem, less than a year old. In pomes it does not bear fruit buds.

(*ii*) The "premature" shoots are young lateral growth which have developed from the lateral buds of a shoot in its first year. These are found rarely in pome fruits but are often developed by stone fruits, sour cherries, plums possessing long internodes and well developed leaves.

(*iii*) Water sprouts are the vertically growing shoots from dormant and adventitious buds in old wood. Clipping makes them develop into ordinary branches.

(*iv*) The brindilles are the 7-30 cm long lateral shoots that end in a fruit (Fig. 8.20).

(*v*) The dard, a shoot lateral short (5-10 cm) ending in fruit bud, has shrtened internodes and very closely spaced lateral buds. (Fig. 8.20). The dards are short fruiting shoots (0.5-10 cm) in some plum and apricot varieties. The dard of plum may end in a thorn which remains for 2 to 5 years.

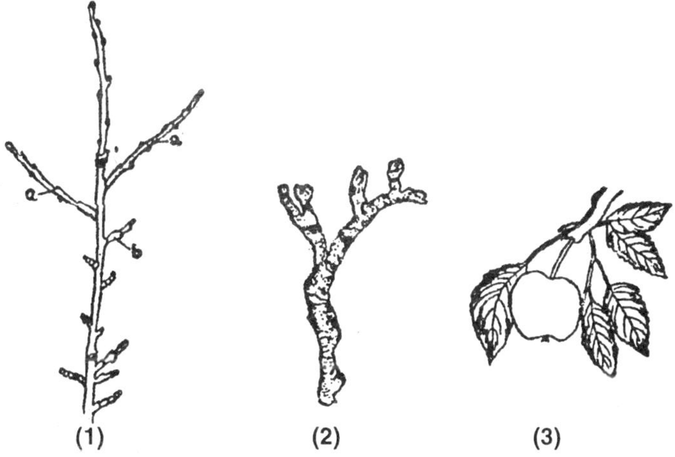

Fig. 8.20. Growth of apple tree (After Kolesnikov) 1. Fruiting branch showing brindille (a) and dard (b), 2. Seven year old spur system, 3. Spur with a fruit and leaves.

(*vi*) The spur (Fig. 8.21) is a short branch with nodes close together often consisting of successive bourses (Fig. 8.22). These occur below dards at right angles to the branch, possess sedimentary lateral buds, end in a well developed wood and fruit bud

(*vii*) The compound spur (also known as spur system, (Fig. 8.20, 2) is a short fruiting branch with several spurs. Spur system and fruit spurs increase in length slowly being covered from year to year by ring shaped scars, where closely spaced buds, scales and leaf petioles were once attached. These annual rings help to make a fail estimate of the age of fruit spur. Best fruiting results from spurs of 6-10 years of age. Spurs of 10 or more years either do not bear fruits or bear poor quality fruits and need be thinned out to encourage new growth.

(*viii*) Bourses are swollen stems (Fig. 8.22 c) at the base of inflorescence in pome fruits. These are persistent and may carry fruit or wood buds. These are formed by the phloem tissue swollen by the food material stored in it.

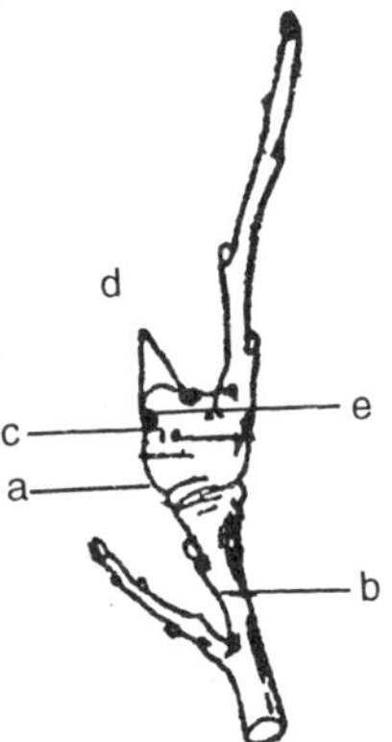

Fig. 8.22. Apple bourse with growth (after Kolesnikov); (a) outer annual ring ; (b) previous year's shoot ; (c) bourse ; (d) large scars at the points of attachment of fruit to stem ; (e) marks of attachment of flowers or dropped ovaries.

Fig. 8.21. Spurs of plum.

Besides the morphologically identical parts *viz.* the shoots, premature shoots and water sprouts in stone fruits, there are fruiting branches possessing numerous fruit buds. There are also certain mixed shoots which have both fruit as well as wood but alternating each other throughout the length of the branches. Shortened fruiting branches (0.5 to 3.0 cm long) ending in cluster of buds containing wood and fruit buds in the proportion of 1 : 4 are known as bouquet branches (Fig. 8.23). Generally speaking no razor-edge difference can be drawn between fruiting and vegetative growth of a plant, these being interconvertible depending on the condition of growth.

Growth and Related Phenomena

All growth initiates with the growth of the cell. Cell growth has been often known to take place in three phases *viz.* division, extension and differentiation. During growth, changes in form as well as chemical composition take place. A fully grown cell may divide, differentiate, or remain latent. Division helps in the formation of more cells while differentiation leads to change in appearance and structure. The cells, when in the latent state, do not exhibit any appreciable change. The division of the nucleus into two nuclei by cross wall marks the first step in the formation of two daughter cells from a single cell. The division is common in the period favourable for active growth, more so in younger regions. When the cells are enlarged to full capacity they generally remain latent when the condition is not favourable. Return of favourable condition initiates the division of the cells leading to the formation of large number of cells.

Embryonic or meristematic tissues as different from the somatic or fully grown tissues are characterised by two peculiarities *viz.* lack of differentiation and capacity to divide and multiply. The meristematic tissue present at the regions of active growth is called premeristem. It gives rise to primary meristem. The daughter cells, of which mention has been made earlier, increase in size till these attain the size of the parent cell. They again divide to form more daughter cells and this continues till the conditions necessary for vegetative growth exist.

This short statement of the division and extension of the cells should not, however, convey that these processes take place one after the other. The two processes proceed simultaneously so that both division and enlargement may be found to occur in the meristematic tissues at any time during the active period of growth. The nature of extension of the cell is determined by the plane of division and orientation. This is generally governed by the physiologica conditions of the tissues and their chemical make up. Cells continue to divide and enlarge when not fully grown. Immature cells divide and give rise to new cells which may either develop from the time of formation or their further growth may be delayed for sometime. The process of cell division in young fruits is very active at the initial stages and often as many cells are formed in short time as the number of cells in the mature fruit. This process of division often ceases when the fruits have attained one-third or one-fourth of the size of mature fruits.

Further increase in the volume of fruits is largely due to cell enlargement which continues and is enhanced thereafter.

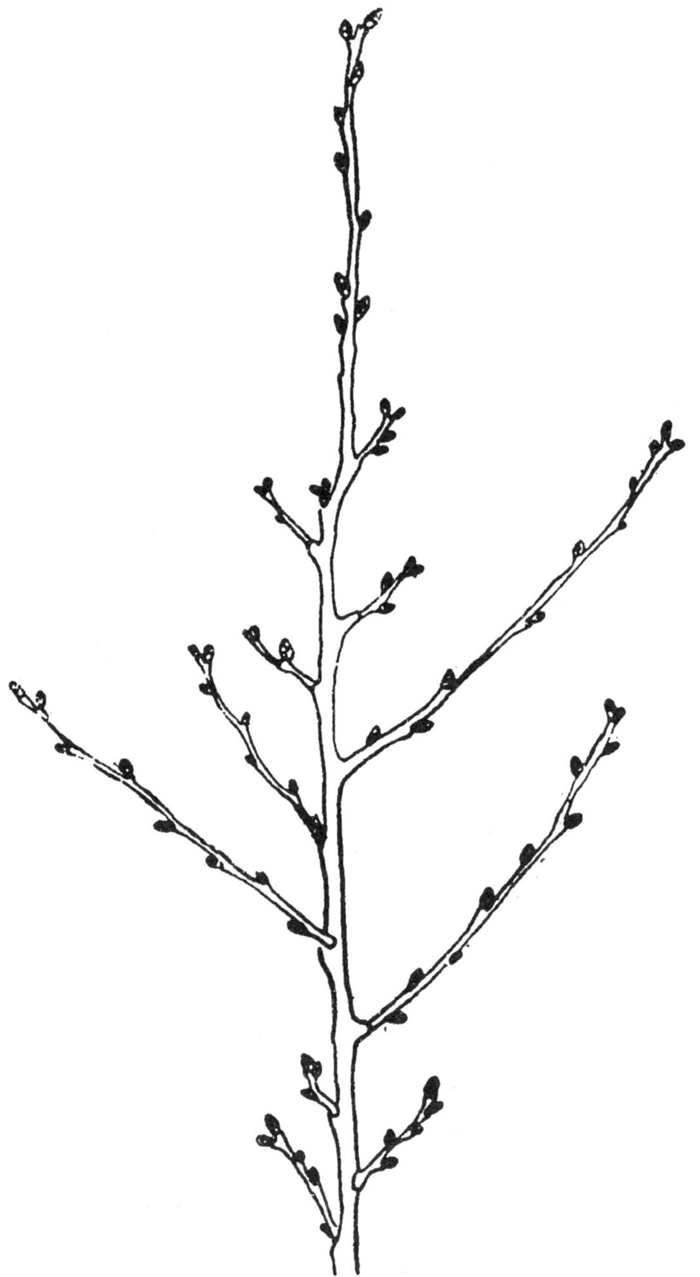

Fig. 8.23. Bouquet shoots of sweet cherry.

Cells take up various shapes by the action of the pressure falling on sides. The pressure varies with the site where the cells are located. Differentiation in shape of the type mentioned here may also be accompanied by the deposition of cell wall substances on the inner side of certain walls resulting in their thickening and consequent rigidity to the cells. This is also one aspect of differentiation. Cell differentiation may not be easily discernible to the naked eye and marked only when viewed under the microscope. Differentiation varies in its nature as well as extent and is materially dependent on the kind of plant and the tissue involved, besides the environmental conditions. The parenchymatous cells exhibit comparatively small changes in the shape with the lapse of time or age of the cell. The formation of fibres from schlerenchyma takes place when the differentiation proceeds to sufficient degree leading to the formation of elongated cells with thick walls. The fibre cells found in the fruit of banana, stem of cactus, or root of radish are thicker than the original parenchymatous cells which lead to the formation of these cells. The difference is easily visible in the length of two states of the cell, the differentiated being longer than the undifferentiated ones. Differentiation imparts characteristics to the cells such that they may be easily recognized. Cells undergo several types of changes during their formative stages, they may show signs of extension or contraction in all directions. Sooner or later, as the protoplasm is absorbed, these may take up the function of vascular tissues which serve as channel of translocation.

Differentiation is often associated with chemical changes as is evidenced by change of sugar into glucose during the thickening of the cell wall or the formation of cork cells in the plants. The change in chemical composition of maturing ears of cereals or the ripening of fruits also form examples of chemical changes during the process. The phases in the life of the cell *viz.* division, extension and differentiation lead to development and growth of plants. As the cells attain maturity these become static in size, shape and functions. In exceptional cases only fully mature phloem parenchyma cells may divide to develop bark or in the case of wounding even apparently mature cells may take up the function of regeneration of new tissues.

Conducting Tissue and Cambium

The non-living cells which play significant role in the plant are of unique importance. The nutrients absorbed through the root hairs are moved to the endodermis across the parenchyma cells of the cortex. The tissue which are specialized for the translocation of materials are the xylem and the phloem. The xylem cells consist of trachieds (Fig. 8.24), the vessels or tracheae, the wood parenchyma and xylem medullary rays. The walls of these cells may be comparatively thin but strengthened by spiral or annular thickenings. In the alternative the wall may be thick and lignified possessing many pits of the bordered type (Fig. 8.25). The trachieds eventually lose their protoplasm and die off. The tracheae consist of long cells of 20, 24, 34, 42, 46, 49 and 50 cm in the case of walnut, cherry, lemon, plum, apricot, peach and grape respectively. The xylem rays are ribbon like strips of tissue that extend radially into the xylem. These strips are composed of parenchyma cells that have their long axes at right angles to the long axes of the trachieds, vessels and wood parenchyma. These are living cells and retain their protoplasmic contents for a very long time. The non-living cells of the xylem are in contact, with some living cells *viz.* xylem parenchyma or xylem rays. Xylem and phloem are the complex tissues concerned in the translocation of materials in the vascular plants. Xylem serves as transport channel for the upward movement of water and dissolved substances. The phloem consists of sieve cells which are closely associated with companion cells (Fig. 8.26). Besides these phloem parenchyma and schlerenchymatous fibres also form a part of the phloem tissue.

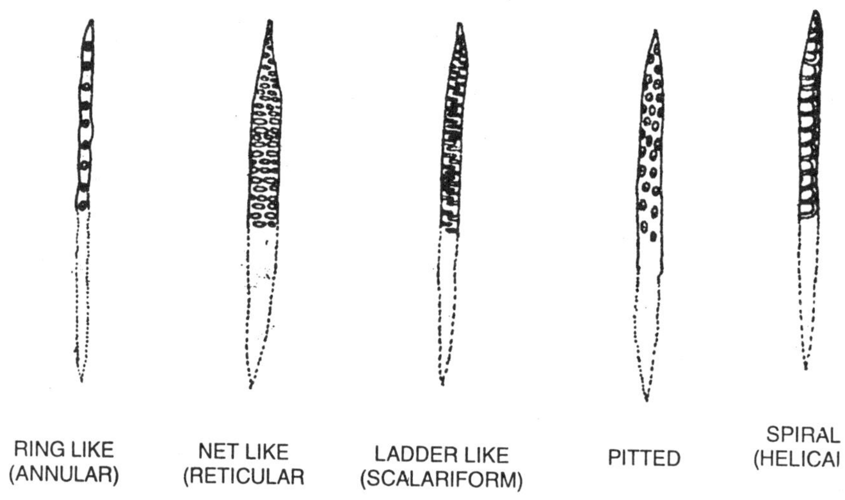

Fig. 8.24. Tracheids showing different types of secondary thickenings.

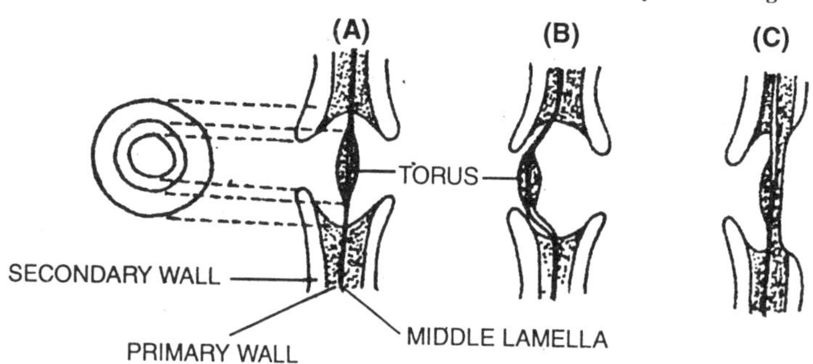

Fig. 8.25. The bordered pit A, its working B, and half bordered pit C.

The elements composing the phloem are living cells in contrast to the elements of the xylem tissue. The phloem consists of sieve tubes, phloem parenchyma, phloem rays and companion cells. The sieve tube is an elongated living cell possessing thin cellulose wall, a large vacuole with large amount of protein content bounded by cellulose on the periphery. Sieve tubes are arranged in longitudinal rows and joined end to end. The protoplasmic contents of the sieve tube are connected through pore or perforation at their end and lateral wall located in definite area round the sieve plates. While the sieve tubes do not have a nucleus, the companion cells are elongated slender, parenchymatous with small vacuole and dense protoplasmic contents and prominent nucleus. Each sieve tube is associated with one companion cell. The phloem parenchyma cells possess dense protoplasmic content and simple pits in both their longitudinal and end walls. Different types of phloem possess different amount of phloem parenchyma. The phloem rays have structure similar to the xylem rays.

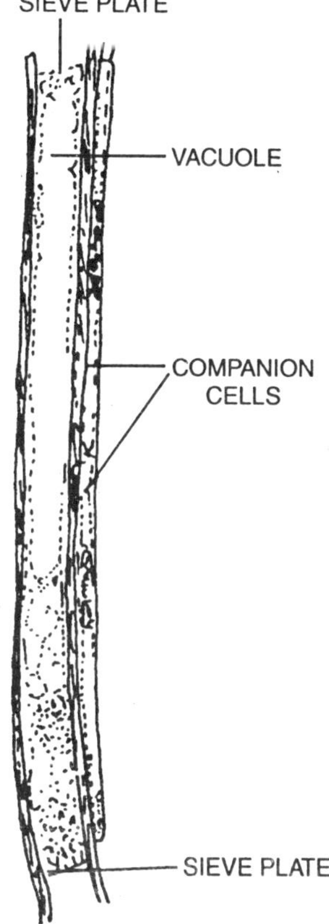

Fig. 8.26. Sieve element of the phloem.

Cambium

The vascular tissue consisting of xylem and phloem are distributed in various organs of the plant in different well known patterns. In woody stems the xylem is largely responsible for the formation of wood and the phloem for the bark separated by layer of cambium which by active division (Fig. 8.27) gives rise to xylem on the inner-side and phloem towards the periphery. Older xylem is largely used for storage and gives strength to the wood. At this stage of development it is known as heartwood and in many cases it has a deeper colour than the sap wood. The older phloem that lies on the outermost side develops corky walls which is sloughed off with passage of time.

Besides the development of xylem and phloem in the case of most woody plants, the cambium may give rise to meristematic cells which divide to form new tissues specially at the tips of growing tissue. The meristematic tissue at the tip is responsible for linear growth and the lateral meristem provides for the side branches.

In stem, maximum cell division takes place in the region where there lies greatest distance between the leaves. Linear growth takes place due to the extension of the cells but as these mature and are fully developed, the distance between the leaves remains unchanged. Such a portion of stem increases in girth due to the single plane division of the meristematic tissues. The tip of the root just below the root cap shows maximum increase in length in unit span of time.

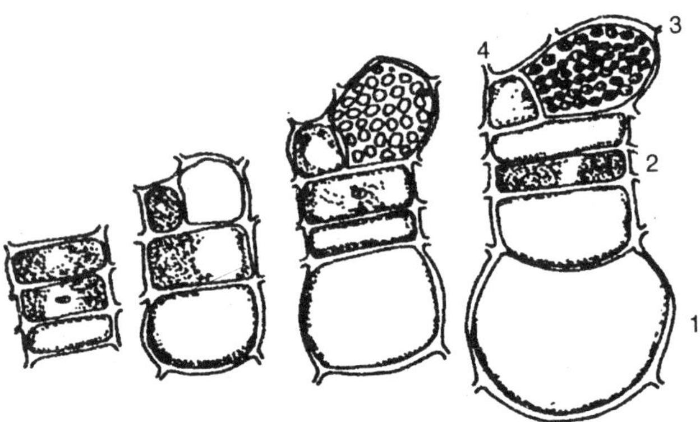

Fig. 8.27. Differentiation of cambial cells. The cambium ring consisting of a single layer of meristematic cells which divide tangentially giving rise to new xylem and new phloem elements. 1. Tracheal cell, 2. Cambial cell, 3. Sieve tube, 4. Companion cell.

PRACTICAL SIGNIFICANCE OF PLANT STRUCTURE AND PHYSIOLOGY IN HORTICULTURE

The study of the structure and functions of stems is fundamental to horticultural practices *viz.* pruning, girdling and grafting. Pruning of undesired branch should be done close to the main branch and parallel to it so that the growth of tissues surrounding the wound may form new tissues to heal the wound. For healing it is necessary that elaborated food material moving down the phloem comes in contact with wound. A cut made such that stump is formed does not help healing of the wound because the leaves were removed with the pruning of the branch. The entry of moulds, bacteria, etc. may be prevented by painting the cut end.

Girdling (Fig. 8.28), the removal of a complete ring of bark (essentially the phloem), is done to check the downward passage of food, which collects above

Fig. 8.28. Girdling (left), ringing (middle) and stripping (partial girdling, right)

the girdle. Girdling may be practised to boost up flowering, produce abnormally large fruits or even kill trees by starving the roots.

Grafting is a vegetative mode of propagation the art and science making it possible rests on the structure and physiology of stem. The stock and scion union must be full and permanent yet the individuality of the two is maintained in that the same stock may bear as many kind of flowers or fruits as the number of scions. Grafting is a horticultural practice of putting together two freshly-cut stem surfaces such that their cells grow together forming an organic union between the two pieces of stem. The preparation of stock and scion should be done in a manner that suits their size, internal anatomy, period of active growth, etc.

Success in grafting is largely based on the placing in contact of the cambial layers of the stock and scion together since these cambium layers cause union. The stock and scion must be firmly bound together so that the cut ends remain together in spite of wind etc. The region of the graft must be covered with wax to prevent the drying out of the exposed tissues and also the entry of parasites. The stock and scion union must be done in closely related species of plants since only in such related species, the growth habits and anatomy are alike to ensure union.

Grafting may be done to propagate seedless varieties of those hybrid plants which have poor quality seeds germinating with difficulty, induce early fruiting, change the shape of the plant, provide better nutrition to the plants by using vigorous stocks and to overcome the environmental difficulties in the normal propagation of plants.

The world production of organic material by green plants is roughly 2,000 billion tons per annum. Besides getting 3,000 calories per day for each individual we need energy for fuel which is being exhausted rapidly. The green plants have a marvellous energy producing process through the use of the pigment chlorophyll to utilize light energy to synthesise sugars and starch to begin with. Today the capacity of the plant to break up the water molecule into oxygen and hydrogen is likely to be harnessed for the production of hydrogen, to be used as fuel. The possibilities are great and tempting but the road is long and winding. Carbohydrate synthesis by chlorophyllous leaves in the presence of CO_2, water and light is termed photosynthesis. The carbon of the CO_2 present in the atmosphere is passed through a series of complicated yet well defined steps and converted to sugars, starch, fats, proteins (Fig. 8.29). The rate of photosynthesis is influenced by light, CO_2, water, temperature, chlorophyll and other factors which influence the number, size or condition of the leaves.

Reduced water and nutrient supply, low availability of CO_2, extremes of temperature and the accumulation of carbohydrates may diminish the rate of photosynthesis. Although theoretically any increase in the concentration of CO_2, should increase the rate of photosynthesis subject to the laws of limiting factor, in practice it is not of much significance. In general, photosynthetic activity proceeds rapidly with increase in light intensity upto a certain level. The optimum light intensity varies with plant species. Although the quality of light has effective influence on the photosynthetic activity of the plants ; it does not have much significance in horticultural practices.

Daylength is an important factor in determining the quantity of photosynthate formed, the control of the type of growth, the onset of the reproductive phase and its march. This aspect is covered under the term photoperiodism and has definite commercial value. Plants receiving continuous illumination in extreme northern regions mature early.

Checking the loss of water from the plants may check the cracking of berries in dry weather. Thin film of oil on leaves helps ripening of oranges. Water supply conditions carbohydrate synthesis directly as well as indirectly through maintenance of the turgidity of guard cells necessary for the opening of stomata to allow diffusion of gases for active photosynthesis. Water has a role in translocation of inorganic nutrients from the medium of growth to the different organs of the plant and that of the manufactured food materials from the leaves to storage organs and roots. Extremes of temperature govern the rate and extent of the manufacture of food in the plants. Above a certain threshold temperature from near about freezing point to about 75°F, the rate of photosynthesis increases. Beyond it the rate declines, the optimum is specific for plant species. Rise of temperature tends to increase the respiratory rate more than the rate of photosynthesis.

Since horticultural practices *viz.* pruning, staking, fertilizing, cultivation, mulching, irrigation influence the rate of carbohydrate synthesis in plants these should be followed not at random but with a consideration of the amount and type of growth needed. Lack of essential elements *viz.* magnesium, potassium, iron, phosphorus etc. also affect photosynthesis.

Plant growth and production is controlled by the relative intensities of the two vital processes-photosynthesis and respiration accounting for the anabolic and catabolic impacts on the plant.

Transpiration

Loss of water, as vapour, from living plants or parts thereof is termed transpiration. Leaves form the main seat of losses due to transpiration and that too through the stomata.

The speed of the transpiration stream is greater if the differences in vapour pressure inside and outside of the tissue is more. Transpirational loss of water is governed by humidity, temperature and movement of air. High temperature as well as low humidity increase the transpiration rate. Movement of air removes the nearly saturated air layer next to the leaf, allows higher gradient, and increases loss of water.

Control of water loss may be desirable at certain stages in the life of the plant ; this may be achieved by providing wind-break or shade to the cuttings and nursery stocks, painting tree trunks white to reflect heat, spraying plants with a wax emulsion, wrapping of cloth round the trunk, use of antitranspirants, judicious use of trace elements *viz.* boron and the removal of excess foliage. Use of antitranspirants might be practical and useful in cutting down the rate of transpiration and consequent reduction in the loss of moisture from the surface of the plants provided proper concentration and time of application is worked out.

Absorption and Translocation of Water

Plants take water and nutrients through their root system and this is conducted via the xylem vessels to the tops. The root tip has a white cap known as the root cap, just before this there lies the meristematic zone and then zone of cell elongation. Above this lies the zone of root hairs, the filamentous outgrowth of the epidermal cell (Fig. 8.30). Root hairs absorb maximum amount of water and dissolved nutrients, in its absence the non-suberized portions of the root carry on the function. The process by which the intake of water is affected is known as osmosis, the diffusion of solvent through differentially permeable membrane from its higher concentration to a lower one. Higher concentration of solutes (lower of solvent) in the cells of the root hair than that in the vicinity of the root hair tend to affect the movement of solvent into the cell. If the concentration of the solute, in the vicinity of the root hairs, exceeds that in the root hair cells exosmosis may take place which means that the solvent (water) moves out of the plant into the medium. Artificially such a condition can be created by applying salts to the top soils, may be to kill the weeds. It is also possible in alkaline soils and withholding of water in the soil due to high concentration of salts leads to a condition known as physiological drought. Evaporation of water from the leaves, known as the process of transpiration, produces a diffusion pressure deficit (DPD) or suction tension (S) in the spongy cells present therein. This tension is exerted on the roots through the vascular system. The roots absorb water from the soil when the DPD(S) is more than that in the soil. Water absorption thus is primarily due to the combined effect of osmotic and imbibitional forces and diffusion-pressure-deficit. The passage of water from the root hairs to the cortex of the roots through the endodermis and the pericycle and the conducting vessels constitute a continuous column into spongy cells of the leaves. Nutrients move from one place to another through diffusion and obey the laws of concentration gradient *i.e.* from higher concentration to the lower concentration.

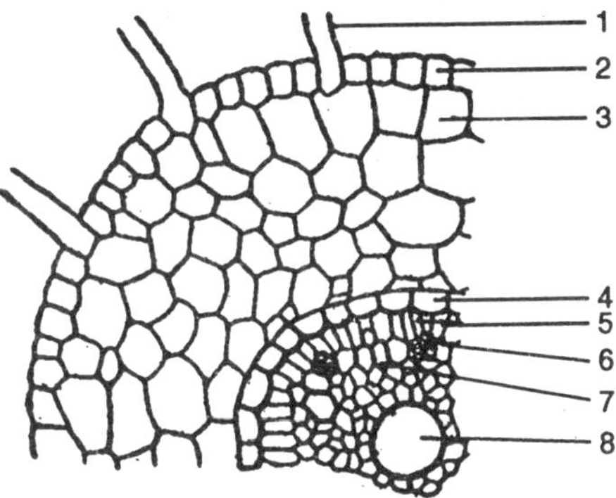

Fig. 8.30. T.S. Monocot root. 1. Root hair. 2. Epidermis. 3. Cortical parenchyma, 4. Endodermis, 5. Pericycle, 6. Phloem, 7. Xylem 8. Central tracheal tube.

Movement of water containing dissolved inorganic nutrients from the medium of growth to the tops and of organic metabolites from the leaves to the growing regions and organs of storage is termed translocation. Xylem caters the need of upward movement while phloem that of downward flow, largely. Translocation is very important as it controls the supply line to the different organs of the plant. Its rate and nature determines the nature of plant products. Many theories have been advanced for the movement of water to the tip of the plants. These may be grouped under three heads *viz.* (*i*) vital theory (*ii*) root pressure theory, (*iii*) cohesion theory. As practical horticulturists one need not go into the details of merits and demerits of the theories. Transpiration is believed to play some role in the translocation of water and minerals also.

Ringing (Fig. 8.28) experiments have shown that the cut in the supply of manufactured food to the trunk and roots lead to starvation and death of the affected parts and eventually the death of the plant. Such devices may temporarily be employed to affect temporary change in the balance of carbohydrates, nitrogen and possibly growth substances to induce more flowering and better fruiting.

Growth Periodicity

Growth has three distinctly distinguishable phases *i.e.* cell division, increase in size, and differentiation. The phases and consequently growth proceed with varying speed depending on environment or other factors giving rise to growth periodicity. Growth rate may be fast, slow or even nil in the life of a plant. Even in a single plant the growth rate may be rapid for particular organs and slow or nil for others. Alternatively the rate of growth may be fluctuating for the entire plant in general. The rapid growth in localized organs of the plant at certain periods may be witnessed and is spoken of as "flushes of growth". Environment or external conditions or genetic makeup control growth to certain extent. This takes away the capacity of the gorwer to control the nature, rate and extent of growth at will. Hormones (auxins), formed at the tips of plants from where they move to other parts also, constitute one of the internal factors that control the direction, type and amount of growth. Control of hormone production or distribution in the plant is useful from horticulturist's point of view. These induce the growth of buds in small amounts only, since larger supplies inhibit their growth. Clipping of shoot tips leads to the loss of the auxin producing part of the plant and this practice induces bushy type of growth and better fruiting afterwards. Hormones are light sensitive. The side of the plant facing light of greater intensity has lower concentration of auxin and slow growth as opposed to the side which has less intense light. This leads to the bending of twigs towards light. Such characteristic behaviour of the plants may be made use of by the horticulturists in many ways. A more detailed account of the effect of growth regulators on fruit production is given in Chapter 14.

The existence of rest period in the case of buds and the after-ripening period in the case of seed (*e.g.* stone seeds) is of value to the plant grower since the knowledge of the conditions which lead to the initiation of growth following these temporary phases is necessary from practical point of view. Stone seeds, for example, need a period of rest to be able to germinate. The rest period may often be shortened by the use of certain mechanical and chemical treatments. Speaking in the scientific language we can count the rest period also as the period of growth because many a time the processes started towards the formation of the seed continue during the rest period also. In certain cases *viz.* tomato seed there is no rest period. Plants or the organs of the plant *viz.* seed, bulb rhizome do not show any external sign of growth when they are in the resting period.

With the onset of unfavourable conditions the growing plants exhibit state of dormancy and resume growth under favourable conditions. If it were in a resisting stage it will not resume growth even under favourable conditions. Dormancy thus signifies simply the cessation of growth activities even under favourable circumstances. Resting period implies inability of cells to grow regardless of the environmental conditions. The realisation of the distinction between the two aspects of apparent stoppage of growth should invariably be made use of by horticulturist to handle the situation for profit.

Growth and Development

A plant grows vegetatively in the beginning, and provided necessary conditions, it flowers and fruits at certain stages. This later phase is known as development. Growth is a prelude to development in plants, in general. Vegetative growth is the precursor to flower production. Flower buds are differentiated without being directly influenced by other buds on the plant. This behaviour suggests the involvement of either nutritional conditions or growth regulators, in the process.

Young seedling with well developed root system, is in a position to supply good quantities of nitrogen to the buds in comparison to carbohydrates. With increase in the size of tops the nitrogen requirement increases and the carbohydrate synthesis also increases causing a stepping up of the C/N ratio. The carbohydrate availability may not be adequate to meet the need of growing plant in all cases. How the C : N balance in the plant is of consequence is some of the horticultural operations like pruning has been discussed in Chapter 6.

The supply of nitrogen should not be limiting since under these conditions neither active vegetative growth nor fruit production proceeds. Even if trees bloom under such conditions they will not set fruit. Ringing may effect accumulation of carbohydrates and consequently the setting of blossom buds. Pruning of roots retard the supply of nitrogen and water to the plant leading to higher C/N ratio and therefore of flower buds formation.

Apical Dominance

Hormones or auxins are produced in apical meristem, their downward movement tends to retard the development of lower buds. On decapitation and removal of terminal portion the phenomenon known as apical dominance becomes evidenced by rapid growth of buds down below on the decapitated stem. Apical dominance may be made use of in the control of the form of a plant and may be utilized for the training of plants to give desirable shape and also control flowering to some extent.

Horticultural plants are subjected to pruning, cutting etc. and the injury so developed is healed up in the course of time depending on the nutrient supply. The wounds which are longer and less broad with tapered bark heal up earlier than the broad cuts with the bark being absent. The cut end of the plant bleeds and dissolved food material is given out. Plants which bleed profusely must not be pruned at a time when the call sap is at high pressure. It is not easy to formulise the extent to which the structure and physiology of the plant and its organs is associated with its total performance *i.e.*, yield in details, and this aspect can best be visualised by the horticulturist in the several routine or day to day operations that the plant yields in response to the stress of heredity and environment.

CHAPTER

PROPAGATION – PRINCIPLES AND TECHNIQUES

PROPAGATION OF FRUIT TREES

Plant propagation, in India, has been a fundamental occupation since the time of Rigveda (1400 B.C.). Civilization throughout the world is largely based upon man's ability to propagate and grow plants which can provide food, shelter, clothing, recreation, aesthetic and other requirements. Plant propagation aims at the reproduction of selected individuals or group of individuals, in this the basic principles of life impose certain requirements which must be met with in practice for successful propagation. Successful plant propagation necessarily requires a knowledge of the science of plant structure and growth. An account of the structure and growth of plants has been presented elsewhere in the book.

Basis of Propagation

Plant propagation requires a complete knowledge of the scientific background of the life processes and of mechanical manipulations and technical skill needed to provide the conditions essential for growth and regeneration. Its perfection needs a certain amount of practice and experience too. The manipulations include the art of breeding, grafting and the like operations. Another important requirement of successful plant propagation is a knowledge of specific kinds of plants and the particular methods by which these must be propagated for best results under prevailing environmental conditions.

The method used must be geared to the requirements of the particular kind of plant being propagated. The propagation of plants can be accomplished by a variety of ways (Fig. 9.1) depending upon the kinds of plant involved and the purpose of propagation. Propagation may be done by seeds, cuttings, bulbils, bulbs, corms, division, layering etc. The common techniques of reproduction are detailed here under the head sexual and asexual reproduction.

See Appendix for Propagation Methods of different Fruit crops.

Seeds may be classified as orthodox or recalcitrant depending on their ability to survive during storage. Orthodox seeds are capable of survival or reduction of their moisture by drying and their storability is increased. On the other hand seeds which lose their viability when dried to certain level of moisture content consequent upon irreversible damage are known as recalcitrant seeds. Recalcitrant seeds are therefore short lived and their stability is low.

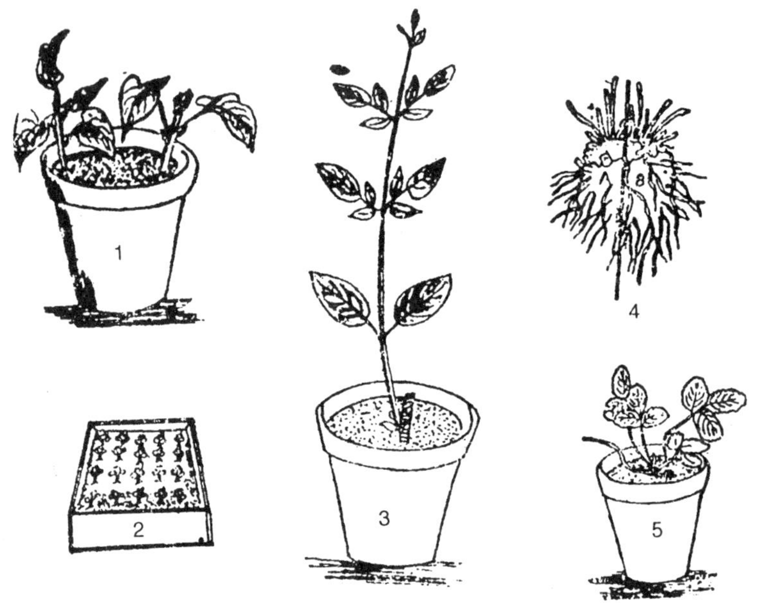

Fig. 9.1. Propagation by 1. Cutting ; 2. Seeds ; 3. Grafting ; 4. Division ; 5. Layering

The longevity and germination of some seeds is indicated below :

Fruits	Longevity (days)	Germination (%)
Jackfruit	30	80
Loquat	180	92
Sour orange	330	22
Rough melon	480	46
Grapefruit	360	88
Mandarin	1450	88
Sweet orange	1450	71
Litchi	30	8
Mangosteen, mango	240	Maximum

The seeds of lime, pummelo and lemon do not show any significant reduction in germination percentage even after being stored for eight months.

Disinfection of Seeds and Soil

Propagation by seed, though common, needs considerable care and caution regarding freedom from disease, health, vigour, vitality and viability of the seed, well drained soil free from fungi, nematodes, bacteria and weeds for success at commercial scale. In practice the seeds may be disinfected and soil rendered free of pathogens and weeds.

Disinfection of seed may be done by placing these in 0.6% acetic acid for 24 hrs at 21°C or in concentrated HCl for 30 minutes to free it from tomato mosaic virus.

Dry seed treatment may be given to protect the seeds from infecting fungi and bacteria. Liquid suspension of organo-mercurial chemicals like Agrosan – GN and Cerasan and the non-mercurial compounds like Brassicol, Captan, Dithane M-45 and Thiram (0.3% solution) are used as disinfectants.

Hot water treatment is very effective against fungi and bacteria infecting the seeds from inside. Dry seeds are immersed in hot water kept at 48-55°C for 20-30 minutes. After the treatment the seeds are spread in thin layers and dried in shade with help of air.

Apply Formalin (40%) : Water : : 1 : 50 @ 2.5 1/30 cm^2 area to disinfect soil. The soil should be covered with polythene for 24 hrs after the application. The soil is then worked and left for 2 weeks before placing of seeds for germination.

Chloropicrin (tear gas) may control all parasitic fungi but it is also highly phytotoxic hence care must be taken to apply it through injection in the soil. No sowing should be done for about 2 weeks after the application.

Methyl bromide. Effective fumigant with deep penetration property. It is selective fungicide. Effectiveness depends on the amount of soil moisture at the time of application. Methyl bromide : Chloropectin : : 2 : 1 mixture is also used.

Vapam (Sodium-N-methyl dithiocarbamate, dihydrate) – When sprayed as solution in water on the soil surface kills the nematodes, fungi, and germinating weed seeds. Sowing should not be done for 2 weeks from the date of treatment.

Seed-invigoration

Several methods have been tried to increase production of horticultural crops. Better seed gives early emergence, dense stand and healthy plants capable of giving higher yields. Some pre- and post-harvest treatments are *in vogue* to invigorate the seeds for obtaining better seed performance.

Some of the pre-sowing treatments *viz.* use of dilute salt solutions, hardening treatment, 0.01 M-Potassium phosphate (Dibasic), Coumarin (100 ppm), Potassium iodide (0.01-0.25%), GA (100 ppm), EDTA (200 ppm) and hydration-dehydration treatments have yielded encouraging results with some seeds. These methods of physiological invigoration of seeds, besides those used for breaking dormancy or making the seed pathogen-free are put under the physical physico-chemical treatments and should be adopted in practice with caution and only when found to be of significant value.

A. Sexual Reproduction

Propagation by seed is common in angiosperms, most annuals, biennials and perennials. Some seeds may germinate after passing a "rest period" to shake off their dormancy. Some seeds need pregermination treatments like scarification while others germinate immediately when placed in proper environment. The plantlet so developed from a seed is known as a seedling.

Seed Propagation

Propagation by seed though common with field crops has certain limitations with fruit trees where the number of seeds to be sown, the quality of the fruit etc. are of significance. Certified, non-dormant, vigorous and viable seeds are needed to reproduce a plant. Seeds essentially need sufficient moisture, adequate temperature and oxygen for germination. Seedling plants need not necessarily be a true image of the parent since the seeds, may at times, do not reflect the true characteristics of the mother plant specially where cross fertilization is possible under natural circumstances. While the cultivars of mango like *Langara, Dashehari* originated as a single plant from single seed, the essential and desired characteristics of these fail to reappear in the seedlings of the seeds got from the plants. Seedlings from cross fertilized seeds are generally heterozygous in contradiction to the asexually propagated plants, which are homozygous and identical to the mother plant.

A variety, which means to the fruit grower, a single plant originated from a seed which reflects its specific characteristics can be safely propagated vegetatively. A clone is described as one plant variety of this nature. A clone can be defined as generally uniform material derived from a single individual and propagated asexually alone. All members of clone behave alike in resistance, susceptibility to diseases etc. although the characteristics may perhaps be modified by environment at times. Plants which generally reproduce by asexual means *viz.* rhizomes, runners or suckers and multiply themselves as a matter of routine may also be included under this. Seedling plants of a "species" are generally not rigidly uniform hence desirable forms are often reproduced by clonal forms. Clonal material may also have some departures in characteristics from the mother plant due to bud mutation caused by cytological accidents. One of many buds may give rise to a shoot which may be of somewhat different nature. This shoot may be able to give distinct cultivar. Rarely a fruit plant produces seedlings resembling itself in all respects. In polyembryony, embryos develop in portion of the seed other than the fertilized egg and give rise to seedlings indistinguishable from the parent plant. In an attempt to raise new cultivars through hybridization, fruit growers must resort to the seed. Apple, pear and strawberry show high degree of variations in their seedlings because of their genetical make up. Many cultivars have arisen either as chance seedlings or as seedlings from a seed of unknown parents.

Methods employed for vegetative propagation may be classified in four categories as given below :

(*i*) Propagation on root system of other plants through (*i*) budding and (*ii*) grafting.

(*ii*) Propagation of its own root system through (*i*) cutting and (*ii*) layering.

(*iii*) Propagation through specialised units *viz.* bulbs (onion), tuber (potato), corm (Gladiolus), offshoot (date-palm), rhizome (banana), runner (strawberry,) sucker (chrysanthemum, banana, pineapple), slips and crown (pineapple) etc.

(*iv*) Propagation by apomictic seedlings which are identical to the mother plant besides being completely free from virus.

Apomictic seedling is one derived from an embryo developed from the diploid cells of the seeds and not as a result of fertilization between male and female gametes. Plants that produce only apomictic embryos are known as obligate apomictics and those producing apomictic and also sexual seedlings are called facultative apomictices.

Apomixis

Apomixis, the development of seeds without the complete sexual process, is a form of non-sexual or vegetative reproduction. Such seeds are known as apomictic seeds. The most significant type of apomixis is where the meiotic cycle is completely eliminated. The seed is produced from a diploid cell which is a non-reduced megaspore mother cell or some cell from the maternal ovular tissue. Due to apomixis a heterozygous cross pollinating plant appears to breed true. It may be employed in the asexual reproduction of grapes, citrus and mango to a limited extent only. Apromixis may be of several types :

(*i*) *Recurrent.* Where the female gametophyte may develop in a seemingly normal manner, without meiosis having taken place, pollination may or may not take place.

(*ii*) *Adventitious embryony.* The apomictic embryos do not arise from gametophyte but from cell or group of cells in nucleus or integument (diploid sporophyte). Sexual embryos may also develop and at maturity the seed may contain one, two, or more embryos enclosed within a single seed coat, only one of which would be true sexual embryo.

(*iii*) *Non-recurrent apomixis.* An embryo arises directly from egg nucleus without fertilization. Since the egg is haploid the resulting embryo is also haploid. It does not consistently occur in any particular kind of plant as is the case with recurrent apomixis and adventitious embryony.

(*iv*) *Vegetative apomixis.* At times (as in *Allium sativum*) vegetative buds or bulbils are produced in place of flowers.

Polymbryony, formation of more than one embryo in one embryo sac, occurs in a number of plants growing under different conditions of climate and soils. Polyembryonic stock induces more scion-vigour than monoembryonic ones but the percentage of germination of polyembryonic seeds is lower than the mono-embryonic ones. The number of sprouts in polyembryonic seeds is much less than the number of embryos. The sexual embryos develop larger seedlings than the asexual ones. Polyembryony and monoembryony are interreversible.

Advantages of propagation by seedlings

(*i*) Seedlings are long lived, hardy, bear more heavily and easier to propagate.

(*ii*) Only method of propagation where large-scale vegetative propagation is not possible as in the case of *phalsa,* mangosteen, papaya etc.

(*iii*) The hybrids are first raised from seed.

(*iv*) Chance seedlings of very superior quality may result.

(*v*) Polyembryonic seedlings raise plants true to the type. These may be used to raise uniformly identical plants.

(*vi*) Root-stock is usually propagated by seeds due to the hardy and well developed root systems of such plants.

Disadvantages of propagation by seedlings

The propagation by seedling has certain disadvantages also. Some of which are given below :

(*i*) Progeny is not always uniform in growth, quality of fruit, yield etc. This method is not safe for the perpetuation of exact replica of the mother plant.

(*ii*) Takes more time for the first bearing than the grafted ones.

(*iii*) Larger trees hence uneconomical to handle.

B. Asexual Propagation : Nature and Importance

Asexual propagation through the use of vegetative organs of the plant, involves no change in the genetic make-up of the offspring and the plants are homozygous bearing all the characteristics of the mother plant because exact duplication of chromosome takes place during cell division.

Asexual methods of propagation involve various plant organs. The advantages that make asexual method of propagation of plants so common are outlined below :

1. Progenies raised by the asexual method are generally true to type, uniform in growth, yielding capacity and fruit quality. Variations inherent with seedling trees are overcome easily by this method.
2. Uniformity of fruit quality. Picking or harvesting becomes easy owing to the restricted growth and early maturity.
3. Plants bearing seedless fruits can be propagated asexually only.

4. Asexually propagated plants come to the fruiting stage early.
5. Budding or grafting to resistant root-stock for vigorous growth, freedom from pests and diseases is made possible.
6. The advantage of better root-stock can be conveniently combined by the method even to suit zonal requirements. Apple on Russian stock or Crab apple does well in very cold regions.
7. Root stock regulates tree size and fruit quality *viz*. Golden delicious apple grafted on malling XVI root-stock may be three times larger than those grafted on malling IX. Pear trees may be dwarfed on Quince A or Quince C root-stocks.
8. Cross pollination may be aided by grafting shoots of pollinizers on branches of self unfruitful variety to encourage better bearing and high yields.
9. Use of graftage in healing of wound *viz*. bridge grafting.
10. Composite trees with different types of fruits may be raised on a common stock *viz*. same stock may be used for several varieties of sweet orange or even different citrus fruits.
11. Methods like top-working improve the bearing capacity of the plants.

In spite of these benefits there are certain disadvantages associated with asexual methods of propagation.

1. Besides the lowering of vigour and hardiness of the new plants, development of new varieties is not possible.
2. Short-lived plant results due to imperfect incompatibility between scion and stock, at times.

General Aspects

Mutations. But for mutation, propagation of plants by asexual method does not ordinarily bring about any genetic change. Only the somatic cells take part and every cell in the new plant has the genetic constitution as the parent cell. As the mutated cell situated in the growing point of a bud divides leading to the formation of a new shoot, the changed characteristics become obvious. Any bud which by abrupt changes in inheritable characters may give rise to an organ of the plant which has the capacity to perpetuate these characters asexually may be termed a bud sport. These bud sports may form the starting point of new clones. Bud sport is a somatic variant. Polyploidy involves an increase in the number of chromosomes of cell. This increase may take place either during mitosis (vegetative cell division) or in the process of fertilization. Polyploid has the advantage of being able to give rise to giant sports in certain cases. Mutation may also be induced artificially *viz*. application of extremes of temperatures, indole-3-acetic acid, colchicine x-rays, radioactive radiations etc. etc.

Chimeras

The growth of a bud involves many meristematic cells. During the course of division of the cells mutation may affect one daughter cell and the cells derived from it while there could be many non-mutated cells. When a shoot arises with some mutated and some non-mutated cells it is known as chimera. Chimera may be chiefly of three types *viz.* sectorial, periclinal and mericlinal (Fig. 9.2). Other types of chimeras (mixed, mosaic) are also reported. Chimeras produced artificially by grafting operation are called graft chimeras.

The stability of chimeras varies with the kind depending on the position of the mutated and non-mutated cells of the portion. As is evident only those which have the mutated cells show variations. Severe cutting back of scions forces adventitious buds which may arise from the callus in the region of graft union and a bud may develop partly with the tissue of stock and partly that of scion. Such a structure is essentially one of the chimeral type in which the cells of two graft components are exhibited. Winkler artificially produced many graft chimeras with tomato and black night-shade as graft material. The adventitious buds which are formed produce plants exhibiting characters of either or combined these in one, the latter group being known as chimera. Production of chimeras artificially has significant horticultural value in certain instances.

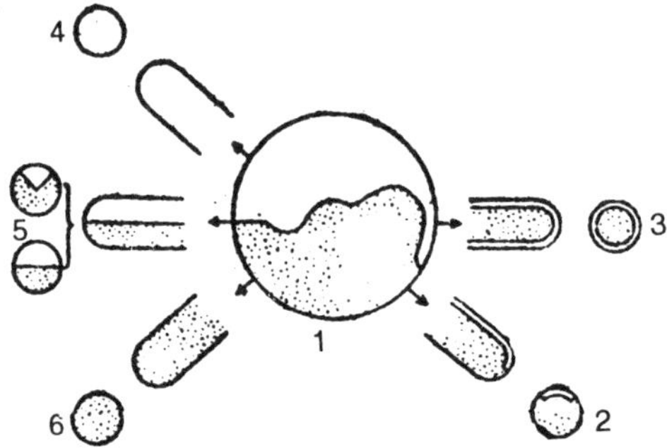

Fig. 9.2. Buds arising at different position on a sectorial chimera may produce sectorial, mericlinal or periclinal chimera.

1. sectorial chimera ; 2. mericlinal chimera 3. periclinal chimera ; 4. complete mutation giving arise to mutated cells only. 5. sectorial chimera ; 6. non-mutated cells.

TECHNIQUES OF PROPAGATION

Cuttage

Leaf, stem, leaf-buds and roots may be used to propagate new plants from existing ones. The stem cuttings may be herbaceous, softwood, semihardwood and hardwood. It forms one of the most important and prevalent modes of vegetative propagation of plants. Cutting of some varieties or species root with ease (grapes and phalsa) while others are difficult to root. Plant parts that are used for propagation may be root (Fig. 9.3), leaf (Fig. 9.4), leaf bud (Fig. 9.5) etc. It is advantageous to have a knowledge of the factors controlling rooting of cuttages specially when the material is difficult to root. Unless the factors are favourable cuttings fail to reproduce totally or partially.

Fig. 9.3. Propagation by root cutting.

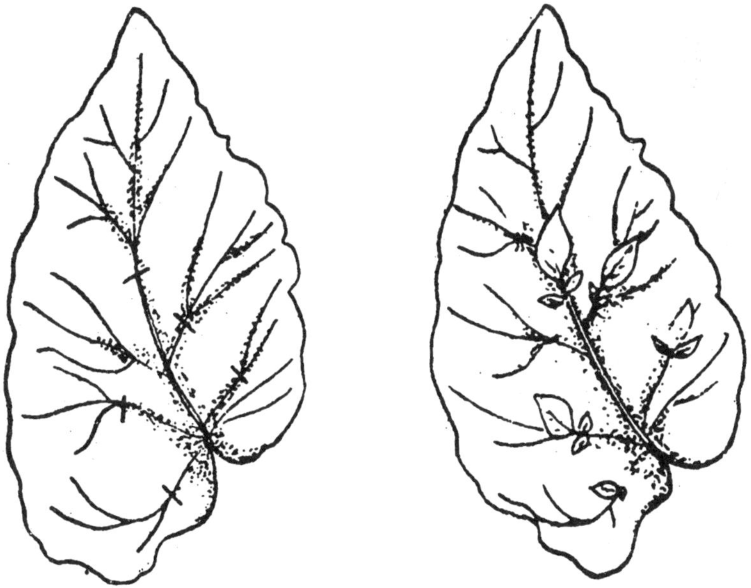

Fig. 9.4. Propagation by leaf cutting.

Material

It is now recognised that the nutrition-status of the stock plants exerts a strong influence on the development of roots and shoots from the cutting. Cuttings from plants with high C/N ratio produce more roots but feeble shoots as against those containing ample carbohydrate and higher nitrogen that produce fewer roots but stronger shoots. Cuttings from succulent stems with very low carbohydrate and high nitrogen do not succeed. Firmness of stem signifies higher carbohydrate content. A test of the freshly cut end kept in 0.2% of iodine in potassium iodide for one minute gives a colour indicative of the carbohydrate content depending on the intensity of the colour. Cuttings from slow growing stock prove better due to high carbohydrate accumulation and low nitrogen content. Deficiency of nitrogen, phosphorus, potassium, calcium or magnesium minimise chances of success.

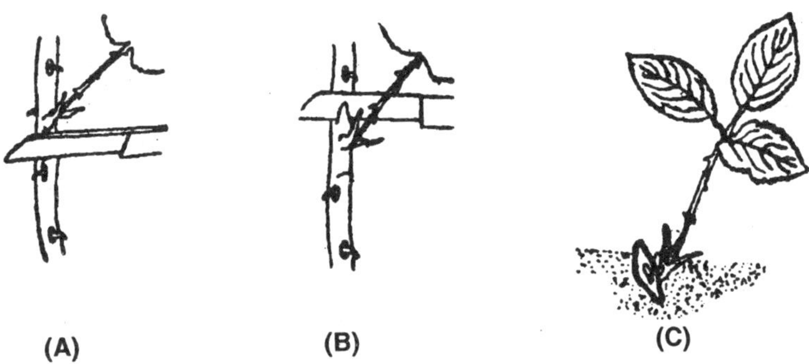

Fig. 9.5. Leaf bud cutting, A and B, removal of leaf bud from parent stem ; C. planting with the bud partially covered by soil.

High C/N balance may be achieved by reducing nitrogen supply to the stock plants for taking lateral shoots when growth has ceased. Cuttings from younger plants often root better since juvenility counts as factor. Type of wood, time of year when cutting is made and treatment with growth promoting substances are also contributory factors governing success with cuttings.

In short the internal factors affecting rooting of cuttings are the age of the plant, nutritional or hormonal condition of the plant, relative position of the shoot on the parent plant, maturity of the tissue, position of the basal cut in relation to the node and the presence of leaves and buds on the cuttings. Among the external factors light, temperature, water, atmospheric humidity and the rooting medium are of universal importance.

The environmental factors that affect rooting are humidity and temperature. Above all, rooting medium is also of significance. Rooting of herbaceous plants may be speeded and definitely increased when treated with growth regulators. Broad-leaved evergreens like the lemon, orange, grape-fruit, olive and the deciduous plants including the apple, pear, peach, apricot, plum, filbert, cherry, mulberry etc. respond to the treatment by growth regulators. Propagation by cuttings is economical, simple, easy and has the advantage that the surplus growth of only a few stock plants may be sufficient to give rise to many new plants. The problem of incompatibility does not arise in the case of cuttings. Variations and problems resulting from the rootstock and scion characteristics do not come into play. It is useful when rootstock resistant to soil conditions is desirable.

(A) Stem Cuttage

Stem cuttings are used for propagation in case of hard wood, semi-hard wood, soft wood and also herbaceous types with some difference in details. Propagation by the use of cuttings of hard wood of deciduous species is inexpensive and easy.

(*i*) *Hardwood Cuttage*. Cuttings usually prepared during the dormant season are generally successful *viz*. fig, quince, mulberry, grape, plums, pomegranate. The cuttings from the previous season's or that from 2-3 years old growth may be employed as in the case of fig. The size of cuttings may be between 10-30 cm. depending on the species and supply of stored food to nourish the developing roots. The cutting with two nodes and the basal cut just below one node and top cut 2-3 cm above the second node is preferred. The cuttings may be entirely free from older wood (straight, Fig. 9.6. A), may contain a small piece of the old wood (heel, Fig. 9.6. B) or have a portion of the old branch (mallet, Fig. 9.6. C). Cuttings may be planted in pots containing a mixture of soil, sand and well rotten leaf manure soon after detachment from the mother plant or treated with rot promoting substances before the plantation or else under unfavourable circumstances placed in well drained sand in upside down position. This period is known as the callusing period when callus formation is encouraged. The practice of putting the cutting in sand in upside down position is useful in making the basal cells develop into roots while the tip does not initiate top growth before the roots. Treatment of the basal ends of the cutting with indole butyric acid or other root promoting chemicals is given and these are stored in moist conditions with relatively high temperature (65° to 70°F) for 4 to 6 weeks to stimulate root initiation.

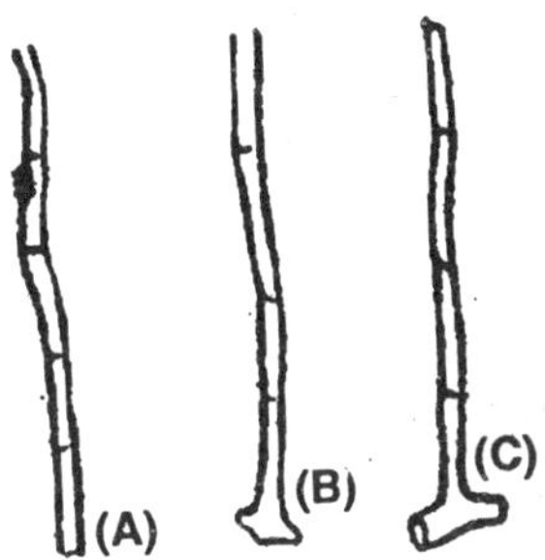

Fig. 9.6. Hard wood cuttings. A, straight ; B, heel cutting with a small piece of older wood at the base ; C, mallet cutting with an entire section of the branch of older wood attached at the base.

(*ii*) *Semi Hardwood Cuttage*. Terminal end of shoots may be used in preference to the more basal portions. About 7-15 cm length of cutting is taken with only the upper leaves intact (not the lower leaves) and rooting is attempted under condition of high humidity and freedom from chill with the aid of growth regulators.

(*iii*) *Softwood Cuttage*. These are easy to root. Cuttings of 7=12 cm length with the leaves intact are used for rooting.

(*iv*) *Herbacious Cuttage*. These are soft, succulent and tender, 7-12 cm length of cuttings taken from the plant and the leaves removed before planting with or without the aid of auxins or chemicals in well aerated sand-leaf-mould mixtures made sterile by chemical additives. Examples of herbaceous cuttage is common in *chrysanthemum* and *coleus*. All types of cuttings respond to treatments with root promoting substances.

(B) Leaf Cuttage

The leaf blade alone or with the petiole intact may be used as starting point of a new plant. Leaf cuttings of *Bryophyllum* and *Begonia* are common examples. They need environmental conditions similar to that required for herbaceous or soft wood stem cuttings. Root promoting chemicals are sometimes helpful in the case of leaf cuttings also.

(C) Leaf Bud Cuttage

Plant species which have the capacity to initiate roots but not shoots from detached leaves may be raised by the leaf-bud cutting. Axillary bud at the base of petiole is initiated to give rise to shoot. Leaf bud cutting must not be taken from leaves unless these are well developed and possess healthy well developed buds. Use of root promoting substances at the cut surface stimulates rooting.

(D) Root Cuttage

Some fruit plants such as blackberry are propagated from root cuttings. The size of the root taken for the purpose is about 5 cm in length and 0.5 cm in diameter and the planting done in sandy soil having good aeration. Root cuttings often give better results than dormant-stem cuttings. In the case of grafted plants the root cuttings below the graft union shall bear the characteristic of the rootstock and not the scion. It must be borne in mind that clone roots do not grow as well as the root cutting.

ROOTING MEDIA

Rooting media may be soil, sand, vermiculate, water or even moisture saturated air. Under Indian conditions, sand is used for the purpose which encourages free drainage while it also retains some moisture near the cuttings. Soil free from organic matter should be used for rooting.

Water is also used for rooting. It lacks aeration but aeration may be done artificially to get results. Running tap water is a good source for supplying oxygen, if need be. Moisture saturated air has been often used to speed up rooting. It has given good results in combination with sand under the conditions existing at Allahabad. Leaf cuttings may be placed in pot containing moist sand and covered with another pot to provide more humid atmosphere.

USE OF GROWTH REGULATORS

Plant growth regulators may be employed to increase the percent success by hastening root initiation and increasing the number of developed roots per cutting. It is realized that though rooting is made easy by the aid of growth regulators the plants do not have better vigour. Attempts have been made for better rooting by treatment with sugar, iron, potassium permanganate, carbon monoxide, acetylene, ethylene, etc. Some of the phenoxy compounds have been reported to be extremely active in promoting the root formation even at low concentrations, 2, 4-5 trichlorophenoxy acetic acid, 2, 4-5 trichloropropionic acid and 2, 4-trichlorophenoxy butyric acid have been used with success to stimulate rooting. Rooting of majority or plant species may be stimulated by indole acetic acid, naphthalene acetic acid and indole butyric acid. The last is supposed to be better due to its non-toxic effect over wide range of concentration. Some of these chemicals are available in the market dispersed in talc as powder and may be used with comparatively less care. All plant species do not respond to the same extent to different growth regulators. In fact the response depends on the nature of the cutting, its age and physiological condition, among other things. The cuttings from difficult to root woody plants need treatment with higher concentration than the succulent tender species that root easily. An account of the use of growth regulators for the benefit of fruit production is also given in Chapter 14.

Dry Application. The freshly cut end of the rooting material should be evently sprinkled with the commercial powder preparation to get good results.

Soaking Method. Solutions of growth regulators of the concentrations of 20-2000 ppm are used to promote rooting. More concentrated solution is used for such species which do not root easily whereas weaker concentrations are used for relatively better rooting species. The lower portion of the cutting is dipped for about 24 hours to a length of 2-3 cm of the basal end just before planting in the rooting medium.

Dip Method. This method consists in increasing the concentration of growth regulating substances for a short period of say about four to five seconds. The concentration of the freshly made solution used in such cases is kept at 500-10,000 ppm depending on the need of the plant. The cuttings are transferred to the rooting media individually after taking out from the solution for best results.

Biological Peculiarities. Asexual method of propagation (cutting, layers, division, grafting and budding) is generally employed to increase planting stock while the sexual method is chiefly important for hybridization. Sexual reproduction is also extensively used to grow root stocks on which cultivated varieties are grafted or budded. Clonal variety is specially useful in fruit growing. The cuttings of buds when taken from a plant in the period of maturity give genetically stable and uniform plants. Plants derived from cutting which have their own roots are very much the same in all characteristics as the parent plant. The plants show an extensive inter-dependence between the root system and the aerial complex as though the plant is a single whole. Plants propagated asexually are true to the type, these retain their varietal characteristics yet the different parts of shoots used in vegetative hybridization do not always give the same type of offspring. In short, even vegetative means of propagation of fruit trees do not ensure against differences in fruit from trees of one and the same variety in the same orchard. In blackberry, raspberry and currants, the plants produced by vegetative propagation produce bad croppers or produce smaller fruits in comparison to the parent tree. Sucker growth close to stem do not produce the same type of plants as those arising from the stem. Vegetative propagation is conditioned by the plant origin, plant part used, age of the parent plant, method employed, nutritional status, environmental conditions, ability of the plant to regenerate, etc. The last depends on the heredity of the plant also in addition to the environmental conditions. Cuttings from old trees are bad rooters as compared to the younger one, plants in juvenile stage are better than mature ones. Regeneration is in fact connected to growth. Terminal buds which are older than the adventitious buds show difference in their capacity of regeneration. Similarity in the development of adventitious buds on roots and the root initial are not accidental. Optimum seasons and the time for taking cuttings is relevant from the point of view of the nutrition content therein.

Root Formation from Stem Tissue. At the intersection of cambium layer and medullary ray in the stem cuttings are situated the root initials which may give rise to roots under favourable conditions. Root formation is certainly related to the anatomical peculiarities of their lenticels, those allowing free gaseous exchange and good transpiration are prompt at rooting.

Root initials are abundant towards base of a shoot specially on the side of the bud and in the upper portion of the internode. Lateral roots arise endogenically from the cells of precambium of the growing root opposite the radial rays of vascular bundles. Trees like apple, plum and sour cherry develop adventitious roots of stem origin only. These arise endogenically but from the endodermis or the pericycle of the stem with the help of newly formed meristems in all probability. Adventitious root formation has been of great importance in vegetative propagation. Roots may be formed from the callus of stem cutting particularly when these are soft wooded.

Basic Principles of Propagation by Cuttage

Cutting is that portion of the plant which is capable of producing a new plant when planted in soil. It should have the potentiality of producing the entire plant when placed under favourable conditions. It may be a leaf, shoot-piece, bud, bulb, corm, root etc. It is possible since these portions are meristematic in nature endowed with the faculty of possessing restorative regeneration. The variability among horticultural plants to root is well known and efforts are on to gain an understanding into causes of hindrances in root formation and the methods to overcome it.

On the basis of rooting behaviour, plants may be placed in four different categories : (*i*) presence of hormonal and nutritional substances at optimum level, (*ii*) presence of all substances except auxin in optimum quantity so that application of auxin helps rooting, (*iii*) deficiency in one or more of the hormonal factors but not auxins so that exogenous application of auxin does not yield positive result, and (*iv*) abundance of inhibitors so that their removal or destruction helps rooting.

Rooting of cuttings has been widely practised for propagation of *Aonla, Litchi*, Avocado, *Ber,* Custard apple, Jackfruit, *Jamun*, fig, mango, mulberry, papaya, *phalsa*, pomegranate etc. The process of rooting can be distinguished in four stages :

(*i*) Development of root initials from some cells,

(*ii*) Differentiation of the cells into root primordia,

(*iii*) Establishing vascular connections with the conducting tissue of the cutting and emergence through the cortex and epidermis as a result of growth and development.

(*iv*) Development of roots in the medium of growth.

In the process of making a cutting cells are always injured and the cut portion develops a layer of suberin, cells situated between the vascular bundles and just beyond it divide to form root primordia consisting of a number of small cells. Root tip is formed due to continuing cell division. A new vascular system develops in the primordium which gets connected with the vascular system of the cutting. Cut surface is suberised and callus formation takes place by the division of cambium and adjacent phloem. It is held that the formation of callus and the formation of roots are independent. It is not necessary that the formation of callus and the formation of roots are independent. It is not necessary that the formation of callus should precede root formation. The two processes continue side by side at the same time. The relationship between root formation and callus appearance is of varied nature. Roots may arise directly from the basal end of the cutting, a cambium develops within the callus and root initials develops from the cells of this cambium. Excessive formation of callus is associated with poor root formation in certain cases.

In some cases adventitious root initials develop on intact stem, get dormant and are embedded in stem tissue due to increase in the girth of stem *e.g.* citrus. These dormant root initials or preformed dormant root initials termed as latent root initials or root germ when present in cuttings resume growth and the roots come to the surface quickly under favourable circumstances.

Wide variation in the rooting capacity has been reported between different numbers of the same family or between strains or varieties of a species. The examples of plants easy to root are *Vitis vinifera, Citrus medica, Old Home-Pear* and Baramasia mango while *Barlette* pear, *Langra* (mango), *Merry Palmer* (bougainvillea), *Vitis berlandieri* and *Citrus aurentium* root with difficulty owing to high carbohydrate level in the former group of plants.

Inadequacy of hormonal factors, mechanical barrier, nutritional ill-balance and accumulation of inhibitors are supposed to be some of the reasons behind difference in rooting behaviour. It is also held that physiological conditions are more pronounced in their effect than anatomical limitations of the strain, variety or species in the difference in rooting.

Rooting of cuttings is influenced by external factors like light, humidity and temperature. Leaves play a significant role in regeneration of roots by supplying nutrients, growth regulating substances etc. to the rooting site. Leaves regulate water loss from the cutting hence in order to strike a balance between the have and have not of leaves provision of intermittent mist keeps the cuttings free from desiccation. Day temperature varies between 21 to 27°C and night temperature of about 15°C is reported to be optimum for temperate fruit crops. The rooting medium should have desirable texture and structure of soil ensuring adequate water holding capacity to nourish the young roots and the cuttings. Water logging should not take place at any stage and the roots forming zone must have adequate aeration.

Cuttings show a diminution in rooting with advance in their age. Juvenile cuttings of apple, apricot, cherry, citrus root easily while mature cuttings from the same plants do not root easily. Mature cuttings, when sprayed with gibberellin, behave as juvenile ones and the rooting is facilitated as well as increased. Lateral branches root better than the terminal possibly due to the availability of more carbohydrates in the former. Basal portion of the shoot gives cuttings which root better than the terminal portion of the same shoot. Cuttings taken during the flowering period of the plant root with difficulty in comparison to those taken during the non-flowering stage of the plant.

Sometimes defoliation or removal of flower buds from the cuttings promotes rooting. Cuttings with a heel or mallet (Fig. 9.6) are better rooters than those devoid of any wood at the base. The advantage may be partly due to higher level of nutrition and/or presence of preformed root initial in the cuttings with heel. Leaves and buds are supposed to be essential in rooting although the season also plays a vital role. The promotion of rooting through the bud during the non-dormant season is supposed to be due to enhanced supply of auxins and the diminution of inhibitors. Leaf area is directly related to rooting. The effect of the position of the basal cut with respect to the node varies with species. There is no uniformity in the outcome of the rooting and its proximity or otherwise to the nodes. Rooting may be from the nodes as in grape but root promoting substances induce roooting from the internodes also. Both photoperiod and photointensity are related to rooting since these affect the carbohydrate status of the plant from which cutting is taken. High nitrogen supply to the mother plant (source of cutting) lowers its carbohydrate status since the formation of proteins is accelerated, making the cuttings less suitable for rooting in comparison to those from the plants supplied with moderate amounts of nitrogen. Jackfruit cuttings which are difficult to root may be rooted easily if the basal portion of the shoot is etiolated coupled with exogenous application of IBA and mist treatment. Rooting is affected by the C/N ratio of the cutting. Adequate nitrogen, phosphorus, boron and zinc level helps in rooting.

Role of Growth Regulators

The meristematic cells of the plant is totipotent *i.e.* it has the inherent capacity to regenerate even when detached from the plant. This capacity is dependent upon several factors. Growth regulators and hormones are known to play a vital role in the process of restorative regeneration. Some of the well known growth regulators *viz*., auxins, gibberellin, cytokinins, abscisic acid (inhibitor) and ethylene influence rooting of cuttings.

Auxin increases the plasticity of the cell wall, the wall pressure around the cell decreases and the turgor pressure caused by osmotic forces in the vacular sap causes water to enter the cells leading to increase in cell size. The loosening of the wall and the intake of water are two distinct processes bringing the increase in cell size. Loosening of the cell wall needs the presence of auxin and oxygen.

The primary mechanism of auxin action is not well understood. Several possibilities have been suggested.

(*i*) Enzyme induction by auxins is operative through the mechanism of regulation of protein synthesis as indicated in the scheme proposed by Jacob and Monod. Without going into the details of the scheme it may be said that either (*i*) auxin combines with regulator gene and prevents the synthesis of the repressor or (*ii*) auxin combines with the repressor making it inactive.

It is obvious therefore that auxin or growth regulator must interact with the regulator gene or its product in one of the two ways mentioned above under the scheme.

(*ii*) Some hormonal responses requiring the involvement of DNA-dependent RNA synthesis might be due to direct effect of the hormone at the level of transcription, other responses might be the result of the action of auxin at the level of translation, while still other hormonal effects probably may not be mediated by any of the mechanisms involving RNA or protein synthesis.

(*iii*) Auxin is supposed to initiate an acidification mechanism in which pH is lowered and enzymes responsible for lossening of the cell wall become active.

Auxins are known to be helpful in rooting. The initiation of rooting is also influenced by gibberellins, cytokinins, inhibitors and ethylene. The ratio of auxin to adenine or auxin to kinetin determines the direction of regeneration. High ratio promotes rooting and a low ratio suppresses rooting and helps bud formation. This suggests the importance of the balance between the internal factors. Auxins play a vital role by interacting with the genetic material of the cell in the living plant causing its increased elongation. The nature of action is possibly through increase in plasticity of the cell wall and increased in flow of water in the cell. The growth is affected by auxin through the synthesis of RNA and increase in protein formation. A repression-depression mechanism controls protein synthesis in regeneration.

As the shoot is detached from the plant the endogenous auxin moves to the base acting as a depressor (inducer) resulting in the formation of specific proteins necessary for regeneration. This postulate gets support from the fact that application of growth regulators induces root formation. It is not known whether the control is at transcription level or at the translation level.

Auxin treatment induces more cuttings to root in much less time, increased number of roots per cutting as well as better branched and healthier root system. Auxin treatment of cuttings induces rooting due to enhanced hydrolysis of nutritional resources. The difference in root under the effect of different auxins is largely due to the difference in initiating hydrolysis of nutritional substances. Gibberellins bring about morphologically and biochemically suitable environment, within the treated cuttings, for regeneration of roots. The effect is therefore considered to be an indirect one. Gibberellins have been reported to retard the auxin-induced root formation. In certain cases of difficult-to-root plants gibberellin induces juvenility. Cytokinins also promote rooting in some cases due to the interaction with different levels of endogenous substances. Zeatin, kinetin and adenine display cytokinin-like properties. Kinetin stimulates the formation of additional root initials. Thiamine (Vit. B_1) promotes the growth of existing roots, it does not initiate regeneration of roots. Pyridoxine (Vitamin B_6) Vitamin K, H and also niacin also encourage root initiation. Endogenous rooting co-factors are present in hardwood cuttings of Crab C, EM 26, *Old Home,* MM106 apple rootstocks but lack in EM_2 and Bartlet pear. There is a belief that chemicals including auxins influence root formation though concentration of inhibitors nullifies the effect of even the exogenous application of auxins.

Certain synthetic organic compounds inhibit, promote or modify the physiological processes of plants in small amounts. Hormone may be defined as "organic substance produced naturally in plants and capable of controlling growth or other physiological functions, at a site remote from its place of production, in minute amounts." Plant hormones have properties similar to auxins, traumatic acid (wound hormone), florigen etc. Auxins are chiefly important in initiation of adventitious roots. IAA is identical in action to natural auxin contained in *Rhizopus* preparation. The use of this knowledge in practice was demonstrated some thirty five years ago. The capacity of naphthalene acetic acid and indole butyric acid to promote the growth of adventitious roots was also demonstrated near about the same time. High concentrations of auxin favour root initiation but inhibit bud formation and that is how the rooting end only is brought in contact with the growth regulator and not the other end. Hormone-like substances were formed in developing buds and transported to the base of cuttings.

Certain varieties of the citrus (*Citrus medica*) root easily while other (*Citrus aurantium*) with difficulty. It is supposed that initiation of adventitious roots was related to inherent factors in the cells to a larger extent than to readily conducted food matter or root promoting hormones.

LAYERAGE

In this method of plant propagation the plant part serves as the nucleus of a new plant which develops while still growing on the food and water resources of mother plant. This, like other asexual methods of propagation, is generally followed in those cases where the plants from the seed do not come true to type. Other methods of asexual propagation *viz*. cuttage, budding, grafting etc. are preferred over layerage if the success with them is satisfactory. Layerage is a natural means of reproduction in black raspberry, filbert, litchi, muscadine grape and trailing blackberries. Propagation by division involves dividing the blackberry root, for example, into a number of pieces each with suckers and then planting in the nursery. If the root is cut into pieces before the formation of suckers and planted to sprout it is known as cuttage. Again the cutting of the tops into many pieces for planting may be known as division. Separation of pieces of plant having rooted nodes and initiated short buds from the stock plant and planted are included in layerage. The chief characteristics of layerage therefore is the independent nature of the layer before detachment from the mother plant. Layerage involves the manipulation of downward movement of photosynthates, auxins etc., which are synthesised in the tops, to the roots.

Factors Controlling Layerage

Success with layerage is not rigidly limited by condition of atmospheric humidity and temperature as is the case with cuttage and graftage since layer remains attached to the plant at least for the period its roots do not establish and begin to feed it. Check in the downward translocation of carbohydrates, auxins or other growth factors from the leaves and growing shoot tips beyond the place of root formation leading to root formation is the secret of layerage. Since the xylem of mother plant is in contact with that of the layered shoot, the supply of nutrients is maintained and therefore layering is independent of nutrition factors which operates in the case of cuttage. Rooting in the layer can also be stimulated by etiolation. Application of rooting substances are helpful in layerage also.

TYPES OF LAYERAGE

1. Simple Layerage

Simple layerage consists in covering the medium portion of dormant year-old twig about 7-15 cm deep under soil in such a way that the tip is exposed and the branch remains attached to the mother plant. Budding of the shoot for the purpose is sufficient to induce rooting although loosening of the bark adds to success. The shoot selected should have vigorous growth at the time of layering. The notching of the stem portion to be covered and application of root inducing substances may give better results.

The layered shoots must not be disturbed till the end of the growing season after which these may be removed and planted at the desired place at suitable period.

2. Tip Layerage

The tip of the branch may be placed below soil 2-5 cm deep for rooting as is done commercially for black raspberries. This is known as tip layerage (Fig. 9.7). The time taken for the development of roots may be 2-3 months when the shoots may be detached

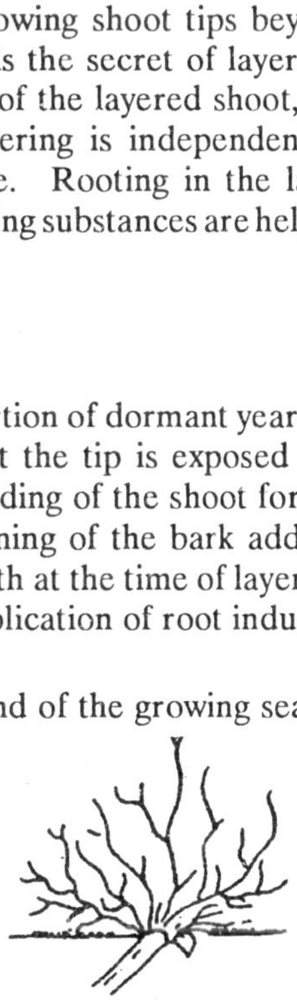

Fig. 9.7. Tip layerage commonly used to propagate black raspberries.

a short distance away from the tip and planted as new plantlet. The method may be employed with blackberries, current and gooseberry.

3. Compound or Serpentine Layerage

When the branches are alternatively covered and exposed over their entire length, the layerage is known as compound layerage (Fig. 9.8). The stem may be girdled at places and covered for better results. The buds present on the exposed part of the stem give rise to shoot while the covered portion to roots. Soon after rooting, portions may be detached containing one rooted portion and planted at suitable place and time for resumption of growth of the portion into an independent plant. It may be followed in Muscadine grape or ornamental vines which have long flexible shoots.

Fig. 9.8. Serpentine or compound layer. Specially suited to plants like Muscadine grape, *Clematis, Smilex* etc.

4. Air Layerage

This method of propagation is also known as "Chinese layerage," "pot layerage", "circumportion", "Morcottage", "gootee" etc. It differs from common layerage in the sense that the branch is not bent to reach the soil, instead the soil is brought near the branch. Rooting is encouraged by placing the girdled branch in contact with moist soil while still attached to the mother plant (Fig. 9.9). Root inducing substances may be applied at the girdled spot to encourage rooting. The soil is held in place in several ways, one of the easiest method and commonly used in India, is to wrap the soil with gunny bag, cheese cloth or plastic. This method has the advantage of being able to affect rooting from any place or portion of choice chiefly due to the fact that the branch has not to be bent to reach the soil. Soil kept in contact with the branch is kept moist to encourage rooting. Plastic wrappers used successfully hold moisture in the rooting media around the part to be rooted. The wrap is at times covered by paper to protect against damage as well as keeping the portion

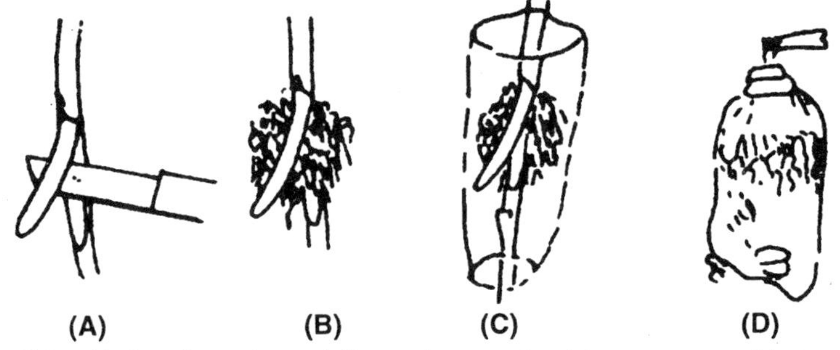

Fig. 9.9. Air layering using polythene. A, upward cut in a one-year old stem ; B, moss (*Sphagnum*) inserted between the previously hormone applied cut surfaces ; C, polythene jacket over the shoot ; D, prepared portion of the shoot stuffed with moss and the end of the jacket sealed by adhesive tape.

from becoming too hot. Excessive moisture is more a problem with air layerage than scanty supply. As the roots are formed, the layer is removed from the plant. Air layering is commonly used in propagation of *litchi*, kagzi lime, mango, pomegranate, guava, jack fruit, persimon etc.

A new technique in which moist grass (moss, *Sphagnum*) is wrapped round the cut portion of scion branch and covered tightly by polythene wraps conserve moisture and helps in rooting of latent buds from the cut-end of the scion wood. Roots are formed in 4-6 weeks in *litchi* gootees, which can be severed from mother tree by giving a half cut after 6-8 weeks, followed by full cut after another 10 days. The polythene wraps are removed and rooted, scion twigs are planted in the field and allowed to grow, flower and fruit. The main advantage of using *Sphagnum* and polythene wraps is that roots can be seen by naked eye without opening the wrap and continuous watering of gootee is not necessary to maintain the level of humidity in the vicinity of the debarked end of the scion.

5. Mound or Stool Layerage

This method is specially suitable where branches are stiff and difficult to be bent, specially in the trees which have perhaps outlived their utility due to some type of defect in the branches. It involves severe pruning of trees like apple, guava, mango, etc. in the dormant season to induce the development of a large number of vigorous shoots from the base (Fig. 9.10). As the new shoots develop these may be thinned down, soil put round them and maintained moist. It is necessary that the tip of the young shoots should not be covered with soil at any stage. The shoots may be covered only half the way. The shoots are severed from the mother plant, after the roots have developed, and planted in suitable season. Girdling the bases of shoots by wiring, about six weeks after their initiation, also stimulates rooting considerably.

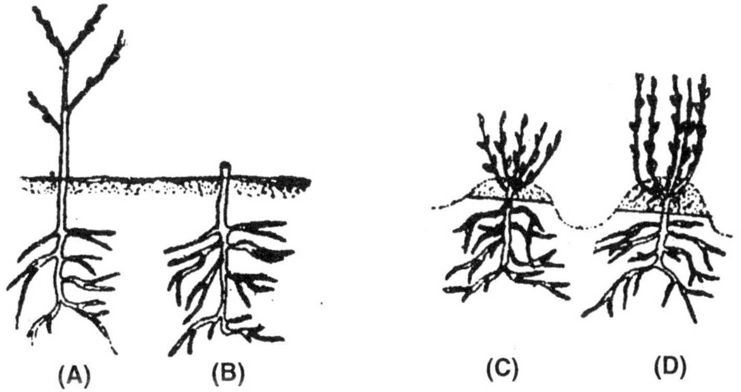

Fig. 9.10. Mound layering. A, established mother plant ; B, top removed, a portion is above the soil level ; C, addition of soil started when three to five inch growth of new shoot is achieved; D, more soil is added and the shoots root which may be severed from the mother plant and planted separately.

6. Trench or Continuous Layerage

A branch may be covered over its entire length to get plants from all nodes (Fig. 9.11). The branches are covered 5-12 cm deep horizontally in the base of trench for the purpose. The method has the advantage of giving out large number of branches at a time but the plants are apt to be weak. The method has been employed for propagation of cherry, plum, and apple stocks although this is not a commercial method. Trench layerage is essentially a nursery method of propagating rootstock of fruit trees which are not easily possible by other methods. It is also useful in the case of shrubs.

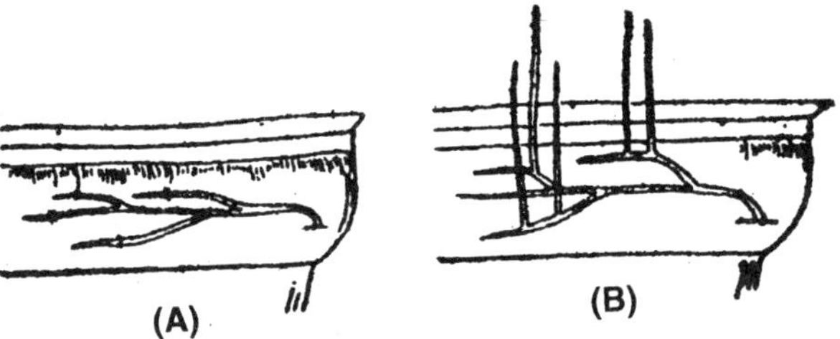

Fig. 9.11. Trench layering. A, plant laid flat in trench just before growth with the tips cut back slightly; B, the shoots emerged out after being covered with soil by the end of the season.

GRAFTAGE

It is the process and operation of inserting a part of a plant on to another plant in such a manner that they unite and continue their growth as one unit. Graftage necessarily aims at cambial union of the two parts. The top part of graft combination is termed the scion, and the other, in contact with the soil, is termed root-stock Graftage encompasses two chief types :

(*i*) Scion not severed from parent plant during process of its union with stock *viz.* inarching, approach graft.

(*ii*) Scion completely severed from parent plant during process of union *viz.* rind or crown graft, cleft graft, wedge graft, splice graft, side graft, notch graft, veneer graft, whip or tongue graft, shield graft etc. and also budding of several types *i.e.* Forkert, T-budding, patch budding, flap budding, chip budding etc.

Success in graftage (grafting, inarching, scion-grafting, bud grafting etc.) depends on compatibility of graft union, cambium contact and closeness between the scion and the rootstock. When the scion is a small piece containing a single bud the method of joining it with the stock is termed budding. Success of a graft depends on the union of the cambial portion of one unit with that of the other so intimately that the sap flows from one to the other. Any two plants having close botanical relationship may be used for graft, one serving as a root-stock and the other as scion. Graft between two plants of different botanical families may also succeed occasionally. The fruit quality, its bearing habits etc.

are derived from the scion while the rooting habit, intake of nutrients, vigour, tolerance to various soil conditions depends on rootstock. Graftage with monocotyledonous plants are successful only in rare cases, hence it is mainly practiced with dicotyledonous plants where the cambium takes active part.

Grafts between members of different genera are possible although intergrafting is of limited significance only. Grafts within a species or genera are more common and successful. The performance of the graft of any two plants must be tested before the resulting union is considered successful.

Grafting or budding is chiefly employed for the perpetuation of clones of apple, pear, walnut, almonds etc. which could not be reproduced by cuttage, division, layerage or other asexual methods with convenience and high degree of success. Clonal rootstocks have uniformity and may be used for the perpetuation of specific characteristics of a particular stock, say plant size. Propagation on large commercial scale may include budding of the desired variety on the root-stock even in those cultivars which can be propagated by cuttage, layerage etc. Graftage gives extra advantages of combining desirable qualities from the rootstock. If the rootstocks are sturdy, of good rooting habit, disease free, tolerant to floods and drought the qualities may be combined in any scion compatible with it.

The method of double working or use of a particular plant which should be compatible to both the rootstock and scion so that the qualities of these come into play and intermediate stocks are available. Top working *i.e.* cutting of trees may also provide for adequate cross pollination throughout the orchard to the proper pollinizing variety. Grafting the scion on older tree cuts down the age of fruiting. It is useful in the repair of damaged parts of the trees. The new cambial sheath formed in the callus bridge start the laying down the new xylem and phloem alongwith the original vascular cambium of the stock and scion establishing vascular connection between these.

GRAFT UNION

Success in budding or grafting by detached scion may depend on the compatibility between the stock and the scion. Some plants respond to grafting or budding (apple, pear, citrus) more easily than othes (mango, walnut) while still others do not succeed. The causes of the failure, total or partial, may be due to inherent difference in their ability to form callus, poor supply of nutrients, tendency to produce excessive quantities of gums, resins, etc. between the callus tissue of the stock and the scion.

Active rootstock and dormant scion ensures greater success in union. The bud should be easily removable from the scion with or without the bark depending on the type of budding to be followed. The season of the operation always plays a dominating role, the early rainy season or the spring being more prone to success.

The damage from insects; pests and diseases should be avoided during the period when union of the rootstock and scion is in the process of being materialised. Grafting wax should be used invariably. It can be made by mixing resin (0.25 kg), bees wax (0.3 kg), raw linseed oil (220 ml) powdered charcoal (30 gm) and glue (45 gm). Bees wax and resin are heated to melt and to it is added linseed oil and charcoal powder. In another vessel glue is dissolved in hot water (100°C), cooled and added to the Bees wax-resin-linseed oil mixture which is cool but still in liquid form.

The requirements of successful graft union may be summarized in brief. The stock and scion both must be freshly cut. These must be held together intimately and securely such that the cambial regions of each are in close contact. This operation of bringing the scion and root stock together must be done only when the environmental conditions (humidity and temperature 45°-60°F) promote the activity of cells. Formation of callus is the first important step followed by intermingling and interlocking of the callus cells of both the graft components for success of the graft. The next phase consists in the differentiation of some of the cells from the callus into new cambial cells. Formation of new vascular tissue by the new cambium developed in the graft components at the point of junction is the stage after which free translocation of nutrients takes place denoting that the union is complete. This process of graft union is the same physiologically as the healing of wound, the slight difference being that in graft union a foreign tissue (scion) is incorporated into the split pieces of a stock as in the case of split graft. For the initiation of the shoot it is necessary that the scion possesses active bud. The region of the graft union must be free from disease. There are very few reports to the effect that graft union is accelerated by growth substances or other chemicals.

Some combinations of grafts may not succeed. It is known that some pear cultivars graft well on quince while other cultivars of pear do not succeed. This is due to incompatibility. Incompatibility is not understood fully but it is the opinion of workers in the field that this is caused due to diseased component besides anatomical and physiological disorders. Structural weakness of either component and inability of the rootstock to supply nutrients to the scion of required quality and quantity are some of the causes of incompatibility. By way of example, Jonathan apple scion when grafted on Malling-9 (nine) develops molybdenum deficiency but the same scion does not show any sign of deficiency on other rootstock. It indicates that the rootstock is not able to absorb molybdenum in good quantity for supply to the scion. *Pyrus pyrifolia* rootstock being prone to virus does not give a healthy plant of Bartlett pear used as scion. Bartlet does well on virus resistant *Pyrus communis* rootstock.

The factors which influence the graft union are compatibility, kind of plant, environmental conditions and growth activity of the stock plant. While attempting a graft union between two stem pieces the morphologically proximal end of the scion should be placed in contact with the morphologically distal end of the stock. If stem tissue is to be grafted on to a root piece (root grafting) the proximal end of the scion (1) should be brought in contact with proximal end of the root piece (2, Fig. 9.12). The distal end of the stem (3) and the distal end of the root (4) is seen in figure. There are several types of grafts in practice. Some of these may be considered here.

Double working. This graft has two graft unions, one between the stock and the intermediate stock and the other between the intermediate stock and the scion. It is generally practised to overcome incompatibility between the rootstock and the scion. This method may also be employed to obtain dwarf or disease resistant plants.

Top working. With the onset of the spring, just before the initiation of active growth by the stock plant all the branches (except 3-5 scaffold branches) are removed. Any type of grafting may be used but the cleft method (Fig. 9.20) has given greater success. After grafting of each stub the grafting area is covered with wax or polythene.

1. Splice or Whip Grafting

Long, diagonal, 3-5 cm. slant cuts of equal length, on both stock and scion of almost equal girth, are made (Fig. 9.13). The cambium of the two plant components is matched and the graft tied firmly in place. This graft is generally used in cases where the union takes place with ease and strong winds do not blow. It is specially suited for plants with a very pithy stem. The method has been followed in cherries, apple and pear. This type of grafting is done just before bud-break in spring.

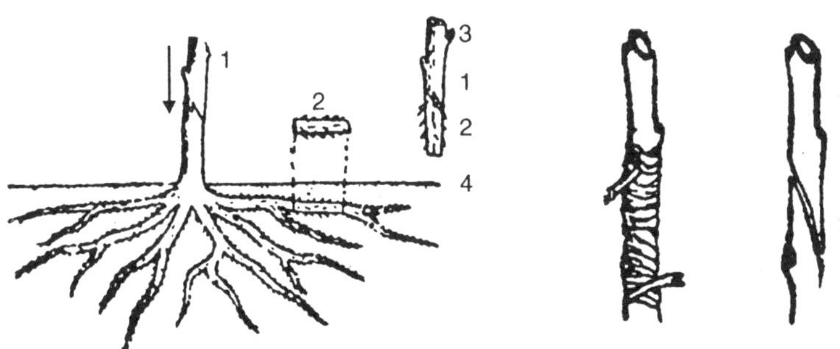

Fig. 9.12. Root grafting : 1, the scion ; 2. the stock ; 3, the distal end of scion ; 4, the distal end of root stock.

Fig 9.13. Splice grafting. A. long slanting cut in stock and scion ; B. stock and scion tied firmly.

2. Whip and Tongue Grafting

The girth of the plants used in this method should not be more than 3 cm in diameter. Both the stock and scion should be of almost equal diameter. Diagonal cuts on both scion and root stock are made (Figs. 9.14 & 9.15). The length of the cut should be about five times the diameter of the plants being used in the graft. The cut made in the stock must be matching to that made in the scion. Long smooth sloping cut is made in one stroke. On these cut surfaces a reverse cut is made downward starting from about 1/4 of the distance from the tip. The scion and stock are interlocked so that the cambium layers match along (at least on one side) preferably on both the sides. The completed grafts are then secured in position with grafting tape until these unite. Wrapping may be necessary for keeping in position and for protection from drying.

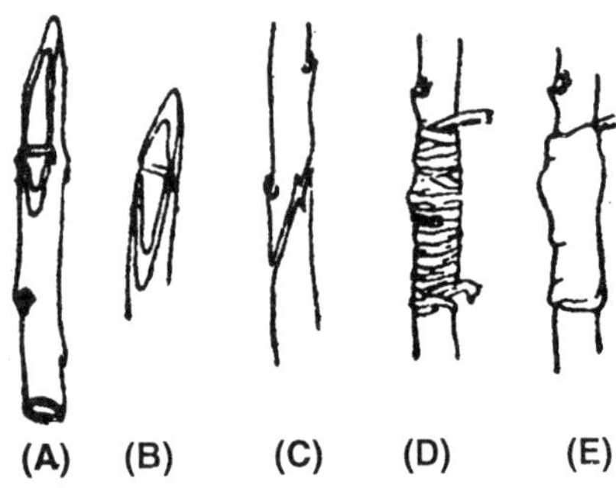

Fig. 9.14. Stages in whip and tongue grafting, A, slanting cut and tongue in scion; B, stock with similar cut ; C, tongues of stock and scion put in contact ; D, stock and scion with tongues in contact finally tied firmly ; E, covered with grafting wax.

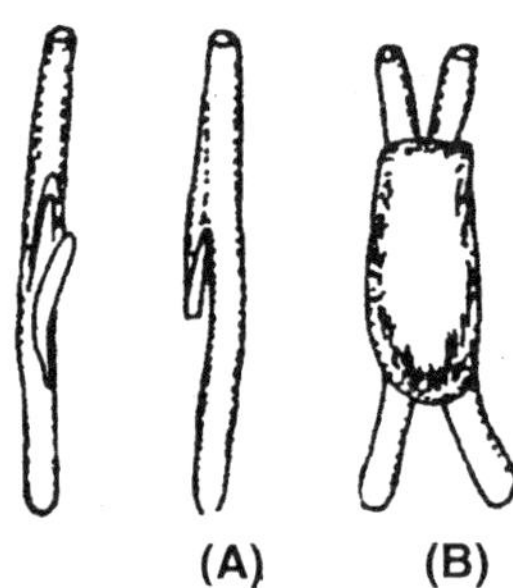

Fig. 9.15. Tongue graft : A, cuts preparatory to the union ; B, the union secured firmly.

3. Saddle Grafting

Diagonal cuts of equal length are made on the scion and stocks of equal girth in the form of saddle to match each other (Fig. 9.16). The components are brought together facing cambium layers, tied securely and coated with grafting compounds.

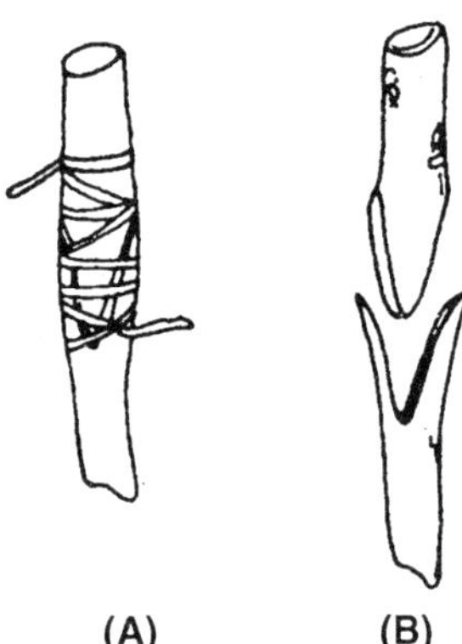

Fig. 9.16. Saddle grafts common in Rhododendron : A, portion at the base of the scion with saddle to receive its counterpart; B, the cambium layers of the stock and scion having matched, the graft is secured in position.

Fig. 9.17. Wedge graft. Wedge shaped scion end is inserted in the split understock.

4. Wedge Grafting

This is specially useful for top-working trees either in trunk of small trees or in the scattered branches of large trees. If the tissues are soft and succulent the components can be united by placing as shown in fig. 9.17. It is necessary to give some strength to the union by tying it with stake.

5. Side Grafting

The scions are placed laterally on the side of an understock. In side graft the scion is put on the side of a stock of larger diameter of about 2-3 cm. The scion is made free of all the lateral branches and growths near the point of graft. The cut made is long, slopping and wedge shaped (B). Soon after bringing together the stock and scion, the whole part of union is tied firmly and sealed with wax (Fig. 9.18 C). The cut end of the stock is painted to protect from infection and desiccation damage.

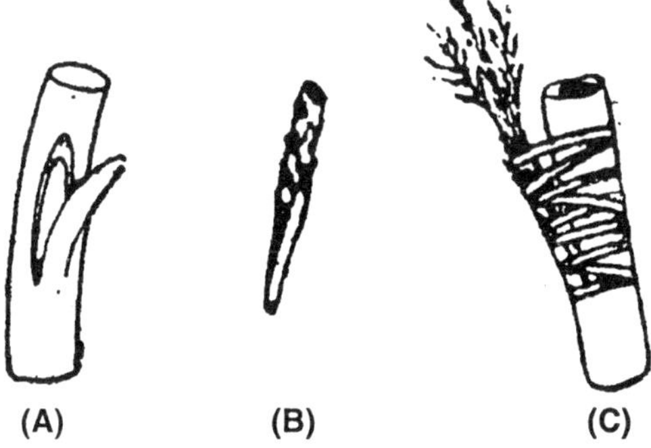

Fig. 9.18. Side graft, A, cut on the understock ; B, the prepared scion ; C, the scion and the stock in position.

6. Veneer Grafting

It is a slight variation of side grafting and is widely used for small potted evergreens. A shallow downward and inward cut of about 3-4 cm length is made 15 cm above the ground on the stock plant (Fig. 9.19). At the base of this cut another short, downward and inward cut is made to join the first cut (Fig. 9.19A). The piece of wood and bark is removed by severing off the separated portions by making a clean cut at its base. The scion (B) is prepared by making a long cut on the side to suit that in the stock. The cut end of the scion and the stock are put together and tightly wrapped with waxed string (C). The graft may be waxed only when necessary and kept in cool humid atmosphere for healing. After healing the stock may be cut above the scion

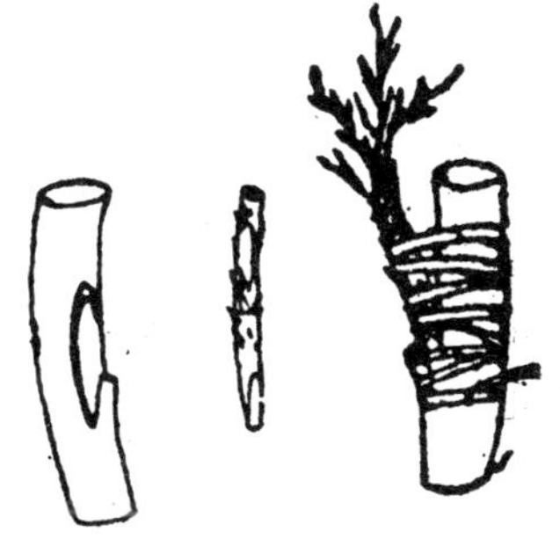

Fig. 9.19. Veneer graft : A, the understock after the second cut ; B, the scion shown separately and C, the completed graft secured in position.

either all at once or gradually. Mango and peach are propagated commercially be Veneer grafting technique. Veneer grafting is also done *in situ*.

7. Cleft Grafting

A method of grafting specially adopted for top worked trees though it may also be used for smaller plants. One of the prerequisites in working with this method is the clean and even splitting of the trunk. It is chiefly done when the plant is in late dormant stage, just before the onset of the spring season and the buds of the stock should be about to start active growth. If cleft grafting is done after the initiation of active growth, the bark separates out from the wood and the union becomes difficult. A 5 cm cut (Fig. 9.20, A) at right angles to the main axis of the branch is given. The cut ends are separated by means of a wedge (W, Fig. 9.20, C) for the insertion of the scion or scions (B) to be placed at the same time on the sides where the cambium is situated (Fig. 9.20 D). The scion receives a long slanting wedge-like cut in a way that the side having the bark is somewhat thicker than the other side of the wedge. The scion should not be more than about 10 cm long possessing more than 3 buds. Proper placement of the scion is very important so that there is good cambium contact between scion and stock. Waxing may be done to check outside contamination and entry of water to the cut surfaces.

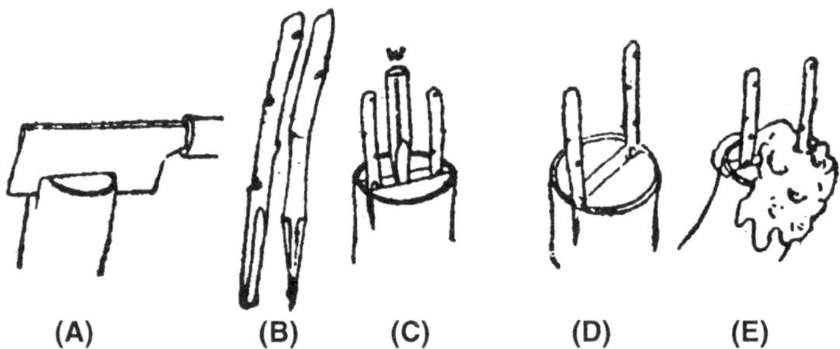

Fig. 9.20. Cleft grafting. A, median splitting of stock ; B, scion with two slanting cuts on opposite faces ; C, putting a wedge (w) to keep the split open to allow the insertion of scions in the splitted stock ; D. scions and stock in the desired position ; E, tying and then waxing, to keep the scion and stock in position.

8. Bridge Grafting

This type of graft is generally useful for the purpose of repairs where the root system of the tree is in perfect condition but the bark of the trunk is injured impending danger to the tops. It is a safeguard against damaged tree trunk. The method is specially useful when there is danger of the roots being starved to death due to injury to the bark. The wounded portion is trimmed up to healthy and undamaged tissue (A). Several one year old scions (B) collected during the dormant season are inserted in the trunk when the bark begins to slip at a distance of 2-7 cm such that the two ends of the scion get in contact with the cambium layer of the tree on both the upper and lower sides of the injured portion (Fig. 9.21 C). The scion must be inserted with right side up. The whole area of the graft union is sealed with wax and kept in a moist atmosphere preferably by putting it under some polythene cover. The buds, if any, on the scion often are pushed into growth indicating that the grafts are successful. These shoots are removed as no branches are desired.

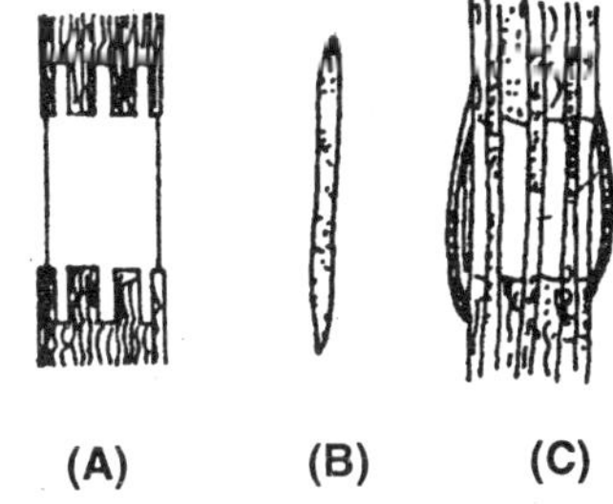

Fig. 9.21. Bridge graft. A cut in the bark to remove the wounded portion. Slots are made for the scion. B, prepared scion with slanting ends and no buds. C, scions nailed in place before covering the unions with grafting wax.

9. Bench Grafting

The scion is grafted on to a piece of root not attached to the soil. The name possibly may apply to any grafting method done at the working table or bench in the laboratory. In this type of grafting the rooted cutting, layered plant or rootstock seedling is taken out of the soil and used as the stock for the graft (Fig. 9.22). The entire root system may be used as in the case of whole root graft or small cut pieces of the roots may be used (piece-root graft). The purpose of using piece root graft is to make the scion develop its own roots. During the period roots from scion develop the piece-root serves the purpose. The development of the root on the scion is encouraged by wrapping a small piece of wire immediately above the point of union. The piece of root which serves temporarily known as nurse root is girdled just below the union and may be separated later. This method is of commercial importance in the case of apples, pears, grapes etc. If the roots are derived from the root-stock the graft should be set in a manner that the union is above the ground level.

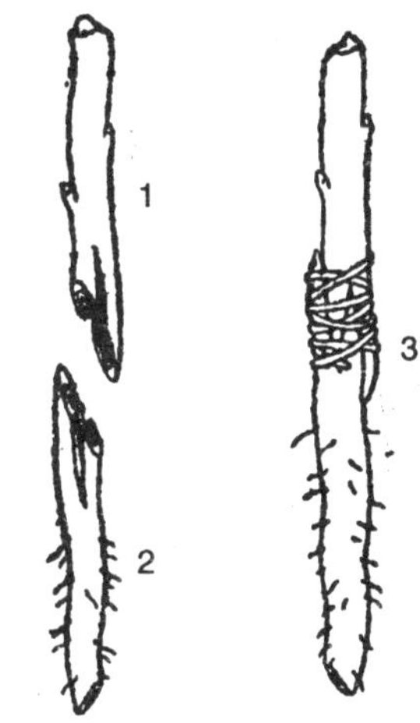

Fig. 9.22. Bench grafts 1. scion ; 2. the piece root. 3. completed graft.

10. Crown Grafting

A union made in the 'crown' region *i.e.* the transition region between the stem and root. Crown graft may be of several types whip, side, cleft or bark graft depending on the nature of the species and size of rootstock. The scion is taken from mature dormant wood of the previous season.

11. Bark Grafting

A method that may be employed on branches of varying growth from 3 to 25 cm. The stock, when in active growth, is taken and the bark removed (Fig. 9.23, A), the scion must be in dormant condition. After the break of dormancy and resumption of active growth the stock is cut at the height where grafting is to be done. The bark is stripped downward. 12-15 cm long scion of 0.6 to 1.2 cm thickness containing 3 buds is prepared by giving a 5 cm long cut. The scion bud is placed in position in the stock and nailed or tied firmly (Fig. 9.23 C) after covering the stripped portion of the bark. The union is very susceptible to injury by wind and needs staking in the first two years even after union has taken place. Many types of bark graft are known.

Fig. 9.23. A. Bark graft method. No. 1. The scions are inserted into the slot made by the removal of bark, the end is slipped under the raised flap of the bark. Left, the grafted stub ; Right, waxing of the stub.

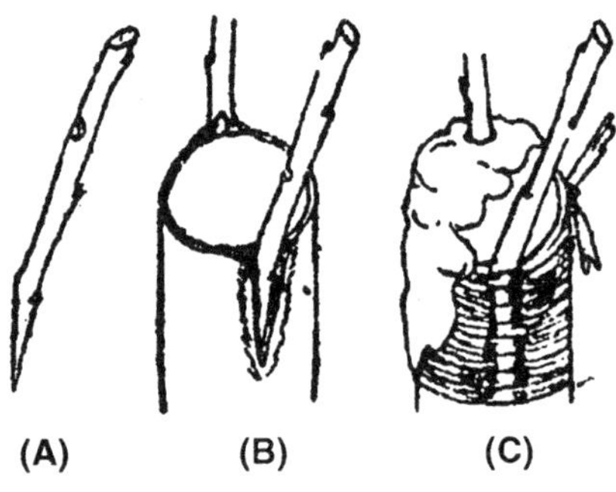

9.23. B. Bark graft method No.2 A, the scion ; B, the scions are pushed downward under the raised bark along one side of the cut ; C, the grafted stub is fastened tightly and waxed.

(*i*) Each strip as several scions (Fig. 9.23 B) which are inserted between the bark and the wood of the stock. The scion may be nailed to the stock for best results. All cut surfaces are thoroughly covered with grafting wax.

(*ii*) The bark is removed from one side while the other side remains undisturbed. A vertical knife cut of about 5 cm length is made at the top end of the stock. The scion is inserted in this cut between the bark and wood. This method is very simple and gives good results.

12. Approach Gafting

The characteristic of the approach grafting is bringing together of the cambium of two components, each of which is independent and self sustaining plant. The spliced approach graft, inlay approach graft and the tongue approach graft are types of this system of grafting. In spliced graft (Fig. 9.24) a slice of bark and 5 cm length of wood is removed from each stem and put together, tied and covered with wax. In the inlay approach graft it is necessary that the stock plant must be actively growing and the bark 'slipping'. The tongue graft is characterised by a thin tongue on each piece and these are interlocked tight and closely fitting. Cut surfaces of the stock and the scion should be fastened together and covered with grafting wax. After the union the top of the root-stock above the graft and of the scion below is severed, so that the top of the stock and the basal portion of the scion are missing. The practice is common in mango.

Inarching

Compatible seedlings are planted near the tree trunk during the dormant season. As active growth starts these may be grafted to the tree according to need. The operations are done by making proper slits in the stock as well as scion, matching and fastening the same to keep in position (Fig. 9.25). Both the stock and the scion are self sustaining at the time of inarching. A seedling plant or one with desired root characteristics growing by the side of an older tree with damaged root system may be grafted by this method. Inarching is useful in supplementing or replacing damaged wood. The root stock material to be inarched (Fig. 9.25) should be situated at about 12-15 cm round the tree. Inarching is common method of propagation of mango and loquat.

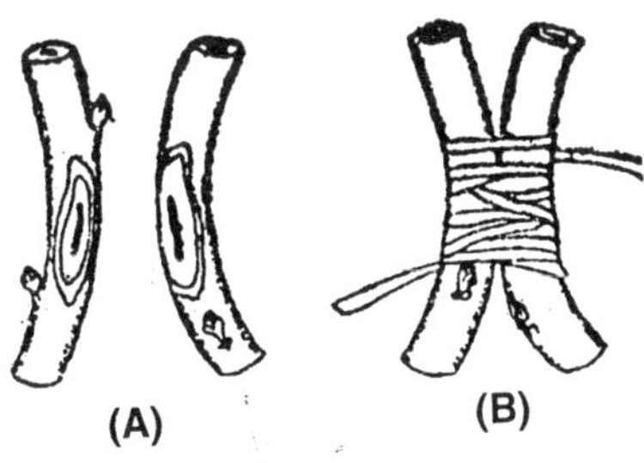

Fig. 9.24. Approach graft. A, identical shields removed from the twigs ; B, twigs tied firmly with the shield removed portion in contact

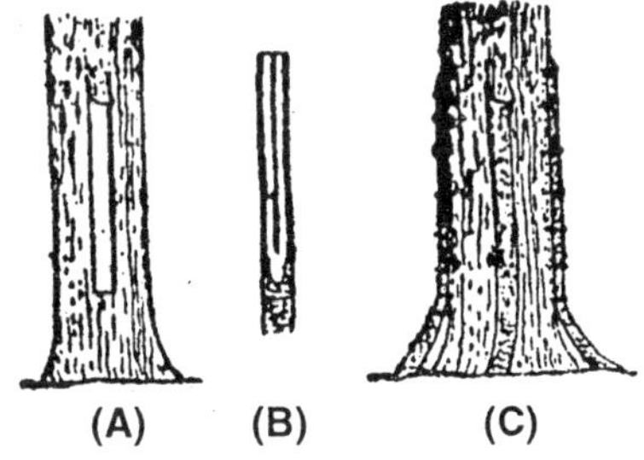

Fig. 9.25. Inarching of a large plant with smaller ones planted around its base. A, removal of about six inch length of the bark equal to the width of the seedling to be nailed. A, short flap of the bark is left at the top. B, the cut surface of the seedling to be placed against the exposed wood of the tree. C, seedling nailed in position, shown before thorough waxing.

Budding

Budding, a rapid method of grafting is of commercial importance specially when tree is the factor. It can be performed under wider range of climatic conditions than any type of graft involving a large number of buds on the scion.

Bud grafting is considered to be a sort of lateral grafting with a single bud scion. The bud (B) is removed (Fig. 9.26) with small amount of surrounding tissues from the plant of desired flowering or fruiting behaviour (A) and inserted in the rootstock (C). The portion of the rootstock above the point of the insertion of the bud is removed to force the bud which is in contact with wood below and the bark of the stock above. The bud must be held in position and these must not experience dry wind soon after budding. Adverse environmental condition *viz.*, drought, defoliation, hail, frost, flooding etc. retard the chances of success.

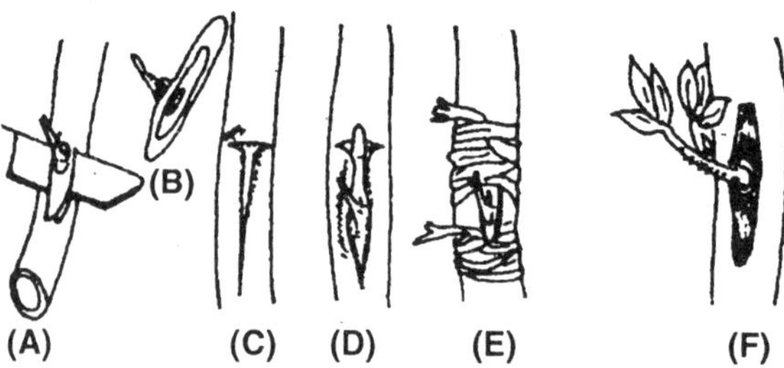

Fig. 9.26. Stages in budding. A, removal of bud from the bud stock; B, separated bud with leaf stalk and shield of bark ; C, T-cut made in bark of stock ; D, bud (B) inserted in T- cut ; E, the inserted bud tied in position ; F, growth of the bud.

The method is specially useful in the case of plants like citrus, plums, peach, cherry, rose etc. where slight injury to the xylem or wood portion resuls in the production of wound gums. There are several types of bud grafting of which T-budding (shield budding), inverted-T budding, ring budding, patch budding, flute budding, chip budding are more common. In budding it is necessary that the vegetative buds are used and not flower buds. Small and pointed buds are prone to be vegetative while large and plump ones are flower buds.

1. "T" or Shield Budding

This type of budding is quick and common. As many as 2,000- 3,000 T-buds may be inserted by rose budders each day only if they have a helper for typing the bud in place. The success with this type of budding is normally over 90%. A matured and rounded, though not old, scion is selected. The leaf at the base of the bud to be taken out is cut off leaving the leaf petiole to serve as a handle. The stock is given a vertical cut (just through the bark and not more) of about 2-4 cm. A small slit is given on the upper end to make a T-like appearance (C). The bud (B) is removed from the desired twig and inserted in the T-slit and there must not be any injury either to the bud or the stock. The bud is kept in water before insertion under the bark of the stock in the T-cut. The bud after insertion (E) is wrapped immediately with plastic tape and waxed. String, raffia, polythene strips or any fibre may be used to keep the bud, firmly in place and to exclude the passage of air, water etc. Buds that remain green and become active (F) after some time indicate that the union has taken place (Fig. 9.26).

2. Inverted "T" Budding

The variation lies only in making the horizontal cut at the base of vertical slit in the stock giving it the appearance of inverted 'T'. Other details remain as in T-budding. It is generally recommended when there is danger of rains during the period.

3. Ring Budding

The bud (2) 1.2 to 2.5 cm long is slipped off from the scion (1) attached to a ring of the bark (Fig. 9.27). The tip of the rootstock is removed and a portion of the bark is peeled off (3). The bud is then slipped over the portion (4) and wrapped. It is practised, though sparingly, with *jujube* and peach on commercial lines.

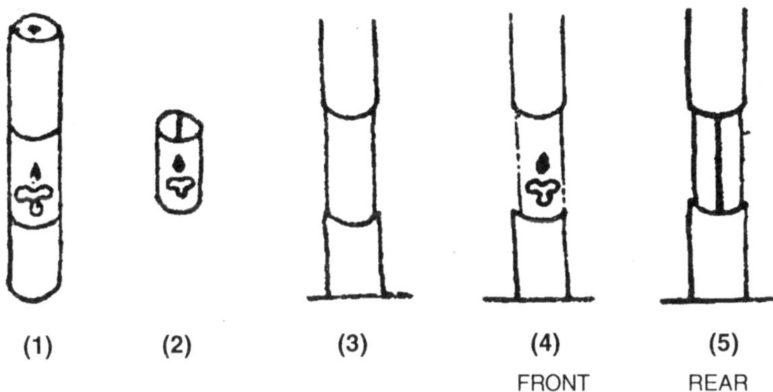

Fig. 9.27. Ring Budding

4. Patch Budding

It is similar to shield budding in all respects except that the section of bark containing the bud is rectangular and the bark of stock removed is also of same size and shape. The bud on the patch is carefully tied in its place (Fig. 9.28). It is commonly practised in the propagation of guava.

5. Forkert Method of Budding

A method common in Indonesia has been recommended for mango in India. Quite similar to patch budding the patch is of the size of 1 cm × 3 cm but the flap of the bark is left intact at the bottom. The scion bud of the same size and shape as the cut in the stock is inserted in place and then covered with the flap of the bark and wrapped by budding tape. The bark may be peeled off instead of being cut after the union. As the bud grows the top of the stock is severed.

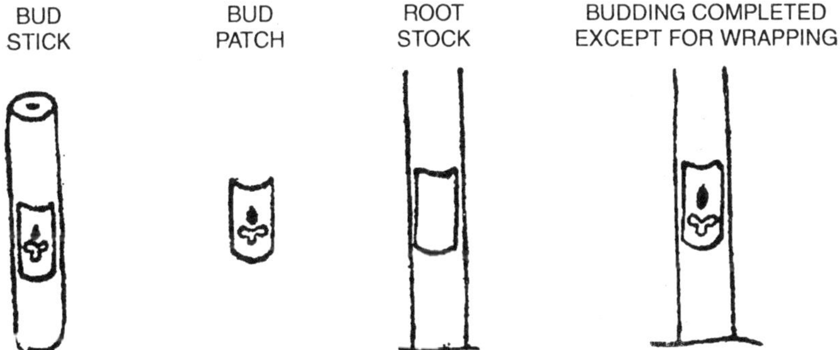

Fig. 9.28. Patch budding.

6. I-Budding

Bud patch may be rectangular or square as in the case of patch bud. The stock is given 1-cut (Fig. 9.29), instead of T- cut. It is useful when the bark of the rootstock is thicker than that of bud stock. It is easier than patch bud system because pairing down of the bark is not necessary in this case.

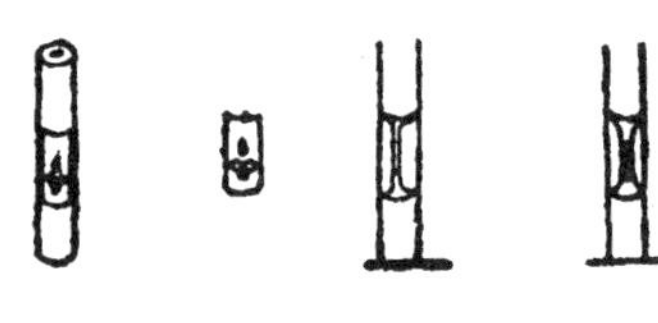

Fig. 9.29. I-budding.

7. Chip Budding

It is practised when the bark does not slip easily. A chip (B) of bark is removed from the stock (A) and replaced by another (D) from the bud stock (C) of same size and shape. The cambium layer of the bud piece should coincide with that of the stock and fastened (Fig. 9.30).

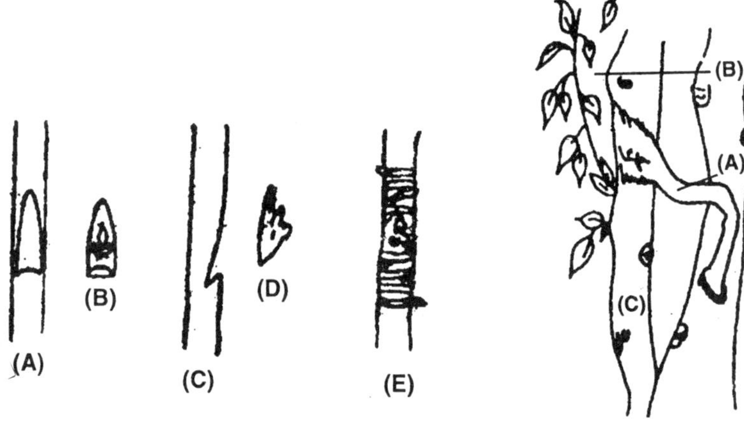

Fig. 9.30. Chip budding. **Fig. 9.31. Bracing.**

Bracing

This is a method to afford "natural brace" in young trees by grafts similar to inarching and bridge grafting. This method is specially used to save the branches from being broken off.

A similar branch (A) about 25 cm above the weak crotch is grafted into the adjacent branch (C) to be supported (Fig. 9.31). The bark (B) of the larger branch is removed and the smaller branch fitted on to it firmly. Grafting is done during the period of active growth of the tree. The graft union after the insertion of the branch, is tied and waxed thoroughly.

MICROPROPAGATION

With the increasing need of having more productive plants the technique of micropropagation has come *in vogue* to meet the challenges of growing population and awareness towards increased use of advanced technology. The meristem tips, cells, protoplasts, embryo, anther etc. may be used for micropropagation in artificial growth medium. It is also known as tissue culture or *in vitro* culture. Rapid multiplication is the chief aim of tissue culture. Another feature of the method is to maintain large number of stock plants in small space for a longer period. Asceptic culture of embryo has yielded useful results in interspecific and intergeneric crosses. Fruit plants like papaya, banana, apple, pear, grape, strawberry and datepalm have been successfully cultured by this method.

Asceptic condition is essential in the entire process of tissue culture. The plant material used in the culture should be sterilised by chemicals *viz.* bromine water (1-2%), ethanol (95%), hydrogen peroxide (10-12%), mercuric chloride (0.1-1.0%), silver nitrate (1%), sodium hypochloride (2%) etc. before being transferred to the culture media which should contain all the elements needed by the plant besides carbohydrate, growth substances, organic nitrogenous compounds, organic acids, vitamins, water etc. The pH of the medium for the culture of plant parts should be acidic, (5.5-5.8 is considered to be optimum pH) $Ca(NO_3)_2$, Na_2SO_4 , KNO_3, KCI, NaH_2PO_4 , $MgSO_4$, $MnSO_4$, $ZnSO_4$, H_3BO_3, KI, $CuSO_4$, Na_2MoO_4, $FeSO_4$, glycine, thiamine-HCl, pyridoxine-HCl, nicotinic acid, sucrose, Myo-inositol, IAA, cysteine, 2, 4-D, kinetin are some of the common ingredients of the culture medium. Sometimes plant extracts (yeast extract, coconut milk, tomato juice) is also added to the medium to increase the growth. Agar (a poly-saccharide from sea weeds) is used to solidify the medium. The concentration of agar is 0.6 to 1.5%. The widely used media for culture of cells and tissues are White's medium, Murashige and Skoog's medium, Gamborg's medium. The ingredients are different so also their quality in different media. The composition of Murashige and Skoog culture medium (1962) is given here in the inset below :

Table 9.1. Composition of Murashige and Skoog (1962) Culture media (mg/l)

Chemical	Quantity	Chemical	Quantity
$MgSO_4.7H_2O$	370	H_3BO_3	6.2
$CaCl_2.2H_2O$	440	$Na_2MoO_4.2H_2O$	0.25
KNO_3	1900	Sucrose	30.00
NH_4NO_3	1650	Glycine	2.00
KH_2PO_4	170	Myo inosital	100.00
$FeSO_4.7H_2O$	27.8	IAA	1.30
$MnSO_4.4H_2O$	22.3	Thiamine	0.5
KI	0.83	Nicotinic acid	0.5
$CoCl_2.6H_2O$	0.025	Ethylenediamine tetraacetate	37.3
$ZnSO_4.7H_2O$	8.6	Kinetin	0.04-10
$CuSO_4.5H_2O$	0.025		

The underlying principle of micropropagation by tissue culture is the aseptic excision of the plant part and its transfer to a suitable medium for development under artificially maintained optimum conditions.

Root formation on callus tissue is promoted by red light and the stem formation is favoured by blue light. When both root and shoot are developing a balanced exposure to these bands of light is needed. In fact light is needed for the synthesis of chlorophyll, initiation of roots, formation of shoots and also growth of the young plant. Generally continuous light for 8-16 hrs/day is considered to be sufficient for most plants. Culture practices and procedures remaining fundamentally the same there is difference in details of culture of callus, cell, meristem, embryo, root etc. Tissue culture is useful in the aseptic culture of embryos to produce interspecific and intergeneric hybrids, for overcoming embryo abortion, induction of haploid plants and overcoming seed dormancy.

Protoplast are cells without any rigid wall, the presence of only plasma membrane makes their fusion from different sources possible. Protoplast-culture opens new possibility of fusion of somatic cells (parasexual hybridisation) and incorporation of genetic material so that the characters of varied plant species and genera may be combined in one. It is also possible to transfer genes from one species to another and from micro- organisms to crop plants through the culture of protoplast. Cell culture technique helps the breeders to induce variability in plant cells and select desirable variants for regeneration of complete plant. Meristem culture affords rapid multiplication, availability of genetically stable virus-free plants to serve as donors. Storage of virus-free germplasm for long periods is also possible by this method. Success in the exercise depends on size of meristem tips, nature of culture medium, type of growth regulators and their concentration, photoperiod and other environmental conditions.

Micropropagation does not suit those species of plants where the rate of multiplication is not fast and plants regenerated by this method fail to establish when transferred to soil. Plants obtained through tissue culture may not be true to type, variation in chromosome number takes place. A decline in organ-forming tendency with successive cultures is also possible. In spite of these limitations micropropagation has been tried in sapota, persimon, grape, banana, *aonla*, citrus, strawberry, papaya, pineapple, *ber*, fig, datepalm, mulbery, avocado, cashewnut, mango, etc. with advantage.

CHAPTER

10

STOCK – SCION RELATIONSHIPS AND INCOMPATIBILITY

The various methods of propagation through the use of stock and scion have been made out in chapter nine. Compatible combination of stock and scion, produce certain unusual effects that may be either utilized commercially or avoided. A combination, of compatible plants with different genetical constitution, has afforded great possibilities in the improvement of the plant including larger yields ; the resultant effect may be due to either partner. One component influences the other, this aspect may be considered separately.

A. Influence of Root-stock on Scion

Rootstock has marked effect on the ultimate plant size, shape, vigour, growth, season of maturity, hardiness etc. besides bearing habits, fruit-set, quality, maturity and yield. The effect of rootstock on vigour and other characteristics can be modified by factors such as scion, soil type etc.

In stock-scion union there does not exist a zone of blended, intermediate or hybrid tissue with mixed characteristics. Although above the point of union the characteristics of the scion are reflected and below this those of the stock are witnessed yet the influences of stock on the scion is well marked specially in the rate of growth and the stature of the tree. Some stock display dwarfing effects ; the stock that dwarfs the growth of one scion may not have similar effect on another.

The size and the growth habits of fruit tree is altered as a result of change in vigour. The classical example of the effect of the rootstock on vigour of plant is that of series of 16 vegetatively propagated apple rootstocks at East Malling. The plant could be grouped into 4 sets depending on the degree of vigour induced in the resulting plant as dwarfing, semi-dwarfing, vigorous and very vigorous.

Malling II, a semi-dwarfing rootstock tends to have vigorous plants with good spreading habit and early bearing though the trees are smaller in stature than the seeding trees. This stock, tolerates variations in soil moisture on either side of the optimum to a greater extent than any of other Malling stocks.

Malling I is a valuable stock with respect of vigour. Malling XII forms a very vigorous rootstock. These produce plants occupying greater area than any other Malling stock and need larger spacing. Plants are very sturdy and strong. This stock has certain drawbacks, *i.e.*, the fruits do not have good colour and cannot adapt to low soil moisture conditions though the trees are large, well anchored, erect with spreading growth habit. Malling XII is some-what resistant to crown gall but is rather difficult to root even by layering. Other Malling rootstocks have different growth habits and maturity. It has been reported that in the case of sweet orange, rootstock influenced tree size, fruit size and also yield. Some rootstock-scion combinations produced dwarf trees also. Sweet cherry roots produce large, vigorous trees of *Prunus avium*, whereas vegetatively propagated variety of *P. cerasus* on Stockton Morello roots produce dwarf trees. *P. mahaleb* rootstock produce trees of moderate size.

In citrus, the rootstock has a marked effect on the size of the scion variety. Mention has been made by Brown that Malta sweet orange budded on the rough lemon (*Citrus limon*) produced trees four times larger than those budded on the citrus (*C. medica*)..

Maturity of Wood

Evidences of the effect on the stock on the maturity of wood is not uncommon. Although climatic conditions and the ultimate maturity appeared to be related to hardiness, it is not invariably true that stock which induces early maturity influences hardiness also. With cherries grafted on Mahaleb stock, earlier ripening of wood resulted with the result that cherris become harder both in orchard and in the nursery. The dwarfing may be associated with earlier bearing of flowers and fruits and often with greater productivity. This effect is not known in case of dwarfing brought about by environmental effects *viz.,* drought.

The claim that dwarfing brought about by stock-scion union is also associated with larger fruit size is doubtful. Buds of grape vines on own roots open much more regularly and somewhat earlier than those of the grafted vines. Vines on the hybrid stock generally blossomed later and more regularly.

Using 5 rootstocks (sour orange, sweet orange, trifoliate orange, grape fruit and rough lemon) it was found that in each case early bearing was initiated when self budded. This suggested that in the case of graft union, some partial chocking of nutrients was affected as though by girdling early and more fruiting resulted. Cases of the influence of rootstock on flower production and fruit set have also been reported in Japanese persimmon (*Diospyrus kaki*). The number of blossoms produced and the per cent of mature fruits was quite pronounced in *Hachiya* cultivar grafted on different rootstocks. *Diospyrus lotus* used as rootstock produced more blossoms with low maturity than *D. kaki.* Trees on *D. virginiana* produced more blossoms and unsatisfactory crops. Stocks with dwarfing influence lowered the vigour and vegetative growth of plants increased carbohydrate accumulation and tended early fruitfulness in proportion to its dwarfing habit. The more vigorous stocks stimulated vegetative growth in the scion and delayed bearing. Studies with apple varieties on Malling stock have revealed that the stocks could be arranged in decreasing order to vigour-producing influences as Malling XII > Malling XIII > Malling V > Malling IV > Malling II > Malling I > Malling VII and the earliness in reverse order *i.e.* Malling VII the earliest and Malling XII the latest. In the case of plums the Myrobalan root produced larger trees and higher yields in comparison to other rootstocks. The influence of Myrobalan roots was due to the soil-water relationship. The vigour and yield was affected to the maximum extent while fruit characters remained unaffected.

Increaed yields of *Vitis labrusca* grapes were obtained when grafted on vigorous rootstocks in comparison to own rooted plants. In general vigorous rootstocks produced larger plants and greater crops whereas dwarfing stock may fruit early. Some grafted mango plants show greater susceptibility to black tip disease than others. The disease is reported to be eliminated by spray of borax showing thereby that a deficiency of boron is caused due to low intake of boron by the particular stock.

Plant species differ with respect to the effect of the rootstock on the fruit characteristics of scion variety. Secondary effects due to dwarfing have been observed to occur in deciduous fruit trees leading to increase in fruit size. It is most probably due to the availability of more carbohydrates for the development of fruits following stunted growth.

Generally the characteristics of fruit of the stock are not carried over to the fruits of the scion variety. Characteristics of pear fruit do not appear in the almond fruit or that of quince on the pear, when the former in each case, is used as rootstock for the later. Sometimes the root characters may affect the scion *viz.* blackend disease in pears.

A number of workers reported that the characteristics of fruits of rootstock appear in the fruit of the scion. Rootstock of sour orange (*Citrus aurantium*) with scion of sweet orange, tangerine and grape fruits produce thin skinned smooth, juicy fruits of very good quality that store well. Rootstock of sour orange (*C. aurantium*), sweet orange (*C. sinensis*) with scion of sweet orange give fruits which are thin skinned, smooth, juicy when factors of nutrition are not limiting. Rough lemon (*C. limon*) as stock produces fruits with thick skin, low sugar, high acid content and of inferior quality. Similarly maximum sized Valencia oranges appear on dwarfing terifoliate orange stock, whereas fruits of smallest size are produced when sweet orange stocks are used. Naval orange fruits are largest in size on sour orange stocks and smallest on the Palestine sweet lime. Tomato fruits were detected to contain nicotine when grafted on tobacco rootstock. No such change in composition occurs with tomato rootstock in union with tobacco scion. The tomato scion produced poisonous fruits when grafted on *Datura strammonium.* Regarding winter hardiness the citrus rootstock influences the quality of the scion though in the case of apple no such relationship exists.

Annual species of plants grafted on perennial species become perennial. The life of scion (that part of the perennial plant which behaves as an annual) is increased by grafting on perennial varieties. Conversely in some instances, scion of perennial grafted on annual stock have died at usual time for stock. Such instances as these are more striking than those observed in fruit plant where the possibility or changes are necessarily more limited.

Spring effects. In the effect of stock on spring growth the behaviour of cherries on chicksaw plum may be taken as typical. The stock starts much earlier and shoots while the cherry graft remains dormant. A delay of about one half the difference between the two blossoming periods is found. The influence of stock on season of scion's activity is wholly regulated by the season and temperature in which the sap moves, independent of the traits of the stock.

Prominent effect of rootstock. Sometimes the quality of the fruit is affected by the stock *viz.*, the acidity of oranges when on rough lemon (*Citrus limon*) is higher by 10-15 per cent than when on sweet orange (*Citrus sinensis*). Another examples is of the coarse, dry insipid fruit of *Citrus nobilis* grown on stock of sweet orange whereas the stock of trifoliate (*C. trifoliata*) stock is associated with fruits of better quality.

Certain stocks influence fruit-bud formation. The dwarfing stock permit sufficient accumulation of carbohydrates to be utilized in formation of fruit buds. The setting of the fruit is influenced somtimes by the stock on which scion is grafted. Vigorous growing stocks fails to set fruit in sandy soils. Kieffer pear grown on wet soil results in better setting if quince roots are used.

So many factors affect the size of the fruit that it is difficult to determine the influence attributable to stock. Fruit size of both the Washington Navel orange and Valencia orange is strongly influenced by the rootstock.

The stock may influence the quality of fruit borne by the scion. Shortened life on the part of scion associated with lack of adaptability with its rootstock is well known. High mortality rate of peach trees after 1-2 years when budded on plum stock after making vigorous growth has been reported. Sudden decline of lemon trees 12 to 20 years after budding on sour orange is also attributed to a lack of stionic congeniality.

Early bearing is necessary for the success of any variety in cultivation. Naturally grafted trees of cultivated varieties should come into bearing early, otherwise the varieties would not be in cultivation. Some varieties come into bearing at an earlier age than other irrespective of being grafted on same stock. This time can also be hastened or retarded by cultural means. Vigorous seedlings are late in bearing, so are vigorous grafted trees.

It has been reported that on certain soils the production from grape vine is larger with compact bunches when grafted on Gloire or Clinton stocks in comparison to the growing on its own roots. Similarly *Rupestris St. George* roots help the concerned variety to yield double than that on its own roots. Rough lemon stocks generally increase the production of citrus fruits than those grown on sour orange roots.

Rootstocks reflect differently on the scion with regard to soil conditions. Almond and Myrobalan plum roots tolerate a higher dose of boron present in the soil than on Mariana plum roots. Both apricot and plum when grafted on Myrobalan plum are affected by nematodes to a higher extent than when raised on peach roots. Myrobalan plum, peach, apricot and almond rootstock used for stone fruits tolerate soil mositure to decreasing degree. It was reported that different rootstocks have high degree of variation in their capacity to tolerate lime induced chlorosis in calacareous soils.

Instance of greater resistance to apple scab (*Venturia inaequalis*) has been observed with Malling XVI rootstock. The effect of rootstock on the variation in germination quality of tomato seeds has been observed with different rootstocks of solanaceae.

B. The Influence of Scion on Rootstock

Scion characteristics affect the quality of the plants to a considerable extent. The interaction of scion and stock determines the characteristics of the plant. The dwarfing predominates in its expression whether from the stock or the scion. Hodgson reported that in the case of citrus the scion determines the vigour and size of the plant and not the rootstock. A vigorous, fast growing scion stimulates the less vigorous rootstock. The scion variety may not affect the root system in the case of Malling IX or XII rootstocks, although these influence the quantity of roots. Seedling stocks are affected markedly by the scion in their root characteristics although the vegetatively propagated clonal stocks do not show such effects.

The rootstock when grafted to a scion which stops growth in early fall induces maturity of the root tissue more than the scion which prolongs root growth so that increased winter hardiness is observed in former case but not in the latter. Sour orange (*C. aurantium*) when budded on Eureka lemon suffered much more from winter injury than the unbudded seedlings.

Effect of Intermediate Stock. The intermediate stock affects the scion as well as the rootstock. The work of Tukey and others has shown that between the rootstock and the intermediate stock the former has greater influence. Dwarfing intermediate stock *viz.*, Malling IX affects the entire tree in the same manner. It is believed that the observed effects the interstock are due to an influence of the length of the piece and not due to the abnormalities at the graft union and that increase in the length of the interstock intensifies the effect.

EXPLANATION OF THE STOCK-SCION RELATIONSHIP

Although not much is known definitely about the fundamental causes of reciprocal stock-scion influences certain theories have been propounded. The chief being (1) predominating influence of stem portion, (2) predominating influence of root system and (3) the physiological factors.

Predominating influence of the stem portion. The mechanism of stockscion relationship is unclear and conflicting ; quite often observed effects remain unsusbstantiated. It is held by advocates of this theory that the effect is primarily controlled by translocation effect and not by the ability to absorb. Robert held that the effects of the rootstock was localized in its stem portion rather than in the absorbing root system. A scion behaves differently when grafted on a root piece than when budded or grafted on the stem portion. Root grafting, therefore, may produce more uniform plants from seedling stock than by budding. If the stem portion of the stock used is uniform the resulting trees from the union must also be uniformly similar. The presence or absence of stem portion in the rootstock also influences the effect of scion on the rootstock. The stem portion of the rootstock dominates in its effect when present on the root system. If the stem piece of the stock is not present and scion is grafted directly on the root piece the effect of the scion is reflected.

Predominating influence of the root system. There is, however, the opinion expressed by certain workers that the root portion of the stock has predominating influence although the stem portion may also exert some influence.

Influence of physiological factors. Physiological factors that control vigour are supposed to affect the reciprocal influence of stock and scion. Since the roots depend for their supply of elaborated food material on the vigour of the shoots and the latter on the supply of elaborated food material on the vigour of the shoots and the latter on the supply of mineral nutrients from the roots the vigour and the supplying capacity of one must influence the other. A vigorous rootstock is able to absorb and supply more nutrients from a soil of relatively poor availability than the other makes the difference. Peach root system grown at low nutrient levels is known to absorb and supply the scion with more of nutrients and water than other less vigorous rootstocks. Such rootstocks become harmful in soils with high level of salt content for obvious reasons. This is how under differnt soil conditions the rootstocks behave differently and their use must be made with full knowledge of the salt content or the minor element availability position of the soil. It is believed that the graft union introduces additional resistance in the translocation of minerals from the root to the scion and of elaborating food materials from the scion to the roots. Stocks with high bark : wood ratio in lateral roots tend to dwarfing while those with low ratio cause increased vegetative vigour in the scion. Marked differences in the anatomy of dwarfing and invigourating rootstocks possibly lead to changes in traslocation and metabolic requirements so as to affect the scion differently. Alteration in growth behaviour may also be induced due to interference in the downward flow of natural hormones to the roots thus causing the observed effect. The interference is caused by the use of intermediate stock or ringing etc. Trees grafted on dwarfing stock have been seen to make vigorous growth for a year or so, the dwarfing effect being initiated later by increasing lack of hormones. The role of growth hormones for the growth and development need not be presented here. Plants in vegetative stage of growth when grafted to a flowering compatible partner or related type being to flower showing thereby that hormonal interchange does occur.

INCOMPATIBILITY

There are instances when the stock and scion do not show a union, this is known as incompatibility. It may be partial, complete or temporary. Here we may consider some aspects of incompatibility. Stock-scion union is possible only between closely related plants. Failures are common between two unrelated plants. The combinations that succeed often show change in the character of uniting partners. The reciprocal influence between stock and scion has been dealt with earlier. The failures in the success of the grafts may be due to several factors of which incompatibility is one. It is sometimes called as uncongeniality as opposed to congeniality or compatibility.

Incompatibility is an extreme case of uncongeniality. Strictly speaking incompatibility refers to causes where either of the two or both fail and graft dies quickly. Uncongeniality relates to the cases where combination does take place but with some modification, *viz.*, swelling at the union or dwarfing in the stock or scion or both. Combinations which do not materialize, *viz.*, failure to unite or result in unhealthy growth, due to distinct difference between stock and scion caused by incompatibility do not include cases of failure resulting from abnormalities or unsuitable environmental conditions. The usual condition of compatibility between stock and scion may be considered under the following heads :

Physiological and structural similarity and dissimilarity. Stock and scion with slight physiological and structural differences make graft unions a success and the tree grows, develops and fruits, normally. The chief criteria of success is satisfactory growth of trees and good fruiting, abnormalities like over growths near about the junction point is not taken into account as to class the union as unsatisfactory.

Wide differences between stock and scion make a satisfactory graft union impossible ; the failure may at times, be delayed for many years. Such cases which exhibit "delayed symptoms" of incompatibility are known as cases of "delayed incompatibility".

Cases where the differences are not great and allow some success are classed as a separate group. The role of environmental conditions which accelerate or diminish chances of incompatibility often weighs in the outcome of such grafts.

Symptoms

Incompatibility is often indicated by the breaking off of trees from the place of the joints after some period of growth, provided, the break be clean and smooth. Such cases have been reported between apricot and almond or apricot and Myrobalan plum roots. Another indication of incompatibility is the presence of sheets of parenchymatous tissue or bark or both at the point of union. The formation and deposition of loose parenchyma cells at the point of union interrupts the normal vascular connection between stock and scion. Swelling at the point of union between scion and graft is often taken to be due to a hurdle in union though not always so. Defective graft unions are many times due to failure to maintain cambium continuity. Some of the symptoms associated with incompatibility may be listed here although these symptoms may also be, at times, produced due to other causes, *e.g.*, lack of nutrients, attack by insect, diseases or defective technique.

(*a*) yellowish foliage, early drop and poor growth of leaves,

(*b*) overgrowth, near about the junction point,

(*c*) visible difference in growth rate or vigour of scion and stock,

(*d*) differences in the time of vegetative growth of the scion and the stock in the season,

(*e*) premature death of the tree, and

(*f*) failure to form a successful graft or bud union in majority of cases.

Combination of peach, plum, pear, cherries and apples as stock and scion with swelling is no indication of imcompatibility. Inclusion of bark tissue upto the point of cambium in the newly grafted or budded trees lead to incompatibility. Development of good xylem connection between the scion and rootstock at the point or union as in the case of peach on *Marianna* plum but failure of the union of the phloem tissue lead to incompatibility. This type of incompatibility is due to translocation difficulties.

Types of incompatibility. Incompatibility between grafted plants is of several types. It may be due to (1) complete lack of union between species or varieties, (2) success limited to small number of cases, (3) union takes place, may give rise to growth but the plant eventually dies, (4) the union gives rise to nutritional disorders or deficiency symptoms, (5) the combination may induce extreme stunted condition, (6) differentiation in growth near about the graft union, (7) degeneration of tissue system, leading to unequal distribution of food material and premature defoliation.

Causes. Compatibility between two plants cannot be tested in advance and foretold. The fundamental cause leading to incompatibility is also little understood. Between the scion and stock the union may meet success in some cases but not in others. It is not possible to determine the ratio between the successful and unsuccessful unions in any set of combination.

The theories that have been put forward to explain the causes leading to incompatibility do not find conclusive evidence in their support. Theories fall into three categories. (*i*) physicology, (*ii*) anatomy, and (*iii*) disease.

(*i*) *Physiological causes.* Marked differences in vigour, nutritional requirement or absorption capacity of the scion and rootstock or time of vegetative growth may be one of the causes of incompatibility. The difference in growth rate and time of active vegetative growth makes the union difficult. This may, however, be not taken as the basic cause of incompatibility because the rate of growth can be influenced by edaphic and biotic factors in many instances. The development of molybdenum deficiency in the case of union between Malling IX rootstock and Jonathan apple scion due to the inability of Malling IX rootstock to supply sufficient molybdenum to the Jonathan portion of the plant is the prime cause of the disorder. Such disorders do not develop when other Malling stocks are used. In the case of Mango it has been observed that black tip disease (which could be corrected by foliar application of borax) develops when certain rootstocks with low intake of boron from the soil are used but not with others. Hormonal deficiencies are also at times associated with causes of incompatibility. It is likely that graft combinations manufacture certain substance or substances in the green leaves of the scion which move down to the point of union to affect the degree of compatibility. The substance possibly helps in the development of antagonism so as to cause disconitnuity along the graft union. It may be true otherwise also if the stock leaves give some substances that promote union.

Cucumis melo grafted on *Cucurbita fictifolia* does not succeed for long though the reciprocal combination is successful. The incompatibility in the former case can be prevented by retaining a few leaves on the stock (*Cucurbita fictifolia*). Similar situation has been reported in the case of peach grafted on *Marianna* plum as sated earlier.

(*ii*) *Anatomical causes.* One of the anatomical factors that possibly contributes to structural weakness in interspecific grafts is the distortion of vascular tissues between the stock and scion causing obstruction in the flow of minerals, water and elaborated food. Likewise any outgrowth between the cambium of rootstock and that of scion also inhibits. Such growth as of parenchymatous tissue or bark and formation of whorls or loops in the vascular or conducting tissues at the point of union are generally responsible for check in translocation. Sometimes a breakdown of medullary ray system leading to the formation of wound gums in the neighbourhood of the points of union also causes structural weakness in many grafted plants.

(*iii*) *Disease a cause.* Virus infected rootstock or scion may produce a toxic substance leading to incompatibility by checking the activity of the meristem as reported in the case apple varieties grafted on *Virginia crab* and U.S.D.A. 227 root stocks or various combinations of citrus stock and scion.

CHAPTER

11

ENVIRONMENTAL CONTROL OF PLANT GROWTH

Fruit production necessitates control of growth and development of plants under the stress of environmental factors which come abundantly in play. Growth as a process is conditioned by ranges of temperature, quality and concentration of nutrients, period, nature and intensity of light besides variation in water supply. The factors that control growth do so in accordance with the law of limiting factor as enunciated by Blackman : "When a process is conditioned as to its rapidity by a number of separate factors, the rate of process is limited by the pace of the slowest factor."

Air-moisture relationships are extremely important in the water economy of plants. Transpiration is slow in humid atmosphere as compared to the hot and dry ones. All water supplied to the soil is not utilized, much is wasted by surface run off, seepage below the root zone, evaporation etc. Humidity, air, light etc. influence the development of fruits, its structure and colour. The fruits deteriorate under the pressure of high wind, frost etc.

The rate of water loss is simple linear function of evaporating power of the air ; water loss in wind is greater than in still air. High wind velocity is harmful for fruits as we find in the mango crop which is adversely affected by high winds specially during the hot months in Northern India. Severe storms in early spring also knock off many of the leaves and fruits from mango trees. Extreme heat or dryness accompanied by winds cause much loss of water and practically restricts the culture of certain fruits in large areas.

Chilling temperature is important in setting limits to the commercial culture of fruits of some kinds besides determining the regularity of crop yield and profits in practically all deciduous fruit growing areas. Frost formation depends to a considerable extent on the radiation of heat by exposed surfaces during the night. It is often seen that frost and hail destroy crop of fruits.

Cold waves in Northern India are associated with the western depressions which enter the country. These depressions cross the plains towards the north-east frontier and are at first warm and moist then cool and dry, with wind blowing from the north and North-west.

Hail is one of the greatest hazards in apple growing regions in the Himalayas. Papaya is also liable to damage, as the fruits are ripening at a season when hail is not uncommon, and the thin skin of fruits are easily injured. The three more important factors *viz.* water, temperature and light are discussed below in detail.

Water

Water has the property of absorbing and loosing heat with minimal changes in temperature due to its high specific heat. As water evaporates it cools the surface due to its high latent heat of evaporation. The high surface tension of water helps it to give high surface energy to surface films. Water is a catalyst and a solvent of universal importance. It forms a major constituent of the protoplasm and also of most of the organic compounds present in the plant tissues. Besides being a solvent carrier of nutrient and chief base of transpiration stream it is essential for photosynthesis, turgidity of the cell and several exergonic processes of the plant cell that bring about many structural as well as physiological adaptations. The carbondioxide intake of the plant is mostly in dissolved state in the thin film of water present in the loose spongy parenchyma as it enters through the stomata. Almost 1% of the water that passes through the plant in the transpiration stream may be utilized in photosynthesis. Loss of water leads to plasmolysis of the cells, dehydration of protoplasm, closing of the stomata and reduction in the rate of photo-synthesis. Most of the water intake by the plant is through the roots. Soil water at field capacity is supposed to be optimum for the needs of most mesophytes. Xerophytes can continue to grow in the normal way even under reduced water supply conditions of the soil. Those plants which grow in water are termed as hydrophytes. There are still others that take up moisture from the humid atmosphere, these are designated as aerophytes or epiphytes. Ordinary land plants are called mesophytes. Soil lacking in available water is not congenial, to the growth and metabolism of plants and thus the moisture status of the soil is considered as more potent factor than any other factor except temperature. Optimum supply of water is a prerequisite for the normal growth of the different categories of plants. Water is essential for the growth of microbes that work towards the well being of the soil, plant as well as the microbes in general.

Since different plants need different amounts of the soil moisture for their optimum growth, variations in plant distribution in the arid, semiarid, sub-humid and humid zones throughout the world is obvious. The difference in the anatomical structure of the plants belonging to the xerophytic and hydrophytic conditions is much more sharp than those growing in temperate, subtropic or tropic zones. The cacti, for example, which serve as a good example of xerophytes shall not grow under the conditions in which hydrophytes *viz.* water lily (*Nymphaea* sp.), water hyacinth (*Eichhornia crassipes*), water chestnut (*Trapa bispinosa*), Hydrialla etc. flourish. Conditions of water distribution and availability are of importance for the growth of the plants in the same way as the changes in temperature brought about by any other cause.

Wherever the plants grow under normal conditions or the conditions existing between those of a hydrophyte and the xerophyte we have the common land plants, the mesophytes. It is the class of plants suited to intermediate conditions that are used for economic as well as ornamental purposes. Paddy, Trapa, Water cress etc. are well known plants which, being partially hydrophytic in nature to a considerable extent, need large supplies of water.

The need of water to plants may also be visualised from the fact that it is present in every tissue or organ or part *viz.* tubers, fruits, flowers, etc. to a great extent (80-95%). Barley root is known to have 93% of water as against only 10.2% in the grain. The leaf, stem and flower of sunflower contain about 80, 88 and 71% of water respectively. Fruits of tomato, watermelon and strawberry have 89-94% of water. The woody tissues of plants may contain as much as 40-50% and dry beans, peas, etc. about 10-15% of water.

The need of the plants for water starts from the time of resumption of active physiological actvities by the seeds leading to germination. Imbibition of water by the seed through the seed coat, most probably, forms the first discernible aspect of the breaking of dormancy or the rest period of the seed. Plants need water throughout their life-cycle continuously. Even short spells of drought bring down dry matter production and shift in protein metabolism of the plants to remarkable degree. With the advent of water supply to young growing seedling the root system as well as the tops gain. With further increase in water supply the root system becomes more restricted and fibrous while the leaves continue to increase in number and dimensions. With increase in water supply more minerals are absorbed but nitrogen is not available in that proportion mostly due to the leaching away of the soluble forms of nitrogen to, depth beyond the root zone. This naturally leads to lowering down of the per cent nitrogen in the plants. Grain-straw ratio in cereals decreases with increase in supply of moisture beyond the optimum. Regulating water supply to the plant is of great economic importance. Maturation is also affected greatly by the amount of water supply. Under conditions of relatively larger supply of water the maturity is delayed and the span of vegetative growth prolonged.

These instances relating to the effect of water supply to the plants show that its excess or shortage must be guarded for economical production. The supply of water should, in general, be sufficient to compensate the loss due to transpiration. Temperature, wind velocity, light intensity, humidity and absence of root hair are factors that should weigh in regulating the extent of supply of water. Besides these, soil characteristics have to be taken into account rigorously for determining optimum water need since different types of soil have different water holding capacity and depending on this as well as the amount of seepage allowed by a soil, the water supply must necessarily vary. Excessive water supply induces the development of large sized cells, leading to a leggy drawn-out type of growth. The first effect of the increased supply of water leads to undue stretching of the cells just below the growing point. The absorption of soil water is definitely more when the soil conditions are optimum and the transpiration pull is at increased rate. In practice the transpiration pull may be minimized by close planting which reduces light supply and impact of wind. With large supplies growth cracks occur due to the breaking of cell walls occasioned by high pressure. Examples of cracking are often visible in guava, lemon, mango, *litchi, bael*, pomegranate, tomato, cabbage, onion bulbs, beets, radishes, sweetpotatoes, carrots etc. The cracks are also seen under conditions of high absorption of water combined by low transpiration due to increased pressure on cell walls.

Water deficit, like its excess, causes reduction in yield and sometimes death of plants due to desiccation causing plasmolysis of cells. Deficits in water may be either due to reduction of moisture content of the soil to the point of unavailability or due to greater loss of water than its intake in the plants. Fruit crops like the apple, plum, pecan, pears and peaches often suffer due to dearth of water. Their stomata are also remarkably sensititve to drought conditions and these check water losses during periods of water deficit. Water deficiency leads to greater susceptibility to injury due to frost. Plants have reduced carbon assimilation rate during period of drought causing the dwindling of the carbohydrate resources. These carbohydrate resources help in the formation of hydrophyllic colloid which bind water. Lack of carbohydrate in plant to low temperature, the physiology of drought and frost resistance being essentially the same. Extreme conditions of drought may lead to permanent wilting of the plants. Under this condition the stomata are closed with subsequent stoppage of the process of photosynthesis. While the process of carbon assimilation discontinues, that of respiration continues leading to carbohydrates, proteins, fats, aminoacids, organic acid etc. into carbon dioxide and water in accordance with the reaction :

$$C_6H_{12}O_6 + 6O_2 \rightarrow 6CO_2 + 6H_2O + \text{energy}.$$

This causes diminution in the dry weight of the plants. Prolonged conditions of this type lead to the death of the plants. Since water deficiency has such consequences of far reaching importance in plants it is always necessary to check the onset of drought condition by proper irrigation. How to determine the stage when watering becomes necessary for the plants is thus a point of practical value to the plant grower. The symptoms of water deficiency may be noted for such cases below.

Dark green colour of the foliage, reduction in the linear growth of the stem and its branches, unfolding of the vegetative bud, reduction in the size of flowers and withering of the youngest tissues of the plants. The loss of water through the stomata of the leaves may at times be so great as to cause reduction in the moisture content of even the fruits so that the volume of the fruit is reduced.

Whenever water losses are great and the amount of precipitation less, irrigation may be resorted to in order to save the plants from the disastrous effet of drought. In the case of transplanted plants the number of root hairs are reduced when the seedlings are taken out from the soil for transplanting. This leads to less absorption of water by the newly transplanted seedlings. The practice of cutting down light intensity and of transplanting during evenings is useful in this regard. Shallow cultivation does damage the feeding roots mostly located at the surface.

Besides loss in yield, water supply position causes other effects of significance to the plant grower. Good supply of water accompanied by optimum temperature and nutrients may lead to larger synthesis of carbohydrates, proteins, etc. in the plant body following conditions that make rapid cell division and extension possible. The available carbohydrates and proteins may be utilized for vigorous vegetative growth of the plant. Restricting the water supply to moderate levels makes discrenible changes in the growth behaviour of plants. Under these conditions stored carbohydrates and proteins are utilized for the development of flowers and fruits and even for the growth of the stem and leaves. Plants receiving abundant supplies of water have increased vegetative growth than those getting moderate supplies with reproductive phase initiated earlier and to a greater extent. Cutting down the water supply to the level which could be classed as scanty reduces cell division and vegetative growth while the carbohydrates and sugars are translocated to the storage organs. In this way we find that the water supply position to the plants may shift the balance between vegetative and reproductive phase of the plants and therefore this may be used for regulating the phase of vegetative and reproductive growth of plants.

Temperature

Any specific temperature has a particular feature mainfested in the organism. Seeds of most plants fail to germinate unless the temperature is warm, some germinate only when exposed to cold before being placed at the proper temperature. Seed cones of *Pinus balsamea* do not open to release the seed unless they have been scorched by fire. The annual growth of *Saccharum spontaneum* are harvested and the cut end of the portion still attached to the soil is burnt to give delayed but vigorous vegetative growth in the next season.

Plants are widely spread throughout the surface of earth having wide fluctuations in mean daily and also seasonal temperatures. All the ranges of temperature found in nature are not conducive to the growth of plants but within a reasonable limit plants do grow. Their maximum growth is only possible with optimum temperature. Any temperature beyond the limit of tolerance of plants inactivates the enzymes that drive the *in vivo* chemical reactions leading to the metabolic processes of plants. Fortunately the temperature that generally exists on the earth's surface forms the tolerance range for plants. The effect of temperature on plant growth may be considered from two viewpoints, the favourable and the unfavourable. The different metabolic stages of the plant have different maximum temperature requirements. Flower buds have higher maximum temperature requirement than leaf buds. The range of temperature which allows successful conduct of plant processes *viz.* transpiration, photosynthesis, respiration, cell division, cell enlargement and cell differentiation may be termed as optimum temperature range. Within this range lies a temperature which is optimum for maximum accumulation dry matter.

The optimum temperature range varies with plants. Some plants have higher temperature requirement than others. The general distribution of plants bears a testimony to this statement. Cotton, sugarcane, oranges, figs, dates, sweet-potatoes require higher temperature than apples, potatoes, sugarbeets, etc. Plants cease to grow at temperature of the freezing of water. Some seeds of hardy and cool weather crops *viz.* radish, oat, wheat have the capacity to germinate at temperatures slightly above the freezing point and continue to grow and produce. It is not possible for seeds of warm season *viz.*, tomato, corn, cucumbers to germinate at such low temperature. Horticultural plants may be classified into two broad groups according to their ability to thrive in temperature range of (*i*) 45-60°F (the cool season crops) and (*ii*) 60-75°F (the warm season crops).

Cool Season Crops

Fruits	Vegetables	Ornamental
Apple, blackberry, cherry, currant, grape, gooseberry, strawberry, pear, plum, peach, apricot.	Asparagus, beet, carrot, cabbage, lettuce, pea, potato, rhubarb spinach	Carnation, sweetpea, snapdragon, violet, calendula, geranium, gypsophilla, daisy, petunia, larkspur, lupin, mignonette, scabiosa, cineraria, schizanthus, zinnia, phlox.

Warm Season Crops

Fruits	Vegetables	Ornamental
Citrus, date, peach, grapes, fig, Annona, persimmon, mango, guava, jack fruit.	Sweet potato, tomato, pepper, egg plant, cucumber, water melon, pumpkin, squash, beans, okra.	Rose, poinsettia, gardenia, euphorbia, lily, hyacinth, amaryllis, orchid.

Cell division, the starting point of active growth is much affected by temperature changes since the rate of division is directly related to temperature. The upper half of the optimum temperature range shows greater cell division than the lower half. This is reflected in pronounced vegetative growth with greater utilization of the carbohydrates. The lower half of the optimum temperature range promotes slow vegetative growth and induces greater accumulation of the carbohydrates favouring reproductive phase of the plant. Temperatures beyond the limits of the suitable temperature range are unfavourable for the growth and development of plants.

Any increase in temperature beyond the optimum range increases the rate of respiration to such an extent that, even with increased photosynthesis, the dry matter accumulation is pulled down. This explains low yields by the plants when subjected to high temperature especially when the temperature increases during the reproductive phase of the plant. Yield and temperature are related positively. Whenever plants are grown in temperatures lower than the optimum range, they again produce low yields because of slackened rate of carbon assimilation and protein synthesis. Such losses are easily seen to occur in the case of warm season crops since temperatures lower than the optimum may be met with in nature. The cool season crops suffer due to lower temperature rather sparingly under Indian conditions.

Plant protoplasm possess hydrophyllic colloids which impart capacity to withstand low winter temperatures. These colloids have enormous surface and store large quantities of bound water which give the property to resist freezing at the ordinary subfreezing temperatures. Mature tissues withstand low temperatures to a greater extent because they hold their water against the force of freezing to a remarkedly greater extent than younger tissues. It is also known that greater carbohydrates content of the tissue increases the ability of protoplasm to withstand cold. Retarded ability of plants to withstand cold during the vegetative phase as compared to the reproductive phase is due to the maturity of the tissue at the later stage. All such conditions which promote vegetative growth and lower carbohydrate accumulation particularly during the later period of their life cycle predisposes the plant to cold weather influence. Any treatment or condition *viz.* withholding of water and nutrients, gradual lowering of temperature, exposure to strong sunlight, which checks cell division and cell enlargement without lowering manufacture of food material by the leaves and their accumulation within the cells is bound to increase hardiness. Different fruit trees have different degree of ability to withstand temperature. While apple and peaches may be safe at 0°F lemon or oranges cannot be grown at that temperature. If the minimum temperature is about 25°F, orange and grape fruit may be tried but not mango and avocado. Again only those cultivars of apple that withstand low temperature should be grown in temperatures below −15°F or so because all cultivars shall not flourish under the low temperature conditions.

Reasonably high temperature tends to promote water intake, development of buds and of opening of flowers. Flowers have less ability to resist low temperatures, sudden fall in temperature upto the freezing point tends to kill the buds and also blossoms. Whenever such conditions are expected, early blooming strains must be discouraged because these are more prone to injury of this type.

Dry wind tends to desiccate the plants rapidly by removing the available moisture through transpiration. The desiccation sets in easily if the the soil supporting the plant lacks capillary water which generally leads to the death of plant, the effect of drought is more evidenced in younger tissues which succumb to the injury early. The young plants and the flower bud suffer heavy losses. High rate of transpiration accompanied by loss of soil water may cause death of the entire plant including the older portions. Winter injury due to drought is often more common in regions where soil moisture is limiting, the atmosphere has low humidity besides intense winter cold. Under these conditions irrigation minimises chances of injury due to cold. Any substance sprayed on the foliage that reduces transpiration rate may also be employed as safety measure. Winter injury is less aggressive if drought conditions are eliminated. Warm temperature after cold weather condition causes flower bud development in peaches. The extent of flower development depends on the proportion of the period of cold and warm weather experienced by the plant. Longer periods of cold temperature followed by warm temperature for short period induces better reproductive cycle than short spell of cold temperature followed by long period of warm temperature. Exposure of peach plants to continuous periods of either warm or cold temperature without any alternation checks flowering since a certain amount of low temperature is required to take the plants out of the rest period ; in the case of peach it has been estimated that about 40 days of cold temperature (45°F) is sufficient to end the rest period. A chilling cold is necessary for the rapid production of rhubarb petioles because it is necessary to break the rest period. In the case of lily bulbs, flourists find it useful to treat these frequently with hot water.

Besides the effect of temperature on growth behaviour, development and injury to plants it has certain specific effects on the quality of the produce at harvest. Carbohydrates synthesized by plants may be stored as starch, sugar, etc. or converted to proteins, fats, organic acids etc. Sweetness determines the quality of many fruits and low temperature helps in the accumulation of sugars while high temperature causes its degradation. Lower mean temperature therefore adds to the sweetness of the fruits. The interconversion of sugar to starch or its final liberation as carbondioxide and water is controlled by the activity of specific enzymes. The reaction proceeds at slow speed when the temperature is low and at increased rates with rise to temperature. The changes in temperature do not affect the reactions involved. The process of conversion of sugar to starch and also its final breakdown to carbondioxide and water are affected adversely by low temperature resulting in the accumulation of sugars with decrease in temperature. Under high temperature increased rate of sugar breakdown lowers its level. Temperature at the time of collection of horticultural products thus greatly influences the quality through adjustment of sweetness. Lower mean temperature during the period of maturity and harvesting retard the rate of respiration and also the conversion of sugar to starch.

High colour of apples develops when the nights are cool and days have bright sunshine. It is important therefore, to know exactly the optimum temperature of the final products of horticultural crops specially at the time of maturity and harvest and grow only those varieties which respond favourably to prevailing temperature. It is not just an accident that the highest quality of *Dasheri* mango comes from Lucknow and *Langra* from Varanasi. Similarly in the case of muskmelon the *Safeda* of Lucknow and *Kajra* of Allahabad have no equal. Temperature ranges are also specific in their action on insects, fungi and bacteria causing disease. Diseases like apple scab and peach leaf curl are aggressive during low temperature period. When the temperatures are higher bitter rot or black rot of apple, smut on corn, blossom end rot on tomato etc. appear. The diseases commonly prevalent in certain areas are many a time due to temperature specificity.

Temperature ranges affect the reproduction and spread of insect pests and thus ultimately control not only the quality but also the quantity of the produce. Different types of winter injury like bark and trunk splitting and crown injury is due to low temperature. Frost canker also occurs in certain cases.

Light

Light is another important factor which influences organisms, more so the autotrophic plants. The arrangement of leaves on the stem in a way as to allow maximum exposure to sunlight is itself a significant proof of the response of plants to light. The amount of light controls the flowering of plants. This day-length effect on plants may be made use of in producing blooms out of season by nurserymen by the supply of extra hours of light or by subjecting plants to period of darkness, as per needs of the plants.

Light influences the closing and opening of stomata and thus regulates the diffusion of gases. In this light controls both the amount of CO_2 entering the stomata as well as the energy supply for the process of photosynthesis whereby glucose is formed in accordance with the endothermic reaction. Glucose is the known starting point for all other higher carbohydrates and in fact for almost organic compounds obtained from plants.

Light has a telling influence on the life of the plants through its action on permeability of roots, germination, enzyme action, dormancy, respiration, photosynthesis, protein synthesis, stomatal movement, reproduction. etc. Light consists of particles known as photons or light quanta. As these strike certain surfaces electrons are ejected, light of greater intensity ejects more electrons.

The velocity of the ejected electron is inversely proportional to the weavelength of light but bears no relationship with its intensity. The energy needed to push an electron from one orbit to another obviously depends on the energy difference between the two orbits. Light is absorbed by the chlorophyll pigments present in the green cells of plants and transformed into chemical energy required for the synthesis or carbohydrates.

Plant growth and development is affected by light in various ways. Light of a definite wavelength and intensity has been known to have formative influence. The formation of flower and time of flowering both can be made to vary considerably by changing light conditions. The effect of light also varies in accordance with duration of light.

A quantum of light of longer wavelength has less energy than that of shorter wavelength. Light is important as radiant energy and its effect is controlled by its intensity, quality and also daily duration or photoperiod.

The light waves are diverging with the speed of 3×10^8 metre per second. The rapidity of the waves can be gauged by the fact that every centimetre length contains about 20,000 of these. When the rays come in contact with the chlorophyll in presence of CO_2 and water these aid carbohydrate synthesis. Plants bend towards light because light retards growth on the side on which it falls and the stem turns to it. In the absence of light both cell division as well as elongation is more rapid.

The light of the sun at noon, on a clear day is about as much in intensity as 8,000 to 10,000 ft. candles. This intensity is reduced by fog, dust, clouds and water vapours in the atmosphere. If the light intensity reduced to 10% of normal sunlight the rate of photosynthesis is reduced by 60%.

A green leaf, in light, generally fixed more CO_2 than released by its own respiration. A stage known as compensation point may be found when carbon accumulation and respiration just equal each other. Leaves on the lower branches of a densely shaded fruit tree receive only a minute fraction of the sunlight so that the photosynthetic activity is reduced considerably pulling down the dry matter accumulation. Pruning and training of plant helps the spreadout of leaves for better utilisation of light. Light of short duration and low intensity lead to slow growth and restricts flower development.

Shaded leaves are succulent with reduced chlorophyll content, fewer palisade cells and larger intercellular spaces in the spongy tissues.

Light of very high intensity destroys chlorophyll, checks photosynthesis, increases transpiration rate, retards cell division and cell enlargement and may also cause sunburn and fading of colour of flower and fruits. High light intensity available to plants grown at high altitude lead to dwarfing of trees, with thick vegetation, slender and weak tree trunks, less branching and so on.

Lack of light may result in etiolation which has a drain on the food material prepared elsewhere by the plant. Plants have differential requirement of light for survival. The effect of shade is not disastrous on plants which have the capacity to tolerate shade. Use of partial shade is made in the production of best type of coffee, fine textured cigar-wrapper type tobacco leaves, betel leaves, ferns, and some crop plants. The need of shade to the cuttings in the beginning to check wilting and then of light to check decay is felt due to limitation of stored food. Supply of light a few days after the initial step of putting the cuttings in the soil and manufacture of food by leaves left on the cuttings is helpful to the plants. The length of time for which shading or light exposure is needed depends on the type of cutting and existing weather conditions. The kind of vegetation which gives better performance in the shade is distinct from those that thrive best under full sun conditions and consequently plantings have to be done with due regard to the available sunlight.

It is believed that under normal conditions 50% is utilized in transpiration, 30% is transmitted, 19% is re-radiated and the remaining (1% only) is used in photosynthesis.

Chlorophyll *a* is considered to be the most important pigment in the absorption of light energy from the rays of light though carotenoid and phycocyanin may also receive energy and pass on to chlorophyll *a*.

The intensity and duration of light are of importance to the plant in many respects. The colour of the light is also important. Light of different colours has different wavelengths denoting change in quality of light.

The clear light that we receive from the sun is composed to violet, indigo, blue, green, yellow, orange and red (VIBGYOR) seven colours. These have different wavelengths. The blue part of spectrum is considered useful for the formation of chlorophyll while both the blue as well as the red lights are useful in photoynthesis. Plants remain stunted in blue light though they look otherwise normal. In yellow light the plants give spindle shaped appearance with long internodes and small leaves. Light beyond the violet on the one hand and red on other is not considered necessary for the plants. Light of shorter wavelength is considered to have more energy as compared to the same quantum of light of longer wavelength. Maximum rate of photosynthesis is seen in the red and next in the blue corresponding to 655 and 440 millimicrons respectively. Photosynthesis corresponds to the degree of the absorption of light by the chlorophyll. The wavelength of different colour of light in millimicrons are 390-422 (violet), 422-492 (blue), 492-535 (green), 535-586 (yellow), 586-647 (orange), 647-760 (red). Waves shorter than violet are known as ultraviolet, beyond these are X-rays, gamma-rays and the shortest are cosmic rays. On the other side of the VIBGYOR beyond the red, the infrared preceded by radiowaves are known to exist in respect of wavelength.

Ultraviolet rays play an important role in the development of colour. More of ultraviolet rays reach the plants when the atmosphere is free from dust. These help to develop anthocyanin pigments to a greater extent. Violet and blue region of the visible spectrum is important for phototropic effect ; photosynthesis is at its best in the blue and the red band (lowest in green yellow) ; the orange and red (to a certain wavelength only) promote seed germination while the longer wavelength of red and the infrared inhibited germination of seed. The red rays of visible spectrum are largely effective in bringing long-day plants into flowering and in delaying the blooming of short-day plants. The lengthening of the day with green or yellow light has no effect.

The energy of sunlight trapped in photosynthesis leads to the formation of carbohydrate, the consumable metabolite which is transformed into energy by the activity of mitochondria for the benefit of the cellular organisation. The shoots grow away from gravity and roots downward towards gravity. The responses to light may be known as positively phototropic for shoots and negatively phototropic for roots.

Photoperiodic influence, as the name indicates, comprises the effect of daily light period on plants. Plants are tuned to specific day-length and their natural distribution bears testimoney of the same. Photoperiod is easily the most dominant factor in fruit and seed production, if other conditions of growth are not wanting. Long day cool-season plants when exposed to high temperature remain vegetative. Besides the effect on flowering, day-length markedly influences the foliage. The horticulturist can make proper selection of varieties with respect to day-length to derive maximum produce of flower, fruit and seed. Longer the exposure to light, greater is dry matter production within certain limits beyond which increase in light duration does not increase dry matter production.

The effect of day-length is of such great importance that the flowering plants have been broadly labelled as long-day, short-day and day-neutral. In fact, according to Lang, nine photoperiodic groups which differ in their response to day-length conditions are known. Seed germination, leaf fall, colour development and the rate of growth of plants may also be termed as light-controlled processes.

A short day palnt is characterized by its ability to form flower buds and bloom when subjected to short-days. The long-day plants, on the other hand, need long continuous exposures of about fourteen hours or more to come to the flowering and fruiting stage. The long-day plants remain in vegetative stage under short day conditions just as the short-day plants fail to flower in long-days. The plants which are not specific in their response to day-lengths are termed day-neutrals and may be grown in any photoperiod.

The characteristic demand of the flowering plant on light determines the season in which these are preferentially grown for obtaining good yields. For better production long-day plants may be artificially supplemented with more light when grown under short-day conditions although the production is more during the long days due to increased photosynthesis.

In plants like *Brassica,* temperature controls flowering; the plants when grown at continuously higher temperature remian in vegetative state, but flowers when exposed to lower temperature. The bolting in the case of onions, beets, etc. may be due to premature exposure to lower temperature. Gibberellin may replace the stimulus due to cold in many cases. The short-day plants when exposed to long-days naturally have greater reserves of carbohydrates and proteins maintaining vigorous vegetative state. The long-day plants, on the other hand, when exposed to short-days have low carbohydrate reserves and continue in a weakly vegetative state. The photoperiodic groups differing in their response to day-length conditions are now recognized to be nine *viz.* (*i*) short-day plants, (*ii*) long-day plants, (*iii*) day-neutral plants, (*iv*) long short-day plants, (*v*) short long-day plans, (*vi*) intermediate or stenophotoperiodic plants (*vii*) amphiphotoperiodic plants which flower in long-day or short-day but not in intermediate day-length. (The species may behave as long-day as well as short-day plants). (*viii*) intermediate species : Light period requirement being variable, these flower under any of the above conditions, (*ix*) Flowering takes place only under continued alteration of long and short-days *viz. Gestrum nocturnum* possible the one-plant group.

Photoperiodic Needs of Some Plants

Light period of 12 hours or less	**Light period of 12 hours or more**	**Indifferent to light**
Short-day or Long night plants	**Long-day or Short night plant**	**Day-neutral plants**
Stawberries, pineapple, salvia, soyabean, prime-rose, sweet potato, multiflora bean, phaseolus, poinsettia etc.	Plantain, castor-bean, cucumber, spinach, radish, lettuce, sugar-beet, onion, aster, calendula, gardenia, delphinium, stock.	Tomato, *Dianthus*, carnation, *Hibiscus*, Violet etc.
Nearly all tropical plants, cereals and spring flowering species of the temperate regions.	Most of the agricultural plants of the temperate regions, wheat, potato, all summer flowering plants.	Independent of day-length, flowering controlled by temperature.

Photoperiodic effects are conditioned by the prevailing temperature. A long-day cool season plant does not flower if the temperature is high and it is believed that temperature masks the effect of day-length. The inter-relationship between temperature and light has been well illustrated by the biennial garden beet. This plant develops the root in the first year of growth, flowers and set seed in the second year. When the garden been plant is grown between 40-50°F they do not flower or set seed in short days ; the higher temperature the longer is the photoperiod needed for flowering and fruiting. During the winter when the temperature and light intensity are lower and the days are shorter the short-day plants are grown. Potato is a long-day plant for flowering but a short-day plant for tuber formation so that the economic growing of potato is possible under short day conditions only.

With rise in temperature the number of staminate flowers greatly exceed the pistillate flowers reducing fruit formation in dioecious plants.

CHAPTER

12

FRUIT PHYSIOLOGY

I. THE FRUIT

Fruits are of immense value for human consumption as well as for seed dissemination. These vary in their morphology and makeup to a considerable extent. The development of the fruit is, to a large extent, dependent on the events taking place in the ovules. Fruits are defined as ripened ovary together with whatever may be intimately attached to it at maturity. Normally, in the angiosperms, a fruit develops after fertilization of the ovary and the seeds lie embedded in the fruit. An enlarged pistil together with closely associated parts may be known as a fruit. True fruits like orange, persimmon, avocado and grape have well developed carpels as the most conspicuous portion. In fruits developing from epigynous flowers such as apple, squash, banana and cucumber the receptacle is a part of the fruit structure.

Several kinds of fruits are known (Fig. 12.1). In peach or tomato it consists of only the ripened ovary giving rise to a simple fruit as opposed to an accessory fruit when additional structures are present. These groups of fruits are generally recognized (*i*) simple, (*ii*) aggregate, and (*iii*) multiple. Simple fruits may be of two types : (*i*) dry, and (*ii*) succulent or fleshy. Sometimes the dry one-seeded fruits *viz.* corn and sunflower are termed as seeds. The type of flowers and stage of development are associated with the type of fruit.

Kinds of fruit. Simple fruits are generally the outcome of the ripening of either inferior or superior, dry or fleshy ovary of one pistil. A fruit developing from inferior ovary consists of other parts of flower besides the ovule as opposed to the fruits arising out of superior ovary consisting of ovary alone.

A simple dry fruit (follicle, legume, capsule, achene, caryopsis, samara and nut) generally has a leathery, papery or woody fruit wall. Some of these are indehiscent *viz.* achene, caryopsis or grain, samara, schizocarp and nut while legume, follicle, capsule, and siliqua are, dehiscent fruits with dry pericarp. A follicle is a dry, monocarpellary, superior, one-chambered fruit splitting along the adaxial or ventral side only. Sometimes several follicles may be associated together to form etaerio of follicles. A **legume,** a dry fruit developed from a simple superior one-chambered pistil, has ovary dehiseing by both sutures as in pea, bean etc. legumes may be indehiscent *viz*. the peanut or monkey nut.

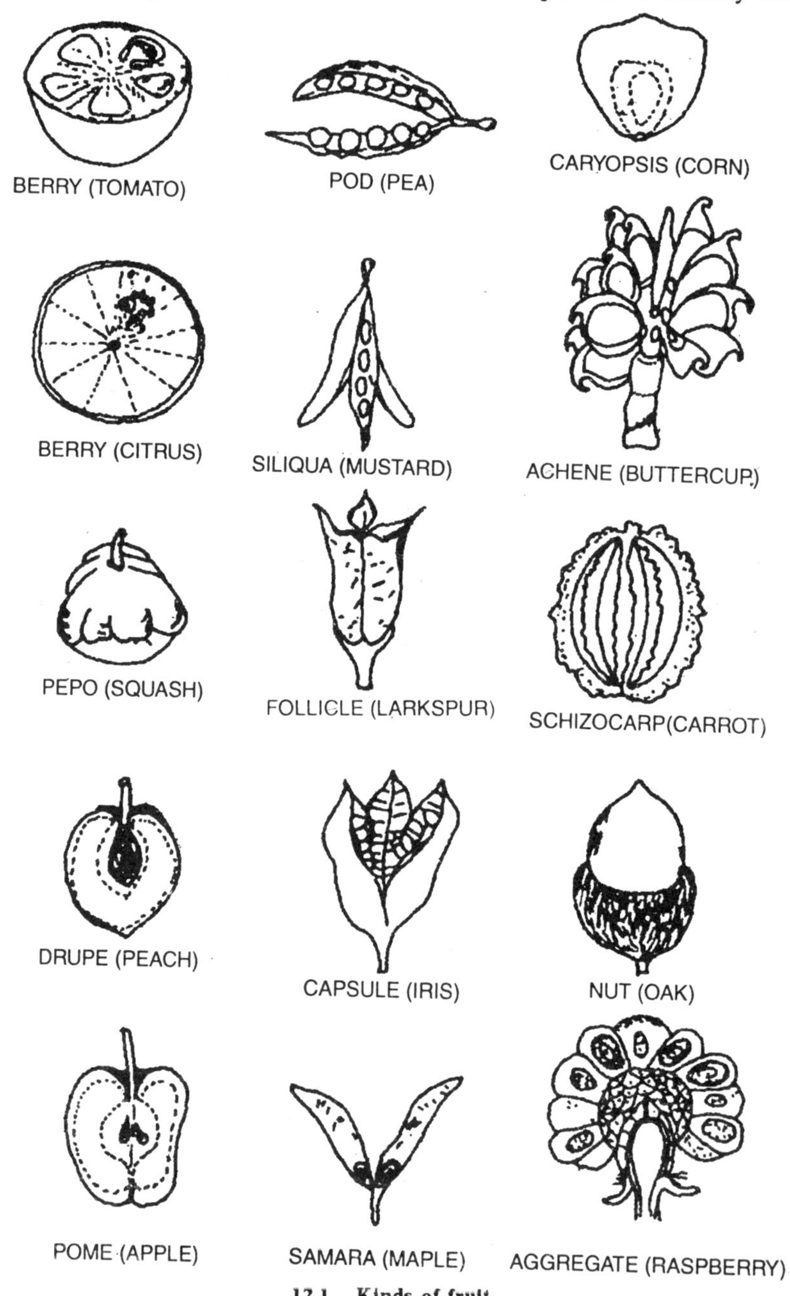

12.1. Kinds of fruit.

Siliqua is long narrow many seeded fruit developed from a superior bicarpellary ovary with two parietal placentae. It dehisces from below upwards by both sutures. The ovary is one chambered in the beginning but later the fruit becomes two chambered due to the development of a false septum the replum, as in cruciferae.

A **capsule** (dry fruit) arises out of a pistil with either superior or inferior ovary. A many seeded uni-or multilocular fruit developing from a superior bi-or poly-carpellary ovary dehisces and may split along the back of individual carpels may be associated together giving rise to syncarpy, *e.g.*, poppy, okra, cotton, datura, etc. **Achene** is a single seeded, small, dry fruit with seed attached to the wall of the ovary at one point. It may develop from a superior or inferor monocarpellary ovary *e.g.* rose, climatis etc. **Caryopsis**, the fruit of the grass family, resembles the achene except that thin seed coat and the pericarp unite to form a single unit. The common caryopsis are corn, rye, oats, maize, grass, bamboo and rice which develop from a superior monocarpellary ovary with the pericarp fused with the seed coat.

Indehiscent dry, 1-2 seeded fruit developing from a superior or tricarpellary ovary with flattened wing like outgrowths are the **samara.** This may be borne in pairs as in the mapple or singly as in elm, tree of heaven etc. Wings develop from the pericarp and the fruit breaks with its component parts each enclosing a seed. Fruits developed from pistil of superior bi-or poly- carpellary ovary where the pericarp is hard and woody are known as **nuts.** These are dry, one-chambered and one-seeded fruit (although sometimes two-seeded nuts also occur). Generally nuts are large fruits *viz.* peanut, almond, cashewnut etc. True nut (oak, chestnut, walnut etc.) has structures formed from bracts arising from the base of one or a number of closely associated flowers. Nut may develop from epigynous flower and may be enclosed in fleshy or dry receptacle. **Cypsela** is also a dry one-chambered and one-seeded fruit developing from an inferior, bicarpellary ovary with the pericarp and the seed coat free as in the compositae, *e.g.* sunflower, merigold, cosmos etc.

In contradistinction to the dry fruits like the nuts, fruit wall remains fleshy at maturity in succulent fruits. Fruits of this category are the berry, pepo, hesperidium balausta, drupe, pome etc. The **berry** is a fruit derived from a superior (sometimes inferior) ovary in which the entire ovary wall develops into a fleshy, often juicy and edible, pericarp. Berry is indehiscent usually many-seeded fleshy or pulpy fruit developed from a single carpel or from a syncarpous pistil with axile or parietal placentation. Date-palm, guava, tomato, grapes, avocado, fruit of the egg plant and persimmon are good examples of berry. In tomato the placentae are fleshy and enlarged containing a number of small seeds, the side walls of the carpels form the interior of the fruit. The date-palm is a single seeded berry. Berry remains succulent. The berry may or may not have seeds and may have one or several carpels. The citrus fruits are regarded as modified berries, **hesperidium**, in which the rind is not eatible.

In the **hesperidium** (orange, lemon, grapefruit, lime and tangerine) the exocarp is leathery and endocarp membranous. Multicellular enlarged, juicy and scale-like outgrowths from the surface of the carpel walls are edible. Carpels form a number of wedge-like segments which have the edible multicellular, juicy and scale-like outgrowths arising out of the carpel walls.

Pepo is also a type of berry in which the outer wall which is the receptacle becomes hard as in cucurbitaceae family. It is fleshy or pulpy, many seeded fruit developing from an inferior one-celled or spontaneously three-celled, syncarpous pistil with parietal placentation. In pepo seeds embedded in the pulp remain attached to the placentae. Capsicum is a type of dry berry. True berries are distinguishable from false berries by the persistence of the remnants of the flower at the top of the fruits in the former. In cran berry, several sepals persist at the apical end whereas the banana fruit has scars from where the floral parts have fallen off.

A **drupe** is fleshy, one or more chambered and one or more seeded fruit developing from a monocarpellary or syncarpous pistil (Fig. 12.2). Olive is an exception where the flowers have two carpels and four ovules although only one carpel develops. The pericarp of the drupe which is mostly derived from a superior ovary is easily distinguishable into an exocarp, mesocarp and endocarp. The exocarp or epicarp is commonly the thin skin of the fruit, the mesocarp is the fleshy pulp and endocarp is represented by the stone. Fruits like mango, peach, apricot, plum, coconut-palm, cherry with well differentiated pericarps are classed as drupes. In almond leathery exocarp and mesocarp are

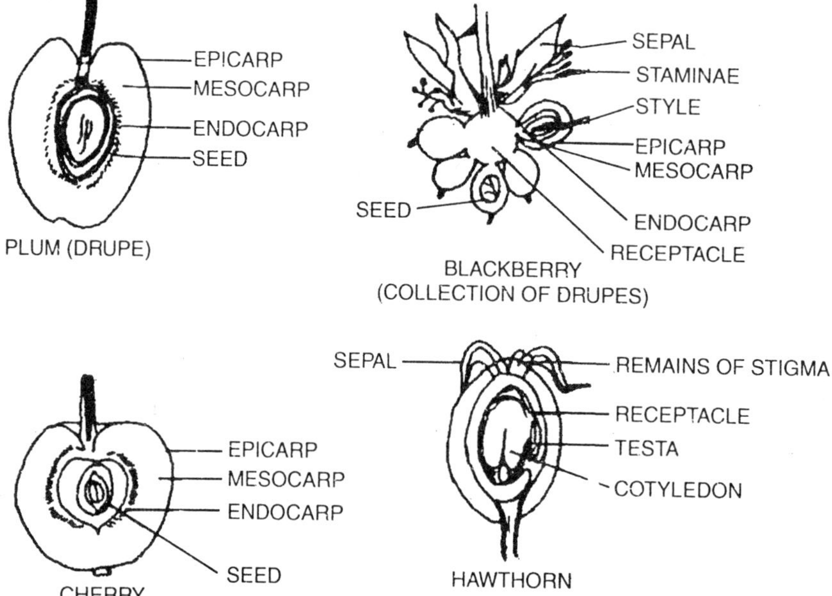

Fig. 12.2. Some fruits (vertical section) classed as drupe.

not edible, only the shell containing the edible seed is available in the market. Dry fleshy drupes with one carpel are mango, olive, almond, plum and peach, those with two carpels include coconut. The hard covering lying over the edible portion, the endosperm, is the endocarp in coconut. The outer fibrous covering is the mesocarp covered by the outermost smooth epicarp. Quince, loquat, apple, pomegranate, and pear are fleshy fruits classed as **pomes** (Fig. 12.3) with the pericarp differentiated into epicarp, mesocarp and endocarp. The pome is an inferior, two or more celled fleshy, syncarpous fruit surrounded by the thalamus. The fleshy edible part is composed of thalamus while the actual fruit lies within. An **aggregate fruit** is a collection of simple fruits or fruitlets developing from apocarpous pistil of a flower. The simple fruit of raspberry has small parts, the **druplets** which may separate out as independent units from the dry dome shaped receptacle. In blackberry the fruit is a little different from that of raspberry where the elongated receptacle is also fleshy and forms a part of the fruit. The blackberry fruit is classed as accessory as well as aggregate. The strawberry fruit has simple pistils developed into tiny **achenes** (also known as seed). In the strawberry the edible portion is the fleshy and much elongated receptacle, a modified stem. This is also an aggregate and accessory fruit.

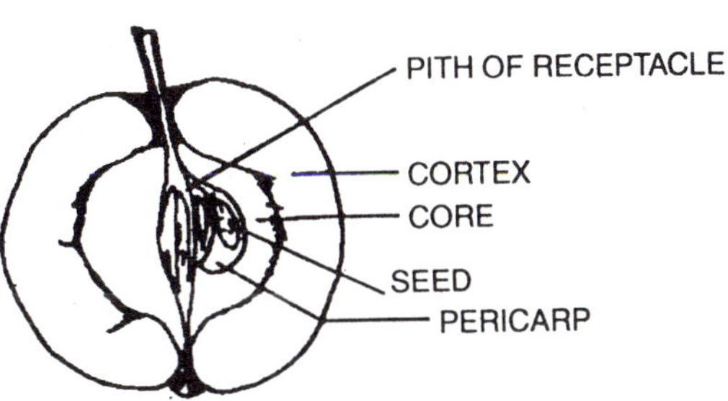

Fig. 12.3. Apple (vertical section) as an example of a pome.

Whenever the inflorescence takes part in the formation of the fruit it forms a **multiple fruit** as in the case of pineapple, mulberry, fig. etc. In these, every flower forms one fruit which remain together as a single mass at maturity. Many sessile flowers attached to the elongated axis of the inflorescence become fleshy with leafy tip (**sorosis**) as in jackfruit, mulberry and pineapple. The **syconus** (fruit of fig.) develops from a hollow pearshaped fleshy receptacle (inflorescence : hypanthodium) which encloses a number of minute male and female flowers. The receptacle grows to form fruit. It encloses a number of true fruits or achenes which develop from female flowers lying within the receptacle of its base as in the fig. banyan, etc.

II. THE PROBLEM OF FRUIT-SETTING

Fruits may be completely self-barren, fully self-fruitful or partially self-fruitful. The fully self-fruitful or automagous varieties may be planted in the orchard without making any provision for cross pollination though it is better to make provision for cross pollination since even varieties which are self-fertile also bear more fruits when cross fertilized. Often the horticulturist grows more than one variety which help in effective fertilization. Proper selection of the varieties is of importance to a fruit grower from the point of view of fruit yield for which besides overlapping of blossoming season over long periods, the time of maturity, productiveness, market value etc. are of importance. The pollinizer must be a heavy pollen producer to ensure better chances of cross fertilization. Although pollenizers are desirable, their number must be restricted so that these do not become uneconomical because the pollenizer trees occupy space, consume nutrition and take away attention of the fruit grower, without any return in yield, which would have otherwise been diverted to the trees capable of bearing fruits.

Table 12.1. Edible parts of some fruits

Fruits (kind)	Edible part
Apple, pear (pome)	Fleshy thalamus
Banana (berry)	Mesocarp and endocarp
Cashew-nut (nut)	Peduncle and cotyledons
Coconut-palm (fibrous drupe)	Endosperm
Cucumber (Pepo)	Mesocarp, endocarp and placentae
Custard apple (etaerio of berries)	Fleshy pericarp of individual berries
Date-palm (1-seeded berry)	Pericarp
Fig (syconus)	Fleshy receptacle
Guava (berry)	Pericarp and placentae
Grape (berry)	Thalamus and pericarp
Jack fruit (sorosis)	Bracts, parianth and seeds
Indian plum (drupe)	Mesocarp including epicarp
Litchi (1-seeded nut)	Fleshy aril
Mango (drupe)	Mesocarp
Melon (pepo)	Mesocarp
Citrus (hesperidium)	Juicy placental hairs
Papaya (berry)	Mesocarp
Pear (pome)	Outer portion of the receptacle bracts and perianth
Pineapple (sorosis)	Fleshy thalamus
Pomegranate (balausta)	Juicy outer coat of the seed
Strawberry (etaerio of achenes)	Succulent thalamus
Tomato (berry)	Pericarp and placentae

The number of pollinizer trees also depends on provision of cross pollinating agents like bees, 2-3 colonies of bees per hectare of orchard is a good aid to pollination. Generally 11% pollinizer trees are considered sufficient for deciduous fruit trees. The pollinizer trees should be systematically placed in the orchard, for instance, the third tree in every third row is considered sufficiently good arrangement. Large blocks of self-unfruitful trees improve in pollination when some of big trees are grafted over with pollinizer varieties than by their replacement by young plants. Sometimes only 2% staminate trees are considered enough.

The placing of bunches of the desired pollinizer's cultivar bearing flower buds ready to open at a time coinciding with the proper flowering time of trees of self-bareen cultivars improves pollination effectively. The larger the period for which flowers are fresh open and shedding pollen the better is the result. This type of artificial pollination has been practised in the remote past for dates and figs. It is a really troublesome and expensive practice.

Though both insects and wind are pollinating agents, the former is of greater importance in the deciduous trees except the nuts. The grapes and nuts are considered to be mainly wind pollinated but the presence of honey bees is apparently more useful in fruit-setting specially because the wind borne pollens may be carried to a distance of about a 1000 metres and rendered ineffective. Honey bee is possibly the most useful insect for pollination and these should be encouraged. Pollination through wind in the primarily insect pollinated fruit trees like apples, pears, peaches etc. is almost negligible. The fruit setting habits of some fruit trees are given here to have an idea of the variations commonly met with.

Almond. Self sterility in cultivars is well known but at times these cultivars develop self-fruitfulness. Self unfruitfulness in almonds is not caused by imperfect pollens but due to incompatibility. Some cultivars are self fertile and others self sterile. It is a good practice to plant pollinizers and also promote insect pollination for better harvest.

Apple. The flowers may have some defective pistil as well as pollen grains in the hermaphrodite flowers of apples. Self-fertile, self-barren, and partly self-fertile cultivars are known. Extent of self fruitfulness is controlled in the cultivars, by internal factors like age and vigour of the trees and external factors like the season, locality, wind, rain etc. The fruits develop parthenocarpically quite successfully but such development is not associated with any particular cultivar. Although some of the commercial apple cultivars have been grouped as self-unfruitful or self-fruitful this character is variable with change in soil conditions. Early fruit drop is associated with apple scab and fire blight which affect the blossoms or the newly set fruits.

Apricot. Both self-fertile as well as self-unfruitful cultivars are known. Self-barrenness is practically universal in this fruit.

Cherry. Cherries may be self-fruitful, self-unfruitful as well as self-barren. Interfruitfulness has been reported among Napolean, Lambert and Bing cultivars of sweetcherry. Sour cherries are also interfruitful to some degree, the former prove better as pollinizer of the latter as compared to the reverse. Self-unfruifulness and barren unfruitfulness in the cherry are due to incompatibilities rather than to any structural defect of pollen or ovule.

Grape. Hybrid cultivars of grape are likely to be self-barren although non-hybrid cultivars may also be self-barren. Grape cultivars may be self-fruitful as well as self-unfruitful, examples of partial unfruitfulness also occur. The cultivars having pseudo-hermaphrodite flowers should be inter-planted with cultivars having perfect flowers. Pollination with impotent or sterile pollen produce a few seedless berries in some grape varieties. Some of the seedless cultivars produce good pollen but the femaleness is not well developed. Deficiency of carbohydrates, natural or caused by heavy pruning, also leads to such a condition.

Gooseberry and current. Commonly grown cultivars are self-fruitful and need no provision for cross pollination. English gooseberry cultivars tested by Hooper were also self-fertile.

Pear. Pear, like the apple, has hermaphrodite flowers. Self-incompatibility makes some of the cultivars self-barren. Good number of cultivars are self-unfruitful although all the cultivars invariably have at least some viable pollens. Pears should be so planted as to secure the benefits from crossing.

Peach. Commercial cultivars of quince as well as peach are generally self-fruitful. J.H. Hale, Minodo, Halberta, June Elberta, Chandoka, Alamar, Saharanpur No. 5. China flat and Maicha cultivars of peach characteristically produced only abortive pollens and are consequently self-barren. Similarly the famous sub-tropical peach cultivar *sharbati* exhibits very high pollen sterility (46%) and is partial pollen fertile. J.H. Hale cultivar of peach with abortive pollen is self-unfruitful. Such cultivars would naturally need cross pollination with some other cultivars with fertile pollen. Except in the case of self-barren cultivars peach growing may be done even with a single self-fertile variety.

Plum. Plum cultivars are also cross pollinated though practically all the commonly cultivated cultivars of the Japanese and American species are self-fertile. European cultivars like Giant, Greenage, Italian, French and Blue Qamson which are partly or completely self-fruitful set better when cross pollinated. Some of the self-unfruitful cultivars are due to the fact that the stemen become petaloid.

Persimmon. The *kaki* or Japanese persimmon includes cultivars being both pistillate and staminate flowers. The flowers in persimmon are pistillate constants, staminate constants and staminate sporadics showing these characteristics with regard to the appearance of male or female parts. Pollination is not important for fruit production in persimmon. Some cultivars set seedless fruits without pollination while others produce fruits with seed after pollination.

Strawberry. Two types of strawberry are found : (*i*) pistillate-flowering and (*ii*) perfect-flowering. The perfect-flowering cultivars are self-fruitful. The flower that mature early are less fertile than those that mature late.

Nuts. Most of the nuts, *viz.*, walnut, chestnut, filbert are monoecious and many of their cultivars show partial dichogamy. Some cultivars of the pecan show complete dichogamy so that the tree or cultivar is markedly self-unfruitful. Pecan and Japanese chestnut cultivars are less self-compatible when cross-pollinated as compared to the cases of selfing that experience heavier fall of the partly grown nuts. Filbert cultivars show both partial dichogamy as well as self-incompatibility. Instances of cross incompatibility are also reported.

Loquat. Experiments carried out at Horticulture Research Institute, Saharanpur and also at Kanpur have shown that loquat cultivars *viz.*, Golden yellow, Pale yellow, Large Agra and Improved golden yellow are self-incompatible. These cultivars should therefore be not planted singly in an orchard. California Advance is a good pollinizer for Tanaka and golden yellow.

Mango. Studies carried out a IARI, New Delhi indicate that *Dashehari* and *Langra* cultivars too have shown less fruiting when planted singly in an orchard. When grown mixed, fruiting increased due to cross pollination.

III. THE PROBLEMS OF UNFRUITFULNESS

The causes leading to sterility are mainly (1) impotence, (2) incompatibility, and (3) abortion of embryo. Impotance relates to conditions when either (*i*) one or both the sex organs fail to develop adequately to the required stage so that there is complete failure of the formation of flowers, or (*ii*) the abortiveness of the male and female organs. In cases where sex organs are fully formed but there is sterility it may be due to incompatibility. There can be yet another case when both the male and female elements are formed and function, yet the embryo may be abortive. The time and stage of the incompatibility of sex elements is important in determining the cause of sterility. Sterility associated with internal functions may be related to (*i*) evolutionary tendencies due to the factors associated with the fundamental constitution of the protoplasm (it is heritable), (*ii*) genetic influence, and (*iii*) physiological reasons.

EVOLUTIONARY TENDENCIES

Defective flower. Self-fertilization is unknown in many plant cultivars due to imperfect flowers, heterostyly, dichogamy and pollen impotence. These are mostly associated with the fundamental constitution of the protoplasm. Plants possess flowers which are complete with both male and female elements present therein. These are known as perfect. On the contrary, flowers which do not have the two essential sex elements are classed as imperfect. Plants may have imperfect flowers containing only one of the essential organs, *viz.*, stamens or pistil. The two types (male and female) of flowers, may be present on the same plant (monoecious) as in litchi, banana, chestnut, filbert, walnut, pecan or on different plants (dioecious) as in papaya, dates. The evolution of unisexual flowers cuts down chances of self-fertilisation completely.

The transfer of a pollen of one flower to the stigma of another (cross pollination) leading to fertilization gives rise to viable seeds only when the male element is compatible with the female element. Horticultural plants vary in their degree of fertility. The cultivars that set viable seeds from the pollen of same cultivar are known as self fertile as opposed to the self sterile which do not form viable seeds with pollen of the same cultivars. Fruitfulness is not synonymous with viable seeds but the two are related rather closely. Generally better fruits are formed with seeds which derive nourishment and protection during the period of latter's development. Apart from the external factors which cause unfruitfulness, *viz.*, nutritive condition, soil type, water supply, cultural practices, wind, low temperature and the like there are certain internal factors also which control fruitfulness. Most of the internal as well as external factors are controllable by the grower.

The pistillate flowers, as in the case of cultivars of mangosteen, oranges, grapefruit and tangerines give rise to fruits parthenocarpically. Strawberry produces both pistillate and perfect flowers. Walnut, pecan, filbert and chestnut are monoecious. Generally cucurbits also are monoecious. There are plants which have both perfect and imperfect flowers. Some of the imperfect flowers of strawberry are pistillate. Sometimes the distribution of sex may be very peculiar, *i.e.* monoecious and dioecious forms being found in the same plants. The *Japanese persimmon* may be (*i*) staminate constant, (*ii*) staminate sporadic, or (*iii*) pistillate constant with respect to its varieties. The staminate constants are those which bear pistillate flowers each year alongwith some staminate flowers. Staminate sporadics are the cultivars which produce staminate and pistillate flowers in different seasons. Pistillate constant varieties produce only pistillate flower. The generally unisexual plant of papaya has many cultivars ranging from being unfruitful to fruitful due to incompatibility, self unfruitful or barren and self-fruitful among those which have the two sexes together. Of the unisexual plants the pure staminate flowering ones are essentially barren.

The presence of short styles with long filaments or long styles and short filaments is dimorphism, a type heterostyly, that may cut down self pollination totally and if cross pollination (which takes place) does not succeed the plants are turned barren though the sex organs are well developed. Heterostyly is relatively less important factors in dioecious plants. Besides this there may be other structural peculiarities of the flower which do not allow fruitfulness. Difference in the time of maturity of the sex elements (dichogamy) is known to be a potent cause of unfruitfulness even in the case of flowers which are otherwise quite potent. Protogymous and protoandrous types are recognized according to the time of maturity of the organs concerned. Dichogamy may be absolute or partial (complete or incomplete), in the former cross fertilization with pollen of some other plants may take place but in the latter some selfing also may be possible.

This aspect of unfruitfulness is very fascinating and prevalent so that it may be well realized that largely all dioecious plants are protogynous while the monoecious plants are protoandrous. One example of monoecious plant, *i.e.*, caprifig that shows dichogamy is well known. The cultivar of walnut which is protoandrous needs interplanting with other cultivars to ensure fruiting. Pecan is both dichogamous and otherwise, *Annona* and plum are dichogamous while chestnut has both protoandrous and protogynous flowers. The flower behaviour in avocado had been termed as protogynous diurnally synchronous dichogamy (see details on page 272 of the book). Dichogamous plants have an opportunity for "illegitimate crossing or hybridization." Completeness of dichogamy varies with environmental conditions.

Abortiveness leading to impotence. Interference either in the development of the flower or in the full development of sex elements and their function may lead to unfruitfulness. Abortion of flowers and flower buds before blossoming is well known. In certain cultivars of pomegranate and plum the pistil is degenerated and unfruitfulness results. In pecan the terminal clusters consisting of pistillate flowers slough off before pollination. This also causes unfruitfulness. In case where complete degeneration of the pistil takes place before maturity no fertilization is possible.

Flower abortion is more common in indeterminate inflorescence as compared to determinate inflorescence. Unfruitfulness may result in certain cultivars of plum and ornamental pomegranate due to degeneration of the pistil. In the case of strawberry the abortion of pistil is late so that unfruitfulness does not result. Falling off of terminal portion of the pistillate flowers before pollination in pecan, grapes and tomato also leads to unfruitfulness. Late flowers of strawberry are as a rule abortive so that no fruits can be formed. There are cultivars of defective pistil (plum) defective embryo (apple), defective embryo sac (oranges) which also lead to unfruitfulness.

Non-viable pollen. When the pollen is not viable the production of fruit is not possible. The cultivars of American grapes may be classed as self-fertile, self-sterile, partially self-sterile when dependent on their own pollen grains for fertilization. The defect being traced to impotence of the flowers. Unfruitfulness in the case of Mascadine grape is due to the fact that the stamens are reflexed and their pollen defective. The flowers are not truely hermaphrodite but pseudo-hermaphrodite. The plants are dioecious hence both male and female plants should be planted side by side in the orchard and at close intervals. Pollen of mango (65%) is defective while that of plums abortive. Pollen abortion is common with blackberries, several cultivars of peach, pear, sweetcherry etc.

GENETIC INFLUENCES

Hybridity is associated with sterility as well as unfruitfulness. It is also responsible for seedlessness in some cultivars. Hybridity and incompatibility are two types of self-sterility which are directly related to genetic factors. The two types are not easy to be demarcated from one another. Self-sterility depends on inheritance but its development is controlled by environment. It can be transmitted to its pure offspring and also to its hybrid progeny as in the case of *Rubus trivialis*.

Self-sterility is a condition determined by the inheritance received although the effect of favourable environment is necessary to make it touch the level of perfection. The degree of sterility increases with wider crossing. Peach-plum hybrids known as Blackman or Mule have completely sterile and barren flowers so also in Kamdesa, a hybrid between peach and sour cherry. Pyronia (pear × quince) flowers and fruits freely but is always seedless. Seedlessness in pineapple is considered to be due to hybrid nature of their ancestors, similar is the case with banana. Unequal distribution of chromosomes during pollen tetrad formation in a number of hybrid raspberry forms is the cause of sterility. Hybridity is not necessarily the cause of sterility although these may be associated. Incompatibility between the pollen and the ovule is also one of the causes of self-unfruitfulness and self-sterility. These sex elements may be fertile in themselves yet incompatible. Apple, blackberry, cherry, almond, pear, olive, apricot, plum etc. are examples of self-unfruitfulness due to incompatibility.

Several of the cultivars of fruit trees though being self-unfruitful give good performance when cross fertilization take place. This signifies that there is nothing wrong with the pistil. Likewise the capacity of the pollen to be effective on other cultivars of the same species tends to suggest that both the pistil as well as the pollen are normal and capable of functioning except when incompatible. There are instances of interfruitfulness and inter fertility where the ability of two plants to bear seeds with each others' pollen is the rule. Dioecious varieties of strawberry, fig and also date palm are not only self-unfruitful but also interunfruitful. Where the pollen is important (some grapes) incompatibility as in the case of cherry cultivars or due to poor pollen production or incompatibility with parent cultivar. Incompatibility may be several type, *viz*., self and cross-incompatibility.

Physiological causes. Poor growth rate of the pollen tubes in the style, possibly due to hormonal or chemotropic control (apple, pears, cherries, citrus fruits) is known. The growth rate of the pollen tube may be slow due to several causes including low temperature.

Unfruitfulness can also result from differences in the stage of maturity of the pollen grain *vis* a *vis* the pistil or ovule. In tobacco, if mature pollen is able to reach immature pistil it may germinate, enter the ovule (which is also not yet mature) and the flowers fall. Pollination of the bud (peach, pear, persimmon) has often led to low fruit setting and in many cases decline in the number of seeds per fruit. Delayed pollen germination is also ineffective depending on the time of delay.

The proper development of the flower, its maturity and fertilization leading to the development of a fruit with viable seeds is controlled by the nutritive conditions within the plant existing at the time of pre-blossoming and post-blossoming stages. The nutritive status of the plants determines the time taken by the pollen to fertilize. The fertilization of dwarf trees, under normal conditions, takes place in shorter time if the trees are in strongly vegetative condition than those in poorly vegetative condition. Nutrition of the plant also controls the percentage of defective pistils and therefore the development of fruits on these. Defective pistils are formed specially on exhausted or weakened plants caused by overbearing, drought, poor nutrition etc. So far as the internal conditions are concerned the deficiency of carbohydrate caused failure to blossom, abortion, and also droping of fruits. Bearing on spurs and shoots (plums and cherry) or on strong and weak spurs are distinctive, the more healthy spurs always promoted better bearing. Shoot-borne plums showed heavy June drops as in caprifig in comparison to the spur-borne fruits. Ringing leads to parthenocarpy only due to the greater concentration of nutrition, at the point of bearing, on tips. Fruit-setting may also be limited by unfavourable nutritive condition in the plant body. Overbearing, overpruning, ill-nutrition or unfavourable water relations lead to defective pistil or pollen grains of low vitality. Flowers borne on weak shoots or spurs or those on normal shoots are no better than those on healthy branches or spurs, the place where the flowers are borne is also of significance in fruit-setting.

External factors. Any external factor that affects the nutrition of the fruit trees, *viz.*, soil condition, water supply, cultural practices are of importance in this behalf. Among the indirect factors, rain and low temperature are well known. Crop production is an adjustment between heredity and environment and the impact of environment is well recognized in fruit-setting just as in growth and development.

Environment. Conditions prevailing at the onset and during flowering have been known to affect fruit setting, sometimes defective pistils, ovules, embryo sac or pollen grains may be formed rendering the plants dichogamous. At other instances the transfer of the pollen may be affected or the development of the pollen tube rendered slow to bring about changes in the extent of fruiting.

Nutrient supply. Nutrient supply affects fruit-settings as well as unfruitfulness possibly through its modifying influence. Grown on rich soil *Agave vivipara* generally produces bulbs but no seeds while in poor soils accompanied by economy of moisture seeds are formed rather than bulbs. Tomatoes supplied with rich nutrients fail to produce normal flowers, the flowers are not fit enough to set fruits. Jonathan apple which has been reported to be mostly self-sterile under the rich soil conditions (Australia) become self-fruitful when grown on poor soils. Certain varieties of grape produce perfect hermaphrodite flowers only when they receive proper cultural treatment. Ordinarily cultivar of a race of strawberry (Hautbois) is perfect flowered and productive but in rich soil the stamens develop poorly and the pollen grain number is rather small so that the fruitfulness is affected adversely. In fact high nitrogen content of the fruit bearing limb of the plant at the time of blossoming encourages fruitfulness. Deficiencies cause disturbances *viz*. boll shedding in cotton, as well as hen and chicken disease of grapes. Excessive nutrition causes sterility in graminae, cruciferae and leguminosae, for example.

Pruning and grafting. Moderate heavy pruning induces good setting and yield in soils having ample available nutrients. Pruning is also known to give rise to normal true hermaphrodite flowers in Hope grapes where absence of pruning leads to sterility through a weakening and abortion of its pistils. Pruning affects flower formation, fruit-setting and productivity. The main idea about the effect of pruning as a cultural practice in the control of reproductive phase of the plant is to affect of nutrient supply through the manipulation of C/N ratio besides other effects like apical dominance etc. Stock-scion effects have been discussed in an earlier chapter.

Locality. A cultivar may set fruits often in a better way in one locality than in another. Factors, *viz*, soil, temperature, humidity, light, etc., possibly play a significant role. How far each of these factors is responsible for the influence on fruit setting has to be determined. Potato, sugarcane and other plants may flower at certain localities but not in others. Intersterility due to incompatibility is often conditioned by locality influences in sweet cherry, almond, plum, and pear cultivars. *Escholtzia* is completely self-sterile in Brazil's hot climate but is fertile elsewhere. Jonathan apple is self-sterile in Australia though not in USA where it is distinctly self-fertile. Lilac a moderate seed bearer in England fails to do so in parts of Germany. Production of defective pistil is known to be a changing character associated with locality.

Season. It is difficult to distinguish the influence of locality from that of season. The fertility of plum is known to vary from season to season so is the case with fruit setting in most of the fruit trees. Seasonal influence on fruitfulness has been observed in figs, of the San Pedro class where the cultivars set freely without pollination, in the early crop, developing seedless fruit. The main crop does not set and mature without caprication. Several other cases of the influence of season may be cited, *viz*., change of end-season fertility to self-sterility in tobacco, mango and grape plants or of end-season sterility to self-fertility in strawberry. Change of sex with season often associated with persimmon, grape vine, date palm, maize etc. is also well known.

Age and vigour of plant. This factor is in majority of cases linked intimately to the effect of nutritive status of the plant or its limb, locality and the season. Change of staminate to pistillate flowers or *vice versa* is often controlled by the vigour and age of the plants. In apple, old and less vigorous trees set fruits better than young and vigorous ones. Age is effective through the vigour of the plant although it must be realized that some internal factors controlling fruit-setting are completely uninfluenced by vigour. Age and vigour influenced possibly through effects on impotence preceeding fertilization and embryo abortion later and not through compatibility factor.

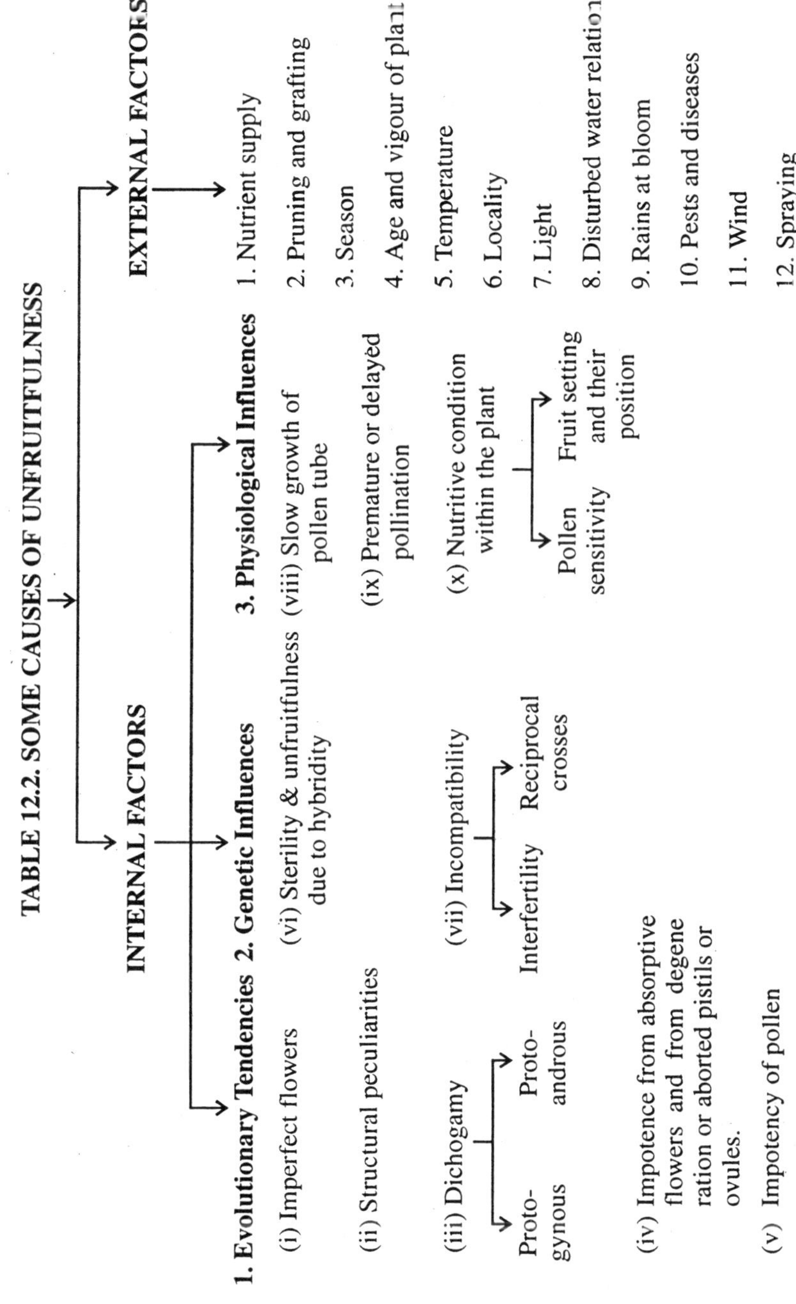
TABLE 12.2. SOME CAUSES OF UNFRUITFULNESS
INTERNAL FACTORS
EXTERNAL FACTORS
1. Evolutionary Tendencies
(i) Imperfect flowers
(ii) Structural peculiarities
(iii) Dichogamy
Proto-gynous
Proto-androus
(iv) Impotence from absorptive flowers and from degene ration or aborted pistils or ovules.
(v) Impotency of pollen
2. Genetic Influences
(vi) Sterility & unfruitfulness due to hybridity
(vii) Incompatibility
Interfertility
Reciprocal crosses
3. Physiological Influences
(viii) Slow growth of pollen tube
(ix) Premature or delayed pollination
(x) Nutritive condition within the plant
Pollen sensitivity
Fruit setting and their position
1. Nutrient supply
2. Pruning and grafting
3. Season
4. Age and vigour of plant
5. Temperature
6. Locality
7. Light
8. Disturbed water relations
9. Rains at bloom
10. Pests and diseases
11. Wind
12. Spraying

Temperature. It is common knowledge that a period of cool yet frostless weather is conductive to better blossoming, fertilization and fruit-set. For tomatoes the night temperatures (50°F is critical temperature) are limiting though not the day temperatures. the abscission of flower bud, fruit etc., is a function of temperature, each has its own critical temperature. Temperatures are quite important for pollen germination in plum, cherry, apple, pear, etc., where 50°F allows free germination but 40°F or lower temperature completely checks the same. The importance of temperature in effecting fertilization may thus be realized. It is in this way that cool and frostless weather is no good for fertilization (although ideal for blossoming). Fruit-setting is sometimes found in 'male' papaya plants under cool climates through the development of the pistils usually present in the male flowers. Temperature influences underdeveloped flowers, more directly the formation of flower buds and flower parts than the fruit setting. Temperature effects may also be influenced through water stress created in the plant.

Light. Low light intensities cause boll shedding in cotton, unfruitfulness in tomato, fruit setting in sour cherries, suppression of male organs in strawberry etc. In delicious apple the light falling on the plant in the fortnight following blossoming determines to a great extent the setting of fruit. It is doubtful if variations in light are important in other fruits. However, it is of interest that some herb plants develop flowers normally and set fruit freely in open sunny situation but when shaded the flower buds abort and fall off before opening.

Water-relations. Moisture deficits created in the plant by any cause whatsoever promote the formation of abscission layers leading to flower and fruit drop. Extremes in the carbohydrate nitrogen balance in plants may lead to the formation of abscission layers in normal flowers or young fruits but this influence is slight as compared to that of water stress during the early stages of fruit development. The dropping of flowers or partly developed fruits or more due to water imbalance than anything else. Irrigation, tillage, use of cover crops, wind breaks and adequate drainage facilities are important to cut down the gap between absorption and transpiration.

Rain and wind. Rain is an important environmental factor that indirectly affects fruit-setting by disturbing the process of pollination, germination of pollen grain and even staminal fertilization. Wind is specially of value in the control of pollination either by promoting wind pollination or by checking insect pollination. Wind may give jirks to flowers causing mechanical injuries. Sometimes it also dries the stigmate fluid prematurely preventing the germination of pollen grains.

Besides the environmental factors affecting fruit formation or fruit-set diseases and pests or the sprays used to check their menace also affect the flower and fruit drops. Sprays, sometimes have direct influence on the insects that bring about the pollination causing blossoms or partially developed fruit to drop. The response of the tree to the several external and internal factors which vary from one variety to another, one orchard to another, and one year to another have a governing influence on the phenomenon of fruit-setting. The manner and extent to which the horticulturists can control this important aspect of fruit production must necessarily be linked with the physiology of the plant.

IV. FRUIT-BUD DIFFERENTIATION

With the growth of shoots, buds are formed which show meristematic activity leading to the successful completion of the vegetative phase. The vegetative cone is formed in 5-10 days. The vegetative cone with three to four layers of meristematic cells marks the beginning of fruit-bud formation. Bearing habits of fruit trees depend to a considerable extent on the inherent nature of the vegetative buds to be converted into a flower, fruit, staminate flower bud, pistillate flower bud or male catkin (Fig. 12.4). These differ widely among fruits.

The tissue leading to the formation of bract are differentiated in several days in the axils where the primordia of future flowers are formed, the centre of each flower primordium assumes a concave shape and on its fringes develop the sepal primordia. A torus is also differentiated simultaneously and on its inner surface the primordia of petals and a little later the carpel etc. develop. This is how by the beginning of the flowering season the primordia of the main parts of a flower are formed. The archesporeal tissue and ovules are usually formed in spring. Many times different stages of fruit-bud differentiations occur on the same tree at the same time. The time of the differentiation of bud depends on the species, strain, age, bearing position, type of fruiting wood, the physical conditions, weather, and cultural practices.

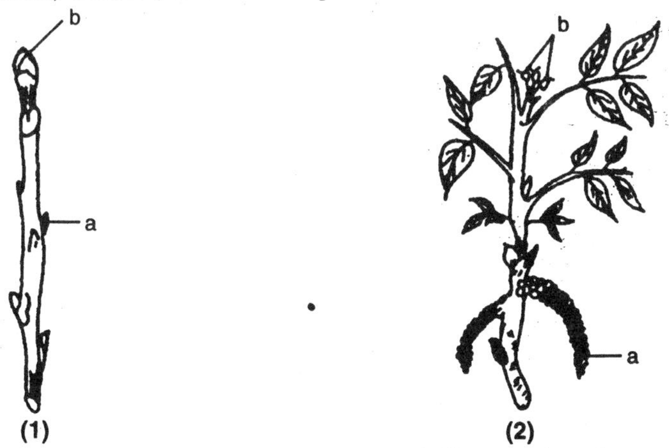

1. Vegetative growth
(a) wood bud (b) fruit bud.

2. Flowering branch
(a) male flowers or catkin, (b) female flowers.

Fig. 12.4. Walnut.

Earliest differentiation takes place of the tip buds of the spurs followed by those of the young bracts and then of those on young yearly wood. The spur which bears the bulk of the fruits completes its linear growth within a few days of the appearance of the bud and after the fall of the flowers, only the growth in thickness occurs. With the rapid growth of the spur the carbohydrate level falls, after some time the leaves again accumulate carbohydrate which coincides with the rapid growth of the spur the carbohydrate level falls, after some time the leaves again accumulate carbohydrate which coincides with the period of bud differentiation. This sequence has been confirmed with apples and pears. Rudenko reported that by the beginning of October the primordia of flower organs are usually formed. The bud development cycle usually takes about two months according to the observations made by Sergeyeva in 1956. Almonds, apricots, apples, pears, peaches, sweet cherries and plum form buds under optimum conditions in July to August. Olive, on the other hand, begins bud differentiation only by the end of April and the Japanese permission towards the end of January. In apricots fruit-bud differentiation may begin in July or beginning of August depending on weather conditions and varietal peculiarities. The spurs develop fruit-buds earlier than the year-old wood. All shoots may have fruit-buds at different stages of development. Zeller holds that the flowering stimulus is generally stronger in the middle portion of a shoot than the base or the tip. After the fruiting shoot is fully grown it takes 48 to 72 hours for the development of the fruit-bud. According to Pilipenko droughts encourage early formation of fruit-buds. The trees growing in north and on higher altitude develop fruit buds later.

Fruit-bud formation. It involves physiological and complex biochemical problems. Flower-forming substances; sugar concentration of cell sap, C : N balance, light intensity, are considered some of the factors controlling the process : fruit-buds are abundantly formed when the period of intensified growth is marked with ample soil moisture and nutrients, while the second half of the year has more sunny days. A knowledge of the kind of buds and their location (Fig. 12.5) helps in profitable pruning operations. The alternate arrangement of dormant fruit-buds is well known specially in the case of apricot and peach (Fig. 12.6). Alongside, it is also of interest to know the time, the steps and the factors in the differentiation of flower-buds. The development of apple fruit-bud (Fig. 12.7) indicates the dormant to mature fruit stages.

The time of the formation of flower buds is important in influencing their distribution by the change in cultural practice or by changing the carbohydrate supply fraction. It is also important because the stage of development of flower buds had great influence on winter injury to plants. The blossom buds may show somewhat lighter packing of tissue although all buds appear substantially alike. There is hardly any difference between a leaf bud and blossom bud upto the predifferentiation stage which is related to the stage of growth of the tree and its biochemical condition at the time.

The buds, on the well grown spurs, at the time of differentiation begin to develop under suitable conditions and are ready to bear fruits. This is possibly controlled by the adjustment of carbohydrate level in the spurs. Both the appearance of the buds as well as their opening seem to coincide with the rise in carbohydrate level. Differentiation of the bud is evidenced by the rounded surface of the tip of the spur surrounded by embryonic leaves having relatively large mass of meristematic tissue. The bud grows more at the point of contact with the spur than at its distal end though there is the absence of swelling at the periphery which preceeds formation of rudimentary leaf. Flower-bud differentiation is manifested in the rapid elevation of the crown into a narrow conical form rounded at the apex. The fibro vascular connections and pith increase at the same time. If a leaf bud develops on the spur it may thereafter form a blossom bud in the growing season.

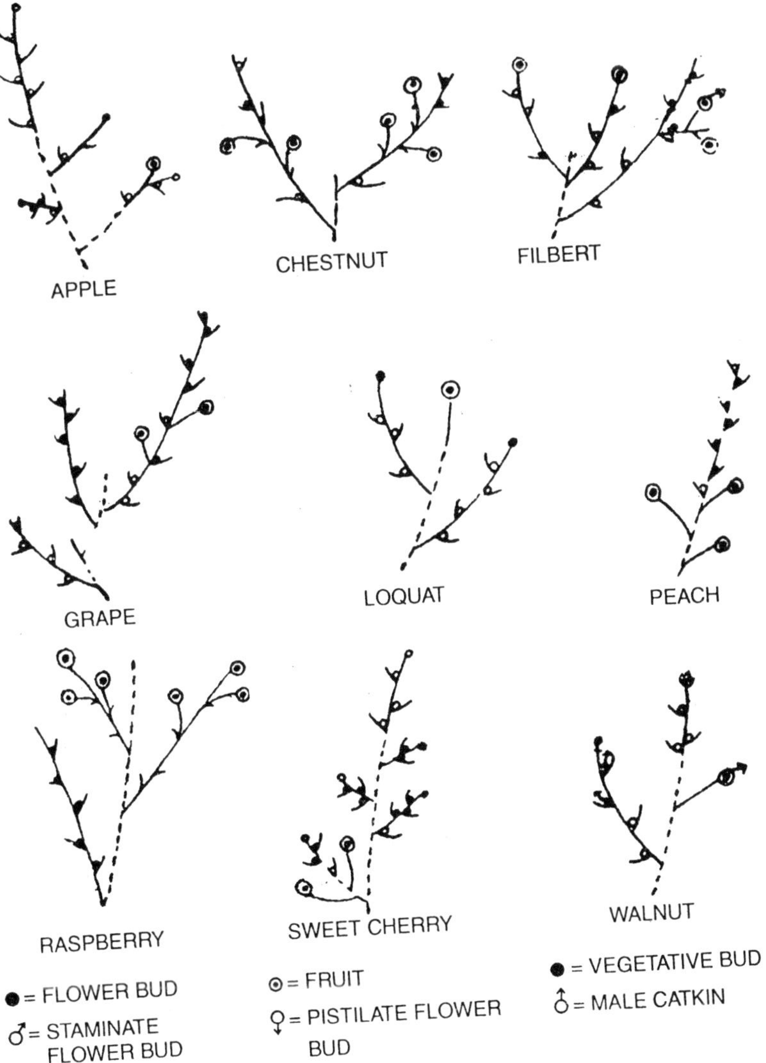

Fig. 12.5. Bearing habits of some fruit trees.

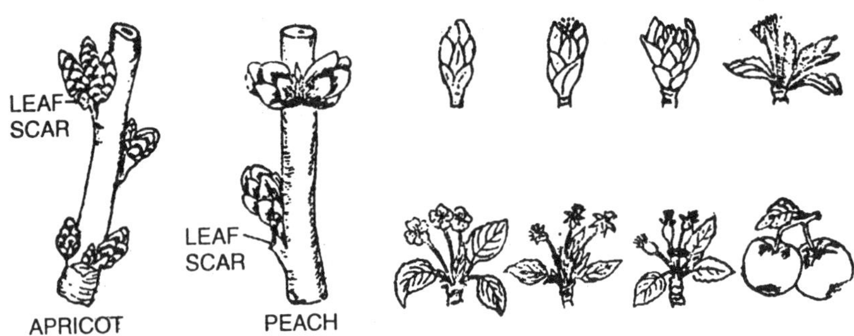

Fig. 12.6. A... ..ate arrangement of dormant fruit-buds.

Fig. 12.7. Certain stages in the development of apple fruit-bud. Top row (left to right)—Dormant, breaking, bursting, green cluster. Bottom row (left to right)—Bloom, petel fall, closing of calyx, mature fruit.

FLOWER BUD DIFFERENTIATION IN DIFFERENT FRUIT TREES

Apple. The broader details of the differentiation and development of various floral structures and organs have been carefully outlined. The differentiation of flower-buds in the apple is very much similar to that of pear. The drupes are different from the apple and pear in this respect that in the former the carpels are not united during development. The flower-buds of apple are formed mostly on the spurs though also on the lateral twigs and may be placed terminally as also in lateral position.

The growing point destined to develop into a leaf bud is not easily distinguishable from that developing into a blossom bud before the differentiation of the floral parts. Both types of buds have a smooth rounded crown of meristematic tissue more or less enclosed by bud scale. The floral organ develops in acropetal succession with the advancing season, *viz*., the calyx, corolla, ovarian cavity and on the placenta of the latter the ovules are borne. The location of the flower-bud and the variety of apple govern the time of its initiation. The appearance of the primordia of flower bud takes place followed by calyx lobes, pistil, ovarian cavity, petals from base of calyx etc. in sequence in about 60 to 75 days. Soon after the flower enters the pollen mother cell stage, tetrad formation takes place and in about a week's time mature pollen is formed. The whole process of differentiation takes place between January 15 to April 1 with the resting stage during the cold winters after the formation of the petals so that the mother cell stage leading to the development of the mature pollen takes place between March-April.

Mango. Unlike apple the blossom-buds in mango are mostly born terminally, rarely axillary buds may also be formed. Mangoes blossom generally on the old spring and summer growths though exceptions do occur where flowers are borne on shoots of the same season. The maturity and age of the seasonal vegetative growths mainly govern flowering in mango which appears in distinct flushes at certain times of the year. There may be three flushes February-March, March-April or beginning of monsoon, and September-October or beginning of winters in Western India while in the Punjab the differentiation is earlier between mid August to November. This is partly due to the temperature effects. The time of differentiation is September in case of *Langra* mango in Bihar and December in case of *Dashahari* and *Langra* in Saharanpur conditions.

Pear. The blossoms appear on the spurs terminally although terminal flowers on long shoots or water shoots may also be found. Bud differentiation in the pear takes place a little later than in apple, the initiation taking place in mid July. The buds are quite large and conspicuous forming a wider angle with the shoots.

Citrus. In mosambi and santara more than one sprout is given off from a single vegetative bud. The growing season in these plants may be July, September-October and February-March as in Bombay. These are flush seasons and are locally known as "bahar" when blossoms appear and fruits are borne. The period of flush in other parts of the country varies with the local conditions.

Grape. Summer is the pre-blossoming period when the flower-bud formation takes place. Each bud consists of a shoot with blossom primordia in its embryonic stage. Differentiation may take place at any time in year depending on the geographical area, climatic conditions, pruning etc. The tendril and the inflorescence primordia differentiate from morphologically analogous structures such that the former may be regarded as a bud of vegetative flower cluster. During late spring, summer and autumn differentiation of anlagen into inflorescence primordia takes place. The anlagen initiated after the winter rest become tendril primordia. Flower-buds differentiate as these open in spring.

Peach. In the peach, flower buds occur in axillary position only on the spurs and differentiate in the month of July.

FACTORS CONTROLLING FLOWER-BUD DIFFERENTIATION

A. Physiological

Flower formation is induced by "flower forming substance", produced in the leaves and passed on to the buds present in the axil of leaves by diffusion. Klebs held that the flower forming substance is of the nature of carbohydrates since the availability of nutrients and light intensity controlled the onset of vegetative or reproductive phase of the plant.

Kraus and Kraybill demonstrated the existence of a relationship between composition of the plant and its fruiting response in which the importance of carbohydrate and nitrogen balance for the initiation of flower primordia and fruit-setting is stressed. The spurs which initiated flower primordia are found to contain greater amount of sugars than non-flowering twigs of the plant showing that carbohydrate has some control on flowering. Weinberger confirmed the findings of Kraus and Kraybill with peach and grapes when he reported that the amount of starch in the annual wood is highly correlated to flower-bud formation. Sugars, that accumulate, serve the need of the increased rate of respiration and in increased production of protein leading possibly to a change in the nature of protoplasm itself due to the increase in concentration of non-sugars. It is believed that the organic compounds in the cell sap might have influence on the initiation of flowering. The flower-bud formation and fruit-setting are two separate phenomena and the conditions leading to the initiation of floral primordia and even blooming are probably different from those accompanying fruit- set. The plants are placed in four classes with respect to the C : N balance by Kraus and Kraybill, details of which are presented later. Pruning of trees deficient in nitrogen induces more unfruitfulness and that of young, vigorous, non-flowering trees delays fruitfulness. This shows that the internal conditions of the plants' composition must be taken into account before deciding the cultural operations. The non-bearing spurs generally are associated with low nitrogen and reducing sugars but more starch and hydrolyzable polysaccharides at the time of differentiation. Carbohydrate deficiency is thus associated with flower-bud formation.

Flower production hormones. Hamner and Bonner asserted that flowering was initiated by certain substances of the nature of phytohormones. The flower forming substances of Sachs may be considered to be a hormone since these substances are also organised in the leaves of the green plants in the presence of sunlight. The interdependence of sugars and hormones is well established and therefore the changes in sugar balance of the plant might be connected with the formation of the so-called flower forming substances. The evidence presented by Sachs, as well as studies on the effect of environmental factors on flower formation emphasize the paramount importance of leaf surface in flower-bud differentiation in fruit plants.

B. Environmental

The supply of concentrated nitrogen – containing fertilizers to young non-bearing fruit trees delays flower formation and accelerates the rate of growth of shoots, spurs etc. If the application of nitrogen is made to bearing trees, flower-bud formation is reduced. Trees of low vigour increase flower formation with moderate nitrogen application. The leaves must be healthy, uninjured containing abundant chlorophyll to make full utility of the nitrogen supplied to the plants.

Carbohydrate-nitrogen ratio. Low early summer temperature does not allow growth to take place to keep pace with the accumulation of photosynthates with the result that the C : N ratio may not be as low as it would have been in plants having active growth forcing utilisation of photosynthates, even with similar amount of nitrogen supply. Similarly root pruning reduces the intake of nutrients, keeps the C : N ratio higher and induces flowering. Plants, in general, have been placed in four categories with regard to the carbohydrate supply. (1) High nitrogen and low carbohydrate supply, (2) High nitrogen and enough carbohydrate as may be found in the case of young trees about to bear, or mature trees which have been severely pruned. Flowers may be formed in both the categories (1) and (2) but differentiation takes place only in the second although unfruitfulness may still be found due to incomplete differentiation. (3) Moderate nitrogen as well as carbohydrate supply leading to good growth and abundant flower-bud formation as may be expected in case of old trees which are planted under good nutritional condition and have access to good solar radiation. There is a slackness on the part of such trees to flower where these have borne fruits heavily in the immediate past. (4) The plants with low nitrogen and high carbohydrate show weak growth, leading to the formation of only few buds. Supply of a minimum quantity of both carbohydrate as well as nitrogen is of primary importance and when the minimum requirement is fulfilled the ratio comes to the forefront in the control of flowering.

Light. Light affects differentiation of floral primordia through intensity, duration, as well as quality. Higher light intensity leads to greater flowering than low intensity. Parts of the tree receiving less light have less flowers or no flower at all. Shading preceeding the pre-flowering period also reduces flowering considerably due to a disturbed C : N ratio. Low light intensity is known to prevent the development of functional anthers in strawberry where shading reduces flower formation and also the development of proper sex units.

Shortened daylength increase flower differentiation in the so called short day plants although it leads to curtailment of the hours of sunshine. Daylength influences the behaviour of fruit plants although temperature also affects to a considerable degree.

Ringing. Ringing relates to the removal of ring of bark from the trunk or branches of tree as opposed to scoring when two or more cuts are made around stem to the depth of the cambium, but no bark is removed. All the available data indicates that girdling results primarily in accumulation of carbohydrates above the ring while the change in nitrogen seems to be of much less significance.

Ringing is of minor importance to the commercial growers since it is seldom practised with mature trees, to stimulate flowerbud formation artificially. Young trees of most varieties also bear soon enough and so the practice is of consequence chiefly in the case of filler trees which are tardy in forming flowers and in hastening the bearing of seedling apple, stripping produces less objectionable wound with results nearly as satisfactory. Whenever ringing is done it must be done well in advance of the time of differentiation of flowers. Ringing results in injury to stone fruits and is not useful for plants which are weakly vegetative, very young, defoliated (top foliage) or alternate bearers.

Root pruning. Severe injury to the roots of fruits trees has frequently increased flower formation. Root pruning has long been employed by amateurs to induce flowering in young vigorous trees which are shy in forming flowers. In order to produce the desired effect the operation, which is usually quite severe, must take place a few weeks prior to the time when flower-bud differentiation is desired. Roots have also been pruned with advantage in the case of rose and fruit (guava and orange) trees where flowering is increased provided it is done at least 10-40 days ahead of the normal flowering time. In principle this practice reduces the entrance of water and nutrients, thus retarding the growth of shoots and spurs. This retardation in growth, as indicated, is associated with differentiation.

It is also known that the effects of bending branches downward so that shoot which were formerly upright become horizontal or even have their tips pointing downward may slightly increase tendency towards flower bud formation.

Temperature. Temperature changes tend to upset the C : N balance in a plant which may influence flower bud formation. Temperatures, lower than 20-25°C favour greater accumulation of carbohydrates in certain plant species and thus force large flowerbud formation. It may however, be borne in mind that there are other factors affecting this type of plant-climate relationship. The effect of two factos, *viz.*, frost and variation in yearly temperature have been studied by R.N. Singh at IARI, New Delhi *vis* a *vis* the mangoes in respect of the start and the periodic amount of bud break.

Defloration. This practice when done in a particular year gives good results in the following year by way of better flower formation. Partial defloration is not as significant. Rigorous defloration of individual branches usually induces the differentiation of flowers with little impact on the fruiting limbs of the plant. The left-over chemical energy due to the non-bearing of fruits in the previous years seems to be of avail in the improvement of flowering and fruiting in the succeeding year.

Defoliation. The leaf subtending a bud bears a direct relationship to the latter, former's removal checks the differentiation of the bud reducing the formation of flowers. Defoliation is detrimental because it brings about a significant reduction in total carbohydrates associated with flower production, if not fulfilled, brings down flowering. Use of chemicals like acetylene and ethylene favour flower differentiation and have been used successfully in pineapple. Any spray that lowers the photosynthetic rate of the plants is not good for flower differentiation.

Fertilization of the ovary makes it swell and in due course ripe fruit is available. At the time of pollination a heavy drop of blossoms and ovaries takes place specially in the spring and early-summer days. Flower and ovary drop takes place in three steps in the season.

First wave coincides with blooming period and involves flower drop. The pistil, in particular may remain underdeveloped or abnormal both morphologically and anatomically. This type of abnormality is rather more common in stone fruits. The second wave begins a weak or two after the petal fall and lasts for one or two weeks. Ovary drop in this case is due either to the lack of pollination or pollination by incompatible pollen. After another fortnight, during the period of rapid embryo development, the "June drop" commences and fertilized ovaries drop. This drop may be sizeable in some varieties and continuc until harvest. It is supposed to be due to lack of fertilization, leading to small number of seeds per fruit as well as poor nutrition. Of the three groups of apple varieties with respect to fruit drop (*i*) Immediately after flowering (*ii*) 2-3 weeks after flowering and (*iii*) almost continuously until harvest, the first is considered best since these provide the best opportunity for coming year's harvest.

Chandler explained the reasons of the drop in apples. The first wave marks the drop of the flowers along with unfertilized ovules, the second has flower drop after pollination with a little growth of ovary and endosperm. The next wave occurs when the endosperm and ovules are developing. Lastly, the drop may take place when the endosperm is fully developed and the ovules too developed considerably though not so much as in the case of undropped fruits. There are varieties which shed most of their flowers in the first drop period and then retain ovaries and fruit till harvest. Only 5-10% of blossoms develop into fruits on a full blossoming apple or pear tree with a more uniform shape, size and colour. Luckwill isolated, in 1953, four different auxins, three of which are acids and the fourth neutral compound. He concluded that growth of apple fruit is regulated by several auxins one of which regulates the growth of fruit, the other fruit-drop, while a third influences seed growth. Healthy leaves are also extremely important in preventing fruit-drop. Auxins are supposed to control fruit drop. Low auxin content favours fruit drop, exogenous application of auxins checks fruit drop but it could not be said that there is an endogenous control of auxin on abscission. Application of 2, 4, 5-trichlorophenoxyacetic acid effectively controls the fruit drop of apricot fruits. Similarly 2, 4-D and 2, 4, 5-TP are effective in controlling fruit-drop in mango.

V. ASPECTS OF FRUIT FORMATION

Fruit-set is an important event in the life of higher plants, yet not much is definitely known about the physiological mechanism of the phenomenon. Fruit-setting is dependent on a host of factor, *viz*., the presence of functional leaves, change in temperature, light intensity, day length, conditions of supply of nutrients and water, the existence of fruits or their removal from the plant. Factors affecting pollination, temperature, light intensity etc. may be taken as of importance in magnitude. Whenever the pollen be sterile or the susceptibility of the female organ be limited by shortness of duration, natural fruit-set is reduced greatly. The number of flowers which are commissioned to develop into a fruit vary widely. In macadamia nut and *litchi* not more than 1% flowers set fruit, in mango the ratio of fruits to flowers may be 1 : 1500. The sterility of pollen affected by rain or chemical treatments is also one of the factors restricting setting of fruits.

Pollination stimulates the development of the ovary and the surrounding tissues leading to the formation of the fruit. The parts of the flower which do not take part in the formation of the fruit, *viz*. petals, stamens etc. are eliminated. The changes taking place from the flower to the fruit stage is known as fruitset. An ovary receptive to the pollen is likely to set the wheel of fruit-set rolling. The receptivity of the ovary initiates before the fruiting, lasts for a few hours (mango) or more (tomato) so that it is essential that the pollen must reach the stigma within that period. It is clear that there is stimulating effect of pollen on ovarian and other tissues (Fig. 12.8c).

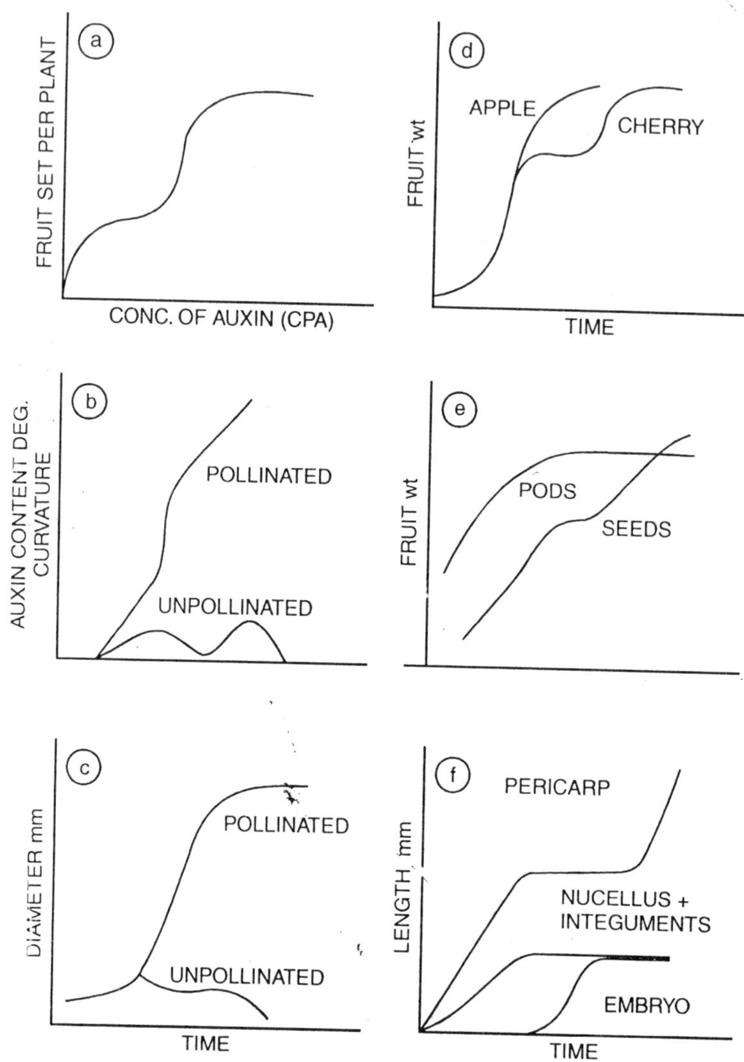

Fig. 12.8. Smothered curves showing (a) effect of p-chlorophenoxy acetic acid on the number of fruit set/plant (Mann and Minges, 1954). (b) Avena curvature response (diffusible auxin from tobacco ovaries) to pollinated and unpollinated flowers (Muir, 1942), (c) rapid increase in cucumber fruit growth following pollination and a decline in its absence (Nitsh, 1952), (d) single sigmoid and double sigmoid growth curves in apple and cherry fruits (Leopold, 1964), (e) seeds follow double growth curve, the pods single growth curve (Carr and Skene, 1961), (f) different parts of the fruit follow different growth curves; pericarp (double growth curve). With the completion of its fruit growth period that of nucellus and of embryo (Tukey, 1933).

Although abscission of flower terminates the period of female receptivity yet in the tomato the ovule remains viable even after abscission of the tomato flower which can be cultured and induced to set fruit.

Fruit-setting and fruit production largely depend on pistil and stamen though other parts of the flower may at times enter into the structure of the fruit. Pollination followed by the germination of the pollen grain and the growth of the pollen tube leading ultimately to fertilization pave the way for growth of the ovarian tissues resulting in the setting of the fruit. Tomato pollen extracts caused wilting and abscission of petals and swelling of the ovary in orchids.

Parthenocarpy encompasses the development of the fruit without pollination or fertilization as in the case of bananas, pineapples, tomatoes, figs, etc. These are usually seedless although seeds may develop therein parthenogentically. Formation of seeds and the development of the fleshy tissues of the fruit are found to be distinctly correlated ; larger fruits usually have greater number of seeds. Successful pollination followed by the germination of the pollen induces ovary growth. The growth of the ovary increases in fleshy fruits even before the lapse of sufficient time for the growth of pollen tubes into the ovules with stimulus and ovary growth (Fig. 12.8c) occurs. In plants where a perceptible rise in ovary growth is seen after pollination, unpollinated flowers usually have smaller ovary. Ovary may even remain static in growth, until pollination takes place, in some species.

The presence of large amounts of auxin in pollen of some species of plants has been recorded. It has been seen that the density of the pollen on the stigma has a telling effect on fruit-set, seed number and fruit size.

It is believed that the auxin stimulus leading to fruit set may not only be from the pollen but also from the ovary. Pollination stimulates auxin formation in the ovary (Fig. 12.8*b*). Unpollinated ovaries have only small amounts of auxin. Auxin production is first maximum at the tip and with lapse of time the style tip and still later the ovary base is the site of maximum auxin (IAA and Tryptophan) synthesis. Though fruit-set in tomato, egg plant, tobacco, okra, fig, cucumber and melon are brought about by auxins almost four-fifths of horticultural species do not set fruits with auxins. The post pollination increase in auxin content occurs 2 or 3 weeks after fruit-set in corn, apple, currant and possibly these contain growth regulators which may be responsible for fruit-set in place of auxins. Of the auxins β-napthoxyacetic acid (2-NOA) and p-chlorophenoxyacetic acid (CPA) affect fruit setting (Fig. 12.8 *a*). It has also been reported that plants grown in suboptimal conditions respond best to auxin application in the early onset of fruit growth as well as increased size of the fruit.

Factors like low night temperature are complementary to auxin action in the production of larger fruits as also increasing the period of fruit production and earliness. Although there is not much evidence in favour of the gibberellins contributing to endogenous control of fruit-set these are known to be effective in almond, apricot, peach, grapes, pears, apples, etc. specially the species which do not give response with auxins. Newly set apple fruits and parthenocarpic banana ovaries contain some amounts of kinins but its exogenous application has not yet been demonstrated to be effective in fruit-set. Fruit development in the absence of pollination may be brought about under the influence of auxins which have been found to exist in the endocarp placenta and flesh of the tomato fruit. Parthenocarpically developed tomato fruits are seedless. The time of the application of axin and also the place of application determine the effect on the development of seed coat.

Many a times seedless fruits may be exclusively parthenocarpic and at the same time several seeded fruits are not incapable of parthenocarpy. High auxin content associated with natural parthenocarpic fruit development are reported from some citrus varieties and other horticultural species. Parthenocarpic fruits have been produced artificially by the use of synthetic auxin like growth substances. Except the embryo which may be empty or filled with undifferentiated cells every kind of tissue irrespective of origin or composition develops in parthenocarpic fruits in a way exactly similar to the development of several fruits following pollination, *viz., Ananas sativus* (swollen receptable), *Ficus carica, prunus, Malus* etc. (receptacles). Receptacle of *Ficus carica* enlarge without the development of the ovary wall when treated with auxin. Cell enlargement and cell differentiation lead to the maturity of the fruit following fertilization that provides required stimulation to the ovary and the surrrounding tissue to continue growth. Growth substances have been used to increase the size of fruits. Exogenous application of 2, 4, 5-T increased the size of *calimyurna* figs, strawberry, seedless grapes, oranges, apricots, pineapple, etc. Low supply of endogenous auxin when supplemented by the synthetic auxins enhances the rate of fruit growth significantly. With agriculture becoming more intensified and specialized the importance of parthenocarpic fruits increases. The flower is believed to contain a small quantity of auxin in the female parts and after fertilization the developing seed becomes the centre of auxin production serving as a controlling factor for the growth of the fruit.

Parthenocarpy has been recognised as setting and growth of fruits without a developing embryo ; vegetative parthenocarpy is parthenocarpy without pollination or any other external stimulus ; stimulative parthenocarpy is that induced by some stimulus (pollination, auxin, low temperature, etc). Natural parthenocarpy represents natural seedlessness and should not be identified with parthenocarpy at all. Parthenocarpy may be principally of three kinds (1) Fruit development without pollination as in tomato, pumpkin, cucumber, citrus, banana, pineapple etc. (2) Full development of the fruit following pollination even when the pollen tube does not reach the ovule and affect syngamy as in orchids and *poa* and fruits of triploid plants (genetically sterile). (3) Abortion of the embryo before the fruit reaches maturity, leading to seedlessness of which chillies, peaches and grapes are good examples.

Intermediate parthenocarpy as in Ginkgo, occurs where pollination stimulated fruit development takes place without syngamy. The pollen tube in this case resumes growth after the fall of the ripe fruit when syngamy takes place. In tomatoes parthenocarpy induced with low temperature and high light intensity conditions while in cucumbers induced parthenocarpy under short photoperiods and low night temperatures have been reported. Among grapes and citrus fruits the seedless varieties show appreciably higher auxin content than those with seeds. According to Luckwill parthenocarpy sometimes may represent a state of auxin autotrophy in the ovaries.

The effectiveness of gibberellines in parthenocarpy suggests that yet another component may be involved. Role of auxins in parthenocarpy is considered to be largely thorough activation of cell enlargement. In case of apples and pears auxin helps fruit-set. Auxin treatment of *calimyurna* fig proved completely effective in causing fruit-set and development of seedless fruits. The pollination of the fruit is not necessary in such cases.

Auxins also bring about abortion of the developing (already fertilized) embryos leading to its use in fruit thinning. Abortion of pollen grains with dichloroisobutyric acid brings down the chances of the setting of fruits. Cresols are able to destroy the flowers and thus minimise fruit-set considerably. At times, the thinning of fruit is practised to improve the size and quality of the produce besides making the alternate bearers regular. In addition to napthalene acetamide (used for thinning) chemicals like napthalene acetic acid (for inducing regular bearing), dinitro-o-cresol have been used for apples and peaches. Malic auxin II a type of regulator produced by seed as antidrop hormone normally does not appear when the plants are sprayed with auxin.

Growth and rest periods in plants. Generally every perennial plant has two periods in a year ; active vegetative growth in which growth and development of the shoots and roots takes place accompanied with blooming and also fruiting, followed by that of rest which is supposed to be an adaptation to tide over a adverse conditions (enforced or organic rest).

The temperature of 20°C is considered adverse for the temperate fruits. Apple, pears, sour cherries etc. need low winter temperatures as a condition to allow the flower initials in buds resume growth. The period of rest undergone by a plant also gives it the necessary strength to survive frost conditions. Rest may be known as enforced or organic and deep rest. During rest stored substances undergo transformation (starch, sugar and oil content declining to the minimum, respiration is at low ebb). The maximum amount of nutrient substances generally present in the trees soon after the shedding of leaves and the termination of root growth (in autumn) declines somewhat during the winters, and significantly during blooming period and still further after vigorous tree growth in spring.

Fruit-bud formation. The views of the formation of specific flower-forming substances stimulating the formation of flower-buds, the accumulation of organic substances and a definite concentration of sugar in the cell sap (C/N balance) or a weak concentration of organic substances elaborated by the leaves in the spring or summer guarantees better fruit-bud formation have received attention at the hands of horticulturists. The formation of fruit-bud is rather closely linked with the total leaf surface of the tree in relation to the amount of growth of the wood and the number of fruitlets formed and not the accumulation of starch. C/N balance hypothesis is not justified and the reserves of carbohydrate to nitrogen is probably of greater value than the balance. The nature of nitorgen transformation that takes place in the plant are of greater importance in fruit-bud formation. Heavier, larger and longer spurs prove better for forming fruit-buds.

The protein nitrogen contents of fruit-forming structures seemed to be related to fruit-bud formation. It is believed that in case of deficiency of carbohydrate, nitrogen remains in organic (NH_3 and NO_3) forms instead of organic forms that are formed in the sufficiency of carbohydrates. Abundant flowering, according to Ursulenko, through an excessive expenditure of carbohydrates halts the formation of proteins. If there is too much accumulation of carbohydrates and also protein too many fruit buds are formed which fall off and the tree is also made functionally important to form a fruit even in the 'on year' following the 'off year'. The factors controlling fruit-bud formation are extremely delicate including the part played by the shoot as well as the root systems.

The florigen concept gained sufficient importance after the work of Chailakhyan. Longday plants accelerated flowering only in excess of carbohydrates and the shortday plants in excess of nitrogen. Hormones affected many aspects of plant growth including the development of fruit so that the formation of fruit-buds is the result of combined activity of all the conditions, internal as well as external to the plant. Annual formation of fruit-buds depend upon correct nutrition, sunlight, moisture and proper cultural practices. The root system, its activity and functioning is also of great significance.

Alternate bearing of fruit trees is dependent on climatic conditions and on the favourableness or otherwise of environmental conditions. High fruit production usually accompanies good conditions of light, temperature, moisture and nutrition. Timely application of NPK in adequate ratio favours fruit formation provided the soil has moderate soil moisture during the 1-3 months of the post-flowering critical period. In fact any lowering of the nutritive content below the critical requirement affects the production activities of the plant adversely.

Heavy supply of nutrition to the medium of growth or timely foliar supplement to the fruit plants is of consequence in defining the vegetative/reproductive destiny of plants and also the annual production of fruits. Adequate ratio of the elements supplied to the plants as well as properly placed irrigation coinciding with the critical periods of water requirement should be helpful in maintaining the yearly productivity of fruits by the formation and growth of the fruit buds. There is a critical period, after flowering in the case of apples and if moderate soil moisture is made available to the plants best fruit-bud formation and development takes place. The cessation of vegetative growth seems to be an important factor affecting fruit-bud formation, differentiation and development which need a freedom from moisture shortage.

The concentration of nutritive substances available to the roots, their ability to utilize it, and the final nature of the metabolites formed seem to be the controlling factors for the growth and development of the buds. The rest period is an adaptive reaction of plants to the unfavourable external conditions. The transition from growth to rest in fruit plants is a highly important ability acquired in the course of evolution. The supply of nutrients and of water to the plants at the time of the active growth period of the plants, soon after the harvest (or depletion of nutrients in the form of fruits), is of great consequence in forcing the fruit buds active.

Fruit production is affected by the extent of pruning, the vigour and the proportion of vegetative to reproductive growth of the plant subjected to pruning. Besides these, another important factor is the distribution of living tissues in the root system. Hard pruning during the dormancy reduces the tendency leading to regular fruit bearing in place of alternate bearing. Orchard care could be able to introduce and guarantee regular bearing in the case of plums (as in Romania). Some trees have their individual characteristics which could hardly be altered permanently and economically. Artificial thinning of the flower and fruits in the 'on' year to conserve energy and reserves for the 'off' years is also a practice which has received support and gives dividends by way regular bearing. Chemical thinning of flowers and fruitlets may be practised to ensure annual fruit bearing. It must be borne in mind that the chemicals used for thinning of fruits must not be a poison for the bees and other insects helpful for pollination and cross fertilization.

Premature fruit drop caused by the lowering of the auxin concentration should be controlled to get larger production. Details on these questions have been furnished at the end dealing with growth substances that may be employed to regulate the physiology of fruit trees.

VI. GROWTH OF THE FRUIT

Enlargement of the ovary and of the receptacular tissues generally lead to nonpolar or marginally polar type of growth of fruits.

Rate of fruit growth. The enlargement of the ovarian tissues stimulated usually by pollination, and the development of embryo and endosperm are the two recognised types of growth of fleshy fruits. The development of the embryo and endosperm of fruit is of greater significance in the later stages of growth of fruits as opposed to the enlargement of the ovarian tissues which predominates in the early stages. Union of one of the germinal nuclei from the pollen tube with the polar nuclei (uncertain numbers) of embryosac leads to the formation of the endosperm. The embryo is formed by the union of the other pollen nucleus with the egg nucleus. All this takes place as a result of fertilization. Diploid tissues from the mother plant, the endosperm and diploid tissue of new embryo compose the fruit. The strawberry is composed largely of parental tissue, the small grains of endosperm tissue, while the composite fruits of embryo tissue.

Apple, pineapple, strawberry, pea, tomato, etc. show simple sigmoid type of growth pattern as in majority of cells, tissues and organs. There is an exponential increase in size to begin with declining as in the case of sigmoid growth. Some stone fruits (peach, apricot, plum, cherry) and non-stony fruits (fig, grape and currant) have two periods of increased growth with a period of slow or suspended growth in between. There is little distinction of the growth curves for the different morphological types of fruits, the there are berries, pomes and simple and accessory fruits in each type. Double growth curves (Fig. 12.8*d*) signify the existence of at least two physiological signals existing during fruit development.

The double growth curve may express the sum total of fruit growth while the separate growth curves may signify the growth of embryo, endosperm etc. The bean pod show growth curve different from that of the seed (Fig. 12.8*e*). It is so because the development of the pod takes place first and that of the seed later. The growth curve for different parts of the peach fruit is different (Fig. 12.8*f*). How the growth of the embryo, nucellus, integuments and the pericarp tissue are related is seen in the figure.

The end of the first growth period coincides approximately with the time of completion of nuclear growth and commencement of the growth of the embryo. This classic example of the existence of separate periods of growth of different parts of the same fruit suggests that fruit growth is reflected by cell enlargement as well as cell division activity, the participation of latter varies widely ; from the time of pollination as in tomato cucurbits, apple, *prunus, Citrus* etc. or to a period much after pollination (strawberry). At the beginning of the growth of the fruit both cell division and cell enlargement seem to be associated with growth.

TRANSLOCATION OF SUBSTANCES TO THE GROWING FRUIT

It is now recognized that the translocation of substances from the leaves to the fruits is not regulated by concentration gradient. Since if it were so the fruit tissue should not have been able to store large quantities of fats, carbohydrates, etc. It is safer to assume that the enzyme activities are largely responsible for such accumulations against the concentration gradient. The manner by which the enzyme activities are controlled is not the subject matter of discussion here.

Dry weight of leaves declines in the case of wheat as the dry matter accumulation in the grains (principally the endosperm) increased, this is later mobilized, by the embryo during germination. Application of kinetin may bring about a mobilization of the nutrients in a manner suggestive of the mobilization of flowers and fruits. It is not unlikely that the growth regulators might be in some way involved with mobilization.

As the fruit grows it needs materials for fulfilling the requirements which must necessarily be supplied by the mother plant. The process by which the transfer of substances takes place is not very well understood. The material synthesized by the leaves are translocated to the fruit which also serve as storage depots of the plant. It has been observed by several workers that the activities of mobilization change in the fruit during its development. Mobilization also takes place from one part of the fruit to another. The fruit may have nutrients for its growth from the nearest leaves or other photosynthetic centers. In the case of wheat (awned variety) supply accounting for almost 10% or the dry weight of the fruit is from the awns, 30% from the rest of the spike and the remaining should naturally come from the leaves. Whenever there is a limiting competition set up by the setting of a large number of fruits the fruit size naturally declines.

Fruit size. The fruit size seems to be related to the cell size specially in the case of fleshy fruits or fleshy parts of the fruit. In some fruits (apple) the number of cells is correlated with the size of the fruit. The development of the intercellular spaces also contributes to the fruit size. All the fruits do not have the same amount of increase in weight during the same period of growth. For instance in the grape there is greater increase in weight than in volume during the later period of fruit growth, it is not so in apple. There is obviously enhancement of the deposition in the apple which has about 20% air spaces.

Auxins and gibberellins increase the growth of fruits either by prolonging the growth span or the diminution of the period of growth. Besides, the time or ripening may be shifted by the application of auxin and growth to a larger fruit be possible. Early application of auxin, to the fruit (plum) increases its size and date of maturity. Application of auxin, late in the life of the fruit, hastens maturity and gives fruits of smaller size. Fruit growth therefore must not be considered as a single event but includes several types of adjustments in cell division, cell elongation, mobility of substances to the fruit and the auxin level inside the fruits.

The process including pollination, fertilization, seed setting, number of seeds formed and the amount of seed material all influence the size of the fruit through the control of development. The effects of the seed can be promotive as well as inhibitive. The influence is generally supposed to be associated with the formation of growth regulating substances so much so that seeds have been taken as growth stimulating signals in the fruits. How do the seedless fruits carry on the functions of the seeds is not clear but it is assumed that some other parts of the fruit may take up the functions of the seeds including hormonal action.

In the developing fruits, the presence of tryptophan has been reported in large quantities and the seed possibly supplied the factor responsible to convert it into auxin. This endogenous auxin is responsible for the growth of fruits. More auxin is produced in the younger fruits and the quantity declined with the age. The time of rapid growth of the fruit synchronised with the maximum amount of auxin present therein. Strawberry, with a normal sigmoid growth curve, showed three different auxins : (*i*) the neutral auxin associated with the first sigmoid phase of growth (*ii*) the acidic auxin more related to the second phase of growth and (*iii*) an auxin concerned more with abscission of the fruit than its growth. These facts justify the presence of double peak of auxin generation corresponding to the double growth curve. At times double sigmoid growth curve did not show any evidence of the presence of more than one auxin increase (grape and fig). Does it mean that only the first growth curve is auxin controlled and not the second ? In instances like that of peach the auxin rise is associated even with the period of suspended growth.

With all this confusion the suggestion came that gibberellin whose presence is detected in first stages of the growth of seedless varieties of grapes stimulated growth of fruits. The gibberellin content of developing fruits fitted very well with the several conditions of double sigmoid growth as well as in case of seedless fruits. With all the available evidences, one is led to believe that some of the growth regulating substances occurring in the fruit control its growth. Although hormones have been detected in apple, banana, tomato no role has been specified.

Delayed harvest is often associated with fruit fall which may account for serious loss as in the case of apple, pear etc. Preharvest drop may be controlled by spray with growth regulators, *viz*. naphthalene acetic acid, and 2, 4, 5-trichlorophenoxy acetic acid. In addition to the check on fruit drop these hormones may hasten the ripening of early-ripening varieties and improve the colour of apples. The check on fruit drop is attained by sprays of different growth regulators in different concentrations depending on the fruit and the time of application. Early ripening can be avoided by spray for the check of respiration rate, lowers fruit drop. Retardants affect the yield performance to a lesser degree.

Auxin application can control drop of fruits arising from self pollinated flowers known to be more susceptible to drop. Phenoxy acetic acid *viz*. 2, 4, 5-trichlorophenoxy acetic acid is known to check fruit drop. Without discussing the merit of the sprays reported for preharvest fruit drop it may be stated that 2, 4-dichlorophenoxy acetic acid or 2, 4, 5, trichlorophenoxy acetic acid as mix emulsion increase the storage life of the fruits.

VII. MATURITY AND POST-HARVEST

The value of post harvest care cannot be under-rated. Proper understanding of the post-harvest life of plant products including fruits is of immense value as the losses running between 5-50% can be mitigated through adequate care and handling. Production-fall and production losses both hit the orchardists economically. High production brought about by the balancing of all hereditary, cultural, edaphic and biotic factors and above all providing ideal marketing facilities are nullified by inadequate or faulty storage. Post-harvest care is needed to safeguard against the injurious effects of the physiological and biochemical changes taking place after the detachment of the fruit from the mother plant. The period immediately following the detachment of the fruit from the mother plant is termed post-harvest meaning after-harvest.

The fruit grows and attains a size after which ripening is initiated. This is the stage known as maturity which bears a distinct relationship with the quality of the fruit. Proper stage of maturity at picking ensures better quality of fruits. Maturity is judged by age, shape, size, colour, appearance, texture, specific gravity, pectins, starch, sugar : acid ratio etc. of the fruit. Generally mature fruits have maximum starch content and highest specific gravity. The orchardist is normally guided by the initiation of falling of ripe fruits to the ground of itself for harvesting.

Presence of high amounts of water in the fruit makes it more prone to deterioration and spoilage. Harvest at proper stage of maturity, regulation of the biochemical changes in the fruit during ripening and protection from microbial infection is the pivot on which the post-harvest life of fruits depends. Matuirty is associated with lenticels and maximum care should be taken to look for the stable characteristics of the cultivar, at maturity. *Alphonso* and *Paire* mangoes, for instance, take 110-125 days, after fruit set, for change in external colour of the rind and also of the colour of the flesh inside denoting maturity. Late harvest lowers the shelf-life of the fruits while early harvest results in poor quality and uneven ripening. Both the conditions lead to losses to the grower.

In pineapple the beginning of ripening is indicated by rapid loss of chlorophyll of the shell. Increase in brix and titratable acidity serve as an index of maturity in pineapple. The pH of the shell declines with maturity. A decline in total pectin and water solution pectic substances is associated with the maturity of orange, which is also characterised by the complete absence of starch. The angularity of individual banana fruits affords a good index of maturity.

Respiration plays a vital role in the changes that take place in the fruit after harvest through biological oxidation of glucose and the consequent production of carbondioxide, water and energy liberated as heat. The process of respiration may be considered to work in two different phases. In the initial phase pyruvic acid is produced from glucose through a chain of reactions, (Fig. 8.29) the glycolytic cycle, representing the anaerobic form of respiration. In the next phase pyruvic acid is converted to carbondioxide and water. This is known as the aerobic form of respiration. The series of events of this phase (Fig. 8.29) are known as Kreb's cycle, tricarboxylic acid cycle or the citric acid cycle.

Respiratory quotient (the CO_2 given out/the O_2 taken in) gives a clear indication of the substrate being utilised in respiration. It is unity or equal (one) when glucose is the substrate, the RQ is less than one when fats and fatty substances are the substrate. In the cases where RQ is more than one the substrate is organic acid. Thus the measure of RQ of a fruit gives a definite clue to the form of plant product being utilised in respiration thereby to the stage of maturity of the fruit.

Lowering of the rate of respiration increases the post-harvest life of the plant. The cycle of metabolic interactions twined with respiration (hence the life itself) is depicted in fig. 8.29 which shows the inter-relationships between the different organic constituents of the plant body or plant cell.

VIII. RIPENING OF THE FRUIT

Fruit development gets impetus from pollination leading to fertilisation and the development of the seed. The germinating pollen and the developing seed provide auxin necessary for fruit development. It is reported that the extraction of the seed from the fruit retarded its growth. Exclusion of pollination and exogenous application of gibberellic acid encourages parthenocarpic development of fruit. Fruit development entails cell division and cell enlargement after which the cytoplasm of the cell shifts to the periphery and the vacuoles thus created are filled with sugars. Thereafter, with increase in the age of the fruit, starch content increases. Mature fruit has the maximum amount of starch in it. Ripening is initiated only after the fruit has attained maturity. Abundance of carbohydrate in the fruit acts as substrate for respiration. Biosynthesis of fats, pectins, cellulose, organic acids etc. also takes place in the developing fruit. The level of carbohydrates, organic acid etc. register a change with ripening of the fruits. The changes are not the same in all fruits even of the same genus *e.g.* oranges show a decrease in organic acid and increase in sugars with maturity. In lemon however, the condition is unlike that in oranges.

Maturation and ripening are distinguishable; the former corresponds to the attainment of full size and the latter refers to the quantitative transformation of the mature fruit. Fruit ripening is one of the developmental phases of its life cycle. It is the natural irreversible process of ageing and has been called as "fruit ageing".

The fruit undergoes a period of ripening soon after the completion of its growth. Ripening includes softening of the fleshy fruit with associated changes in pigmentation and flavour. The non-fleshy fruits may not apparently experience ripening. Ripening in some cases may be stimulated by picking of the fruit (Avocado, pear) while picking hastens the process (banana, apple, papaya).

Changes with Ripening

Growth substances often control fruit drop in apples, pears and peaches. They also hasten maturity. Ripening of some detached fruits can also be successfully controlled by growth regulators under suitable conditions. Not much is understood about the biochemistry of these effects and their relationships to naturally occurring growth hormones and to ethylene.

The loss of chlorophyll and appearance of caratenoids is one of the visible changes in fruits during ripening. The fruit becomes soft, as it ripens due to the breakdown of insoluble protopectins and formation of soluble proteins. There is increase in sugar forms and in volatile compounds adding to the flavour of the fruit.

Ripening of fruits involves physical and chemical changes like decrease in total carbohydrates content, formation of sugars from polysaccharides, decrease in acid content. Nitrogen status remains almost the same although a very small increase in the protein content takes place. Colour changes occur with the alterations in pigments including chlorophyll. Carbon dioxide and ethylene production increases, the latter is believed to be a ripening hormone. Generally ripening involves increased activities of malic enzyme and pyruvic carboxylase leading to climacteric rise in carbon dioxide production that takes place in all the fruits showing climacteric.

Fruit ripening is a process well connected with the condition of the fruit at the time of detachment from the plant, the state of the plant and also the earlier stages undergone by the fruit before the ripening phase. Basically ripening may be taken as the stage in the development of the tissue and the organism under the direction of factors affecting other stages. Ethylene possibly acts by its action on nucleic acids and protein biosynthesis. Ethylene given off by the fruit itself may be a by-product of the ripening process without doing anything in ripening.

Softening of the fruit flesh, hydrolytic conversion of storage material in the fruit and changes in the pigments' aroma, consistency and flavour associated with ripening have been attributed to the energy provided by respiration. There are a number of evidences in favour of this concept. Respiratory inhibitors prevent ripening. The existence of a linear correlation between ripening processes and metabolic activities and the essentiality of respiratory energy for ripening changes is well settled.

Changes in pectic materials, cementing cell wall, hydrolysis of starches (squash) or fats (avocado) are associated with softening of fleshy fruits during ripening. The softening of apples during the ripening is ordinarily brought about by solubilization of pectic substances from the middle lamellae accompanied with rise in soluble pectins. Making the pectic substances in fruit soluble is possible by polygalacturonidases. The losening of cell cements of a ripe apple gives a sense of tastelessness as if the flavour had vanished with advanced ripening. Tastelessness may result due to the extreme case of cellular separation leading to a slippage of the cells during mastication. Consequently cells remain unruptured and thus not releasing the flavour.

Hydrolytic changes. Hydrolytic changes during ripening usually lead to the formation of sugars. Fruits vary in hydrolytic activity in rate as well as extent.

Ripening usually accompanied by hydrolytic changes, leads to the formation of sugars, the banana ripens quickest, apple slower and citrus fruits (orange, lemon) the slowest extending over months. During this activity starch, fat etc. may give rise to sugars. Pectic substances may be hydrolysed forming soluble pectins causing softening of the fruits. Proteins are synthesized at increasing rates, during the ripening of apples.

During ripening of fruits marked changes occur in pigmentation (decline in chlorophyll content and increase of carotenoids) production of flavours and usually depletion of astringent substances. In banana, loss of chlorophyll from fruits accompanied by formation of carotenoid pigments may be synchronous with ripening. In the oranges, it may occur only in the earliest stages of ripening. Sometimes carotenoids increase with process of ripening. It is also possible that the fruit may have the specific pigments *viz.* carotenes in papaya and anthocyanins in strawberry. The pigments may be restricted to the outermost skin (apple) or the entire fleshy part within (peach). Anthocyanins may be formed in response to sunlight. Accumulation of sugars may accentuate pigmentation. Pigmentation may be affected by temperature in a manner quite distinct from the ripening changes indicating that colour changes may not be proportional to fruit ripening. With ripening the aromatic esters, alcohols, carboxyl compounds which are basically responsible for flavour become pronounced. Amyl acetate is responsible for flavour in banana and terpenoids and cumarins are responsible for the smell in citrus fruits. Many esters, aldehydes, ketones, and a series of saturated and unsaturated hydrocarbons have been identified in apples. It has been observed that light inhibits the formation of some flavour substances in apples apparently due to the diversion of the necessary substrates for the formation of pigments.

RESPIRATORY CLIMACTERIC

During ripening the rate of respiration increases—this sudden upsurge in respiration is known as climacteric rise which is followed by deterioration and senescence. Fruits like banana, guava, mango, fig, chiku, jackfruit, papaya etc. show climacteric rise. Certain other type of fruits *viz. litchi*, lemon, lime, orange, grape, pomegranate, pineapple are non-climacteric in nature where there appears to be a single gradual decline in respiration. Ripening has been known to be associated with well noted changes in the rate of respiration, lowering in the case of the mature fruit followed by a significant rise. Typical climacteric fruits include apple, avocado, bananas, guava, jackfruit, papaya and peas (Fig. 12.9*a*). The fruits enter the senescent stage after it has attained a climacteric peak and the respiration rate is lowered. Refrigeration prolongs the storage life mainly by suppression of the intensity of respiratory climacteric and also with increase in its duration. The climacteric peak synchronises with the time of optimal eating quality of pears. In the case of banana and apple it slightly preceeds the optimal eating quality of pears. In tomatoes the climacteric is achieved much before the fruit is fully ripe. Association of climacteric with the hydrolysis of reserves in the fruit, has been suggested, for fruits like lemon and fig, while cherries and strawberries generally do not ordinarily have either a climacteric after harvest or any sizable hydrolytic activities. Climacteric fruit contain carotenoids and nonclimacteric fruits contain anthocyanins. Avocado and banana show rapid climacteric while apple, pear and mango are intermediate. A distinct respiratory decrease in the mature fruit preceeds a rise in climacteric peak after which again a drop is registered (Fig. 12.9*a*). There are plants (peanut) that show a decline in respiration during the ripening period. A decline in temperature reduces the chances of reaching the climacteric and this is perhaps the basic secret of the refrigeration industry. With the climateric, free fructose disappears from cytoplasm, possibly due to phosphorylation. A change in the permeability of the tonoplast permits fructose movement from the vacuole to the cytoplasm leading to the increase in respiration. The rate of respiration is also regulated by the concentration of phosphate receptors (*viz.* ADP). Whenever the ATP : ADP balance is high the respiration declines specially when the energy requirement of the system is at low ebb. When the energy requirement is high, with the initiation of the ripening phase, ATP is broken down to ADP and inorganic phosphates so that the rate of respiration increases. 2, 4-dinitrophenol inhibits ripening by its uncoupling action of oxidative phosphorylation. This is true in tomatoes but not in avocado.

Storage of fruits is affected by nitrogen in atmosphere, low oxygen tension or control of the concentration of carbon dioxide and oxygen both at the same time mostly because these factors could prevent the climacteric rise. Similarly rise in carbon dioxide and oxygen-deficiency could be controlled by storage of fruits in polythene.

Ripening mechanism. Exposing the fruits after picking to certain gases and conditions and providing them with certain metabolites helps to initiate the process of ripening. Picking of fruits initiates the ripening process (avocado and some pome fruits). Thus the possibilites are that there is a trigger in the fruit itself which is able to operate when it is detached from the tree and which may otherwise be serving as an inhibitor to ripening. Unsaturated hydrocarbons (ethylene) can stimulate the ripening process. Ethylene has a direct bearing on the rate and extent of ripening. Peaches and apricots, that respond poorly, while oranges and lemons which do not normally experience a climacteric during ripening are easily forced to such a change through the use of ethylene. The discovery that the climacteric in apple could be stimulated by the presence of exogenous malate suggested that the addition of certain distinctive respiratory substrates could aid ripening. Additions of methanol to persimmon fruit produces a similar ripening response. All this leads one to conclude that there exists a chemical regulation of ripening by metabolic transformations.

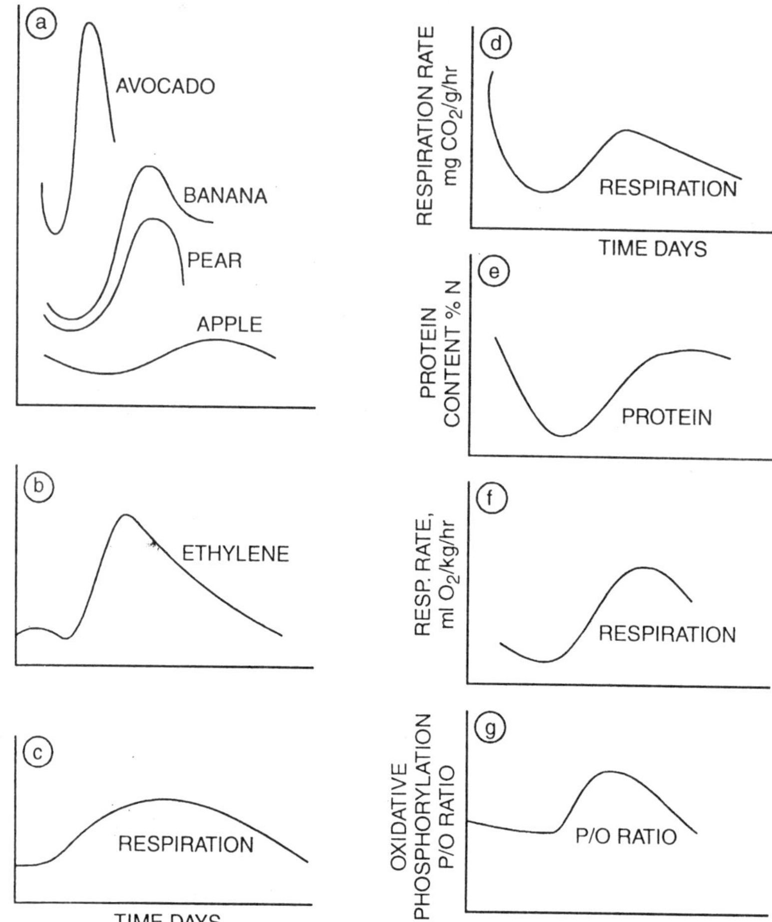

Fig. 12.9. Smothered curve showing (a) the relationship between the rate of ripening and the climacteric peak (Biale, 1950) ; (b) and (c) relationship between the climacteric rise and evolution of ethylene under storage (Hauson, 1942); (d) and (e) rise in respiration associated with rise in protein content after the preclimacteric dip in Brambley's seedling apple (Hulme, 1954); (f) and (g) Avocado fruits showing respiratory climacteric and oxidative phosphorylation (as indicated by rise in P/O ratio ; (Romani and Biale, 1957).

Ripening may be taken as a type of ageing. In some fruits like apple, banana rich in caratenoids it is associated with a steep rise in respiration known as climacteric. Fig and cherry do not have a climacteric and are rich in anthocyanin. The onset of climateric marks the disappearance of fructose from the cytoplasm as a result of phosphorylation. More fructose movement from the vacuole takes place and the respiration rate increases once again due to low ATP/ADP ratio. It is suggested that a high energy requirement of the ripening fruit causes the respiration to increase.

The early part of climacteric is associated with an increase in the synthesis of RNA and mRNA and also the appearance of the enzyme necessary for ripening.

The synthesis of new RNA is considered to be essential for the ripening process and also increase in production of protein. Chlorophyllase causes breakdown of chlorophyll and lipase of free fatty acids. Hydrolytic enzymes change starch into sugar. Other enzymes *viz.* lipoxidase and peroxidase are involved in the biosynthesis of ethylene which possibly regulates the translation of mRNA that initiates ripening. Placing ripe fruits along with unripe fruits under a cover hastens the ripening of latter. Ethylene brings about climacteric even in fruits which are non-climacteric by nature.

In the case of apple fruit it has been reported that with the rise of the climacteric, fructose begins to disappear from the cytoplasm of the cells and the respiratory rise continues as long as fructose is being made available from the vacuoles to the cytoplasm. This observation led to the conclusion that the climateric is a consequence of metabolism of an "activated" fructose. It has been suggested that such a role could be taken up by phosphorylated fructose. All this yielded strength to the concept that substrate transformations are the fundamental steps regulating the mechanism for climacteric rise.

Fruit ripening constitutes complex physiological events. Wolfgang Holl, in 1977, suggested the ripening mechanism (Fig. 12.10). It is of importance in the sense that the relationship between ethylene and ABA remains obscure. Ethylene treatment raises the level of ABA which may initiate ripening by stimulating ethylene production.

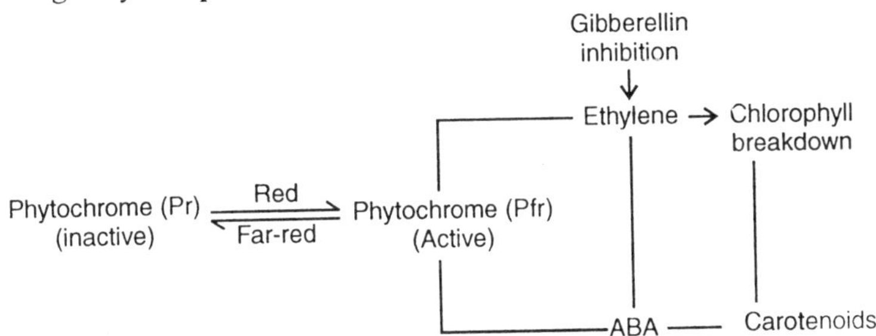

12.10. Changes associated with ripening.

During the course of ripening of tomatoes colour changes from green to white, yellow-orange or red-orange and finally red caused by the formation of chromoplasts from chlorophyll *via* carotenoids. Chromoplasts are formed by the disruption of grana thylakoids. In fruits like apples, berries, cherries, there is a synthesis of anthocyanins during the process of ripening, which accumulate in vacuoles. Cartenoid as well as anthocyanin formation is controlled by phytochrome system through its action.

In fruits, RNA synthesis has been generally associated with their maturation; there is a rise with beginning of the climacteric and a fall at the height of the climacteric. The RNA synthesis has exhibited a rise in protein content. Many hydrolytic enzymes show increased activity during ripening and the starch is degraded to sugars adding sweetness to the fruits. Sometimes the lipase activity increases the free fatty acid content. Enzymes undergo qualitative changes during the process of ripening and also have a role in the production of ethylene.

Wardlaw and Leonard, working with tropical fruits, noted that as the size of the fruit increased certain physical limitations to free exchange of carbon dioxide and oxygen developed and as these limitations became acute, carbon dioxide accumulated inside the fruit and oxygen level declined considerably. This led to the belief that the respiratory climacteric may be anaerobic respiration. Accumulations of carbon dioxide and depletion of oxygen did occur during the climacteric. Some fruits actually ripened from the centre outward so that climacteric rise may be taken to be a case of anaerobiosis positively influenced by temperature. This theory lacks support on two counts firstly anaerobiosis, inside the fruit, does not preceed the climacteric (rather accompanies), and secondly anaerobic conditions actually prevent rather than promote ripening as experimentally demonstrated by Blackman and Parija. Dinitrophenol, an uncoupling agent, increases respiration of ripening apple fruits by causing an increase in the availability of phosphate acceptors at the expense of phosphate esters. It was so, provided the dinitrophenol was applied in the preclimacteric state of fruits and not in climacteric or post-climacteric ones. Ripening in avocado fruit behaved in a manner observed in the case of apple under the influence of dinitrophenol. It was further gathered that extracts of climacteric fruits caused respiratory uncoupling in mung bean mitochondrial respirations and it was suggested that the respiratory rise was due to the accumulation of naturally occurring uncoupling substances which obstruct the effectiveness of oxidative phosphorylation.

These experimental evidences indicate that climacteric rise causes a deficit in available energy, a fact which is not supported by the continued increase in protein synthesis during ripening of apple fruits (Fig. 12.9 *d, e*). Further the uncoupling agent like dinitrophenol inhibited the ripening of tomato. Significantly enough it was later shown that the efficiency of phosphorylation increased with ripening instead of the decrease as envisaged earlier. The respiratory climacteric in the avocado fruit leads (Fig. 12.9 *f, g*) to an increase in oxidative phosphorylation (rise in P/O ratio).

Fruit ripening, in view of the modern concept, is a positive constructive and endothermic metabolic activity.

Ethylene's role. Ethylene, the 'ripening hormone' is considered to be a must for ripening probably through the initiation of the climacteric. Ethylene is not a hormone in the strict sense of the word. It is produced in high concentration during the preclimacteric stage to induce the process of ripening.

Ethylene has been used exogenously to induce ripening for over eight decades now. Ethrel, an ethylene liberating compound, has been used for hastening the process of ripening in mango, banana, citrus, etc. In mango ethrel induced ripening is accompanied by attractive colour and lowering of acidity. Ethylene treatment @1.0 ppm accelerated the onset of climacteric in banana. Ethephon treated citrus showed a decrease in chlorophyll and development of yellow colour by inducing caratenoid synthesis. Calcium carbide is also used for obtaining attractive colour, better palatability as well as improved quality of fruits. Early ripening has commercial value besides reducing the chances of losses due to incidence of diseases during storage over long periods.

Methionine an aminoacid is the immediate precursor of ethylene in the fruit.

Mature but unripe fruits ripe under the influence of 0.1% ethylene applied exogenously. Sometimes the concentration of ethylene as low as 0.02% is considered sufficent to induce ripening of fruit.

Some fruit may be arranged in decreasing order of ethylene production : chiku, and passion fruit > avocado and papaya > banana, fig, guava, mango > pineapple > citrus, grape and pomegranate. Exposure of fruits to incense and myrrh helped in ripening. The unsaturated hydrocarbon gases coming out in the fumes of kerosene stoves were used to improve the colouring of lemons some six decades hence. Ethylene could strongly promote ripening activities by stimulating the climacteric rise. It has been observed that as the fruit reached the climacteric peak these become insensitive to ethylene treatment. At times fruits, ripe bananas for instance, could stimulate the climacteric in the ethylene way. The emanations, from the ripe banana fruits, have ethylene in detectable amounts, its production generally followed a pattern consistent with the intensity of the respiratory activity.

Ethylene found in fruit emanations is known to trigger the ripening processes of fruits. In general respiratory climacteric is controlled by the metabolic activity and also the metabolic products accumulated therein. Ethylene, for instance, is a product of ripening that enhances the process of ripening. Whenever its production is limited due to temperature control (freezing or above 40°C) or lowering the oxygen level, ripening is checked. That ethylene production is directly correlated with respiration rate in climacteric (Fig. 12.9 *b, c*) suggests the importance of temperature in the storage of fruits. In mango, avocado and bananas preclimacteric formation of ethylene caused respiratory rise. It has been observed that induced climacteric in leaves, and also developed changes in the pigmentation. Ripening fleshy fruits undergo endothermic chemical changes in it accompanied by the production of ethylene and other unsaturated hydrocarbons.

Ethylene is a powerful stimulator of respiration and even fruits which do not naturally experience a climacteric such as oranges and lemons can be induced to do so by its treatment. When ethylene is used to effect ripening (no effect on ripening of tomatoes) softening is more rapid than in air but the taste remains unaltered. The effectiveness of ethylene treatment on ripening at ordinary temperatures decreases when the preliminary cold storage period increases. The effect of ethylene on respiration and ripening are rather limited at low temperature.

Ripening after picking is notably influenced by (*i*) temperature, (*ii*) radiation, (*iii*) air humidity, (*iv*) ethylene, (*v*) volatiles, (*vi*) growth substances, and (*vii*) chemicals.

Temperature. Fruits picked up at right time generally ripen at any temperature between two critical limits. In certain cases fruits may require a cold treatment before being placed in the temperature limit for ripening. The effect of ethylene is the same as of pre-cold treatment.

Radiations. These may act as inhibitors or stimulators of ripening. Grapes ripe more quickly under treatment with infra-red radiations. Bananas irradiated with X-rays exhibited a decrease in softening but an increase in skin blackening.

Air humidity. The relative humidity and velocity of the air in the vicinity of the fruit influenced the maturity, especially in the evolution of the flavour ; saturated air hinders the development of good flavour in pears. Apples show blackening of the core.

Volatiles. The role of volatiles is most controversial. Non-ethylenic volatiles can stimulate ripening. Air purification with activated carbon, H_2SO_4 and NaOH slowed down the ripening of preclimacteric apples in a recirculation system. Depressing volatiles are also reported. Carbon (activated) reduces the effect in both the cases. Whether volatiles from the ripe fruit can stimulate preclimacteric in fruit in cold storage condition is a problem still to be solved.

Growth substances. Growth substances sometimes stimulate ripening of gathered fruits. It seems that the treatment is effective specially when the application is made very early soon after the picking. Stem of bananas immersed in solution containing 1000 ppm sodium 2, 4-D, 2, 4, 5-T or parachlorophenoxyacetic acid showed that ripening was accelerated. Reducing sugars of treated bananas are higher than the control. 2, 4, 5-T and to some extent 2, 4-D, when sprayed in a wax emulsion delayed the development of yellow colour in the rind of lemons during storage. The storage life increased.

Auxins may slow down (generally) or even (sometimes) accelerate ripening process. Ethylene formation is inhibited by auxin and therefore auxins have to be broken down by peroxidases (IAA oxidases) to control fruit ripening. Ripening is accompanied by a rise in auxin-degrading enzymes. Gibberellins inhibit degradation of chlorophyll so also do cytokinins. Gibberellins also stop colour change in fruits (banana). Cytokinins degrade protein when applied exogeneously. Accumulation of abscisic acid (ABA) is also associated with ripening. ABA increases markedly the degree of ripening of tomatoes in harvested fruits but not in those still attached to the plant. Auxins, gibberellins and cytokinins delay senescence by blocking the point of attack of ethylene.

Chemicals. Chemicals like calcium chloride, calcium nitrate, maleic hydrazide, isopropyl-N-phenyl carbamate, 2, 4-D, 2, 4, 5-T have been reported to affect the shelf-life of fruits. The storage life of citrus fruits was prolonged by spray of 2, 4-D and 2, 4, 5-T. Coorg Mandarin experienced a decrease in the loss in weight and also vitamin C content when sprayed with 2, 4, 5-T, maleic hydrazide hastened the maturity of chiku but delayed the ripening of mango. Chiku experienced reduction in the rate of respiration, ripening and spoilage when treated with isopropyl-N-phenyl carbamate. Pre-harvest treatment with 0.6% calcium chloride and 1% calcium nitrate enhanced the storage life of mango and guava. Ionic radiation also retarded the process of ripening of mango fruits. Ripening of banana fruits is retarded by gibberellic acid treatement given after harvest. In order to keep the ethylene below the threshold level, potassium permanganate or vermiculate has been used with success.

CHAPTER

13

PICKING, STORAGE AND PRESERVATION OF FRUITS

The benefits of production are lost unless wastage is effectively reduced during storage and distribution of fruits. Although this statement chiefly relates to fruits it does apply to vegetables, as well, in principle. The gains to the producer are possible only when the picking, packing, storage, transport, processing of fruits etc. are done in the right way at the right time. The agency of distributors, many a time derives larger profits than producer only through proper picking and disposal. Besides guarding carefully the conditions leading to diminished demand and low prices, the proper quality of the produce must be maintained. Picking of the fruits needs adjustments with nature of the fruit, stage of maturity and of ripening, the purpose for which required and the distance it has to traverse before consumption. The fruits should not be damaged in any way to avoid spoilage losses. Fruits may be harvested fully ripe if they are needed for on the spot consumption. All fruits cannot be picked fully ripe because of the dangers of post-picking losses mounting up. Some fruits give the indication of full maturity only when the tree-ripe fruits fall to the ground while others may do so earlier by colour changes. Pears are spoiled when over-ripe. Softening of the flesh (grape, fig) or change in colour (papaya) are some of the external indications which may guide the orchardist in individual cases. In general, fruits must be harvested firm enough to stand handling and to keep for a number of days and also to allow long distance transport, if so required.

The fruit picked green does not equal the tree-ripe picked ones in quality and it is necessary to strike at a balance as to when the picking is to be done. Off season fruits fetch better price than those sold at the height of the season. Picking is done with ease in the morning hours. Some considerations have also to be made for the prevailing environmental conditions for instance citrus should be harvested 2-3 days after the rains. Most fruit have the largest content of sugar when picked up at 16 hrs and therefore fruit needed for home consumption may be picked from the tree near about that time. Fruits should be properly graded, stored and packed keeping the market condition and the stage of maturity of fruit in mind. Grading mainly takes into account size, colour, freedom from blemishes. The over-ripe or damaged fruit which may decay in transit and damage those in contact with it should be removed. Immature fruits should also be removed so that the impression about the fruit lot is not undermined.

Packing. Many types of packing are used in the country *viz.*, baskets, boxes with perforations, fibre-board cartons, sacks etc. Baskets or containers have various shapes and sizes to suit the convenience of transport.

Packing of fruit is of importance. These could be packed nearly with or without wrapping in tissue paper, cellophane or similar paper to retain their attractiveness. Grapes, strawberry or other delicate fruits are generally kept in small packings which are kept in larger ones lined with some wood, wool, dry grass or leaves for padding at the base, sides and top during transport.

Storage. Proper storage to ward off post-picking spoilage is a problem of importance and it should be done in accordance with the needs of the fruit and the farmer. Proper storage has the advantage of keeping the supply of perishable fruit even in off season fetching high price to the orchardists.

Factors Affecting Storage

Respiration. Higher respiration leads to rapid breakdown of glucose and all other forms of organic compounds metabolized in the plant tissue. Higher temperature aggravates the losses while low temperatures diminishes. Respiration leads to certain events in the life of the fruits etc. which can be depicted in the two equations below :

$$\text{Stored food} + H_2O = \text{Soluble foods}$$

$$\text{Soluble food} + O_2 \xrightarrow{\text{Enzymes}} CO_2 + H_2O + \text{heat and energy.}$$

Respiration is affected by factors *viz.*, amount of sugar available, tissue moisture, environmental temperature, oxygen supply, transpiration etc. Higher the concentration of sugars within the living tissues higher is the rate of respiration. Taking the case of potatoes it may be mentioned that at 0°C these become more sweet than at 5°C with the result that these should be better stored at 5°C and not 0°C to cut down respiration losses. The fruits or vegetables which have higher proportion of living cells respire at a higher rate than the one having a lower proportion of living : dead cells. Tissue moisture bears a direct relationship with the rate of respiration. Succulents respire at a higher rate than the non-succulents ; a cabbage head respires more vigorously than the same quantity of potatoes and that the immature fruits respires at a higher rate than the mature ones.

Loss of moisture. Fruits have waxy cuticle (apple) while tubers have corky epidermis (potato) in order to cut down the losses of water from them. Whenever water losses exceed water intake there is withering leading to loss in weight and quality of the produce. Special care need be taken to cut down water losses from fruits, more so, after harvest and in transit.

Enzymic activity. Processes leading to respiration, ripening of fruits, decomposition and disintegration of the tissues are brought about by enzymes. The activity of these enzymes are controlled by temperature and humidity conditions of the environment.

Micro-organism. The nature has provided the fruits with a skin as in apple which normally checks invasion by pathogenic micro-organisms like moulds and bacteria. Whenever the fruits have cuts or bruises accompanied by lower resistance the attack of the organisms and their entrance is facilitated leading to rotting. Fruits may become unfit for human consumption as physiological disorders develop. Such cases are specially aggravated by unsuitable and unhygenic storage conditions.

Spoilage

Spoilage in fruits is caused by various factors or agencies some of which are discussed here.

1. *Mould.* It causes spoilage in fresh fruits as well as in the preserved products. It is non-poisonous but spoils the taste of the product. It may be destroyed by heating to a temperature of 165°F for 30 minutes.

2. *Yeast.* Yeast ferments the sugar, present in fruits, into alcohol and carbon dioxide. The bubbles of carbondioxide may be seen rising during fermentation. Yeast cells are destroyed by heat (150°F) for half an hour. Certain strains of yeast are also used for producing wine and vinegar.

3. *Bacteria.* Bacteria are of different kinds, some of them being highly heat resistant. Certain bacteria are poisonous and therefore greater care has to be taken to destroy them. Bacteria found in fruits are destroyed by exposing to a temperature of 212°F for 15-30 minutes, while bacteria found in vegetables are destroyed at a temperature of 240°F for 30 to 60 minutes.

4. *Enzymes.* Enzymes present in fruits cause them to ripen and then decay by their activity. Enzymes may be destroyed either by heating or by soaking the fruits in 2-3% brine.

Sanitation, temperature variations beyond the limits of tolerance, addition of chemicals, changes in tissue moisture, fermentation, carbon dioxide content of storage atmosphere are factors that affect the preservation of vegetables and fruits.

Temperature. The temperature is probably the most important factor but its action is controlled by the type, variety, stage of maturity at the time of picking, time spent between picking and storage, climacteric, seasonal variations and several other factors. Good handling of fruits at picking, firm yet airy packing are the important factors effective in preservation. There are several methods entailing low temperature in *vogue viz.*, cool temperature storage, freezing storage, and gas storage which are useful in the preservation of fruits and vegetables for storage. Any rise of temperature increases the rate of respiration, storage requires minimum rate of respiration. The optimum storage temperature, relative humidity for some fruits is given in table 13.1.

Table 13.1. Storage schedule for some fruits

Fruits	Relative Humidity (%)	Temperature (°C)	Average life (weeks)
(1)	(2)	(3)	(4)
I. Tropical and sub-tropical fruits			
Banana (Green, for ripening)	80–85	11–13	1–2
Ripe	85–90	11–13	3
Chiku (Sapota)	85–90	3–4	6–8
Citrus			
(*i*) Mandarin (*Citrus reticulata*)			
(*a*) Coorg orange	85–90	6–8	10–12
(*b*) Nagpur Santra	85–90	5–6	10–14
(*c*) Punjab Santra	85–90	2–5	4–6
(*ii*) Sweet orange (*Citrus sinensis*)			
(*a*) Malta, Blood Red	85–90	2–5	16
(*b*) Malta, Common	85–90	5–6	16
(*c*) Valencia Late	85–90	5–6	20
(*d*) Mosambi	85–90	7–8	16
(*e*) Sathgudi	85–90	7–8	16
(*f*) Pummelo, Lime	85–90	8–10	3–6
(*iii*) Lemons	85–90	8–9	8–12
(*iv*) Grape fruit (marsh seedless)	85–90	8–9	12
Custard apple	85–90	5–7	5–6
Dates (*i*) Ripe (fresh)	85–90	7–8	2
(*ii*) Cured	65–70	1–0	52
Figs (*i*) Fresh	85–90	0–2	4
(*ii*) Dry	65–70	0–2	52
Grapes	80–85	0–2	6–8
Guava	85–90	9–10	3
Jackfruit	85–90	11–13	6
Litchi	85–90	0–2	10
Mangoes	85–90	8–9	4–7

Contd. on next page

(1)	(2)	(3)	(4)
Mangosteen	85–90	5–7	6–7
Papaya	80–85	9–10	1–2
Passion fruit	80–85	7–8	4–5
Pineapple			
(*i*) Mature (green)	85–90	11–13	3–4
(*ii*) Ripe	85–90	8–9	4–6
Pomegranate	80–85	0–2	4–6
II. Temperate fruits			
Apple	85–90	0–2	16–32
Apricot	80–85	0–2	2
Cherry	85–90	0–2	2
Persimmon (*Diospyros kaki*)	85–90	0–2	7
Peach	80–85	0–2	2–4
Pear	85–90	0–1	12–26
Plum (*i*) Mature (for ripening)	85–90	13–15	1–2
(*ii*) Ripe	85–90	0–2	4–8
Strawberry	80–85	0–2	2–3
III. Other fruits			
Cashew apple	85–90	0–2	4–5
Dried fruits	65–70	0–2	41–52
Melons			
(*i*) Watermelon	80–85	2–4	2–4
(*ii*) Musk melon	80–85	0–2	1–2
(*iii*) Persian melon, 'Sarda'	80–85	2–4	6–8
Quince (*Cydonia vulgaris* Pers)	85–90	0–2	3–4
Star Gooseberry **(Kamarakh)**	85–90	0–2	4–5

Temperature fluctuations of the storage chamber must not vary by more than 0.5°C at any time. Other important factors are good air circulation, device to have lower temperatures and constantly high relative humidity. With loss in humidity the fruit loses weight and also quality. The relative humidity requirement of the fruit (Table 13.1) shows that most of the fruits do well with 85-90% while, some may have a little variation percentage of ± 5 on either side. Soft fruits with higher moisture content need high humidity of 90 to 95% to prevent serious withering. Increasing the insulation of the store rooms and the cooling surface area with reduction in the intake of outside air by preventing too frequent opening of the doors help to preserve the existing level of relative humidity.

The accumulation of carbon dioxide evolved as a result of respiration of fruits is bad for storage. Most of the fruits are harmed due to the concentration rising over 1%. The notable exception to this rule is that of pears and some varieties of apples. Storage temperature has a relationship to the period of storage besides the type of fruit, the stage of maturity and the purpose for which it is kept in storage. Let us examine the storage particulars of a few fruits.

Apples. The variety, crop and size of fruits are responsible for storage life. Small to medium sized fruits store better than larger fruits. Overripe apples do not store well ; only those picked at the correct stage of maturity can be stored. Most of the apple varieties like Delicious; Jonathan, Baldwin, Winter Banana, Rome Beauty, Yellow Newton, etc. keep well for 16-26 weeks. Of the Indian apples **Ambri Kashmiri** keeps longest almost double the length of time.

Banana. Ripe bananas when kept at 11°C are chilled and spoiled. Mature green bananas when kept at 13°C develop yellow colour and fine flavour.

Citrus fruits. Fully mature Nagpur santras from the February harvest just developing a tinge of yellow colour, ripen uniformly and improve in juice flavour, when stored between 11°C and 13°C. Fruits of ripe yellow colour can be stored for a further period of 8 to 12 weeks at 5°C to 6°C, Coorg oranges develop chilling effects if stored at a temperature lower than 7°C. Orange picked early in the season are most susceptible to chilling with shorter storage life. Covering separately with tissue paper gives a better look to grape fruits. The best picking time for lemons is the turning of skin colour from light green to yellow ; when deep yellow, these do not store well. Low temperature storage does not suit lime well since these are prone to chilling.

Grapes. Potassium metabisulphide (1 : 2000) contained in muslin bags when introduced in the crates of grapes help in warding off decay and increasing their storage life. 1-2% sulphur dioxide is also beneficial. If grapes are required for storage these must not be picked immediately after the rains.

Litchi. In cold storage litchi assumes a dull brown colour but the pulp gives a good taste and flavour.

Mango. Green, hard, uninjured mature fruits store better. Temperature below 8°C chill the fruits. Some are more susceptible than others. Mangoes, when fully ripe, do not store well. Wrapping individual fruits with tissue paper before packing always gives good results. Relative humidity and temperature conditions which are optimum are indicated in Table 13.1.

Pear. The storage must be done soon after picking and the temperature of the storage cell must be brought down very quickly between 0 – 1°C.

Plum. Fully mature plums assume characteristics colour, flavour and softening of flesh when stored between 13°C and 15°C within a fortnight of storage.

Micro-organisms. Omnipresent micro-organisms cause spoilage, impair flavour, odour, appearance and above all the nutritive value of the fruits. Most micro-organisms *viz.*, common yeast, moulds and bacteria fail to proliferate, grow and multiply at temperatures in the neighbourhood of 5°C. At 0°C most micro-organisms which cause spoilage in fruits do not grow. Low temperatures cannot be used to sterilize the fruits at temperature below –18°C which arrests the activity of the organisms but fails to kill these. As the temperature rises their activity is witnessed again. Thus such low temperature subdues the activity of the enzymes responsible for respiration and ripening processes. It has been seen that apples can be preserved for a period 6 to 9 times longer when kept at 0°C than at temperatures which are not able to destroy the enzymes. High temperature treatment can destroy the enzymes. Although freezing is a good means of preserving some fruits and vegetables by reducing the rate of respiration considerably, the method does not always find favour as it leads to loss of desirable flavour. Fruits may be exposed to sufficient heat to destroy enzymes as well as micro-organisms so that decomposition can be checked effectively. This aspect is used in preservation ; 100°C for processing and 120°C for canning according to the requirement of the product. The hermetical sealing of the sterilized products ensures safety against further contamination by micro-organisms. Spoilage of fruits, vegetables and foods caused by micro-organisms can be offset, temporarily by the judicious use of some compounds which are harmless to human beings, *e.g.*, sugar, salt, alcohol, vinegar and organic acids. Water losses caused by temperature may be recouped by the water formed in the process of respiration, at least partly in certain cases. Water loss is controlled by factors like tissue moisture, nature of the outer cover (cutinized or waxy), temperature of the atmosphere, humidity etc.

Fruits need relatively high humidity in storage in the absence of which they shrink, loose weight and also quality appreciably. It will not be out of place to mention that the storage life of fruits is increased if a humidity of 85–90% is maintained in the storage cells. In all probability cold storage achieves humidity values of this range when loaded to full capacity and high efficiency.

Fruits continue to respire during storage and the carbondioxide produced, if allowed to remain in the vicinity of the fruits, cuts short the storage life of fruits. The safe concentration of carbondioxide for storage is below 1%. During the course of respiration the supply of oxygen leads to the formation of carbondioxide with the release of H_2O and heat.

$$C_6H_{12}0_6 + 60_2 \longrightarrow 6CO_2 + 6H_2O + 677.2\,\text{K cal}$$

In the absence of oxygen, glucose undergoes changes due to respiration as below :

$$C_6H_{12}0_6 \longrightarrow 2C_2H_6OH + 2CO_2 + 25\,\text{K cal}$$

There is a loss of heat in the anaerobic condition to almost 1/26th of that under aerobic condition. Therefore, a restriction in oxygen supply is not economical in the energy balance besides the production of certain chemical compounds which are not useful. Apples, for instance, develop storage scald. Therefore, a sufficient amount of oxygen supplied to the storage cell is beneficial. Bulk storage has the disadvantages of poor aeration as against storage in containers and this method is really useful in bringing down the temperature in storage.

PRINCIPLES AND TECHNIQUES OF PRESERVATION

Preservation of food is a substitute for storage in that the material kept in a condition in which it is consumable and it gives all the other advantages of satisfaction and energy to the consumer. It has the added benefit of being more palatable and of keeping materials in proper condition which was not possible under storage. Food preservation has been practised in one form or other by sun-drying, cooking, smoking, pickling, in salt etc. since the dawn of civilization. That sterilization, coupled with vacuum aided preservation, helping to destroy the enzymes and the micro-organism associated with the fruits and vegetables is well known. It also minimises contacts with fresh micro-organisms. All that inactivates the enzymes as well as micro-organisms forms the basis of fruit preservation techniques. Preservation for limited period may be achieved by retarding the activity of enzymes and micro-organism by refrigeration, large addition of chemicals like salt and sugar, exclusion of moisture and also of air (to check the growth, activity and multiplication of aerobic micro-organisms).

I. Temporary preservation

Fruits can be preserved for 1-24 weeks in the following manner.

(*a*) *Low temperature*. The rate of growth of bacteria is slowed down if fruits and vegetables are kept at a low temperature. Refrigerators, cold storage etc., are used on this principle.

(*b*) *Use of antiseptics*. A mild solution of sugar, salt and vinegar may also be helpful for temporary preservation.

(*c*) *Cleanliness*. Washing fruits and vegetables thoroughly and keeping them in a clean place will help reduce the number of bacteria and micro-organisms present and preserve them for a longer time.

(*d*) *Pasteurization*. Fruit juices and drinks may be kept for 3-6 months by pasteurizing them at a temperature of 80°C.

II. Permanent preservation

Fruit products are preserved for a period of 1-3 years by the following methods.

(*a*) *Use of sugar, salt and vinegar*. Used in certain quantities these act as preservatives. Any product containing over 65% of sugar can be preserved permanently. Jam, Jelly, etc. are preserved by sugar.

(*i*) *Salt*. (15-20%) is required for permanent preservation. Certain pickles, *e.g.*, lime pickles are preserved on this principle.

(*ii*) *Vinegar.* Acetic acid (1-4%) preserves vinegar-pickles permanently.

(*b*) *Use of preservatives.* Potassium meta bi sulphite, sodium-meta-bisulphite and sodium benezoate are generally used for preservation.

(*c*) *Drying Dryhydration.* Fruits and vegetable may either be dried in the sun or dehydrated by machines. The absence of moisture destroys all bacteria and helps in permanent preservation.

(*d*) *Fermentation.* Fruits juices are fermented by good strains of yeast, converted to wine or vinegar and preserved.

(*e*) *Canning.* Properly canned fruits and vegetables can be kept for 1 to 3 years.

(*f*) *Deep freezing.* It is a recent yet popular, handy and useful method.

Preservation of a permanent type may also need pasteurization by heat to destroy most of the harmful forms of organisms which evade destruction, *e.g.*, spores. It is essential that the pasturized product be kept away from the micro-organisms of air. The temperature range for this treatment for fruits, is generally considered to be in the neighbourhood of 115- 120°C. Sodium benzoate (1%) and H_2SO_4 (0.07%) help in the preservation of most fruits. The use of large quantities of sugars (70% or so) is not uncommon in the preservation methods of fruit juice, jellies, etc. against most organisms. Sugar acts on the principle of osmosis whereby it inactivates the air borne micro-organisms which happen to reach the layer of sugar-cover over fruit. Almost 15% salt and 1% acetic acid also help in the preservation of fruits.

Drying reduces the moisture content to a level that micro-organisms fail to grow. Freezing stops the activity of micro-organisms and bacteria and therefore quick freezing or deep freezing is also useful in preservation. Sometimes the decomposition of carbohydrates by the activity of enzymes and micro-organisms into substances that prevent the activity of spoilage organisms also helps preservation. One common example of alcoholic fermentation whereby simple sugars are converted to alcohol is through the activity of yeast. Further contamination by aerobic bacteria, mould etc. must be checked for success with such methods of storage.

The methods employed for the preservation of fruits are : (*i*) storing, (*ii*) canning, (*iii*) drying, (*iv*) pickling, and (*v*) quick freezing.

The modes of preservation of some fruits (Table 13.2) are outlined on page 226. In order to meet the requirements of successful preservation various ways are *in vogue* to process fruits. These are discussed below :

HERMETICALLY SEALED FRUITS

Fruits may be preserved by heating either before or after packing in hermetically sealed containers to ensure complete sterilization of the product and also check re-contamination by micro-organisms. It is known as canning and gives unspoiled wholesome food. There is no loss of minerals, carbohydrates,

Table 13.2. Preservation modes of some fruits

	Anola	Apple	Apricot	Banana	Cherry	Fig	Grape	Grape-fruit	Guava	Jack fruit	Lime	Mango	Orange	Malta	Papaya	Peach	Pear	Pineapple	Plum	Strawberry
Barley water	×	×	×	×	×	×	×	√	×	×	√	√	√	√	×	×	×	×	×	×
Canning	×	√	√	√	√	√	√	√	√	√	×	√	√	√	√	√	√	√	√	√
Cordial	×	×	×	×	×	×	×	√	×	×	×	√	=	=	×	×	×	√	×	×
Jam	√	√	√	=	√	=	√	√	√	√	×	√	√	√	√	√	√	√	√	√
Jelly	×	×	×	=	×	×	=	√	√	√	×	×	=	=	√	×	×	√	×	√
Marmalade	√	√	×	×	×	×	×	√	×	×	×	×	√	√	×	×	×	√	×	×
Preserve	√	√	√	=	√	=	×	×	=	=	√	√	√	×	×	√	√	√	×	√
Pulp	×	√	√	√	×	×	×	×	√	=	×	√	×	×	√	√	×	√	×	√
Pulp juice	×	√	×	×	×	×	√	√	×	×	=	×	=	√	×	×	=	√	=	=
Squash	×	×	×	×	×	×	×	√	×	√	√	√	√	√	√	×	×	√	√	√
Syrup	√	×	√	=	×	×	√	√	√	√	√	√	√	√	√	×	×	√	√	√

√, suitable ; =, partly suitable ; ×, not suitable

proteins etc.. In canning the digestibility of the product is increased. The essential aminoacids are retained in addition to good amounts of vitamins, Canned fruits can be easily taken as the next best to the fresh ones. The frozen foods are even better than the canned ones but this is a more expensive and less convenient method needing electric appliances and electricity round the clock.

CANNING AND BOTTLING

Fruits may be canned or bottled and preserved in a more or less fresh state. Steel cans polished to prevent rusting or those coated with acid or sulphur resistant lacquers may be used. The cans are made air tight by exhausting the air before sealing. Any bacteria present are later destroyed by sterilization. The requirements and procedures are given here in bare details.

Bottles. Specially made glass jars can be used for bottling. They are made of strong glass and the stoppers have a rubber ring around them to give a good seal.

Can sealing machine. A small compact machine is used to seal the cans. In India, this machine is generally available as Dixi-Can-Sealer.

Pressure-cookers. Vegetables need sterilization due to the fact that these grow near or underground and are contaminated with large quantity of heat resistant bacteria which are not affected by the heat of boiling water (100°C) and require a temperature of 115.5°C or more for their destruction. High temperature may be obtained in a pressure cooker at 10 lbs pressure. The cooker should be large enough to hold the cans or bottles. Whenever fruits come in contact with the soil, sterilization is again useful and desirable.

The method of canning and bottling fruits and vegetables is almost the same excepting that (1) the vegetables are blanched before filling into cans or bottles. (2) 2-4% brine or 3% salt solution and 3% sugar solution is used for vegetables. (3) These need sterilization under 10 lbs pressure.

Selection. Select fresh, firm and fully sound fruits. For the best flavour and appearance use only choice fruits of uniform maturity, colour etc.

Preparation. Wash the fruits 2 or 3 times in running water and peel either by the help of knives or machine. Fruits with thin skin, *i.e.*, peaches, oranges, etc. may be peeled in the following manner. Boil a solution of 2-3% caustic soda in an iron container. Put fruits into cloth and dip for 1/2 to 2 minutes. Remove and dip in 2 to 4% citric acid solution. Wash several times. Blanch vegetables to soften, cut fruits into uniform pieces.

Filling. Fill cans 2/3 and bottles upto the neck. Prepare a syrup of 5 to 50% sugar according to the required sweetness. Boil, strain and pour into containers leaving about 1/8 inch head space on top. Use 2% solution, boil and fill in a similar manner for vegetables.

Exhausting. Place filled cans in boiling water bath. For bottles use a false bottom. In about 10-15 minutes when the centre of can records temperature of

77 to 86°C remove. This process drives out the air present inside the can and fruit tissues and helps prevent discolouration and loss of flavour during sterilization. Seal cans with the help of the can sealer and screw on tightly the lids of bottles.

Sterilization. Heat to 116°C for 30 to 60 minutes in a pressure cooker at 10 lbs steam pressure. Fruits may be sterilized in boiling water at 100°C for 15-30 minutes. The time of sterilization should be increased at higher altitudes. After sterilization cool the cans in running tap water and place the bottles on a wooden table. Afterwards wipe thoroughly and store in a cool dry place.

Spoilage. Spoilage is caused in canned goods due to leakage, metallic contamination etc. This may lead to food poisoning. Special care is required in the use of canned fruits. Some precautions are listed here.

(*i*) Keep fruit or vegetable can in boiling water for 10 minutes before opening.

(*ii*) With the first opening of the well sealed can whistle like sound of air rushing in should be heard.

(*iii*) Bulged can should not be used.

(*iv*) Discard if brine or sugar solution sprays while the can is being opened.

(*v*) After opening if the product gives an offensive flavour and smell then it should not be consumed or even tasted.

(*vi*) Discard products giving a flat sour taste, offensive odour or smell.

(*vii*) Do not purchase the cans if the date of manufacture is more than 3 years old.

Mustard oil is commonly used in North India while most south Indian pickles are preserved in til oil.

Fruits of good variety are selected at proper stage of maturity while these are still firm yet fully ripe though in the case of vegetables tender and young ones are more desirable. These must be washed well to get rid of dirt and reduce microbial population. In fact the fruits and vegetables are treated in the way as one would prefer for cooking. All contacts with iron or copper (knife, containers, equipments) should be avoided and keeping in 2% acid brine (good quality salt) followed by thorough washing with water is recommended to avoid discolouration of the products needed to be preserved.

Delicate materials like guava, orange segments, peaches, etc. may be prepared by dipping in boiling hot 1% caustic soda solution (Lay peeling) for 15-30 seconds only, while dipping 2-4 minutes is recommended for those with tough and thick skins. After this the skin can be removed easily. Washing off all the alkali in running water is necessary. The excess water is drained off. Other method for peeling may also be employed depending on convenience *viz.* scalding which consists in keeping the fruits etc. in boiling water for 1-3 minutes and then dipping in cold water. The skin shrinks and can be then removed by hand.

After peeling, precooking (heat treatment) is done for shorter period in steam or boiling water followed by a dip in cold water for not more than 10 to 20 seconds. This process known as blanching is chiefly done to admit closer filling, significant elimination of micro-organisms, inactivation of enzymes, getting rid of acid and a stringent taste, improvement of texture and fixing of natural colour. The processing is conditioned by pH of expressed product and the types and numbers of the organisms. The fruits are generally of two types : (*i*) those with pH more than 4-5 which cannot be processed in boiling water or steam (banana, jackfruit, some mango varieties, papaya, etc.), (*ii*) those which can be processed in the usual way without any difficulty *i.e.*, orange, pineapple etc. having pH lower than 4.5. Substances of category (*i*) could be processed in boiling water after making the material more acidic.

Filling is then done in clean cans or jars in a way as to permit sealing. Boiling hot canning medium is then poured. The medium consists of sugar 150 gm + salt 78 gm dissolved in 5 litres of water, boiled, strained or filtered before use. It follows exhausting of cans for creating a vacuum to check oxidative changes, loss of vitamins etc. After sealing the can hermetically, following the exhausting process, sterilization is done by heat treatment so as to destroy the organisms. Several types of juices (apples, grape, pomegranate, pineapple, lemon, tomato, etc.) are preserved by this process.

FRUIT DRINKS AND BEVERAGES

Various kinds of drinks are prepared from fruit juices. These may be (1) alcoholic (wine, perry, cider, vinegar), (2) none-alcoholic (squash, lemon barley water, cordial syrup) or (3) synthetic.

Drinks from Fruit Juices

The juices are rich in nutrition with respect to minerals as well as vitamins. Besides the alkaline nature of the juice it is favoured by the medical science. The vitamin C constituent expressed as mg/100 gm in juice of malta, mandarin, grape fruit, lemon, lime generally ranges between 30-50.

Squash. It is prepared from the pulp and juice of fruits. It is a concentrated drink and used with 3 times its weight of water. In any squash 10% to 25% of fruit juice should be present.

Select fully ripe and juicy fruits. Wash peel and cut if necessary. Take out juice either by squeezing pressing or with the help of a machine. Measure and take sugar, water and acid (if necessary) accordingly. Dissolve sugar in water by heating, or boiling, add acid and after a while remove the strain. Cool, add juice to sugar syrup. Mix thoroughly. Add preservative and bottle. In order to preserve squashes and other drinks chemicals like potassium-meta-bi-sulphite or sodium meta-bi-sulphite or sodium benozoate may be used. For lightly coloured and hot drinks use only the last, as the first two contain SO_2 which bleaches coloured drinks and escapes when added to hot drinks. Apart from the use of chemicals fruit drinks can also be preserved by either using sugar which should be 65% or by pasteurization. For pasteurization heat juice to 80°C and bottle, then wrap juice bottles in a thick cloth and place in warm water. Heat the water to required temperature for 30 minutes, remove and allow to cool on a wooden table. Pasteurization helps to preserve the juice for 1-3 months only.

Cordials. When the juice is totally clear, free from all pulp or parts of the fruit and moderately sweetened it is known as cordial.

Syrup. The juice of the fruit is extracted, the concentration of sugar is raised to over 65% by addition of sugar. The syrup may usually have a low acid content and a small quantity of fine pulp of the fruit.

Lemon barley water. Take lime juice 1 kg, water 1000 ml. sugar 2 kg, barley powder 30 gm, potassium meta-bi-sulphite 2.5 gm. Make a smooth paste of the barley and boil for 10 minutes and slowly put in the remaining quantity of water. Add sugar and one teaspoonful juice-dissolve, strain, cool and mix with juice, add preservative and bottle.

JAMS, JELLIES AND MARMALADES

Jams. A mixture of fruit and sugar, cooked to a consistency of a gel, to be able to hold the fruit pulp is generally unclear. Proper gel consistency with evenly distributed fruit particle, bright colour, true fruit flavour, and free from crystallization of sugar marks the quality of Jam.

Method. The method of preparing jam is generally the same for all. Select ripe fruits, wash thoroughly in running water and peel, if necessary. Cut into small pieces and weigh. Before adding sugar, it is essential for the fruits to be softened. This is done in two ways.

(*i*) *Cold process.* For very soft fruits take half the quantity of sugar to be added and place the fruit and sugar in alternate layers in an aluminium or stainless steel utensil. Sprinkle sugar on top and leave overnight. Next day add the rest of sugar and cook.

(*ii*) *Hot process.* Harder fruits are softened by cooking with water. The quantity of water to be added depends on the hardness of the fruit. For very hard fruits add an equal amount of water, for softer fruit, add 120-300 ml water per kilogram, cook and when the fruit pieces are softened add the needed quantity of sugar. For very sour fruits add 1 kg of sugar per kg of fruit and for sweet fruits add 1/2 kg of sugar per kg of fruit. Addition of sugar is important for sweetening and also because it helps the gel to get well set in the presence of adequate proportions of pectins and acid. Sugar must not be added until the fruit is softened. The tests for pectin may be made and if the pectin content is less than 58 gm per kg of the weight of fruit it is added @ 10-15 gm per kg. After the sugar has been added increase the heat and stir continuously for 5 minutes to dissolve the sugar. Acid is added as citric, tartaric acid or lime juice. Addition of acid improves the colour, taste and flavour of jam. Also the sugar will not crystalize when the jam sets. Add 4-12 gms of acid per kg of fruit according to the sweetness of the fruit. Keep stirring and test for the end point as follows :

(*a*) *Temperature.* 105°C at sea level (for places above sea level decrease 0.5°C for every 1500 cm). (*b*) *Spoon test.* Cool a spoonful of jam and allow to drop slowly, when the jam starts dropping in big pieces, remove from fire. (*c*) *Plate test.* Place a little jam on a plate and on cooling tilt slightly. If water does not separate from pulp, remove. (*d*) *Water test.* The prepared jam must weigh one and a half times the weight of sugar used. (*e*) *Refractometer test.* 68-69° Brix. Remove the jam from fire and fill into clean, dry open-mouth jars. When cool, seal with paraffin-wax and store in cool place.

Jelly. Fruit jelly is prepared by boiling fruit with an equal quantity of water and cooking the obtained extract with sugar and citric acid. An ideal jelly should have a good set, be transparent and of an attractive colour. The proper setting of jelly is very important and depends upon the correct proportion of the three constituents, pectin, sugar and acid.

Pectin. This carbohydrate is found in varying quantities in all fruits. It is found mainly beneath the skin of the fruits and around the seeds. Pectin is present in raw fruits in the form of pectose and in over-ripe fruits in the form of pectic acid. Of these only pectin has the property of jelling and hence only ripe fruits are used for jelly with the exception of mango and papaya. However, the pectose in semi-ripe fruits may be hydrolysed into pectin by adding a little citric acid while boiling the fruit. Pectin is only soluble in warm water which makes it essential to boil the fruit with equal quanity of water for 40-45 minutes. The amount of pectin extracted may be tested in 2 ways.

(*i*) *Spirit test.* Take one teaspoonful of extract, cool in a glass tumbler and slide 2 teaspoonful of spirit gently down the side. A big clot of thick, firm, spongy (if the mass is rich in pectin), white ppt will be formed (a high grade) ; if two or three lumps are formed a medium grade is indicated and if very small pieces are formed then the pectin is of a poor grade.

(*ii*) *Jel-meter test.* Pectin may also be tested with the help of a jel-meter which indicates the exact quantity of sugar to be added.

Sugar. It is very important to add sugar according to the pectin present in the extract. For a good grade of pectin add equal quantity of sugar, for medium grade add 3/4 kg of sugar per kg of extract, and for poor grade of pectin add 1/2 kg of sugar for every kg of extract. In case the sugar added is excess of the pectin then the jelly will not set at all. This condition is known as weeping jelly. On the other hand if the sugar is proportionately less, a very hard set is obtained.

Acid. 1% citric acid helps in strengthening the pectin fibres and thus is essential for obtaining a proper set. Usually 4-6 gms of citric acid is added for every kg of extract.

Method of preparation. Fruits rich in pectin such as guava, apple, apricot, plum, grape, *karonda*, raw mango, raw papaya, orange, *kaitha*, lemon, etc. should be selected. The fruits should not be peeled unless the peels and seeds are bitter as in papaya. Wash and cut in 1.5 cm thick slices. Add an equal quantity of water and boil for 40 to 45 minutes. Also add a little citric acid to convert the pectose into pectin. Test for pectin by above methods. Strain the extract through slip cloth. Do not squeeze pulp while straining and if possible allow to drip overnight. Decant the clear liquid from top, measure and add sugar according to the grade of pectin obtained. Place on fire and when the sugar dissolves, strain again to remove dirt from sugar. Now keep on a hot plate and do not stir jelly, with spoon or else the pectin fibres will be disturbed and proper setting will not be obtained. After 2 or 3 boils or when the jelly reaches a temperature of 102°C add citric acid according to need. Now test for end point as follows :

1. *Temperature.* 105°C (at sea level). The temperature may be lower by 0.5°C for every rise of 1500 cm in altitude.

2. *Sheet Test.* Dip spoon in jelly, cool and allow the jelly to drop. When in starts falling in sheet remove from fire. If it runs off as thin separate streams the end-point has not yet reached and further boiling is necessary.

Boll-test. Add a little cold jelly to one glass of cold water. If it settles down at the bottom of the glass without dissolving, remove from the fire.

Marmalade. It is made from all citrus fruits. It is slightly bitter in taste as shreds from orange or Mosambi peel are added to it. Otherwise the method of preparing marmalade is the same as for jelly.

Select fully ripe citrus fruits such as oranges, *mosambi*, maltas, grape fruit etc. and for every 4 of them add 1 big lemon. Peel and cut into slices. Add double the quantity of water and boil for 40-45 minutes. The white portion between the peel and pulp known as albedo contains considerable pectin and should be added along with the fruit while boiling. After testing for pectin strain through muslin cloth and measure by Jelmeter. Add sugar according to the grade of pectin obtained and boil.

Meanwhile soak the rind of orange or *mosambi* in water scrape albedo with spoon or knife and cut into 2-3 cm long and 1/8 cm wide shreds. Boil shreds tied loosely in cloth bag for 18 minutes in fresh water. In case lemons have not been used citric acid should be added in the proportion of 2.4 gm per kg of extract, when the temperature reaches 102°C. Keep boiling and test for end point as in jelly. 5 minutes before removing from fire spread the shreds on a tray. After removing from fire, allow to cool a little, then remove scum carefully and when the marmalade is luke-warm, sprinkle the shreds and bottle. Shreds should not be added to hot marmalade otherwise they bunch up and float on top. A good marmalade must have the shreds evenly distributed in the whole mass, besides all other characteristics of a good jelly.

Preserves and Candies

Preserves are prepared by cooking whole or cut fruit with syrup. The concentration of sugar in the finished product is 75%.

Method. Apple, mango, papaya, *aonla*, carrot, ash gourd, ginger etc., are used for preserves. Wash and peel if necessary, cut into large piece or leave whole. Before immersing in syrup, the fruit must be prepared in the following way :

1. *Pricking.* Prick fruit thoroughly with a fork. This will facilitate the assimilation of syrup within the fruit.

2. *Soaking.* Before or after pricking the fruit should be soaked for 2 to 24 hours in one of the following solution.

(*a*) Brine : 2-4% for sour fruits.

(*b*) Alum : 2-4% for bitter fruits.

(*c*) Lime. (2-4% solution, dissolve lime in water and after half an hour decant clear liquid). Used for soaking very hard or very soft fruits.

(*d*) Water soaking is an important step as it reduces the sourness or bitterness of certain fruits, makes them clear, transparent and firm.

3. *Boiling.* After washing fruits 2-3 times in water soften by boiling. The length of time for boiling depends upon the softness of fruits. Hard fruits may be boiled and softer ones blanched by placing in cloth bag. Immerse in boiling water for 2-3 minutes, remove and place instantly in cold water. Ginger is softened by boiling in pressure cooker, under 15 lb pressure for 1 hour.

Candy. Prepare preserve by the slow process. After 2 weeks remove fruit pieces from syrup and place on a perforated or ordinary tray. Allow to dry in the room.

Crystallized candy. For making crystallized candy, wrap the fruit pieces in sugar before drying.

Glazed candy. Prepare a little syrup containing 70% sugar, boil and dip the dry candy for 1 minute in syrup. Dry again. This will give the candy a glazed appearance.

Syrup. Now prepare the syrup by either of the two methods.

1. *Layer method.* For soft fruit, place alternate layers of fruits and sugar (the thickness of the sugar layer should be about half of fruit layer) and keep over night. Next day warm slightly, remove fruit pieces, add the rest of sugar and water to prepare syrup.

2. *Ordinary method.* Take sugar, fruit and water in 2 : 3 : 4 ratio by weight. Place water and sugar on fire when dissolved add 2-4 gm citric acid per kg of syrup. Boil and strain through muslin cloth.

I. *Slow Process*. After this the preserve can be made by boiling the syrup daily for 10 minutes for 8-10 days. Eventually when the syrup starts forming 2-3 threads it is considered as ready. Soft fruits are not boiled with syrup but harder fruits should be boiled initially with the syrup for 2-3 days.

II. *Quick process*. The syrup can also be concentrated in one day until the end point is reached. The slow process gives better results. Allow the fruit pieces to be in the syrup for 1-2 weeks before using.

Dried and Dehydrated Fruits

An economical and perhaps the earliest method which is still employed in some form or the other. This method depends on the simple rule that there is a minimum level of moisture necessary for the activity of spoilage organisms and the moisture content must be reduced considerably so that the micro-organisms either remain dormant or perish. Further reduction of moisture in certain types of foods is sometimes necessary to avoid other undesirable changes during storage. The drying of fruits and vegetables is one of the most economical means of preserving them. Compared with canning, drying is much less expensive and the properly dried product is often as palatable and nutritious as the canned one. Dried product, however, cannot be used quickly and requires preliminary treatment and cooking before serving.

Dehydration has certain advantages over other methods of storage *viz*., quickness, economical, check of enzymic activity, no need of special equipment. The product is of a very uniform quality.

CHUTNEYS, PICKLES, SAUCES, KETCHUPS

Chutneys are a form of spicy matured preserve made from a variety of ingredients. The main ingredients are salted overnight and then smeared with spices. Some vinegar is added. The whole material is kept in a jar and covered. It is better to seal them to check drying. Let us consider the preparation of chutney by taking the example of tomato and apple. The requirements : apples or tomato (red) 1 kg, 2 onion, 1 pint vinegar, 1 teaspoonful pepper, 1 level dessertspoon ground ginger, 1/2 kg brown sugar. Peel cone and slice apples or tomatoes. Pour vinegar and add pepper and ginger. Cover for the day. Turn all, together into a preserving pan with the sugar. Smear well and put in a clean sterilized, dry jar and cover closely with the grease-proof papery lined metal cap when cold. Only aluminium and stainless steel utensils be used, copper and iron must be avoided. It seems clear, therefore, that the pieces of fruits etc. need not be cooked unless it is not taken raw. Cooking just enough to make the material soft and eatable before the addition of spices and vinegar is also possible. Sometimes it is necessary to do the cooking with sugar to improve taste. In the preservation of chutneys, pickles, sauce and ketchups vinegar, sodium benzoate, oil, or lime juice work as preserving agent by warding off infection from outside.

Tomato sauce is prepared by cooking tomato with sugar, spices, salt and vinegar. It is preserved with the help of sodium benzoate.

Method. Select fully ripe and red tomatoes, wash and remove the stem portion and any blemished parts. Add 30 gm of water for every kg of tomatoes, after cutting into small pieces. Place on fire and cook until tender when the skin of the tomato separates from the pulp, remove from fire and pass through an aluminium seive or a thick double net. Weigh the strained juice and take the following ingredients in the given proportion.

Tomato juice	5 kg
Salt	40 gm
Garlic **(lahasun)**	6 gm
Cloves **(lavang)**	3 gm
Cumin seed **(Zeera)**	2 gm
Red chilli powder	4 gm
Aniseed **(Saunf)**	4 gm
Cinammon **(dalchini)**	4 gm
Sodium benzoate	5 gm
Sugar	600 gm
Onion **(Piaz)**	60 gm
Ginger **(adarak)**	30 gm
Cardamum **(elaichi)**	2 gm
Mace **(Jaiphal)**	1/3 gm
Pepper **(Kali mirch)**	4 gm
Correiander **(dhania)**	4 gm
Vinegar **(Sirka)**	250 gm
Acetic acid	15 gm

10-12 ml of glacial acetic acid may be added in place of vinegar. Add 1/3 of the sugar to juice and boil. After 12 minutes add the spices. These may be added in three ways :

1. Direct addition : as in chutney.

2. Cloth bag method : Grind coarsely and tie loosely in a cloth bag and add to sauce.

3. Extract method : Boil spices with water for half an hour, strain and add extract to sauce.

Of these the cloth bag method is considered to give the best result. Concentrate juice until quite thick, then add rest of the sugar. In case vinegar is used, it should be added with the sugar. Remove cloth bag after squeezing. When the sauce is one-third of the quantity of the original juice add the salt and remove from fire. If acetic acid is used then add the acid after removing the sauce from fire. Finally dissolve the sodium benzoate in little warm water, mix thoroughly with sauce and bottle.

Deep freezing. Freezing arrests the activity of the enzymes and micro-organisms and helps in preservation. This is now becoming popular in the country. The foods which are given a deep freeze treatment may be either cooked or raw. These have to be packed in polythene bags, containers or cartons to suit the convenience. Raw foods should be frozen as soon as picked. Before putting down for freezing, vegetables should be blanched by putting in fast boiling water for 1-2 minutes, draining and holding in cold running water for 2 minutes. Drain again, pack in boxes and bags. Deep freezing can be done on home scale by keeping the food in refrigerator cabinet usually maintained at *minus 8°C-20°C. It takes about 6 hours to freeze.*

Increasing per Capita Availability of Fruits

Although India ranks third largest producer of fruits (24 m. tons) in the world yet the per capita availability of fruits works out to only 30 g. Which is far short of the minimum dietary needs. Major constraints in the development of fruit industry in India are factors such as lack of statistical data, non-availability of quality disease-free planting material, low productivity, poor quality of produce, heavy initial investments, high rate of interest, inadequate budgetary support and weak post-harvest management infrastructure.

Besides adequate know-how coupled with the willingness and capacity of the grower to translate the technique into practice inadequate post-harvest management outstands as the main cause of low availability of quality fruits in the country. There is dire necessity of setting up of grading and packaging houses, provision of cheap transport facilities, opening of retail outlets and construction of small cool chambers for avoiding glut in the market. Strengthening of post-harvest management and transfer of technology should help increase in per capita availability of fruits.

In India, the post-harvest losses have to be minimised. This can be achieved by adjusting the picking time, resorting to proper storage and checking spoilage in all its forms. The processed food industry hardly utilises 0.5% of the produce in India as against 60-70 per cent in other countries. In a nutshell integration between production and processing should be achieved for better vigour and health of the people.

CHAPTER

14

GROWTH SUBSTANCES IN FRUIT PRODUCTION

Substances which play a significant role in the control of growth (growth regulators) include auxins, gibberellins, kinins and inhibitors. Natural growth hormone and also the synthetic materials constituting the auxins behave like-wise in regulating cell enlargement. Gibberellins are known to regulate growth in the presence of auxins. Kinins stimulate growth through cell division, these also need the presence of auxins for being effective. Inhibitors are substances which inhibit growth. Endogeneous growth inhibitors actively participate in growth and development of plants. These are specially associated with the mechanism responsible for dormancy. Light and temperature, the two environmental factors, specially affect the action of inhibitors on plants. Kinins affect light responses, apical dominance, leaf growth etc. through cell division, stimulation or otherwise. Kinins elevate the level of DNA, RNA and also proteins giving message that these are very well connected with many aspects of growth and development. Kinins are not acidic as the auxins and gibberellins, the latter seems to be responsible for the control of dwarfism in plants. These also stimulate stem, leaf, flower and fruit growth, flowering, breaking of dormancy, the processes under the control of temperature and photoperiod and other environment-regulated phenomena. Auxins are chemical messengers influencing many patterns of plant development. While endogenous auxins affect growth and differentiation activities of plant, exogenous application also has modifying capacity. Processes like tropism, cell elongation, differentiation, abscission, flower initiation and development, pollen tube growth, fruit-set, fruit growth and development are affected by growth regulators. The physico-chemical properties of the cell like plasticity, elasticity, cytoplasmic viscosity, protoplasmic streaming, respiration, and metabolic pathways are auxin controlled. Most auxins not only stimulate growth of stem and coleoptile but also effect differentiation and abscission.

Widespread applications of growth regulating substances in the service of fruit industry may be grouped under several heads. A few may be listed here :

(1) propagation of plants through cuttings and grafts, (2) prevention of pre-harvest fruit-drop, (3) artificial induction of seedlessness, (4) increased fruit-set, (5) continuance or otherwise of dormancy, (6) control of post-harvest developments (sprouting and root formation), (7) inhibition of growth of buds of nursery stock in storage or delaying growth temporarily after planting, (8) regulation of flowering and fruit size as in pineapple (through ethylene, acetylene, and napthalene acetic acid), (9) defoliation and prevention of leaf fall, (10) thinning of flowers and fruits as in mangoes, apples and peaches, (11) selective weedicide. A very authentic and uptodate account of use of plant growth regulators and gibberelins in horticulture has been presented in ICAR Bull. (Agric.) No. 34, 1971

Uses in Plant Propagation

In propagation are involved root promoting growth substances and/or compounds which regulate (inhibit or delay) shoot growth. Much of success in plant propagation depends on the ability of cutting to root under favourable environmental conditions. Root primordia are ordinarily not present in stems of most plants but may be induced.

Vegetative propagation is preferred to reporductive mode of multiplication due to ease of maintenance of hereditary characters of the mother plant. Cuttings of fruit trees do not always root readily under normal conditions. Growth regulators have been used to increase the efficiency of propagation by cuttings, layering etc. Generally growth substances in extermely minute quantities are effective (not below the optimum) whereas high concentrations kill the base of the cutting, cause excessive proliferation of cells and excessive callusing, or check further growth of roots and shoots.

Once the concentration is of the effective dose is known the duration of treatment is to be adjusted. A high concentration may be used for a short time or a lower one for a longer period. Growth substances or regulators accelarate the rooting of cuttings, produce a larger root system increase the per cent of rooting and make it possible to root cuttings of some species that normally root poorly. Inherent factors other than those affected by growth substances are involved in the propagation of plants by cuttings and therefore comparatively only few species which root with extreme difficulty under ordinary conditions of propagation are benefitted by the treatment.

Since the discovery of Indole-3 acetic acid (IAA) as an active plant growth regulator, Indole-3 butyric acid (IBA) and a-naphthalene acetic acid (NAA) have been freely used to boost vegetative propagation of plant specially rooting of cuttings. Success in rooting with growth substances like IAA, IBA, NAA, NAD, etc. differs with species, the rooting medium the concentration of growth substances, their duration and the physiological state of the cutting among other factors. IBA is more effective, causes no damage to cutting. Treatment with vitamin B_1, B_6 etc. after the application of growth regulators yield better response.

IAA + IBA produces better rooting as against either singly. Carbon monooxide may also induce root initiation and cause stimulation of pre-existing root primordia but it is not an economical feasibility. Carbon mono-oxide gas induced definite rooting responses in 27 plant species. One species produced roots only from flower stump, 7 from the nodes and one species formed epinasty of the leaves (curl downwards) and unusual proliferation of lenticular tissue with retarded elongation of stems.

Unsaturated hydrocarbon gases, *viz.* ethylene, acetylene and propylene initiated and stimulated adventitious roots by exposing growing plants. The concentrations varied from 0.001% to 1.0%. Of 27 species and varieties tested, the gases caused definite rooting response in 15 species Since these did not induce shoot formation it was believed that these gases are specific for adventitious root formation. For use in commercial propagation, these gases are yet of little practical utility.

Methods of Application of Growth Substances to Help Propagation

Growth substances may be applied to plants or plant parts in the form of solution as dip/immersion/spray or as powder or lanolin paste. Watering the rooting medium with aqueous solutions, application from aerosol bombs, injection of dilute solutions into internal tissues and volatilization of crystals at high temperatures are generally followed.

Strong solution : Quick dip method. Whenever prolonged soaking is not desirable or causes injury to the cuttings, dip in strong solution containing growth regulator equivalent to 1-20 mg/ml has been found to be useful. The dip duration may be a few seconds only of about 1 to 1.5 cm length of the basal end of the cutting. Sometimes re-treatment with growth substance 3 weeks after the initial tratment improves rooting of cutting.

Immersion method. Encouraging responses have been reported from 80 species of plants tested by immersion ; 3 cm of basal ends dipped in the dilute solutions of the growth regulators for desired period kept in non-metallic containers in the dark regulators for desired period kept in non-metallic containers in the dark during the period of treatment. Of the hormones IAA disintegrates early while IBA, NAA, NAD etc. keep for a longer period and may be used over again several times.

Both the dip and immersion methods have the disadvantage of facilitating the transmittance of systemic and surface-borne fungi and bacteria from one cutting to the others, *via* the stock solution. The growth regulators in aqueous solutions have been generally found to be more effective than the solid powders. Generally crystalline forms of growth-substances are first dissolved in a little alcohol and then diluted with water to make the required concentration.

Spray method. First three methods are easy to apply to cuttings but the spray method is generally applied to the mother plant before taking the cutting for the propagation. Spraying of stock plant with solution of 2, 4, 5-TP in concentrations varying from 25 mg to 100 mg/litre before taking the cuttings between 10 to 40 days after spraying yields good results.

Powder-dip method. This is a widely employed method 2 cm length of basal end of the cutting mositened in water is brought in contact with an adequate propagation of substance, mixed and the excess shaken off before the cuttings are placed in the rooting medium.

Lanolin paste method. It is generally confined to air layering now.

Stem cutting. A growth regulator does not necessarily influence the cutting of even a single plant alike much less to talk of those from others of different varieties of the same species. Certain factors *viz.*, age of the plant, age of the propagation, wood used from the parent plant, nutrition status of the stock plant, and time of propagation are of importance in the response of cuttings to growth regulators.

The morphologically older end of the cutting may be moistened with water and then brought in contact with the growth regulator placed in shallow container or on waxed paper. A dust automizer can also be used to dust the powder to the cuttings or may be sprinkled in liquid form.

Mixture of more than one growth regulator has been often used to produce more roots and faster rooting on stem cuttings. Indole butyric acid, naphthalene acetic acid, and nephthalene acetamide are popularly used in most commercial hormone preparations either alone or in combination. Carriers such as talc, powdered charcoal, bentonite or flour are used to dilute the concentration of the commercial hormonal preparations. The ratio of the talc : growth regulator may be varied from 1000 : 4-50 (for coarser talc) or 1000 to 0.5-12 (for fine textured talc) so that with finer talc lower concentration of the growth regulator is used.

Hard wood cuttings generally need a higher concentration of growth regulators to affect rooting than those which root easily. The low (1 : 100) ; medium (4 : 1000) and high (10-20 : 1000) hormone : talc ratio is generally considered successful with easy (soft wood), intermediate (ordinary wood) and difficult (hard wood) rooters. Composite powders carrying 500 to 2000 parts Indole-3 butyric acid, 3-naphthalene acetic acid and napthalene acetamide per million of talc effectively stimulate roots in soft wood or green wood cuttings. Concentrations from 10,000 to 20,000 parts per million may be used to root more difficult cuttings. Mixtures of these growth-promoting substances may be more effective. A low concentration of IBA + NAA brings about a better overall performance than either applied singly to cuttings through initiation of more prolific rooting in a larger percentage of the cuttings. Besides, the role and/or specificity of hormones, the physiological condition of the wood counts in the promotion of root initiation and growth in soft wood cutting more than in hard wood cuttings.

Besides the rooting of cuttings, the role of the growth regulators has been immense in different types of layering [trench, chinese (or pot)], aerial etc. and in the propagation by root and leaf cuttings. In the latter case the success is not spectacular possibly due to the inhibitory effect that accompanies growth regulators.

Mode of action. The interaction of an artificially raised auxin level (by exogenous application) and an adequate nutritional status can force rooting in any living plant tissues, *viz.*, cambium, epidermis, pericycle, cortical parenchyma, pith-rays etc. The optimum concentration of auxin required by the plant for cell division, cell elongation and cell differentiation are not alike, more so in different plant organs. Levels that initiate rooting are higher than those needed for cell elongation and cell division in root cambium or pericycle forming lateral roots. Cells divide to give rise to a growing point in the root pericycle preceding root initiation. The roots formed are similar in origin irrespective of the use or otherwise of growth regulators.

It is suggested that there is an interaction between exogenous hormone and rhizocaline, an intermediary auxin which is specific for directing root initiation from stem cuttings and evenly distributed throughout the entire length of the stem. Rhizocaline is mobilized to the morphologically basal end of the cutting for root formation. Since the number of roots developed is directly proportional to the length of the stem cutting, the hypothesis of rhizocaline controlled root formation has some support.

It may be, however, borne in mind that the examples only give an indication of the possible uses which one can derive and no guarantee can be assured for either the concentration or the growth substance being always promotive of any physiological apsect of the plant.

Nepthalene acetic acid (NAA), indolebutyric acid (IBA), 2, 4-dicholorphenoxy acetic acid (2, 4-D), other substituted phenoxy acids Maleic hydrazide (MH) etc. have proved of immense value, among the growth substances, to the pomologist.

Herbicidal activity. α-NAA ; 2, 4-D ; MCPA ; 2, 4, 5-T ; 4-(2, 4-DB) ; 4 (MCPB) ; 4-(2, 4, 5-TB); 2-(2, 4-DP); 2- (2, 4, 5-TP); 2-(MCPA) ; 2, 3, and 2-6 phenoxy thioacetic acid, 4-fluro and 4-bromophenoxy acetic acid and 2-6 phenoxyalkyl carboxylic acid are some of the important chemicals known for their herbicidal activity.

Herbicidal action of phenoxy growth regulators depends on their absorption, translocation and accumulation. Absorption in molecular form, entry through cuticle, selectivity without being dependent on stomata and thanslocation from the leaves to other parts of the plant *viz.* phloem, the food duct, and also from the roots through the xylem vessels causes injury and death at the site of accumulation. Meristematic activity being activated to cause blocking of the translocation channels leading in both uptake of nutrition as well as photosynthesis, carotene level and also the activation of the phosphate system has been associated with the activity of 2, 4-D as a herbicide in plants.

Selectivity of action is dependent on biological mechanism. It has been seen that the herbicide acts on the plant only after its entry and translocation. The effects may be visible as the arrest of plant growth, twisting of the tops, change in shape and size of the foliage, formation of adventitious roots, splitting and swelling of roots in about 4 to 8 hours of the application of the herbicide although the death of the plant takes place only later. It is related to the structural peculiarity of the plant. The movement of 2, 4-D is slower in monocots than in dicots. Differences in selectivity have also been mostly due to absorption, translocation and accumulation of growth regulators. Protoplasmic resistance to absorption and translocation is considered to be due to electric charge of the particle, the H-ion concentration of the cell sap and the nature of proteins in the protoplasm. Moreover in dicotyledons the potential cambial layers are abundant, close to the phloem in a ring. Proliferation of these cells destroys the phloem-system and therefore the tanslocation of food. The monocots are devoid of the cambium the meristem is also very much reduced so that the phloem activity is not disturbed. 2, 4-D generally is more effective on broad-leaved plants alone. CIPC on the other hand is toxic to grasses but not to plants with broad leaves. 2, 4-DB and MCPB are biologically inactive and their action is attributed to the present or otherwise of some enzymes. In the presence of the enzymes the two are converted to 2, 4-D and MCPA respectively and consequently, these may be toxic only to some plant species and not all. Selectively is never absolute. It is relatively dependent on concentration, method of application and crop species among other factors. In thin leaved weeds, once absorbed the 2, 4-D is quickly translocated particularly to the meristems which leads to the death of the harbaceous broad-leaved dicotyledonous annuals (cereals, sugarcane and other monocotyledonous grasses, remaining more or less unaffected.) Generally the concentrations of 2, 4-D which is suited to weed control is harmless to humans and animals, 2-methyl, 4- chlorophenoxyacetic acid (MCPA), 2, 4, 5-trichlorophenoxy acid (2, 4, 5-T), various benzoic acids, *e.g.*, 2, 3, 5- tridobenzoic acid (TIBA), 2, 4, 6 and 2, 3, 5-trichlorobenzoic acid (TCBA) have been successfully and economically used as herbicide.

The 2, 4-dichlorophenoxybutric acid is one of the substances of an interesting large group of compounds, phenoxy alkylcarboxylic acids which are being increasingly tested for their suitability as weed killers which do not in any way damage the particular crop species in the field.

Flowering. Iodine substituted benzoic acids, *e.g.*, 2, 3, 5-triodobenzoic acid (TIBA) is good for including profuse flower formation in many crops. The part that the growth substances play in the endogenous control of flowering has received wide attention, production of auxin by the petals in the flower bud, pollen and ovary in the developing fruit-bud and by the seed in the fruit suggests that auxin in intimately connected with flowering and also fruit-setting. The growth substances may promote or inhibit the pace of the activity of the plant or its part. There is a well established relationship between the auxin content and the natural inclination towards seedlessness ; seedless cultivator have appreciably higher auxin content definitely correlated to flower initiation. The growth hormones are produced under the influence of day length due to the high production of the growth regulators under optimum photoperiod. Plant growth regulators have been largely used for causing as well as inhibiting induction of flowering, fruit set and fruit-drop.

Exogenous application of NAA caused flowering in pineapple @ 0.05 mg/plant and higher strength *i.e.*, 0.25 mg or 0.5 mg per plant caused 100% flowering, 2, 4-D also induces flowering but NAA application induces heavier fruits. NAA is widely used for crop control in the pineapple. β-hydroxyethylhydrazine also controlled flowering depending on its concentration. For litchi SNA (sod. nephthalene acetate) caused maturation of the stem terminal leading to blossoming in the following spring. In dry months it resulted in inhibition of vegetative growth.

Flower thinning. Thinning which is useful for consistent fruit production can be achieved through the use of growth substances in many ways like NAA at flower stage through (1) check of natural pollination, (2) forcing abscission by change in auxin gradient at the point of abscission, and (3) the abortion of young embryo.

Both the prevention of natural pollination and the abortion of the embryos are most probably the outcome of a disbalance in the production of auxin in the concerned part of the flower of fruit. Alteration in auxin gradient across the abscission zone, diminution in the auxin supply to the distal end or increase in that of the proximal end. After a series of investigations Struckmeyer and Roberts indicated in 1950 that auxins may cause increased fruit-set by thinning of excessive number of flowers so that the nutrient supply may not limit. It was also seen that the exogenous auxin application seemed to be non-operative in the absence of leaves.

DNC (dinitro compound) is effective for flowers and NAA (20 ppm) for fruit thinning. 20 – 60 ppm of naphthalene acetamide has been found effective with Jonathan apple. NAA (10-20 ppm), NAA mide (25 – 50 ppm), NAA + Tween-20 and DNC (0.2 – 0.4%) have been found to be effective in the thinning of fruits specially at full bloom or a little after that. It is not necessary to indicate in detail the effect of various compounds on different varieties of fruit which vary with concentration and also the stage of application and the reader may refer to the ICAR Technical Bulletin (Agric.) No. 34 by G.S. Randhava, 1971.

Ample flowering is also promoted by the application of NAA to induce thinning in the 'on' year and thereby getting better blossoming in the otherwise 'off' year. NAA is also effective in causing fruit set in apples and pears.

Parthenocarpy. 2, 4-D; 2, 4, 5-T, naphalene compounds, β-naphthoxy propionic acid, 4-CPA, BNOA, BNOP, GA Gibberellin + auxin have been associated with parthenocarpy. Concentration, time of application, environmental conditions, etc. are well knit. Parthenocarpy, speaking broadly, seemed to be associated with higher degree of fruit-set, advance of maturity, light coloured and astringent fruit (until fully ripe and soft), sugar content, vitamin C content increased at least in some cases.

Sod. salt of α (2-naphthoxy) propionic acid at 100 mg/l and 2, 4, 5-TP has been found to be able to induce parthenocarpic fruit development in frost damaged fruit-buds of apples and pears. Maleic hydrazide stepped up flowering of strawberry, delayed in pear but did not show any effect in apples and grapes.

GA, NAA and also IAA + GA are used for the production of parthenocarpic seedless fruits *viz.* pears, peaches, apricot, almonds, etc. Cherry seems to require both GA and IAA to induce parthenocarpy. GA has been found to be more effective in setting fruit parthenocarpically, increasing fruits size etc. The effect of GA application on citrus fruits varies according to strength and the time of treatment. 4-chlorophenoxyacetic acid has been most successful in unpollinated grapes and NAA is pineapple. Generally spray method and application of hormones and the lanolin paste are common.

Fruit-setting. Fruit-setting has been achieved in several fruits and vegetables by the use of NAA, NOA, MCPA, CCP, 2, 4-D, 4-CPA etc. in different concentrations. In mango 10 mg/l NOA caused better fruit-setting and the fruit development was parthenocarpic. For pineapple, 0.0015% NAA produced 100% fruiting. Fig, citrus, apple and pears all have been found to bear fruits by the use of auxins or GA or both in different concentrations.

Pre-harvest drop. Fruit-drop in no menace if it is done for the normal and regular annual bearing of the tree. Abnormal means of fruit drop of immature apple fruits are found synchronising with the critical periods of stress of environment and physiological conditions. The abscission, brought about by the development of the abscission layer can be regulated by the use of growth substances as 2,4-D controlling fruit-drop of citrus orchards in California. Factors such as species of the fruit, variety, occurrence of fertilization, position of flower and other environmental conditions weigh in the fruit-drop wave.

The effect of growth regulators in the control of preharvest drop of apple fruits follow the manner as given here. IAA > IBA > IPA (acid, sodium, potassium, calcium and ammonium), NAA > β-naphthaoxyacetic acid β-(1-napthoxypropionic acid salts). The effect of 2, 4-D varies with the cultivator (Delicious, Golden delicious and McIntosh give negative response). 2, 4, 5-T and 2, 4, 5-trichlophenoxy propionic acid give good result with apple but Bramley seedling is insensitive. 10 mg/l of growth regulators is effective in conjunction with spreading or wetting agent. These are effective for 5-6 days before the fruit drop while 2, 4, 5-TP (20 mg/l) is influencive for 15 days earlier. Rains develop acceleration of ripening process and accumulation of application of 25 mg/l of NAA or 2, 4-D in water. Spray of 2, 4, 5-TP, 3 to 6 weeks after blossoming, hastens maturity of *Amini* mangoes. 30-40 mg/l NAA controls fruits drops of *Fajri* best. 2, 4-D (100 mg/l) also lowers fruits drop by more than 25%. 10 mg/l 2, 4-D and 20 mg/l NAS reduce abscission in *Dashehri*. Fruit drop in *Langra* is checked by 2, 4-D to the tune of 15%. Fruit drop is checked in apple, pear, stone fruits, citrus, sweet orange etc. by the use of auxins like NAA ; 2, 4-D, 2, 4, 5-T ; 2, 4, 5-TP etc. in different concentrations. Time of applications also affects the effectivity of the compounds. Growth retardants such as B-9, CCC and phosphon seem to be valuable in fruit production by retarding pre-harvest drop.

Breaking of dormancy. Both 2, 4-D and IAA have been found to be effective in breaking the dormancy of seeds, buds and cuttings. Growth retardants contribute to the formation of sprouts in potato. Premature potato sprouting may be prevented by the application of MH, CIPC and TCNB. Ethylene is known for breaking such dormancy. Some volatile essential oils, coumarin and other hydroxyl acetones have been successfully used for prolonging dormancy of seeds and cuttings.

Abscission. The abscission layer is a thin zone of small cells produced by newly divided cells, the walls of which becomes soft and gelatinous due to weakening of the middle lamellae of the cells so that eventually the slightest air movement or mechanical disturbance is sometimes sufficient to break the stalk at this point. Abscissic acid is responsible for the formation of abscission layer. Spray of dilute solutions of IAA, NAA and 2, 4-D neutralise or delay prevent the production of separation layer thus prolonging the life of the fruit of the trees. The exogenous IAA is much more effective in retarding abscission in summer than in autumn or early winter. Phenoxy auxins, *e.g.,* d-2, 4-DP (d-2, 4-dichlorophenoxy propinonic acid) and its optical isomer, 1-2, 4-DP and also 2, 4, 5-T have been widely used in the inhibition of abscission.

Alternative bearing. Auxin can reduce the alternate bearing of certain fruit trees such as apples. Triodobenzoic acid (TIBA) has been used for this purpose and it can also increase yield of soyabeans by 30%. TIBA is considered to be an antiauxin.

Mango, apple, Valencia orange, pear, olive, coffee are often found to bear irregularly or in alternate years termed as biennial bearing. The simple physiology of cure or prevention is the check on large drain on the nutritional resources of the plant in the 'on' year made by excessive bearing such that the normally coming 'off' year may be offset to guarantee full bearing. DNO (1-2%) at full bloom ; 20-60 ppm of napthaleneacetamide at the petal fall stage ; 25 ppm NAA three weeks after full blooms and N-1 naphthalene thalmic acid (NTA) have given encouraging results in the elimination of alternate bearing.

Vigour of the plant, level of nitrogen, the age of the plant, the time of spraying and elements of the environment affect the success with the chemicals. A vigorous tree needs higher concentration of the chemical to be effective than a weak, ill-fed or diseased one. The vigorous trees have chances of less leaf injury due to auxin application. Larger nitrogen availability to the plants makes them more resistant to auxin thinning. Susceptibility to the sprays for thinning increases with the age of the bloom in apples where there is the June drop (4 to 6 weeks after fruit formation). Thinning treatments become ineffective after the June drop. In the case of peach the thinning treatment becomes more effective with the lapse of 3 weeks after full bloom. Fruit thinning improves yield and quality of fruit, 2, 4, 5-TP (40 ppm spray) accelerated fruit ripening when applied 6-7 weeks before harvest.

Gibberellin. Plant activities like pollar transport, root initiation, inhibition of lateral buds, delay in leaf abscission are negatively correlated with GA. Gibberelins (1) overcome genetic and physiological dwarfism, (2) promote seed gemination and break dormancy, (3) induce flowering in long day plants and biennial plants, (4) affect elongation of cells, and (5) induce parthenocarpy in certain unfertilized ovaries.

Gibberellins may effect elongation of cells of young tissues only, *viz.*, significant elongation of internodes of autotrophic plants. Its effect on root initiation is considered to be supported by some and opposed by others. The gibberellin effect is both through cell division as well as cell elongation. Early flowering under the effect of gibberellins has been reported in plants irrespective of their nature of photo-sensitivity. Gibberellins may increase flowering in some species of plants. It is supposed to replace photoperiodic requirements of flowering in case of Hyoscyamus. Long day plants under non-inductive pnotoperiods can be induced to flower by gibberellin treatments of 100-1000 mg/litre. Obligate short-day plants grown under long day-length have not been induced to flower. In some short day plants gibberellin ingibits flowering, while in others it increases flowering but does not initiate it.

Gibberellins spray fulfills the cold requirement of many plants and helps them to reserve their growth even in the absence of cold treatment. The requirements like after-ripening or cold treatment or exposure to light or other factors before being able to germinate is nullified by only one gibberellin treatment in several instances.

Parthenocarpic fruit development in peach has been reported by the application of GA which also affects fruit-setting and fruit development. GA brings about maximum change in carbohydrate fraction of the fruits. Some fundamental dislocation in nitrogen metabolism has also been reported. GA increased nicotine in tobacco plants.

In fact the inter-relations between gibberellins and auxins are believed to be more influential than either alone. The effect of GA is complicated and that of auxin and GA still more complex. GA is known to effect leaf expansion leading to more area for photosynthesis and consequently higher yield.

Gibberellic acid has been known to be effective or reduce the setting of open pollinated flower blossoms of some cultivator grapes. GA is also used as a thinning agent for excessive flower blossoms in vine. GA acts as a pollenicide at anthesis in grapes. The dual action of GA in reducing pollen germination at the same time inducing parthenocarpy results in fruit drop which mainly consists of fruits, which are seedless but reduced in size. Gibberellins cause increase in length of cells although some strains of corn do not respond to gibberellin. The relative length of light and darkness control flowering in several plants. Some plants flower only if the light period exceeds a critical length. GA can substitute long day requirements in some species. It also overcomes the need of some species for an inductive cold period to flower. The effects of GA on the flowering of long day or cold requiring species can be made use of to induce bolting in plants (sugarbeet). GA also activates the cambial activity as in apricot and may act as auxins to prevent the development of lateral buds.

Cytokinin. It prevents the breakdown of chlorophyll and causes a few effects similar to responses by light. Certain seeds that require light to germinate need cytokinin for their germination. Cytokinin stimulates the formation of seedless fruits in *Ficus carica* and *Vitis vinifera.* Cytokinin also affects the abscission of leaves and transport of auxins. Cytokinin may prove to be valuable in extending the storage life of lettuce. Cytokinin can also initiate roots from African violet leaves if petioles are placed in moist medium.

Ethylene. It affects root formation in many ways, hastens blanching, accelerates abscission, inhibits elongation of roots and shoot and sprouting of potatoes, increases the rate of respiration of potatoes and proves effective in securing the germination of some seeds. Generally potassium salt of NAA is more effective than that of IAA in preventing the sprouting of tubers. Me NA, isopropylester of 2, 4, 5-T, TCNB, MH, gamma irradiation (8,000 r for 48 hours) are also widely effective in this regard. Ethylene (100 mg/l) also induces flowering in pineapple plant. It is used for colouring lemons, loss of chlorophyll of oranges and acceleration of ripening of banana (climacteric fruit) and also citrus (non-climacteric). In the banana (climacteric fruit) 0.1-1000 mg/l ethylene gas applied before the preclimacteric encourages ripening. Ethylene in conjunction with IAA encourages rooting. A resume of the success achieved with different fruit plants is presented (Table 14.1) to afford the nature of experiments conducted all over the world.

Table 14.1. Some Effects Induced by Growth Substances on Fruit Trees

Fruits (1)	Growth Regulators (2)	Effects induced (3)
Fig	IBA, NAA, 2, 4, 5-T and CPA	Early maturing parthenocarpic figs
Grapes and currant	4-CPA	Greater fruit set in grapes, synthetic auxins replace girdling
Grapes	4-CPA (5 mg/l)	Larger bearer
Grapes	GA	Improved yields (excellent for seedless varieties)
Cranberry	NOA, 2, 4, 5-TP and NAA (alone or in various combinations)	Increased fruit-set
Strawberry	GA and 4-CPA	Increased fruit-set
Phalsa	GA (10 mg/l), 2, 4, 5-T (5 mg/l)	Markedly increased fruit-set, the latter considered better for yield and quality
Pear var. Precoce de trevoux	NAA and its sod. salt (Lanolin paste) IAA spray (water solution)	Induced fruit-set to frost damaged unopened buds otherwise ineffective.
Bartlett pear	2, 4, 5-TP (100 mg/l) singly, spray before flower opening	20-26% increase in yield, wide varietal difference
Bartlett pear	Aqueous spray of 2, 4, 5-T and its alkanol, amine salt at 100 mg/l applied at bud stage.	Significant increase in yield
Pear	GA alone or GA + 2 naphthoxy propionic acid	Fruit set more in GA alone, GA + 2 Napthoxy propionic acid better than control
Pome	Sod. salt of (2-naphthoxy) propionic acid (100 mg/l) water solution, spray several times	Induced fruit-set.

(1)	(2)	(3)
Apple (millers seedlings)	Frost damaged fruit sprayed with water.	100% parthenocarpic-fruits developed (Cox's orange and Pippin showed no effect)
Plum (*Alpha* var.)	NAA, CIPA and GA	Negative results
Valencia orange	Aqueous spray of 2, 4-D	Better fruit-set, better size
Valencia orange	2, 4-D (16 mg/l) spray at flowering	Increased fruit size but no improvement in fruit set
Valencia orange and Washington Naval	2, 4-D (5 mg/l)	Reduction in fruit-drop, higher yield
Valencia orange	2, 4-D and 2, 4, 5-T (spray)	Significant increase in size
Washington Naval orange	2, 4-D	Increased fruit-set and size
Washington Naval orange	2, 4-D (8 mg/l) spray	Increased fruit-set and fruit size
Clamentine mandarin	GA (100 mg/l)	Increased fruit-set
Bearess lime	GA (100 and 1000 mg/l)	86 and 79% fruit-set as against 56% in control
Eureka lemon	GA (1000 mg/l)	Significant increment in fruit-set
Eurkea lemon	2, 4-D (8 mg/l), spray	Favourable
Sweet lime	GA (10 mg/l) spray, 2, 4-D (10, 15 mg/l), 2, 4, 5- T (10-15 mg/l)	Increased fruit-set but not as much as by GA
Apple	NAA (5 mg/l), 2, fl. oz/plant	7-14% increase
Pineapple	High concentration of NAA	Good
	Low concentration of NAA	Not effective
Pineapple orange	2, 4-D (25 and 50 mg/l)	Effectively reduced the fruit-drop

Common Abbreviations

Ga, Gibberellic acid ; IAA, Indole-3-acetic acid ; IBA, Indole butyric acid; 2, 4-D, 2, 4-Dichlorophenoxyacetic acid ; TIBA, 2, 3, 5-Triodobenzoic acid ; NAA, Naphthalene acetic acid ; MCPA, 2, methyl-4-chlorophenoxy-acetic acid; 2, 4, 5-Trichlorophenoxy acid ; TCBA, 2, 3, 6-Trichlorobenzoic acid ; Chloro-IPC, 3-Chloro isopropyl-N-phenylcarbamate ; CPAC, Chlorophenoxyacetic acid ; CIPA Chlorophenoxyisobutyric acid ; TCPA, 2, 4, 5- Trichlorophenoxypropionic acid ; ABA Abscissic acid ; NAD, Naphthalene acetamide ; KNAA, Potassium salt of naphthalene acetic acid ; TCNB, Tetra chloronitrobenzene ; CIPC, Isopropyl N-(3, chlorophenyl) carbamate ; CCC, 2-Chloromethyl trimethyl ammonium chloride : Phosfon D, Tributyl-2, 4-dichlorobenzyl-phosphonium chloride ; AMO 1618, 1-(-N-Piperidine carboxy)-2-isopropyl-5-methylphenyl-4-trimethyl-ammonium chloride ; KN, Kinetin.

(6-furfurylaminopurine) ; IAN, Indole-3-acetonitrile ; IAE, Indole-3 acetic acid ethyl ester ; ICA, Indole-3-carboxylic acid ; IPYA, Indole-3-pyruvate ; ILA, Indole-3-lactic acid ; CK, Cytokinin ; B-9, Succinic acid, 2, 2-dimethylhydrazide; MH, Maleic hydrazide, CEPA, 2-chloroethylphosphonic acid ; CH, Cycloheximide; BA Benzylademine ; IAdo. N (2-isopentenyl) adenosine ; Ade-Co-thr, N-(Purin-6-ylcarbamoyl) threonine ; Ado-Co-thr, N-(9-B-D-ribo-furanosyl-9H. Purine-6-ylcarbamoyl) threonine ; BAP, N^5-benzylaminopurine; RZ, Ribosylzeatin ; Z, Zeatin ; 2 ms, 2-methylthio ; NTA, Naphtalene thalmic acid. TMTD-Tetra methyl thiuram disulphide ; TEPP-Tetra ethyl pyrophosphate ; EDCT-Ethylene dichloride and carbon tetrachloride ; EDB-Ethylene dibromide; DD-Dichloroprapane dichloroprapine.

□□□

PART–II

SCIENCE AND PRACTICE OF FRUIT CULTURE

1. AONLA
2. AVOCADO
3. BANANA
4. BER
5. BREAD FRUIT
6. CHIKU
7. CITRUS
8. CUSTARD APPLE
9. DATE-PALM
10. FIG
11. GRAPE
12. GUAVA
13. JACK FRUIT
14. JAMUN
15. KARONDA
16. LITCHI
17. LOQUAT
18. MANGO
19. MANGOSTEEN
20. MULBERRY
21. PAPAYA
22. PASSION FRUIT
23. PHALSA
24. PINEAPPLE
25. POMEGRANATE
26. APPLE
27. APRICOT
28. CHERRY
29. JAPANESE PERSIMMON
30. PEACH
31. PEAR
32. PLUM
33. STRAWBERRY
34. ALMOND
35. CASHEW
36. HAZEL NUT
37. PECAN NUT
38. WALNUT

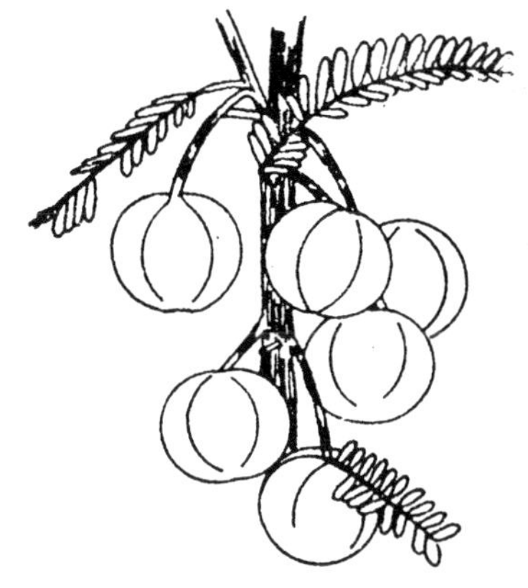

CHAPTER 1

AONLA

(*Emblica officinalis*)

Euphorbiaceae

Origin and Distribution

The aonla tree is indigenous to tropical Asia and is widely grown in tropical India as well as on the hill slopes upto an elevation of 1350 metres. The 'aonla' (*Emblica officinalis* Gaerth) is called by several names in different parts of the country, *e.g.*, Indian gooseberry, *'myrobalan', nelli, 'aonla', 'malacca tree'* and *'amlaki'*. *Phyllanthus acidus* Skeels (*Averrhoa acida* Linn, *Cicca disticha* Linn, *Phyllanthus distichus* Muella., Arg.), well known as Otaheite gooseberry or Star gooseberry is mostly grown for ornamental purposes. The fruits are round, deeply ribbed, pale green and highly acidic and attractive. Male and female flowers appear in spring when hot winds appear in North India. *Phyllanthus fischeri*, a wild species is found in South Indian forests and its fruits are used in pickling.

Emblica officinalis is referred to as an evergreen tree although its leaves are deciduous. The main trunk looks like that of guava. The bark is smooth and comes off in thin strips or flakes. The fruit is nearly stalkless and smooth divided into six lobes through pale linear grooves. The surface is shiny, quite hard with a thin and translucent skin. The hexagonal stone containing six small seeds is embedded tightly in the flesh. The fruit tastes bitter and extremely acidic. As a good source of ascorbic acid (500 to 750 mg/100 gm. of pulp) it is used in several medicines and also utilized either in fresh form or as a preserve, like *murabba* and *chutney*. It is richest natural source of Vitamin C and works better than synthetic ascorbic acid in the cure of deficiency diseases.

Climate

Essentially a subtropical fruit but can be successfully grown in tropical as well as dry conditions. It is not much affected either by desiccation through hot wind or frost. Under the North Indian conditions, the tree starts shedding all its leaves from February onwards and by the third week of March it is almost devoid of foliage. New branchlets emerge at nodes from the side of the scar left after the abscission of previous season's branchlets in the third week of February and continue till the first week of April. The process of shedding of branchlets overlaps the appearance of new branchlets on the naked branches.

Aonla is characterized by very small leaves attached to branchlets in such a way that each branchlet looks like a compound leaf.

Propagation

Soaking of seeds in gibberellic acid solution (500 ppm) for 24 hrs. increased germination as well as the root system. Cuttings from branches yielding good crops, when treated with IBA (1500 ppm) also succeeded well. Propagation from endosperm by the micropropagation technique also gave good results. Aonla plants when raised from seeds bear fruits of inferior quality. Shield budding has been successfully tried at Saharanpur (U.P.) and also in Madhya Pradesh and it is reported to be cheaper and better than inarching. One-year old aonla seedlings having girth of about 1 cm. are shield budded in early June. For top-working, the inferior trees are headed back to a height of $1\frac{1}{2}$ metres in March and the shoots produced from that are shield budded in May with scions of improved cultivars.

Patch budding is being tried with equal success. Budding done in the month of September gives almost 100% success and most of the plants survive after transplantation.

Cultural Practices

Planting of aonla seedlings should not be done during Oct. -Feb. A distance of 10 metres all round is recommended for good cultivation of aonla grafts which are planted in pits. Fortnightly irrigation is required in summers. Aonla plants may be given 10 kg. compost, 100 gm. nitrogen, 50 gm. phosphorous and 75 gm. of potash in the first year. This quantity is doubled in the second year, trebbled in the third year and so on till in the 10 year the quantity becomes 100 kg. compost, 1 kg. nitrogen, 500 gm. phosphorous and 750 gm. potash. This quantity is given each year thereafter. Entire quantity of compost and phosphorous along with 1/2 of nitrogen and potash should be given in June-July and the remaining after a few months. 3-4 kg. of superphosphate is necessary preferably at the preflowering time to boost up the bearing capacity of the plant. Aonla cultivation is also possible in soils of pH 8.5 to pH 9.5. The rise in pH needs some change in the quantity of additives for proper production. While the quantity of FYM remains 50-60 kg/cu meter and sand 15-20 kg/cu meter in the pits the quantity of gypsum should be raised from 5 to 15 kg. and of pyrite from 4 to 12 kg/cu meter with increase in soil pH. In addition it is recommended to add $ZnSO_4$ @ 50 g/pit along with the FYM-sand-gypsum-peyrite mixture in the pits. Plants bear fruits, generally, after 8 years, although the newly evolved cultivar N.A. 7 has the reputation to start bearing in three years.

Several methods have been employed by horticulturists to get good yield of attractive fruits. The more common are as under :

(1) Spray in September 0.5% urea, 0.4% Agromin and 0.5% potassium sulphate

(2) Spray at fruiting stage Dithane M 45 @ 0.3% and methyl Demiton 25EE @ 0.03%.

(3) Spray borax @ 50 g/tree to ward off boron deficiency.

(4) Plant 5% desi variety of Aonla in the orchard.

(5) Spray cultar @ 3 ml/l water at the collar zone in April

(6) Spray GA @ 150-200 ppm to check fruit drop.

(7) Prune 10-15 cm. of the terminal portion of the old branches from top.

Bearing Behaviour

The budded plants commence bearing from the fifth year. Vegetative growth of the tree continues from April to July. Alongwith the new growth in the spring, flowering also commences. Flowers are of two types : (1) male and (2) female. Both the male and the female flowers are located on the branchlets, the male in the axil of the leaf in clusters throughout the branchlet while the female flowers are located at the apex of only a few branchlets. The ratio between the male flowers vs. female flowers has been recorded as 307 : 1 and 197 : 1 in two years of observation. Male flowers predominate female flowers. Sometimes there might arise branches in aonla which have almost all male flowers. If buds are taken from such a branch for propagation the resulting plants may turn out to be unfruitful although the tree may look very healthy.

The trees remain dormant from April to July and resume growth from the last week of July to first week of August subsequent to dormancy break. The seedling trees flower first. Flowering periods continues for about a month. Aonla trees have been reported to flower twice in south India. February-March and June-July, with a very poor crop in the latter case. Under North Indian conditions, removal of shoots from the tree between July to October induces flowering but an economical crop is not obtained.

The flowering periods in aonla is less affected by temperature fluctuations than by changes in day-length.

Pollination. Wind, honey bees and gravity—all three plays an important role in the pollination process of aonla flowers. Common visiting hours for honey bees is late evening and morning. The initial percentage of fruit-set may vary from 12 to 18 per cent in Banarasi. Mechanical pollination or incorporation of honey bee in aonla orchard may improve percentage of fruit set. The poor fruit set in Banarasi and Francisi may be due to low sex ratio or lack of pollination.

Three waves of flower and fruit-drop are observed in aonla. More than 70% of flowers drop within three weeks of flowering which is normally due to lack of pollination. The seccond drop is that of young fruitlets at the time of dormancy break. The third drop of fruits due to embryological and physiological factors is spread over the entire period of the developing fruit starting from latter half of August to harvest. Spray of GA @ 100-200 ppm checks fruit-drop.

Fruiting problem. In a survey conducted to study the problem of aonla growing in Uttar Pradesh it was observed that amongst the cultivars grown in compact blocks Francisi suffered most from fruit necrosis (60-80%) whereas the fruits of Chakaiya were not affected. Fruit mecrosis symptoms start with the browning of the mesocarp in the fruits extending towards the upper surface and resulting finally into brownish black areas on the fruit surface in the second and third week of October. Depending upon the severity of the disease mesocarp of the affected fruits turns black from brown which later turns into corky and gummy pockets. The problem has been considered to be non-pathogenic in nature.

Control. Borax spray (0.6 per cent) twice or thrice in the months of September-October with about 10 to 15 days interval.

CONSTITUENTS OF THE FRUIT

The fruit has following contents and is considered very useful for human consumption :

Contents	Per 100 gm. of the edible portion
Moisture	80 gm.
Minerals	0.5 gm.
Carbohydrate	13.7 gm.
Calcium	50 mg.
Phosphorus	20 mg.
Iron	1.2 mg.
Protein	0.5 gm.
Fat	0.1 gm.
Fibre	3.4 gm.
Vitamin B	30 mg.
Nicotinic acid	0.2 mg.
Carotene	9 micrograms
Vitamin C	600 mg.

Traces of other vitamins such as thiamin (Vitamin B_1) and riboflavin (Vitamin B_2), are also found. Among the fruits, aonla is the richest source of vitamin C except Barbadose cherry. The fruit of Aonla has a high content of polyphenol in the pulp which protects is from oxidation for long periods. Besides it has high percentage of tannins also. The presence of six gibberellin–like factors and six cytokinin–like substances has also been reported in aonla fruits.

Harvesting and Marketing

The aonla blossoms late in spring and the fruits ripen during winter. In southern India the trees bear the fruits throughout the year. Two quintals/plant may be the annual production of well looked after trees of a good cultivar.

Cultivars

The aonla fruit may be distinguishable on the basis of their colour.

Banarasi aonla. The fruits are fairly large in size, shining, thin skinned, transparent, and of yellowish colour. These are good for preserves. This is considered to be the best cultivar for *murabba.* Av. wt. 46.6 g fruit.

Low yielder, sensitive to User conductions.

Green-tinged cultivar. The fruits are fairly large, nearly green in colour and utilized for the preparation of pickles and *murabba. Red-tinged cultivar* : Although the fruits are small these have attractive red flush near their base. *White streaked cultivar :* The tree bears medium-sized fruits which have white streaks on them.

The well known cultivars of Aonla may be put in three categories according to their maturity (1) Early maturity (mid Oct. to mid Nov.) Comprising Banarasi, Krishna, NA9 and NA10. (2) Mid-early (maturity-Nov. end) *viz.* chakaiya, Kanchan, NA6, (3) Late : (Maturity late December) *viz.* francisi, NA7 (Nadia ke par).

Hathijhool or Francisi is a commercial cultivar of *Aonla* of U.P. The trees are fairly large, spreading and heavy-bearing. The size of fruit is good (Av., wt. 41.5 g) with shining, conspicuous glands. Egg-shaped large fruits of cultivars Hathijhool are prone to necrosis and turn black. Two-three sprays of 0.6% borax every 10 days is recommended to check blackening of the fruits. The sprays should be given as soon as signs of the blackening begin to appear.

Chakaiya is also a commercial cultivar which bears heavy. The tree size is large, upright with heavy bearing shoots. The fruits are small, fibrous and flattened in appearance. It is free from blackning (necrosis) disease of Aonla to a good degree. This cultivar is very common in U.P. Av. wt. 30-40 g.

Besides the above well known cultivars NA-7, *Kanchan* (NA-5) and *Krishna* (NA-4) are newly evolved and promising ones. *Krishna* a chance seedling of Banarasi. Fruit size medium to large ; colour yellowish ; skin smooth with red flush. Flesh fibreless, hard and semi transparent. Av. wt. 40 g/fruit.

Kanchan a chance seedling of Chakaiya. Fruits medium to large, Av. wt. 32 g ; skin smooth, suitable for pickles, profuse bearer. NA7 (fruit 47.5 gm wt.) starts bearing in three years and the fruits are of good quality. *Kanchan* and *Krishna* both have fruits of good size. *Kanchan* is a heavy yielder and *Krishna* is good for making *murabba*.

Pests. No insect pest of significance is reported. Caterpillars bore the shoot, gall is formed and the growth of the shoot is checked. Among the pests *Betousa stylophora* and *Inderbela* sp. the shoot gall maker and the bark eating caterpillar respectively are of importance. Shoot gall maker caterpillars bore into the shoot and get into the pith region during August and September. The damaged tissue develops into a gall-like structure. Removal of infected portions followed by spray of monocrotophos 40 EC @ 1 ml/lit. of water during July-August controls pests.

Scale insect. A tiny insect having covering on it is transmitted by ants. The insects secrete honey which is responsible for sooty mould. Spray of monocrotophos, 0.05% is useful to check damage.

Aphid

Both nymph and adult damage the crop during April-September. Methyl dimaton @ 1 ml. or Endosulphan 0.1% in water spray can be end to control the pest. The use of natural predator coccinalidis is also successful in the control.

The bark eating caterpillar gets into the main trunk and makes tunnel. The pest can be controlled by spraying 0.5% monocrotophos or injecting kerosene oil or petrol in the holes and plugging them with cotton and waxing. Inserting powdered celophos tablets and plugging the holes is also effective.

Diseases – Blue mould (*Penicillium islandicum*). Brown patches with water soacked areas are formed on the fruit which are finally covered with bluish green pustules. Spray of borax or sodium chloride solution of low concentration has been found to control the mould.

Aonla Rust (*Ravenellia emblica*) causes brown pustules on fruits and leaves which turn dark brown after some time. Control of the rust is possible by spray of dithane Z-78 at 0.2% concentration. Banarasi and Chakaiya cultivars are supposed to be relatively free from the disease.

Boron deficiency symptoms have been reported causing internal necrosis. Three sprays of 0.6% borax at 15 day intervals from early September removes the deficiency. Application of borax @ 50 gm/tree in the soil is also recommended.

Rains in the month of september cause untimely flowering in some parts of Uttar Pradesh. Many times the increase in insect pest and also incidence of disease have been reported. To overcome plant protection method and agronomic practices need attention to reduce the damage.

□□□

AVOCADO

(*Persea americana* Mill.)

Lauraceae

Source and Distribution

Orginated in America, avocado (butter fruit), a staple food and native to tropical (central) America was introduced in India almost 1000 years ago. It has a high nutritive value, rich in fats (5-7%) and proteins (1-4%). Avocado is also a rich source of carbohydrate, minerals (K, P, Mg, S, Ca, Cu, Fe, Mn, Na) and vitamins A, Thiamin, Riboflavin, Niacin and C). This fruit has high calorific value but the sugar content is low making it useful for persons suffering from diabetes. It has twice as much energy value as in bananas. Avocado is common in West Indies, Hawaii, Philippine Islands, Cuba, Sri Lanka and India. Evergreen subtropical fruit, culture limited to areas with relatively mild winters.

Climate and Soil

The West Indian race of avocado (*P. americana*) enjoys a humid tropical climate. Under India conditions the minimum temperature ranges between 15-22°C and maximum between 27- 34°C in avocado growing tracts. The avocado prefers mild winters of the tropics and sub-tropics. The trees are tender to frost and do not stand hot dry winds. *P. americana* is successful in south India specially in areas with rainfall of 75 to 180 cm. The avocado loves a loam of medium texture overlying a porous sub-soil with proper drainage. Some of the loams and sandy loams of alluvial origin are reported to be excellent for avocado. Poor drainage and water logging are injurious.

Propagation

Commonly propagated by seeds. Seeds of avocado have short viability and must be sown within two weeks of extraction from the ripe fruit for success in germination. Vegetative cultivation is often done by cuttings from Mexican race which roots easily. The west Indian race is difficult to root. Leafy cuttings from young plants root easily. Supplying bottom heat (26°C), continuously soaking the basal end in IBA solution and spray of NAA + cytokinin every week to check leaf shed also facilitate rooting. Vegetative propagation methods such as layering, shield-budding, Forkert-budding or side-grafting are followed, Budding (chip and shield methods), grafting (approach, veneer, cleft and whip) have been successfully employed for the propagation of avocado. Girdling of the alternate bearing cultivars increases the yield. The plants become ready in six months and may be planted in the field after another six months and may be planted in the field after another six months. Purple avocado serves as a root-stock. Rootstock of Mexican race is preferred. West Indian race does well in salin soils and warmer areas. The rootstock from the Gautemalan race does well in saline soils and warmer areas. The rootstock from the Guatemalan race does not compare well against that of Mexican race. Care must be taken not to have the rootstock as well as the scion from the Maxican race because it makes the plant most susceptible to salt. Some rootstock selections are reported to be resistant to phytopthora infection. Rootstock of the West India race of the Nachlat origin is highly tolerant to lime.

Planting

The seedlings are removed from the nursery beds with a ball of earth round the root system for transplanting in the orchard. Hand weeding and hoeing are important operations. Weedicides like simazine and monouron are recommended to get rid of weeds. The tree is planted over pits (1 × 1 × 1 metre) filled with top soil mixed with well rotten F.Y.M., compost or leaf-mould. The distance between plants should not be less than 10 metres. The graft-joint should be well above the ground while planting supported by stakes against possible damage from strong winds.

Structurally bisexual flowers are functionally unisexual in avocado. Dichogamous flowering male and female parts of the flower mature at different times *i.e.* all the open flowers on the tree are female (♀) at one and the same time and the male (O) at another time. This means that the flowers are functionally female at one time and male at another time. The pistil matures earlier than the stamen. Self pollination is ruled out. The avocado cultivars are of two groups A and B with respect to their floral behaviour. The flowers of group A open first in the morning and the second opening is in the afternoon of the next day. The flowers of group B cultivars open first in the afternoon and the second opening is in the morning of next day. Thus it is possible that every morning the pistils of A can be fertilised by B pollen and every afternoon the B-pistils can be fertilised by A pollen. Based on the behaviour and nature of flowers in avocado the planting in any orchard must be done so that plants of the two groups A and B are present side by side.

Cultural Practices

Preflowering period is the fittest for cultural operations like ploughing and harrowing in the orchard. An application of fertilizer and manure mixture *i.e.*, 60 kg. of F.Y.M. 5 kg. of castor cake + 1 kg. of superphosphate should be given to a plant at the age of about 7 years. This is followed by immediate watering, second watering after 3 days subsequent waterings every week. Watering by sprinkling is preferred since flooding promotes root rot. While no definite recommendation can be made with regard to nutrient supply to the plants it is believed that nitrogen, P_2O_5 and K_2O should be supplied in equal amounts to young trees. With increase in age of the plant the amount of nitrogen and K_2O be doubled, phosphorus remaining the same. Zinc deficiency symptoms have been reported to avocado plants. Sprays of 150 g. of zinc oxide in 100 litre of water in the nursery is recommended. This is even more necessary in neutral and alkaline soils.

Purple and Green cultivars flower and set fruit in January ; the Purple avocado fruit matures in June and July, whereas the Green cultivar matures later, in August and September. Besides unfavourable temperature, frost injury, poor fruit-set, alternate bearing and foot-rot are factors limiting the fruit yield from the orchard.

Harvesting and Marketing

The fruit is clipped as soon as it changes colour (purple to deep maroon and green to greenish yellow) so that it may be marketed before becoming soft, with ease. The fruit (berry) is something like a pear in shape with a single large seed surrounded by buttery pulp and a hard skin. Mature fruits ripen early with ethylene treatment. Softening of the fruits takes place only after picking, it is not known in fruits still attached to the plant.

Mexican race (*Persea drymofolia*) has small fruits. It has leaves with very sharp apex and small flowers. It does not thrive in tropical regions. The Guatemalan and the West Indian race (*P. americana*) bear larger fruits which are less resistant to cold. Hybrids between the two races are commonly more adaptable to California conditions than the Mexican race.

Cultivars

Cultivars commonly grown in India are purple, green, fuerte, paradeniya purple hybrid and pollock. Yulia and Ardith are cultivars selected from Californian material while Hass and Lula are from Guatemalan race.

The cultivars of Avocado may be put in four different races :

(1) *Guatemalan race.* Low oil percent (8.15%). Hass, Lula, Green belong to this group.

(2) *Hybrid race.* These have been developed by crossing Guatemalan and mexican races. Prominent cultivars of this race are Fuerte, and Winter Mexican.

(3) *Mexican race.* Most resistant to cold temperature yielding highest percentage of oil (30%). Grater, Gopatopa, Dube and Herman and the e..amples of this race.

(4) *West Indian race.* The cultivars in this groups have lowest percentage (3-10%) of oil. It is least resistant to cold temperature. Purple and Pollock are examples of this race.

Fuerte. Believed to be hybrid between the Guatemalan and Mexican race. The fruit is pyriform. The flesh has a buttery texture, a rich nutty flavour and has about 18 to 26% oil. Common cultivar with alternate bearing habit.

The Green. It is believed to be a member of the Guatemalan race. The fruit is large (250-400 g), oval to obovate. The seed coat thin, closely united and adherent to the cotyledons throughout. The seeds are embedded tightly in the cavity. The 2.5 cm. thick flesh is soft greenish yellow changing to green close to skin.

The Purple. The West Indian race includes this cultivar. The fruit is pear shaped with a long neck and about 250 g in weight. The surface is smooth and deep crimson in colour, the skin is leathery in texture and easily removable from the pulp. The seed is large, conical and is held loose in the seed cavity. 2 cm. thick flesh is firm, smooth, fine in texture, deep, yellow changing to yellowish green close to the skin. Leaves are elliptic in shape.

Ardith. Pear shaped dark green fruit with long neck, rough rind, av. wt. 250 g, good flavour.

Yulia. Small fruits of about 200 g. Good yielder but bieniel bearing.

Hass. Belongs to Guatemalan race. Earlier than Fuerte. Small fruits becoming purple on maturity. Useful for subtropical region.

Pollock. A cultivar of West Indian origin with large fruits (1000 g) and low oil content. Grown in Florida, Alternate bearer.

Lula. Large pyriform fruits (400-700 g) suitable for tropical climate of Florida.

Diseases and Disorders

Among the diseases of avocado *root rot* is possibly the most damaging and extensive. It may even finish the tree in soils which are not well drained. Soil treatment with metalaxyl @1 g a.i./10 kg. soil before planting of seedling or spreading 1 g a.i./10 litre round the plant wards off the attack for about 4 months. Metalaxyl applications in higher doses is also able to cause recovery of plants infected with root rot.

Cercospora spot and scab attack the fruits as well as leaves. Fusarium infection causes accelerated softening of fruits. Some cultivars are susceptible to anthracnose and to stem end rot *e.g.*, fuerte.

Fruit spot diseases caused by *Collectotrichum gloeosporiodes* leads to fruit shed and deformity and arrest of increase in size of the unshed fruits. Spray of lime-sulphur in water checks the disfiguring of the fruits. A disorder of the avocado in which there is a drying out and shriveling of surface areas on the fruit pedicels is also reported. It is commonly known as Ring Neck.

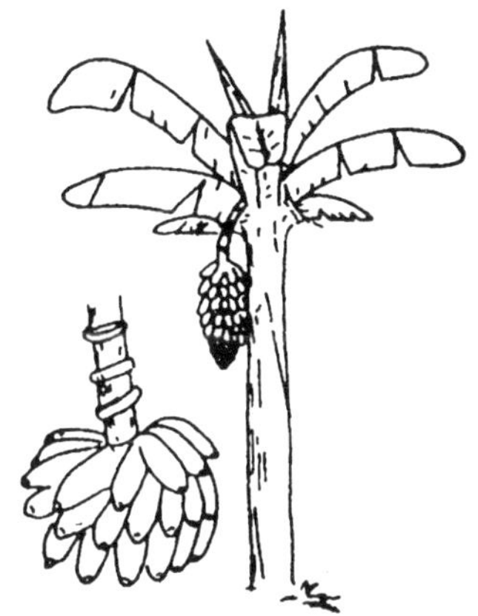

CHAPTER

3

BANANA

(*Musa paradisiaca* subsp. *sapientum*)

Musaceae

Origin and Distribution

Banana (*Kelaa* H.) believed to be one of the oldest fruits probably originated in the warm moist tropical Asia where Assam, Mayanmar, Thailand and Indo-China meet. Its name 'Adam's fig' speaks of its antiquity. Besides Andhra, Andaman, Assam, Bhopal, Tamil Nadu, Kerala, Maharashtra, Karnataka, West Bengal, Bihar, U.P. and Himachal Pradesh, Orissa, Madhya Pradesh in India, other notable producers are Philippines, Thailand, Hawaii, Fiji, Congo, Egypt, Israel, Ghana, Sri Lanka, Taiwan, Indonesia, West Indies, Florida, Southern California. In Uttar Pradesh economic cultivation of banana is feasiable in the districts of Basti, Baharaich, Gonda and Gorakphpur due to favourable climatic conditions. Stray plants have been fruiting in places like Kanpur and Allahabad but the quality, size and yield of fruits is not good. A fruit of major importance occupying nearly 1/5th of the area under fruits in the country. Its cultivation is confined to regions between 30°N and 30°S.

The commercial bananas are classified by Linnaeus into 3 species (*i*) dwarf sp. *i.e. M. cavendishii,* (*ii*) tall ones *i.e. M. sapientum,* whose fruits are edible raw, and (*iii*) *M. paradisica* whose fruits are cooked. Cultivated forms of bananas are classified in a way which is confusable. On the basis of cytogenetical studies, it has been concluded that *Musa paradisiaca* developed from the wild species. *M. acuminata.* Another group seems to have been derived from the seedy *M. balbisiana* while *M. sapientum* appears to contain those developed from hybrids of the two wild species. Chatterji and Randhawa consider *M. nana* the correct name for the dwarf banana rather than *M. cavendishii.*

The banana plant has good height 3.5 to 7.5 M or more, the stem consists of a column of sheathing petioles of spirally arranged dark to yellowish green leaves which are variable in size having an obtuse and entire but easily torn margins. Plant, a perennial shurb, has a large, starchy, subterranean rhizome studded with buds. The inflorescence, a spike has three types of flowers, *viz.*, neutral, staminate, and pistillate. Parthenocarpic fruits are also formed on the plant.

Culture

The seed propagation in edible banana. Off shoots popularly known as suckers and the rhizomes may be used for propagation. Where the climate is not dry and appreciable humidity exists in the atmosphere whole sucker is used to propagate the banana. Sucker may be of two types : (*i*) Sword sucker, which are considered superior to the (*ii*) broad leaved waver sucker. Three month old 50-60 cm. tall sword sucker gives better performance.

Banana can also be propagated by a bit from rhizome possessing at least one bud. Corms or rhizome are easy to handle. It is held that the cut portion of the rhizome must be allowed to dry for several days before planting. This practice is common in West Indies. Corms must be selected from high yielding plants giving excellent performance in every way. The suckers or rhizome pieces are prepared for planting by trimming off all roots and leaves, drying in the open for about ten days and then treating with a 5% solution of formaldehyde or diaflotan (0.2%).

Micropropagation. The need to multiply a new clone of desirable character in large numbers in a short period prompted the scientists to resort to micropropagation or tissue culture. Shoot-tips, adventitious buds, determinate floral buds, meristem tips, apices and other plant tissues have been employed for the purpose in different media with or without the aid of chemicals, growth regulators or physical means.

Cultivation

Planting may be done during the rains and when the night temperatures are not low. Planting may be done in winters or heavy rains. While plantings are quite common in September-October and Feb.-March in Bangladesh. In the western coast the months of April, June and August are considered best. A deep rich, moisture retentive soil containing plenty of organic matter is needed. Good manuring is a must for good planting results. The methods of placing the suckers in places where planting is desired is not as important as abundance of manure and adequate moisture. The soil must not be more shallow than 45 cm. On light soils the usual planting distance is about 1.8 × 1.8 M to 3 × 3 M to give an average of around 3000 to 860 palnts per hectare ; the best spacing on heavy soils is 3.3 × 3.3 M. Tall cultivars need more space than the dawrf. The planting holes are 50 cm. in diameter and also 50 cm. deep. Earthing up the interculture operations are welcome.

In the case of *Basrai* variety which is grown on soils with about 10,000 to 15,000 kgs. of manure/hectare the field is prepared and the suckers are planted in July in deep ploughed furrows 1-2 M apart. Each bed consists of 4 plants and between the beds there are channels which are filled with water every 3-days on an average in the months of November to May. Top dressing of 0.2 kg. N/plant in the form of groundnut cake is given. At times some sulphate of ammonia may also be used to supplement the nitrogen need of the plant.

Application of fertilizers should be done 30 cm. away from the plant and at a depth of 15 cm. in a circle round the pseudostem. Nitrogen may be supplied in five equal doses in the months of August, September, October. February and April. Application of phosphorus preferably in two split doses, the first at pre-planting stage and the second two months after transplanting. One foliar application of Agromin 4 gm + Urea 10 gm + muriate of potash 10 gm per lit. of water three weeks after flowering gives good harvest of attractive fruits.

Being a heavy feeder it needs good quantities of fertilizers which varies from place to place and with variety to variety. 450 gm. nitrogen as sulphate of ammonia, 270 gm phosphate as superphosphate, and 560 gm potash as muriate of potash per hectare is considered sufficient in three split doses 2, 4 and 6 months after planting. A banana plant needs 200 gm N_2, 100 gm P_2O_5 and 200 gm K_2O every year for good growth and high yield. The entire quantity of phosphorous and 1/3 of nitrogen should be given at the time of planting. Of the remaining half should be given in September and the rest in June. Compost, neem cake, bone-meal or any other organic form of nutrient is always helpful in increasing the yield. The rapid growth, high yield and shallow root system of banana necessitates heavy fertilisation, the per unit are intake of nutrients is more than for any other crop. Foliar application of B, Cu, Mn, Mo and Zn ensure healthy growth and higher yield. The deficiency of nitrogen turned the colour of the leaves to pale green and that of potassium reduced the number of fingers in the bunches. Magnesium deficiency is related to Mg/Ca ratio of the soil. The Mg/Ca value lower than 0.2 made the deficiency symptom more acute.

Though at no stage the banana plant should be devoid of soil moisture yet stagnation is undesirable. Plantations are irrigated in summer. Irrigation to be given according to the needs of the crop. Sprinkler irrigation yields better results. Tall cultivars may be grown under barani conditions but dwarf cavendish needs regular irrigation.

Application of about 15 kg. of silt, 2 lbs of mustard cake and 25 gm. of $(NH_4)2SO_4$ to a plant placed at distance of 1 metre has given good yield.

Select planting material free from weevil, Panama disease, bunchy top and mosaic, add 10 gm. carbofuron or 5 gm phorate granules to the top soil at the time of planting. For leaf spot disease spray with 60 gm Blitox in 18 litres of water. Remove and destroy plants affected by bunchy top or mosaic. The bunches bend in the direction of the sun and to protect against scorching, these should be covered with banana leaves.

Intercropping is done in Banana in various parts of the country in the first 4-6 months of planting. Sometimes it is grown to provide shade to young plants of melon, mango, orange, coffee cocoa etc.

Weeding is necessary and paying. It can be done mechanically or through the use of chemcials *viz.* diuron, simazine, paraquat. Other cultural practices include desuckering, earthing up during rains and regular vigilance against pests and deseases.

Climate and Soil

A tropical humid region crop which may be grown upto elevations of 1200 M or so. But for dwarf banana these are not grown above an elevation of 100 M above sea level. Well fertile as well as productive top soils are needed for profitable commercial growing of bananas. Rich deposits of silt, deep cotton soils with no water stagnation favour the growing of banana. Alluvial soils (pH 4.5 and 7.5), more or less evenly distributed rainfall over the entire year and protection against heavy winds should be considered as ideal. It does well in areas of high rains. Warm humid and rainy areas near the tropics suit banana. Arid zones with strong hot wind and freezing temperature during the winters do not find with this fruit crop.

Removal Floral Remnants in Robusta Banana

In Cavendish banana, the floral parts namely style, stamens, the perianth comprising the compound and free petals remain adherent on the finger tips during fruit development. As the fruits develop, the parts become completely twisted, blackish, and desiccated, the style alone remains erect and hard, detracting the appearance of fruits. Such withered floral parts on the tip of fingers get covered with fungal spores and form the seat of initial infection for fruit tip disease which causes much loss and bunches become unsaleable. The removal of withered floral parts by hand, has proved effective not only in the control of the fruit tip disease but also in increasing attractiveness of the bunches.

In Robusta, it has been found that the withered floral remnants could be removed between the eighth and twelfth day after the bunch emergence. At this stage of development, the base of the floral parts on the fruit tip becomes soft as jelly. In fact, at this stage, the floral parts fall off by a light brushing of the fruit tip by the hand. Because of such a condition the elimination of these floral parts becomes simple and rapid. However, certain precautions have to be observed to avoid as far as possible the damage to fruit tips. The damage to the fruit tips not only detracts the fruit appearance but also provides access to pathogenic organisms.

Once the soft jelly like stage is passed, the removal of withered floral parts become difficult and can only be done fruit by fruit, because with drying off, the floral parts become more firmly attached to the fruit tip.

Tipping in Robusta banana has induced earliness in fruit maturity. The fruits are attractive, well developed and give a distinct market appeal with better returns.

Ripening of the fruits. For ripening bananas, the following methods are adopted :

(*a*) In Tamil Nadu, Kerala, maharashtra, Karnataka and the Punjab the bunches are smoked. Smoking has to be carried out for 2 to 3 days or even longer. It takes about 5 to 6 days for ripening. Bananas so smoked are likely to have black specks and other skin blemishes besides producing a smoky flavour.

(*b*) In other instances bananas are stored in warehouse and ethylene gas air : : 1 : 1000 mixture is let in.

(*c*) Robusta bunches take about one to two weeks for ripening without any aid. Experiments carried out revealed that Ethrel solution (200 ml of 5000 ppm) with 5 gms. of NaOH pellets kept in a beaker in the ripening chamber ripened the bananas within 48 hours. The colour is more attractive, the fingers glossy, and golden yellow fruits sweeter as against the dull green colour of ordinarily ripened fruits. The aroma and flavour of Ethrel treated fruits are pleasant and delightful.

Harvesting

Banana flowers in about 9-12 months and the fruits mature in another 3-4 months. As the fingers become round and plump with change in colour to light green these are harvested. Crop is ready for harvest in 12-16 months after planting, depending on the variety. The yield is roughly 30-40 tons in Cavendish dwarf 35-45 tons in Robusta and 20-30 tons in the case of other varieties per hectare. The fruit may be stored to ripen at a temperature of 16-21°C for a week or two or smoked for 14-23 hours during summers and not less than 48 hours during winters.

Storage

Temperature above 13°C and relative humidity over 85% enables storage of banana for about 3 weeks. Ripening of banana takes a week or 10 days at a temperature range between 17-21°C. Storage life of banana can be prolonged by keeping the fruit in relatively high concentration of CO_2 and a low concentration of O_2. Many investigations have been conducted with the aim of increasing the storage life of banana. Some of these are mentioned here—

5 minutes dip in 2500 ppm triforine, spraying 1% calcium chloride before storing, treating with benzimidazole fungicide for control of *Aspergillus niger,* keeping a cement block soaked in $KMnO_4$ solution to remove ethylene, dipping in 200 ppm TBZ as a post-harvest treatment etc. The harvesting of banana needs kill to judge the right stage since it make all the difference in the ripening and, after ripening processes. The colour and form of the ribs on the fruit determines the proper time for harvesting the bunches.

By-product. The by-products of banana and various sorts of paste or flour sold under name like "bananina."

Varieties

Bananas may be grouped into five main categories :

(*i*) *Dessert banana* consisting of table cultivars for consumption as such after they are ripe, *e.g.*, Champa (Poovan), Rasthali (Morthaman), Robusta (Harichal), Dwarf Cavendish, Cheenia, Malbnog, Alpaan, Banarasi Dwarf etc.

(*ii*) *Culinary bananas* can be used only after cooking, such as Monthan, Bankel, Kabuli, Kothia, Hazara, Compeirganj etc.

(*iii*) Cultivars suitable for making chips, *viz.* ; Nendran.

(*iv*) Cultivars for jam making *e.g.*, Virupkshia.

(*v*) Cultivars fit for making powders (babyfood) such as Kunnan.

Regional Distribution

The cultivars commonly grown in different regions are as follows :

Western and Central India. Basarai, Harichal.

Southern India. Poovan, Dwarf Cavendish, Robusta, Nendran, Rasthali, Chakkarakeli, Virupkshi and Monthan.

Northern India. Champa, Rasthali, Bankel, Batheesa and Kothia.

Eastern India. Kabuli, Champa, Kothia, Batheesa, Martaman and Amrit Sagar.

Species

The banana family Musaceae has two genera (*i*) *Musa* and (*ii*) *Ensete : Ensete ventricosa* cultivated in Ethiopia is probably the one important species of the genera. This genus is supposed to have originated in Asia from where it spread to Africa.

Musa genus has several species being grown in the tropics where temperature, relative humidity and light intensity are high and the drainage adequate. *Musa paradisiaca, M. cavendishii* and *M. sapientum* have been derived from two wild species (*i*) *M. acuminata* (*ii*) *M. balbisiana.* The basic number of chromosomes is N = 11 and the cultivars have 22, 33 or 44 chromosomes so that these are diploid, triploid or tetraploid. Genus *Musa* is divided into five sections : *Eumusa* (X = 11 ; $2n$ = 22), *Rhodochlamys* (X = 11 ; $2n$ = 22), *Callimusa* (X = 10 ; $2n$ = 20) 5 *Australimusa* (X = 10 ; $2n$= *20) and Incertae sedis* (X = 7 ; $2n$ = 14).

Cultivars

Many cultivars are grown in different parts of the sub-continent, some with different names in different places. The more common cultivars are mentioned below. 'A' represents the genome with 11 chromosomes from *M. acuminata* and 'B' stands for the genome with 11 chromosomes from *M. balbisiana.* The AAA group signifies the best bananas of the world belonging to *M. acuminata*. The B genome from *M. balbisiana* carries greater resistance to drought and diseases. Banana with B genome are more suited for culinary purposes, for making chips, jams and powder for baby food.

Poovan belongs to Musa (AAB) group. It is known as Poovan in Tamil Nadu and Champa in West Bengal. It is also cultivated in Andhra Pradesh, Maharashtra, Kerala and Nilgiris. It is common in India as well as West Indies. The pseudostem is medium-tall, mid rib has rose-pink colour on the outside. The fruits are yellow skinned, small with sub-acid taste and good keeping quality. It is resistant to Panama disease and fairly resistant to bunchy top. Syn. Lalvelchi, Chakrakeli, Kadali, Dora vazhai, Bangalow vazhai, Fill basket etc.

Harichal of Musa (AAA) group. A semi-tall sport of dwarf cavandish cultivated in Maharashtra, Andhra Pradesh and Tamil Nadu. Fruits large with greenish-yellow, thick skin and sweet taste. Good keeping quality. Similar to Basrai except that the fruit is more green and straight. Syn. Bombay green, Robusta, Peddapachaarati.

Dawrf Cavandish belongs to Musa (AAA) group. Dwarf, suited for dry interior areas under irrigation. Fruits thick skined—green, sweet, yellow coloured if ripening in winter others remain green. Susceptible to bunchy top leaf spot but resistant to Panama wilt. Poor keeping quality. Very popular in Maharashtra. Syn. Basrai, Loton, Kabuli, Vamankeli, Jahaji, Bhusawal, Pacha vazhai, Mauritius, Hirvi. Extensively cultured in Maharashtra, Tamil Nadu, Orissa, Gujarat and Andhra Pradesh. Does not tolerate severe winters. Grown in East Khandesh for export to the north under the name 'chittidar'.

Martaman [Musa (AAB) group] known as Rasthali. Commonly grown in West Bengal, Tamil Nadu, Maharashtra, Bihar, Andhra Pradesh and Karnataka. Best table variety of Karnataka and Tamil Nadu. Tall plants with green stem, reddish margin of petiole and leaf sheath. Fruits of medium size, skin thin of ivory yellow colour, tastes sweet. Av. wt. of bunch 12 kg. Susceptible to Panama wilt. Syn. Rasthali, Mutheli, Malbhog, Amruthapani, Rasabale, Sonkel, Ripe fruits drop from the bunch. At times hard lumps are formed in the pulp of ripe fruits.

Kanchkela [Musa (AAB) group]. It is known as Monthan in Tamil Nadu and Kanchkela in West Bengal. It is tall, light green, very hardy, robust capable of being grown in unirrigated areas. Chiefly a culinary variety. Big bunches of 100-110 fingers weighing 15 kg. Fruit big, thin shoe shaped. Good tolerant of low temperature. Syn. Bainsa, Aunda Bale, Mangakai, Khasdi, Banket, Ponthan, Batisa, Monthan, Bontha.

Nendran [Musa (AAB) group]. A dual purpose cultivar of Kerala. Fruits are large but the bunch is sparse. Av. wt. of a bunch is 15 kg. Fruits good for preparation of banana fig. Very good keeping quality. Grown in Malabar, Karnataka, Travencore and Cochin. It is generally known as plantain throughout the world. Syn. Ethakai, Rajeli, Kochikehel, Plantain (Trinidad).

Hill Banana [Musa (AAB) group]. High quality perennial banana, speciality of Tamil Nadu. Two types, Sirumalai and Virupakshi are grown, the fruits of former has better taste. The plants are tall, fruits sweet. Fruits easily break from their neck when fully ripe. Can withstand low temperature. Syn. Malavazhai.

Giant Governor [Musa (AAA) group]. Large, greenish to dull yellow fruits sweet in taste. The plant height is medium-dwarf. It is susceptible to bag spot but resistant to wilt. Popular in West Bengal.

Lal Kela [Musa (AAA) group]. Common in Maharashtra, also grown in Bihar and Tamil Nadu. A cultivar grown throughout the world. The visible parts of the cultivar is purple red. Fruits are of medium size with pleasant flavour. Bunch is almost 20 kg. as massive as harichal or Dwarf Cavandish. Syn. Chenkadali, Anupan, Savazhai, Red Banana.

Safed Velchi [Musa (AB) group]. A delicate, medium tall plant with slender stem. Fruits are small. The fruit-skin is papery thin and the pulp very firm, milky white. The plant looks like Martaman. Syn. Ney Poovan, Sonery, Kadali, Nhali Poovan. Very common in Tamil Nadu, Malabar, Kerala and Maharashtra.

Gros Michel [Musa (AAA) group]. The cultivar is possible the best desert banana with desirable fruit characters in size, quality and attractive colour of the skin. It is susceptible to wilt. This cultivar is common in America, Sri Lanka, England and Hawaii. Syn. Jamaican Banana, Blue Field, Anamalai.

Kunnan [Musa (AB) group]. Slender, medium-sized plants. Fruits tasty. Syn. Jirige Bale, Patti, Mokiri.

Thella Chakrakeli. Grown in Andhra Pradesh, good table cultivar, large fruits with thick yellow skin and soft, very sweet, pulp of good flavour.

Diseases and Pests

Panama disease (Fusarium oxysporum). It is wilt disease caused by presence of soil-borne fungus in vascular tissues. The affected plants should be removed and burnt. Use disease-free suckers and follow rotation. Grow in soils unaffected by the disease. Cerasan or 0.1% solution of HEMe supplied in irrigation water controls the spread of the disease. Growing resistant cultivars like Poovan or Champa and Basrai dwarf which is immune to the Panama disease is recommended. Flooding the ground with water 60 cm. deep also kills the fungus. Inject 0.2% solution of carbendazim into rhizome for control.

Shoot rot, anthracnose, leaf spot, black finger, main stalk and finger stalk rot are other disease of banana. To minimise and combat these certain measures have been recommended *viz.* : removal and destruction of affected parts, spraying of 1% Bordeaux mixture and planting of resistant strains in disease- free localities. Sprays of 0.3% **Nuvocon/Thiodon** also controls the disease.

Blood disease (Xanthomonas celebensis). Wilt disease caused by presence of fungus in vascular tissues, dark-red colouring of vascular tissues causing complete destruction of plant.

Banana leaf spot (Cercospora musae = mycosphaerella musicola). Longitudinal oval brown spots. Spraying with Bordeaux mixture and breeding resistant cultivars proves useful. Application of bayleton granules in soil spray of dithane M-45 @2.5 gm per l water or 8 aerial sprays of 6 benzimidazole formulation in oil has proved useful in control of leaf spot. Besides dithane Z-78 Cuman may be used as prophylactic.

Bunchy top (virus disease) (Vector : *Pentalonia nigronervosa*) Stunted plants. Destruction of infected trees. Planting of healthy, disease-free suckers has been recommended and roguing of the infected plants. The treatment is the same as for other mosaic infections viral in nature. Destruction of affected portions with the application of 2, 4-D and spray of Methyl Demiton 25 EC @ 2 ml/lit. water is recommended.

Brown spot and diamond spot affect the fruit and the bunch. Dwarf bananas are highly susceptible to bunchy top disease. Rouging and complete destruction of plants, plant parts including rhizomes should be done. Planting material should be taken from disease-free areas. Aphids vector *(Pentalonia nigronervosa)* can be controlled by spraying with insecticides like metasystox (0.1%- 0.5%), dimecron or parathion to stop the spread of virus. Any stray plant infested with the virus causing bunchy top disease must be removed. Monocrotophos solution in water (1 : 4) injections at 30-day interval 2-3 times after planting is also effective. Herbicide like, 2, 4-D may be applied on stool, after cutting down, for effective killing, 2, 4-D : water : : 1 : 30 conc. is recommended.

Bacterial soft rot of rhizome and pseudostem and bacterial wilt are two common diseases of banana caused by bacteria. Soft rot is apparent by soft odorous rot of rhizome reaching the pseudostem in course of time and destroying the growing point fully. Leaves show yellowish and finally they wilt. Good drainage and drenching of the soil with bleaching powder @ 2 g/l every ten days could control the disease.

Moko disease *(Pseudomonas solanacearum).* Bacterial wilt, thus far of little importance in the Indian Subcontinent. It is characterised by the presence of yellow fingers on the green stem. Check the insects visiting the flower and kill the plant *in situ* by the use of 2, 4-D or 2, 4, 5-T and apply chlorpyriphos on the cut surface as a safeguard against the visit of insects to other plants.

Bonnygat disease *(Sphaerostilbe musarum).* Attacks base of stem. Eradication and destruction of affected plants helps in the control of the disease.

Heart rot of Pseudostem. The heart leaf is affected, lamina decays and later in serious cases, the crown is killed. Sanitation, clean cultivation, good drainage, more spacing and the use of captan or dithane M-45 or Dithane Z-78 sprays have proved useful in checking the disease. Dithane M-45 sprayed every 7-10 days.

Other diseases *viz. Heart rot, Sooty mold, Stem-end rot* and *Colombia stem-end rot* damage banana. The latter attacks growing or ripening fruits. Sanitation in cultural practices, harvesting, handling and shipment, use of antiseptic solutions or pastes are useful. In cigar-end-rot disease of banana the pistil, stamens, and the remains of perianth are removed carefully 8-12 days after bunch formation.

Insect and Pests

Besides the diseases like bunchy top, Panama wilt, main stalk rot, cercospora leaf rot etc. there are certain harmful pests also. *Banana stem weevil, Red rose thrips, Banana aphid, Banana beetle* are insects that cause maximum damage to the crop.

Root stock weevil. The grubs are creamy white, stout, fleshy and highly wrinkled while the adult is a stout reddish brown weevil. Both damage the rhizome Pseudostem and suckers. Maximum damage is done during April-October period.

Clean cultivation of pest free suckers and spray, dimethoate 30 EC @ 1 ml or Fenitrothion @ 1 ml/lit water, if attacked, around the base of the plant. Precaution may be taken to treat the pits before planting the suckers with phorate 10 gm or Carbofuran 3 gm. Placing celphos tabs/sucker inside the pseudostem and plugging the slits with mud gives effective control of the insect.

Banana beetles. This beetle may be seen in the fields between April and end of rainy season. It causes severe infestation causing burnt up appearance of the leaves and spotted fruits. Clean cultivation and spray of Methyl demiton 25 EC or Quinolphos 20 EC @ 1.25 ml/lit. water help in the control.

Banana stem weevil (causes plants to tip over). Red rust thrips (discolour fruits), flower thrips (disfigure fruits), Florida red scale (discolour green fruits), coconut scale (discolours green fruits), Red spider (sucks leaves), caterpillar (chews leaves), Oriental fruit fly (infests ripe fruits). Grass hoppers (chew leaves), Banana aphids (vector of bunchy top virus) need be removed by clean cultivation.

Banana Aphid. The aphid is brownish with black-veined wings. It transmits the banana bunchy top virus disease. High humidity and moderate temperature favour increase in aphid population. Spray of methyl demiton 25 EC @ 1.25 ml or monocrotophos 40 EC @ 1 ml/lit water is recommended for control.

Nematodes. Borrowing nematode (*Radopholus similis*), root-knot nematodes (*Melodiogyne* sp.) Spiral nematode (*Helicotylenchus* spp.), Lance nematode (*Hopolaimus* sp.), meadow or root lesion nematode (*Pratylenchus coffeae = P. musicola*) etc. attack the root and rhizome in banana. The plants affected by nematodes do not respond to additives or cultural practices. These have to be eliminated.

Control. Hot water treatment of planting material and soil fumigation.

Application of DBCP @ 40 l/ha at planting (May-June) and 25 l/ha in the month of October has proved effective against nematodes. Use of EDB @ 150 kg/ha is also effective in the control of nematodes. Neem cake, *Mahua* cake or castor cake also help in checking nematode infestation.

CHAPTER

4

BER

(***Zizyphus mauritiana*** **Lamk)**

Rhamnaceae

Origin and Distribution

Indigenous to the area stretching from India to South Western China and Malayasia this fruit has been grown in India since antiquity. It is knwon as *ber, bor, bogari, boroi* or *kool* and *regu* in Hindi, Panjabi, Marathi, Assamese, Bengali, Telugu respectively. The per capita consumption for ber is lowest in India. In certain areas it has stepped in as a substitute for mango and citrus cultivation with more profit than growing of cereals. The gross income from one hectare is round about Rs. 25,000 per annum. In America it is called jujuba or chinese date. Its cultivation in India is restricted to Rajasthan, Hayrana, Madhya Pradesh, Bihar, U.P., Punjab, Maharashtra, Assam and Gujarat. It is also called chinese date.

The Indian ber Z. *mauritiana* with spreading vine-like branches have dark green leaves on the upper side and dense felted on the underside. It is grown commercially in north-west plains. Certain other species of Ber *viz. Z. nummelaria* and *Z. rotundifolia* are also found growing in the Indian subcontinent. *Z. jujube* is commonly grown in temperate parts of the world. *Z. mauritiana* plants are small, spreading with drooping branches. The fruits are oval, round and the colour greenish-yellow to brown.

Climate and Soil. it prefers a hot dry climate of the tropics and subtropics. Ber, a member of Rhamnaceae family has two types of tree habits the spreading and the erect one. It is a hardy fruit requiring hot and dry climate with adequate watering during the fruiting season. The tree has a deep tap root system which responds to a larger number of soil types including those exposed to droughts as well as water logging. Neutral or slightly alkine deep sandy soil seems to be ideal for this fruit tree.

Propagation

Propagation by seed, root cuttings, grafting, forkert, budding (ring, shield or 'T') etc. have been *in vogue* under different conditions and circumstances. Generally the seeds collected from fully ripe ber fruits are viable although in many cases seed treatment by H_2SO_4, their stratification with sand, keeping in warm place, or extraction of seed by breaking the hard shell (endocarp) helps in early germination. Seedlings may be germinated in small containers to 4 leaf stage and may be planted in the field at distance of 10-12 M between them. Clonal multiplication through the tissue culture technique from a single bud has also been tried with the aid of NAA for differentiation of shoot with success.

The technique of 'I' budding standardized at CAZRI, Jodhpur has proved very convenient with good success in the propagation of Ber.

Budding is considered useful in the improvement of the ber fruit. The ber seeds are sown in the field at desirable distances and budding is done *in situ* since success in transplanting of seedings is difficult. Whip grafting has been practised successfully for ber. Both ring and shield budding can be done successfully during July, whereas only shield budding is more feasible during August-September. Budding during March, April gives very little success as there is little sap flow in the mother plant. The root stocks may be taken from the wild species of ber *viz.*, *Z. oenoplia, Z. nummularia, Z. zylocarpa* and *Z. rugosa.* Seedlings trees of *Z. mauritiana* are generally used as root stock although *Z. rotundifolia* (Malha, Jharberi) is also used.

Top-Working of Seedling Ber Trees. There are numerous wild species growing on neglected land and along the borders of the fields, which bear poor quality fruit and fetch nominal income to the cultivator. Top working these trees with improved varieties will undoubtedly adds to the income of the grower. In North India the trees are headed back during March-April leaving about 1 to 1.5 metre of the stem above ground. New sprouts arise from below the cut during May. One strong shoot is retained and others removed. One or two buds of an improved cultivar are inserted on the selected shoot. The budding time is dependent on the atmospheric humidity and temperature. It is best done in the month of June (Haryana), August (Punjab), July (Rajasthan) and just before the onset of rains in Gujarat. As soon as the buds sprout, the portion of the shoot above the point of bud union is cut off. Within a year, the top worked tree develops sufficient head growth and starts bearing.

Manuring

Addition of N, P, K at the rate of 10, 5 and 10 gms. per plant in the first year 20, 10 and 20 gms. in the second year ; 40, 20, 40 gms. in the third year ; 60, 30, 60 (fourth year) and 80, 40, 80 (fifth year and every year onwards is recommended). In addition 20 kg. of well rotten dung must be given to each plant every year.

When both manuring and irrigation are wanting immature fruit-drop takes place. During the early period the ber plants are benefited by intercropping with castor.

Beer trees are liable to be affected adversely by heavy winds and in order to avoid damage to the plants pruning is desirable. It is also customary to make a mound of earth round the trees to give them support against the uprooting effect of high winds. Regular annual pruning ensures better growth for providing better fruit bearing condtions. Too much crowding should be avoided by thinning out of branches specially during the hot months of the year when the plant is dormant.

Training and Pruning. Training of ber is highly desirable during early years to build up a strong framework of the tree. It is a fast growing tree and the commercial cultivars are spreading in nature, therefore staking with strong bamboo sticks is absolutely essential to avoid breakage of bud union during first two years and to train the tree properly. The trunk should not have branches upto 90 cm. height and above this height 4-5 laterals most favourably located around it should be selected to make the scaffold limbs of the tree.

The ber fruit is borne in the axil of leaves on the growing shoots of the current year. Therefore, a regular pruning is necessary to induce a good healthy growth to provide maximum fruit bearing area on the tree. Depending upon the tree vigour, removal of 25 to 50 per cent portions of the previous season's growth gives good yield and also improves fruit size and quality. Lower branches should be pruned to the extent that new growth remains little above the ground and does not spread over it. The best time of pruning in northern India is mid-May to mid-June. It is advisable to complete the pruning before mid-June, when the plant sheds its leaves and goes to rest.

Fruiting Behaviour. The ber tree bears its inflorescence in the axil of leaf on current season's growth. Growth regulators like 2, 4, 5-T and NAA could reduced fruit drop when applied in proper concentration. These regulators increased fruit size, TSS, sugar and acidic taste.

Spray of $ZnSO_4$ (0.4%), $FeSO_4$ (0.4%) and boric acid (0.2%) in August and September proved beneficial by encouraging the fruit set.

Harvesting and Marketing

One deep ploughing in the pre-flowering period increases the yield remarkably. Plants start bearing from thrid year onwards. Ber fruits are ready in February and April when no other fruit is available in the market. The yield of grafted trees is heavy and regular too. The Umran (cultivar) in the Punjab may yield as much as 4 quintals of fruit per tree. Yields as high as 6 quintals have also been reported although 0.75 to 1.25 quintals is the average.

There may be as many as 40-60 fruits to a kg. Grafted-tree bears fruits earlier than the seed-tree. The fruits must be picked when crisp with a light yellowish scarlet colour on the outer skin. It can be stored in open container for 45 days at 32°F.

The fruits are dried or preserved in the form of Jam or Jelly. The powdered fruit is mixed with molasses and Jaggery are made into a drink. Unripe jujube is said to be good for making Jelly provided some other fruit was also added alongwith to supplement the pectin.

Cultivars

More than one hundred cultivars are grown in India. Some of the important ones are given below :

Chonchal. Cultivated throughout Haryana. Large fruit with a long distinct bear (Chonch). The fruit is sweet and juicy.

Dandan. Mid-season cultivar grown in Punjab, Medium sized fruit, soft skin and juicy. The fruit-end is pointed.

Gola. It is a spreading early cultivar grown in Punjab, Rajasthan, U.P. and Delhi. It is early, round fruited with poor keeping quality. Fruit quality is excellent. Fruits ripen in January. Gola (early) tolerates usar.

Kaithli. A cultivar with poor keeping quality not standing transport well. Sweet taste, soft pulp and nice flavour. Mid season cultivar of Punjab and Haryana. Well suited to Usar.

Kheera. A good cultivar for Delhi area.

Sanaur-2. Out of the six No. 2 is most promising cultivar. It is midseason, ripens during second fortnight of March. Fruit is oblong in shape, golden yellow in colour, large in size, apex slightly pointed, surface smooth and shining. Pulp sweet in taste with typical flavour. Resistant to powdery mildew.

Umran. Of the forty cultivars tested Umran has been found to be the best. A heavy yielder, originally from Rajasthan, developed at Fruit Research Station, Bahadurgarh near Patiala. Umran is well blended for sugars and acids. It can be processed and dried and can be used as 'chuharas' (dry dates) throughout the year. It can grow on usar soils where other fruits trees do not grow. It needs very little irrigation. Sweet fruits, fairly big in size (approx. 30 gm) and round at the apex, good yielder and a good keeping quality. Fruits ripen 15th March to 15th April. The pulp content is about 95% as against 105 in indigenous cultivars.

Z.G.2. It ripens in March. It has good blend of acidity and sweetness. Fruit is almost round, light green colour but turns to reddish brown in advance stage of maturity. Fruit size small with round apex. Resistant to powdery mildew.

Banarasi Karaka. Fruits globose. Shape oblong to long with tapering to pointed styler end. Unripe fruits green, greenish-yellow when ripe glossy with smooth skin. It can be grown in usar soils also.

Jogia. Fruits oval with tapering styler end. Unripe fruits green with red pigments, assumes a greenish yellow colour with deeper yellow spots when ripe.

Chuhhara. Fruits ovate-oblong, stone sharply pointed. Resistant to powdery mildew.

In U.P. the important cultivars are Banarasi Gola, Delhi Gola and Narma Varanasi (Early) ; Banarasi Karaka, Muthia, Muriya and Pewandi (Mid-early) as well as Jogia, Aliganj (Late), Ponda and Umran.

Gujarat has Ajmeri, Chameli and Randeri as late cultivars and Mehrun as mid early. For Rajasthan Gola, Seb and Seo (Early) ; Jogia, Mundia and Tikadi (mid-early) ; Ketha, Maharwali and Bagwadi (Late) are common. Punjab is credited with largest number of cultivars of which Umran, Ilaichi, Pathani, ZG2, ZG 3 (Late) ; Nazuk, Noki, Seo, Rohataki Gola. Selected safeda amd Sandhura Naranaul (Early) ; Banarasi Dandan, Kaithali, Wilaiti and Thornless (mid-early).

Bihar cultivators favour Banarasi, Thornless ; Nagpuri, Kurnool and Cuddapah are of commercial importance in Andhra Pradesh while Dodia and Banarasi are also grown. Dodhia is resistant to the attack of white fly. In Bengal Narikeli, Ghughudanga, Banarasi, Prolific and Baruipur and popular.

Pests and Diseases

Carpomyia vesuviana (fruitfly). It damages the crop (fruit) extensively throughout the country. Fruits containing the maggots fall to the ground and enter the soil. Best remedy is to stir the soil occasionally specially during the summer. For control of fruit fly application of methyl parathion 2% dust @ 20-25 kg/ha. Pick and destroy the fallen infested fruits. Spray methyl demiton @ 1.5 ml/l water or Dimethoate 30 EC @ 1 ml/l water at the pea stage of fruits. Melathion 50 EC @ 1 ml/l water mixed with 2.0-2.5 gm of *gur* should be sprayed at maturity of fruits.

Katha, Tikadi, ZG 3, Dodhia, Mehroon, Sanaur-l, Michia and Umran are resistant to white fly while Kaithali, Gola, Dandan and Ilaichi are susceptible.

Adoretus pallens (ber beetle). It has a special preference for foliage of ber wherein they make hole and destroy, specially druing the night. It can be controlled by dusting the trees at dusk with sodium fluosilicate and ash in the ratio of 1 : 8. Endosulfan 500 ml or malathion 300 ml. in 500 l water may be sprayed to control the beetles. Adults can also be trapped by keeping source of light surrounded by water containing some kerosene oil so that these are attracted towards light and have a watery grave.

Bark eating caterpillars (*Inderbela quadrinolata*) can be controlled by painting the exposed portions with monocrotophos. The cultivar Desi Alwar is resistant while Dandan, Kaithali, Gola and Ilaichi are susceptible.

Safeda selected, Chinese, Glory, ZG 5, Chuhhara, ZG 2, ZG 3, Sanaur-2 Nazuk show resistance to powdery mildew.

Other pests which may sometimes need attention are : *Thiacidas postica* (hairy caterpillar), *Tarucus theophrastus* (hair streak), *Myllocerrus transmarinus* (leaf weevil), *Xanthochelus supercilious* (leaf weevil), *Porthmologa paraclina* (Moth).

Control. Hand picking or a preventive treatment with malathion.

Powdery Mildew caused by *Oidium* sp. is possibly the only fungal disease in India. It causes shedding of the fruits. It can be controlled by lime-sulphur wash, Bavistin 0.05%, wettable sulphur 0.2% or karathane 0.2%. Although reported from some places *Mitterellia ziziphina* and *Cercospora* sp. are not serious.

Black leaf spot (Isariopsis sp.*)* The leaves have black spots on the lower surface. The other side is brownish in colour. The leaves start falling after some time. Spray the trees with copper oxychloride @ 3 gm/l or Dithane Z-78 @ 2 gm per l of water 2-3 times at 15-day intervals just after the appearance of the disease.

Ber may contain 50-150 mg of ascorbic acid/100 gm of fruit besides 12-18.7% sugar. The ber plant is also used for rearing *Tachardia laccad* which yields shellac commonly known as lac.

□□□

CHAPTER

BREADFRUIT

(***Artocarpus altilis*** **Fosberg)**

Moraceae

Origin and Distribution

The breadfruit (H. *Khadel* or *Barhahal*)), a staple food just like bread and a native of Malayan Archipelago is also cultivated in West Indies. It is a rich source of carbohydrates, calcium, vitamin A and B ; vitamin C is present in minute quantity. It is grown in the foot hills of Nilgiris in Tamil Nadu. Used for culinary purposes. Besides the nutritious fruit the milky juice from the tree may be used as sealing material for making the boats air tight, wood for buildings and leaves for livestock feed.

Climate and Soil

The breadfruit requires a warm humid tropical climate with plenty of precipitation for optimum results. In general the breadfruit requires an annual rainfall ranging from 240-300 cm. distributed throughout the year, a temperature range of 64°F to 100°F and a humidity of 70 to 80 per cent.

A high humus content coupled with high fertility are the requisites of a suitable soil for breadfruit. Good drainage of the soil is essential to maintain the trees in proper condition. Restricted drainage stunts the trees and leads to premature fruit drop. Lateritic red loams are well suited to breadfruit.

Propagation

Seeds may be sown immediately after removal from the fruits to get the seedlings. The propagation of seedless cultivars, where the fruit is developed due to stimulative parthenocarpy, is possible through root cuttings and air layering of root suckers. Root cuttings, about 2.5 cm in diameter and 22 cm. long, should be planted horizontally. These plants bear fruits in about 6 years as against 8-10 years needed in the case of the seedling raised from seeds.

Cultural Practices

The breadfruit is normally planted in pits (1.5 × 1.5 × 1.5 metres) with a spacing of 20 metres. At the time of pit-filling apply 30 kg of FYM along with single superphosphate @ 500 gm/pit before planting. Good growth is obtained by the addition of FYM (10 kg), nitrogen (75 gm), phosphorus (75 gm) and K_2O (100 gm) to each tree every year. The application of FYM is recommended during the months of June-July. Fertilizers may be given during September-October and January February. Watering of the breadfruit trees is a must in the first two years to enable them to establish well. The tree is generally not manured but rich soils have given high yields. Shallow inter-culture to eradicate weeds and to keep the soil open may be profitable. Inter-cropping and cover-cropping of breadfruit orchards may be necessary in certain situations.

Harvesting and Marketing

The fruit is ready for harvest 2 or 3 months after blossoming. Green premature fruit turns yellowish green when fully mature. Ripe fruits are insipid in taste. Fruits borne at the extremities of branches must be harvested when still unripe. the ripening process quickens by keeping the fruits in closed packing for 4-6 days. A fully grown tree may produce 50 to 100 fruits weighing about 25 to 50 kg. every year.

Cultivars

The seeded and the seedless cultivars cannot be distinguished by their plants. The seeds of the seeded cultivar are small and resemble chestnut. These are relished when roasted. The seedless fruits possess only a mass of whitish mealy pulp. In Jamaica, Yellow Heart is said to be the best cultivar.

Insects, Pests and Diseases

Soft Rot may be caused by the fungus *Rhizopus artocarpi.* It is controlled by spraying of Bordeaux mixture at intervals of a fortnight during the fruit season.

Fruit rot caused by *Phytophthora palmivora* and Pink disease caused by *Corricum salomonicolor* in breadfruit can be controlled by their treatments detailed under jackfruit.

□□□

CHAPTER

6

CHIKU

(*Achras sapota* L.)

Sapotacease

Origin and Distribution

Sapodilla plum, chiku, sapota or bully is tropical tree common in Mexico. It is considered to be indigenous to Mexico and Central America. It is extensively grown in the moist coastal tracts of peninsular India. Maharashtra, Karnataka, Gujarat, West Bengal, Andhara Pradesh, Punjab, Haryana, Tamil Nadu and submontane tracts of Uttranchal are specially suited for the fruit.

Climate and Soil

Achras sapota L. (or chiku) prefers a warm and moist weather. Coastal climate suits it and an annual rainfall of 125-250 cm. is welcome. Cloudy weather does not harm chiku crop. Optimum temperature range for chiku is between 11°C to 35°C. Winter injury is met with in certain parts of north India. Absence of hot, dry summer is its climatic requirement. Immature tree is easily injured by frost in North India but mature plant can withstand mild frost of very short duration.

Well drained soil is necessary for the cultivation of chiku. Alluvial loams, sandy loams near the sea, red lateritic soils, medium black soil overlying murum substrata and deep alluvial, loamy soils of the Indo-Gangetic plains are suitable for growing chiku. The chiku is a shallow-rooted tree. It can tolerate saline soils to some extent. Soils underlaid with hard pan or lime, yellow in colour or sticky are not suitable for chiku. Wherever the lime content of the soil is high chiku plants do not survive for long.

Composition

The edible portion of chiku fruit has 21% carbohydrate, 0.7% protein, 1.1%

fat, 74% water besides phosphorus, calcium, iron and ascorbic acid as 27 mg, 28 mg, 2 mg, 6 mg respectively per 100 gm.

Propagation

Seedling trees are being replaced by vegetatively propagated trees obtained by inarching with Khirnee (*Mimusops haxandra*) or 'Mahuwa' (*Madhuka latifolia*), Star apple (*Chrysophyllum cainito*), Mee tree (*Basia longifolia*), Sapota seedling (*Achras zapota*), Adam's apple (*Manilkara kauki*) root stock. Sapota or Khirnee proved superior as rootstock. For raising sapota seedlings the seeds must be washed well to get rid of the pulp and dried in shade before sowing. Self sown seedlings of *Khirnee* are collected and used as rootstock in Ratnagiri and Surat with success. the seedling is ready for transplanting when it has attained 15 cm. length. Planting pits need exposure to weather for 2-3 months and should be made in April or May for planting in monsoon. The chiku is commercially propagated by air layering or gootee layering and inarch grafting. Air layering is a popular method of propagating chicku in West India. In other parts of India, inarch grafting is generally practised. Forkert method is cheaper, easier and more efficient method of propagation. Budding is the best method of vegetative propagation. A single bud is employed and the suitable period is the month of May for the coastal regions.

Planting

Well prepared levelled land with windbreaks of tall growing trees should be selected and the plants be planted at a distance of 9.0 metres in area of heavy rainfall and 7.5 metres in dry regions during the rains in pits (1 × 1 × 1 metre) filled with soil-compost mixture. The young sapplings need support to keep erect.

Cultural Practices

Manuring gives better yield, 25 kg of F.Y.M. and 1/2 kg of castor cake are recommended to a year-old tree. The dose is increased with increasing age of the plant. Well established trees require 55 kg F.Y.M., 8 kg castor cake and 2.5 kg bonemeal per tree. The fertilizer quantity recommended for a 3 years old plant is N : P : K : : 50 : 27 : 75 g which is doubled for every 3 years age of the plant till the age of 10 years. Thereafter the quantity remains static. A full grown chiku tree needs 1.0 kg of nitrogen, 0.45 kg of P_2O_5 and 0.58 kg of K_2O for good growth and high yield. Sometimes banana which is manured heavily is inter-planted with sapota so that the sapota plants also get the advantage. Superphosphate improves the fruit size and therefore the income also. No pruning is needed for at least a period of 7 to 8 yeras. Self unfruitfulness may be a problem in case of stray plant in a locality. Irrigation is needed depending on weather conditions.

Harvesting and Marketing

Grafted chiku begins to bear in the third year after planting. Fruti production increases with age upto 30 years followed by decline. Fruits mature in about 4 months after flowering. Sapota blossoms continuously in several flushes, at short intervals, throughout the year and consequently the fruits also mature in bits. The time for harvesting the fruit is indicated by the development of good bloom on the surface while it is still firm. The fully mature fruits are individually picked by hand with their stalks attached. Mature fruits should not be allowed to ripen on the tree. Fruits ripen and soften within 5 days after picking if kept in a warm room. During storage the milky latex disappears completely and the fruit becomes sweet and edible. The fruits that ripen in summers are sweeter. The yield may be between 1500 to 3000 fruits a year from a well looked-after adult tree. The fruits have usually good keeping quality and can be transported to long distances without loss.

Several types of chiku cultivars are known to exist in India. These are distinguished on the basis of characteristics of tree : (*i*) erect growing habit, (*ii*) drooping habit (*iii*) spreading habit. The branching may be regular, in whorls and irregular. Further differentiation is by the colour of the foliage which may be light green or deep green. Fruits may be small brown or large yellow.

Cultivars

The fruits may be round, oval or conical in shape, 5-8 cm in diameter and almost like a smooth potato in appearance. The skin is russet-brown, thin and scurfy. The flesh of ripe fruit is honey-brown, soft, mellow, granular, very sweet and delicious. There are a few cultivars of chiku in India.

Bangalore. Fruits have nine ridges running from base upwards. Large, oval, mildly sweet. Grown in Andhra Pradesh.

Calcutta round. Light green foliage, fruits large, medium quality, susceptible to leaf spot. Commercial cultivar of West Bengal and Karnataka.

Pot Sapota. Very early fruiting while still in pots. Fruits small, oval with painted apex without ridges on the surface. Very sweet, pleasant flavour.

Gavarayya. Small fruits with one shoulder drooping and the other raised, 8-10 marked ridges of the fruit. Soft, melting, sweet flesh. Grown in Tamil Nadu and Andhra Pradesh.

Thagarampudi. Medium sized fruits, round or oval with a flat base, thin skin, no ridges, rough. Flesh buff coloured, juicy and sweet. Good for export. Grown in Tamil Nadu.

Ayyangar. Fruits large, round, rose scented, sweet with pleasant flavour. Rough, thing skin. Grown in Tamil Nadu.

Bhuri or *Bhuripatti.* Tnick foliage, medium bearing fruits large, quality good.

Vanjit. Sterile or male type. Slow growing nature. Knots appear on branches and roots emerge from these knots. Fruits of good quality. A shy bearer.

Jonnavalasa Round. Fruits round, small to medium size. Jonnavalasa I and J. II are also grown in Tamil Nadu.

Baramasi. It is commonly grown in West Bengal, Bihar and U.P. and has roundish fruits of medium size. It is not available throughout the year and the name is misleading.

Chhatri. It is commonly grown in Maharashtra. The plant has a distinct drooping umbrella like whorl of branches. The leaves are of light green colour and the fruit is similar to Kalipatti in appearance but not in quality. Good cropper.

Cricket Ball. It is a large round fruited cultivar but the number of fruits from any individual plant is not more than 500 in a year. The fruits are not very sweet. It is considered to be the same as Calcutta Large. Arid climate may also suit the fruits.

Dwarapudi. It has a small sweet fruits, resembling, cricket ball, popular in Andhra.

Kalipatti. It is leading cultivar of Maharashtra, North Karnataka and Gujarat with dark green leaves, spreading habit and oval fruit of outstanding quality. The flesh of the fruit is juicy, sweet, and mildly fragrant. It contains 1 or 2 seeds per fruit. Heavy cropper.

Kirthabharthi. The fruits are egg shaped, very sweet with rough and thick skin, 4-6 ridges present on the rind. Good transporation value. Popular in Andhra Pradesh.

Pala. Thin skinned oval fruits borne in clusters. Cultivated in Tamil Nadu. Small egg shaped fruits. Sweet with good flavour. Heavy bearer.

Dhola Diwani. Common in Maharashtra, light green foliage, whitish oval fruits of good quality harvested in summers. Long leaves, narrow and small. Fruits thin, long and very sweet.

Besides the above culivars Pilipatti, Jumakia and Mohan Gooti (Maharashtra), Badam (Tamil Nadu) and Gede and Guthi (Andhra Pradesh) are grown on small in restricted areas.

Insects, Pests and Diseases

Leaf Miner. It is an extremely tiny caterpillar of a greyish moth which mines into the surface of young chiku leaves and makes them curl. The infected leaves get distorted, dry up and fall. One or two sprays of dimethoate (100 ml in 60 litres of water) or malathion (100 ml in 50 litres of water) control the pest.

Mealy Bug. Mealy bugs are found sticking to the underside of chiku leaves and at base of the fruit, near its stalk. Control measure may be the same as for leaf miner.

Stem Borer. It bores into the bark of the trunk in circular galleries and feeds upon the living tissue of the inner bark. The borer can be killed by thrusting a stiff wire into tunnel. Where the borer has entered the wood and consequently cannot be traced, the hole may be plugged with a wad of cotton wool dipped in kerosene or diseal oil and plastered with wet mud. The borer gets suffocated and killed.

Leaf spot. Caused by fungus may be controlled by spray of 0.2% dithane Z-78 or 0.5% Blitox.

Sooty mould. It can be controlled by spray of 100 g starch in 18 litres of water. On drying it forms thin flakes which drop off, spraying with 40 g Zineb in 18 litres of water checks the diseases.

Flattening of branches due to the attack of a pathogen *Botryodiplodia theobromae* has been reported. It affects the yield adversely. No control measure has been recommended.

CHAPTER

7

CITRUS

(*Citrus species*)

Rutaceae

Oranges and allied citrus fruits are extensively grown owing to their medicinal and nutritive values so as to form an industry of importance. Although citrus is essentially a crop of the subtropical regions it is also grown under favourable conditions of temperate regions. The citrus fruits probably rank third among the subtropical fruits of the world.

Cirtus is a member of the family : rutaceae, subfamily : Aurantioideae ; Tribe : Citreae, Subtribe : Citrinae. The subtribe includes 6 genera which are indigenous in an area extending north-eastern India and north central China to eastern Australia and New Calidonia. These are characterized by having fruits with slender-stalked pulp vesicles filling the space in fruit segments not occupied by seed and by having at least 4 times the stamens as the petals. The genera citrus has 16 species. It has two subgenera *Papeda* and *Eucitrus*, the former with 6 wild forms, inedible due to the presence of droplets of acrid oil in the pulp vesicles and long petioles while the latter include all the 10 commonly cultivated species free from the oil droplets with narrow winged or wingless petioles.

Related genera (1) Trifoliate orange with only one species. *P. trifoliata*, very thorny plants, fruits inedible. Used as rootstock and for intergeneric crosses, seeds polyembryonic, susceptible to blight but tolerant to most of virus diseases, Phytophthora root rot and citrus nematodes.

(2) *Fortunella – (Kumaquat)*. Several species *viz. F. japonica, F. hindsii, F. margarita* and *F. crasiflora*. Trees small and shruby. Plants used for ornamental purposes. Fruits used for candying.

Origin and Distribution

The cultivated species are believed to be natives of tropical and subtropical regions of south-east Asia. May be that they originated in drier monsoon areas rather than the tropical rain forests. According to De Candolle *Citrus* originated somewhere in the north-eastern parts of India, Siam or Southern China from where the varieties of *Citrus* slowly spread to Europe and America. They are now grown in Italy, Spain, Palestine, California, Florida, Central and South America, South Aftrica, Japan and China. Bhattacharya and Dutta hold that two species in *Eucitrus, Citrus indica* Tan. and *Citrus assamensis* and three species of *Papeda viz., C. ichangensis* Swing, *C. latipes* (Swing) Tan. and *C. macroptera* Moutr. are definitely indigenous to Assam. According to them *Citrus aurantium, C. magaloxycarpa* and *Citrus reticulata* may also be added to the list. Problems involved in the history of the citrus fruits are discussed by Tolkowosky. According to Hu, Citron has commonly been considered a native of India.

Stress has been laid on the cultivation of Citrus in the past 100 years or so. Man took into cultivation those having the best quality of edible juice and desired flavour. Natural hybridization seems to have accrued between cultivars and species producing and array of complex hybrids. Some of the species have been found in a wild state, others got naturalized in restricted localities. Most of the commercial production is now in subtropical regions having Mediterranian climate.

In India the chief Citrus growing areas are Tamil Nadu, M.P., Punjab, Karnataka, the southern slopes of Cherrapunji, hills of Assam, Northern Orissa, Maharashtra and U.P. The acreage under citrus is almost 10% of the total area under fruit trees. Mandarin oranges, sweet oranges and acid limes are grown extensively in India citrus may be grown on a variety of soils and under several environmental conditions all over the world. Israel, Egypt, Spain, Morocco,USA, India, Algeria, Australia, Mexico, Brazil and South Aftrica are chief growing regions of citrus. USA contributes almost 40% of world production of citrus. Spain is the largest exporter of fresh fruits followed by USA, Israel and Morocco in descending order.

Ecology

Citrus species are grown in the subtropical regions of the world from 45°N to 35°S between sea level and 700 M. Most of the cultivars cannot withstand frost for long period. The susceptibility depends on the dormancy of the trees. Any temperature below 55°F reduces growth markedly. In the rainfed areas the minimum rainfall should be about 100 cm per annum. The degree of readiness to frost damage is in the following order trifoliate orange, kumaquat, mandarin, sour orange, sweet orange, grape fruit, lemon, lime, citron. High humidity induces diseases ; mandarins can tolerate maximum amount of wet atmosphere. Citrus is also intolerant to high wind and needs wind break for protection. Citrus is not sensitive to photoperiod. Light loams are preferred though wide range of soils have been found to be used for citrus cultivation. Citrus plants are very sensitive to water logging. The soil pH should be between 5-8, a wide range from moderately acid to slightly alkaline. Citrus trees are seen to thrive well in good drained soil such alluvial, medium black or loamy soil. Orange trees are injured in the absence of calcium salts though not of potassium salts and therefore calcium seems to be necessary for slightly heavier subsoil. Citrus is also sensitive to boron, sodium carbonate and sodium chloride. Maximum temperatures are not very important, except in desert where exposure to very hot sun may be harmful, but the total heat during the growing season is extremely important. Little growth takes place when the mean temperature is below 55°F or above 97°F.

Modern System of Classification

Two comprehensive and widely recognised systems of calssification of *Citrus* (Swingle, 1943 and Tyozaburo Tanaka, 1945) are available. Both worked for more than four decades. Swingle (1943) appears to have followed the classification of Engler. Swingle travelled in *citrus* growing areas of Japan, China, the Philippines and the Mediterranean and also collected *Citrus* varieties from all over the world. Tanaka was associated with Swingle during 1915-30 yet the collected more *Citrus* material. Swingle (1943) divided subtribe *Citrinae* into 3 sub-tribal groups (*i*) True citrus, (*ii*) Near citrus, and (*iii*) Primitive citrus ; 13 genera and 65 species in all.

(*i*) True *Citrus* group includes 6 genera and 29 species (*ii*) Near-citrus fruit trees include 2 genera and 22 species and (*iii*) Primitive citrus fruit trees consists of 5 genera and 14 species. The six genera under the true citrus fruits are (*i*) *Fortunella,* (*b*) *Eremocitrus*, (*c*) *Poncitrus,* (*d*) *Clymenia,* (*e*) *Microcitrus,* and (*f*) *Citrus.*

True citrus trees have orange or lemon like slender, stalked fruits with highly specialised, usually more or less fusiform pulp vesicles filling entire space in the segments of the fruit not occupied by the seeds. The group is characterised by the number of stamens being 4 times that of the petals. All the six genera except the monotype genus *poncitrus* have persistent, unifoliate or simple leaves. The leaves of *poncitrus* genus are trifoliate and deciduous.

The genus *citrus* is sub-divided into two sub-genera *viz.*, *Eucitrus* and *Papeda.* Sub-genus *Eucitrus* is characterised by having pulp vesicles free from acrid oil droplets ; petioles being wingless or with narrow wings and if broadly winged, always shorter than the leaf blade, stamens usually free (Swingle 1943 ; Bhattacharya and Dutta, 1956).

Tanaka treated the genera *Fortunella* and *poncrius* approximately similar to that of Swingle, but he had given a liberal treatment to *Citrus*. He recognised two sub-genera (*i*) *Archicitrus* and (*ii*) *Metacitrus* ; comprising of 8 sections and 144 species in *Citrus*. This classification is given on the next page :

Genus	Sub-genus	Sections	No. of species in each section
Citrus I	*Archicitrus*	*Papeda*	12
	(98 species)	*Limonellus*	16
		Caphalocitrus	21
		Aurantium	28
		Citrophorum	21
II	*Metacitrus*	*Osmocitrus*	9
	(46 species)	*Acrumen*	36
		Pseudo fortunella	1

Thus Tanaka's classification of *citrus* fruits involves a total of 3 genera and 151 species of which 144 are *citrus*.

hodgson (1961) held that Swingle's classification was not very comprehensive as numerous ancient forms of orient *e.g.* Attani, Gajanimma, Kichili, Amilbed, Kimb and Sadaphal which are of economic importance have been ignored. Secondly, it has denied species standing to many ancient and highly distinctive forms of economic importance, *viz.* rough lemon, Indian (Palestine) sweet lime, Satsuma and Kunembo mandarins, Rangpur lime, Natsudaidal and yuzu. Thirdly, its treatment to certain distinctive forms is complicated and speculative. Thus rough lime has been named as *Citrus* lemon hybrid ; the king mandarin, *C. reticulata sinensis* hybrid ; sweet lime, *C. aurantifolia* hybrid ; the Rangpur lime, *C. reticulata* var. *austera* hybrid and the Calamondin, *C. reticulata* var. *austera-Fortunella* sp. hybrid.

According to Hodgson the classification proposed by Tanaka had identified an excessive number of species in the genus *Citrus*. He recognised 35 species in mandarin group alone which obviously seems to be based on very minor differences. Secondly, Tanaka has granted species standing to numerous cultigens, some of which are of questionable validity. Likewise species standing has been assigned to some natural hybrids. In view of the extreme divergence in species concept in the two systems Hodgson is of the opinion to recognise an intermediate system. The names of the existing common cultivars given against the species.

(A) Acid Members

(*i*) *Citrus bargamia* Risso ; Bergamot, Limetta.

(*ii*) *C. jambhiri* Lushington (Rough lemon) ; citronelle, jamburi. Important root-stock, tolerant to virus. Better yield of fruits but poor quality. Susceptible to root rot and blight. Tolerant to Tristeza.

(*iii*) *C. latifolia* Tanaka ; Tahiti or Persian lime.

(*iv*) *C. limmettoides* Tanaka ; Palestine sweet lime, sharbati or mitha nimbu of India. Good as root stock.

(*v*) *C. limon Burman* (lemon) ; Important source of lemon oil. Eureka and Lisbon (USA), Femminello and Monachello (Italy), Bernia (Spain).

(*vi*) *C. medica* L. (Citron) ; Monoembryonic species of India. Acidic pulp, juice scanty. Susceptible to frost.

(*vii*) *C. aurantifolia* (sour lime) ; Kagzi nimbu or nimbu lie a cultivar of India. Susceptible to Tristeza.

(*viii*) *C. karna* Raf ; Karna, Karna Khatta, Khatta nimbu, Soh-Sarkar of Assam. Polyembryonic, used as root-stock.

(*ix*) *C. limonia* Osbeck ; Rangpur lime, carton or cravo lemon, mandarin lime. Susceptible to root rot, tolerant to Tristeza and salinity.

(*x*) *C. limetta* ; Limetta of mediterranian, Pani jamir of Assam.

(B) Orange Group

(*i*) *C. sinensis* (sweet orange) ; Mosambi, Malta, Blood Red, Sathgudi of India, Valencia, Pineapple, Washington Navel of USA, Shamouti of Israel, Succari of Egypt, Dobla Fina of Spain etc. are some of the important cultivars. Polyembryonic species of Chinese origin. Fruits of good quality. Used as rootstock also which allows the accumulation of more nitrogen, phosphorus and copper in the leaves. Resistant to Tristeza.

(*ii*) *C. aurantium* (bitter or sour orange, known as Khatta (H.)) Susceptible to Tristeza. Citrus nematode and burrowing nematode. Flowers fragrant may be used for making perfumes. Not used as rootstock now. Fruit orange-red skinned at maturity.

(*iii*) *C. myrtifolia* Raf. ; Chinnoto, chinois, hazara, myrtle leaved orange.

(C) Mandarin Group

(*i*) *C. deliciosa* Tenore ; Polyembryonic, Mediterranean origin, Willowleaf mandarin, Kinnow and Wilking of USA, Blinda (Algerian selection).

(*ii*) *C. nobilis* Lour ; (True mandarin orange) King mandarin, Kunembo of Japan, tangors like Temple, Canton orange. Orange coloured or reddish fruit, sweet pulp, juicy, excellent quality. Native of Indo-China.

(*iii*) *C. reshmi* hort. Tanaka ; (spice mandarin) Cleopatra mandarin, susceptible to root rot but tolerant to Tresteza.

(*iv*) *C. sunki* hort. Tanaka ; Rootstock of ponkan in China.

(*v*) *C. reticulata.* Blanco ; Polyembryonic species of China. Fruits sweet loose skinned, orange coloured. Nagpur, Coorg, Khasi mandarin of India and Ponkan of China.

(*vi*) *C. tangerina* hort. Tanaka ; tangerine, Daney's of USA, Obeni mikan of Japan.

(*vii*) *C. unshiu* Marc ; Satsuma mandarin, economic citrus of Japan, Owari, Kara, Silver Hill are important cultivars. Seedless fruits of very good quality. Petiole narrowly winged but long cold hardy.

(*viii*) *C. madaraspatana* hort. Tanaka ; Kichili, Valdapudi, Guntur sour orange.

(*ix*) *C. medurensis* (Calamondin).

(D) Pummelo-Grapefruit Group

(*i*) *C. grandis* L. (*C. maxima* Merill) ; Pummello or shaddock monoembryonic species indigenous to Malaya. Large leaves broadly winged petioles. Fruit sweet, moderately juicy, may be white fleshed or red and pink fleshed. Kao Pan of Thailand, Buntan of formosa.

(*ii*) *C. paradisi* Macf. (grapefruit or pummello). Polyembryonic species originating in China. Petioles less broadly winged as compared with pummello. Fruits juicy, sweet with better aftertaste. Josts, Ruby, Marsh, Duncan (seedless) are cultivars or commercial value.

(E) Others

(*i*) *C. junos* Steb ; Yuzu (known in China for over 2000 years).

(*ii*) *C. macrophylla* Webster ; colo, alemov of Cebu.

(*iii*) *C. natsudaidai* Hayata ; Natsudaidai, dai dai miken. Japanese summer grape fruit.

(*iv*) *C. pennivesiculata* Tanaka ; Gajanimma (identical to C. moi).

KEY TO CULTIVATED SPECIES OF CITRUS

A. Petioles wingless without apparent articulation at top ; fruits large with very thick peel and scant acid juice. — *C. medica* (citron)

AA. Petioles clearly articulate with lamina usually winged or margined.

B. Petioles normally margined, flowers, buds, tinged reddish, stamens usually more than 5 times the number of petals. — *C. limon* (Lemon)

BB. Petioles with broad or narrow wings ; flower bud white, stamens usually 4-5 times the number of petals.

C. Flowers small, usually 2-3 cm or less across; fruits usually 4-6 cm in diameter, juicy, very sour. — *C. aurantifolia* (Sour lime)

CC. Flowers large, fruit usually much bigger, juicy, sweet, mild or sour.

D. Leaves rather broadly winged.

DD. Leaves normally winged.

E. Leaves subcordate, twigs and under-surface of leaves sporting pubescent ; flower very large, usually 10-12 cm in diameter ; pulp vesicles large, separating easily, seed monoembryonic.	*C. grandis*
EE. Leaves no subcordate, glabrous beneath; pulp vesicles coherent, seeds polyembryonic.	
F. Flower large, fruits usually 9-13 cm in diameter, pulp vesicles rather large.	*C. paradisi* (Grape fruit)
FF. Fruits medium sized, 4-6 cm in diameter with hollow cone and small pulp vesicles, sour and bitter.	*C. aurantium* (Sour orange)
E. Leaves usually small and narrow, 4-6 cm long, fruit with loose peel, embryos green.	*C. reticulata* (mandarin)
EE. Leaves larger and broader ; fruits with adherent peel, embryos white.	*C. sinensis* (Sweet orange)

Colour of fruits is unreliable in tropics as full colour of ripe fruits seldom develops and may remain green. With mature fruits keys should not be necessary. Size of the flowers, adhering of petiole is also very variable. Interspecific hybrids, have not been included in the classification.

Generally the common varieties recognised are sweet lime, Kagzi lime, Lemon, Eureka, Kagzi kalan, seedless lemon and Pant lemon. Lemon is a wondeful detoxifier and antitoxidant.

1. SOUR ORANGE (*Citrus aurantium* L.)

Sour orange a highly polyembryonic species has been extensively used as a root stock. It is considered immune to foot rot and all the gum diseases but is very susceptible to citrus scale and to quick decline diseases. It's very abundant and deep root system makes it thrive best on heavy and wet soils. The place of origin at the species is south-eastern Asia, probably in Cochin/China and then spread eastwards. It was introduced in Europe in the 11th century, earlier than sweet orange which was introduced almost 5 centuries later. It is now grown in the entire tropics and subtropics.

Height of trees may be upto 10 M, thorns slender, leaves medium sized. Petioles 2-3 cm long, rather broadly winged, lamina ovate or elleptical, upto 10 × 7 cm, sometimes crenulate. Flowers axillary, large, white, very fragrant, with 5-12% staminate only, stamens 20-25, ovary with 10-12 loculi. Fruits sub-globose with thick rough strongly aromatic peel, often white, orange red at maturity. Pulp very sour and bitter, central core usually hollow, seed numerous, polyembryonic with a proportion of nucellar embryos.

Used for making marmalade, too bitter to be used fresh. Volatile oil used in perfumary may be available from fruits, flowers and leaves. The peel may be used as a medicine. These are too bitter to be used as fresh fruit. Sour orange is extensively used as a root stock for lemon, sweet orange and grape fruit. It is resistant to gummosis but in comparison with sweet orange and grape fruit it is susceptible to tristeza virus not preferred as rootstock. Predominant glucoside is aurantamanin.

C. aurantium has at times been compared with *C. sinensis* but they are distinct in morphology, chemical as well as physiological properties and also smell and taste. It is harder than sweet orange. But mutation of this has given rise to var. *Myrtifolia* ker Gawl. Hybrid with *C. sinensis* perhaps give the bitterest oranges.

2. SWEET ORANGE (*Citrus sinensis* L. Syn. *C. aurantium* L. var. *sinensis* L.)

This is a very susceptible to Malda Goma foot rot and is unsafe to grow except on light and dry soils. It has shallow root system but produces shapely trees. In California sweet orange root stock reported to be gaining popularity. It is not found in wild state and is a native of southern China. It does not do well in the wet tropics or on high altitudes.

It is perhaps the most widely grown of all *Citrus* sp. Used principally for making orange juice and squashes and as dessert fruit. Oranges are used for marmalade and also for flavouring. The plant provides essential oils, neroli oil, petit grain oil and orange oil from flowers, leaves and peels, respectively. The peel is also used for making pectin. The pulp is used as a cattle feed after taking out the juice. It provides hesperidin as the prominent glucoside from the mesocarp.

The trees are 6-12 M high at maturity, often with angled twigs when young, and stout spines. Leaves ovate or oval elliptic, glabrous, dark green. Petiole 1-2.5 cm long, normally winged, articulated. Flowers axillary, borne singly or in small groups. Petals 5, stamens 20-25. Fruits ripening to an orange colour but often remaining green in the tropics. Fruits subglobose. Pulp juicy, subacid, seeds nil to many, polyembryonic. Cultivars divided into (*i*) those with normal fruits, (*ii*) naval oranges with a second row of carpels, opening at fruit apex in an umbilcus, (*iii*) blood oranges with red or redstreaken pulp. Aberrant forms may occur within the group.

Important cultivars. 'Valencia' a vigorous grower and prolific late cultivar. Fruits of medium size with excellent flavour and 5-6 seeds. Of importance in California, South Aftrica, Australia and West Indies.

'Washington Navel' possibly originated as a bud mutation in Brazil. Fruit larger with naval and rather thick peel, juice sweet and of excellent flavour, seedless. It is a mid-season cultivar. Other prominent cultivars are 'Hamelin' or Hamlin, 'Parson Brown' ; early cultivars : Pineapple, Midseason. 'Jaffa', 'Shamouti', 'Maltese' and Ruby are blood oranges.

Jaffa. One of the famous sweet oranges from Palestine not of much commercial value in India.

Shamauti. This is again an orange of world fame grown extensively in Palestine deserving trial in the arid Indo-Gangetic plains in the north.

3. LEMON (*Citrus limon* Burm. F. Po. *Citrus trifoliate* Raf.)

This is used as a rootstock in China and Japan. It does not do well under tropical conditions. Its value lies in its being the hardest of all citrus rootstocks and is therefore used in frosty localities. It has dwarfing effect on the scion.

C. limon originated in south-eastern Asia, may be east of Himalayas, in north Burma and south China or in north western India. It was known to the Arabs in the 10th century. 'Eureka', 'Lisbon' and 'Villafranca' are the most widely grown cultivars. Rough lemon is probably a hybrid with C. medica. 'Perrine' is a cross made by Swingle between 'Mexican' lime end 'Genoa' lemon.

Widely used in the preparation of lemonade, squashes, etc. Used in the preparation of cosmetics and for the production of lemon oil, citric acid and pectin. Lemons are less important than limes in the tropics but they are extensively used in temperate countries.

Short trees 3-5 m high with stout stiff thorns, petioles short margined but not winged and with a distinct articulation and a petiole. Flowers found all the year round, axillary, solitary or clustered. Petals pink in bud, on opening white above and pink below. Stamens 20-40, fruits oval with a terminal nipple, sour taste.

4. PUMMELO OR SHADDOCK (*Citrus grandis* L. Osbeck Syn. *C. maxima* Merril. ; *C. decumana*)

A valuable rootstock on dry lands but the plants budded on shaddock are badly mottled, in Philippines, *C. grandis* is supposed to be a native of Thailand and Malaysia from where it reached India, China and Persia. It was only towards the 12th century or the beginning of the 13th century that it was taken to Europe as a curiosity. Captain Shaddock introduced it in Barbados in the 17th century. It is known as Pomelo in the East where it is esteemed as a dessert fruit. Best Pomelo crop is raised in Thailand on the ridges surrounded by brakish water. The fruit may be used in making marmalade also.

The trees are spiny, spreading 5-15 m in height. Leaves large petioles broadly winged. Flowers large, borne single or in clusters, petals cream coloured, stamen, 20-25. Fruit very large globose or pear shaped, 10-30 cm in diameter, thick peel, pulp pale yellow or pink with sweetish juice, monoembryonic seeds. Predominant glucoside in naringin. Important cultivars are Walter, Chakaya, Large red fleshed, Large white fleshed and Gill's Jeolikote.

5. GRAPE FRUIT (*Citrus paradisi* Macf.)

The root system is more fibrous than that of either sour or sweet orange but these have a small and crooked tap root. It is susceptible to root rot and certain kinds of gummosis. Grape fruits are closely related to pummello. In Barbados and Jamaica it was known as "the forbidden fruit" or "small shaddock" in 1750 and 1789 respectively. The first record of the name grapefruit in Jamaica is in 1814. It was introduced in Florida in the 19th century. It is used as a breakfast fruit, the juice has a characteristic flavour with wild bitterness. It is commonly grown in all tropical countries. Important exporting countries are Brazil, West Indies, South Africa and Israel.

The plants are spreading with a length of 10-15 m, petioles broadly winged. Flowers axillary, single or in clusters, 4-5 cm in diameter, petals white, usually 5, stamens 20-25. Furit large, usually globose, 8-25 cm in diameter, greenish or pale yellow when ripe ; rind thinner and pulp vesicles smaller than those of *C. grandis*. Seed white, polyembryonic. The prominent glucoside is naringin.

Most widely planted cultivar is 'Marsh' (Syn, Marsh seedless) with only 4-8 seeds per fruit. 'Thompson' (Syn. "Pink Marsh") has pink flesh and 0-5 seeds per fruit. It is a bud mutant. Bud mutation of 'Thompson' are 'Ruby' and 'Webb'. The interspecific hybrid with *C. reticulata* is knwon as Tangelos. Parnambico, Triumph and Saharanpur special are also important varieties.

6. LIME (*Citrus aurantifolia* Swing)

It is used fairly extensively as a root stock for Malta oranges and Santara, for the latter it is the best. It grows wild in North India and may have originated there. Lime is indigenous to the East Indian Archiphelago from where it moved to the main land. It reached Europe in the 13th century and from there to the New World. It is now cultivated throughout the tropics as a common species of cultivated acid *Citrus*. Its production is extensive in Mexico and Egypt.

Although lime is not closely related to other *Citrus* species it is known to hybridize with them. Lime is used as fresh juice, a flavouring agent, lime juice cordial, marmalade etc. The peel may be used for extracting lime oil. Citric acid is also made from the juice. It is transported in small quantities. Since it does not transport well its place is taken by lemons.

The tree is small, much branched upto 5 m in height, heavily armed with short, sharp spines ; leaves small, oval elliptic, margin crenulate, petiole narrowly winged, inflorescence axillary, 1-7 flowered, petals 4-8, stamens 20-25, ovary with 9-12 loculi. Fruit small, oval or globose, peel thin, adherent, greenish-yellow when ripe, pulp greenish, very acid ; seed polyembryonic, cotyledons white. Larger fruited 'Tihiti' is almost immune to whither tip but susceptible to citrus canker. Mandarin lime and sweet lime are interspecific hybrids probably with *C. reticulata* and *C. medica* respectively.

7. CITRON (*Citrus medica* L.)

A tender variety known as galgal (in Hindi) is largely raised by seeds and stem cuttings in Uttar Pradesh, Punjab and also Egypt. In the initial stages it is very easy to work upon and gives quick growth but is not a good root stock because of its reported susceptibility to diseases and short orchard life.

C. medica is believed to have originated in India by some but there is a greater possibility of its having originated in south-east Asia where it has been cultivated since remote antiquity. It first reached Europe in 300 B.C. and ranked first of all the citrus cultivars. Theophratus has mentioned it as the Persian apple. Pliny named it citrus for the first time and this name was adopted later by Linnaeus for the genus. Commercial planting is limited to Italy, Greece, Corsica. Romans used it for flavouring and medicine, Fingered citron (*C. medica* var. *sarcodactylis* Noat Swing) is used as medicine and for ornamental purposes.

3 m high shrubs or small trees with stout spines. Leaves elliptical, serrate, petiole short wingless and not articulated with lamina. Inflorescence racemes, 5 petioles tinged pink, stamens 30-40. Fruits large, oblong, 10-20 cm long, peel usually bumpy, yellow, very thick, sour taste. Hesperedin is the prominent glucoside present.

8. MANDARIN (*Citrus reticulata* Blanco ; Syn *Andrews*-non Lour.)

It is also known as "fancy fruit" and is taken as a dessert fruit. It originated, in all probability, in Vietnam or China and has been cultivated in China and Japan over long periods. It reached Europe in 1805 and U.S.A. later in 19th century. It is said to be immune to scale and resistant to gummosis. Mandarin is used as a root stock for its own scion. It is most compatible rootstock for orange, lemon, and grapefruit. It is resistant to root rot and nematodes. It is hardiest of all citrus species and is grown in most tropical and subtropical countries.

The terms mandarin and tangerine are used indiscriminately. Mandarin should be rightly used for yellow fruited cultivars and tangerine for the deep orange ones. The cultivars can be classified as below.

Satsuma group. Commonly grown in Japan, very hardy and cold resistant.

Mandarins. Fruits yellow or pale orange, *e.g.*, cultivar Emperor. Coorg, Khasi, Nagpur and Srinagar and exclusively grown in India.

Tangerines. (Citrus tangerine hort. Tanaka) Fruit deep orange-red *e.g.* 'Clementine' and 'Dancy' grown in Algeria and United States respectively.

Citrus reticulata var. austera Swing is sour mandarin, tangelo (*C. reticulata* × *C. paradisi*) is a interspecific hybrid with fruits resembling oranges with a flavour in between the parents. Tangors (*C. reticulata* × *C. sinensis*) is a natural hybrid from Siam and is the type of *C. nobilis* Lour.

Small tree 2-8 m in height, leaves small, ovate, narrow, elliptical or lanceolate usually crenate, flowers white, fruits depressed-globose, 5-8 cm in diameter, contains tangererine as prominent glucoside. Pulp sweet and juicy, orange in colour, seeds polyembryonic.

Propagation

In the past seeds usually formed the chief source of propagation but now it is propagated vegetatively by budding. Seeds with zygotic embryos do not reproduced true to type, whereas those with nucellar embryos produce seedlings with the same genetic constitution as buddings but they tends to be rather more vigorous, more thorny and late bearing. Seeds of good, fully matured fruits from vigorous, mature, healthy trees may be used to grow root stocks. The seeds need partial drying in shade after being removed from the fruits since complete drying in sun makes them nonviable. Storage of the seed, if required, may be done in charcoal. Optimum temperature for germination is 26° to 32°C whereas any temperature below 12°C completely checks germination. Seed-rate for raising the seedlings should be 400 gms for every 100 sq. m of seed bed. 20 cm tall seedlings are transplanted in nurseries at a spacing of 1 m × 30 cm generally 6 months after sowing. They are ready for budding after another 6 months. Small seedlings denote zygotic origin and bent roots the off types. Both should be discarded.

Root Stocks

The cultivar to be budded and the area in which the crop is to be grown determines the type of root stock. The stock and the scion must be compatible and the union should produce long lived trees of good productivity. The effects on the size, shape, juiciness, grade and yield of fruit, the length of the productive life of the tree, resistance to cold, drought, wind, diseases etc. should be considered before deciding about the root stock to be used.

The common root stocks used are those of sour orange, rough lemon, sweet orange, grape fruit, trifoliate orange, Karna Khatta etc. (See Table 7.1).

Citrus aurantium (sour orange). Very good root stock for damp heavy soil, deep tap root, not susceptible to drought and also hardy with slow growth in the beginning but gives high yields of excellent quality fruits later. The fruits are held on the tree to full maturity in good condition. This stock is resistant to gummosis but not scab, scion compatibility is good except with lime (*Citrus aurantifolia*).

Citrus sinensis (sweet orange). Suited best to well drained rich loams, shallow rooted and less hardy than sour orange. The growth is rather slow in the beginning but the trees are long lived giving fruits of high quality. It is resistant to tristeza and scab but susceptible to gummosis, qualities just the reverse of sour orange. Compatible with oranges and lemons.

Citrus lemon (rough lemon). Good drained loamy or sandy soils are best for this stock which does not have tap root but abundant widely spreading lateral roots. Fruits on this stock are of poor quality, more acid thick-skinned and large but are not held for long period on the trees in good condition. It gives rise to early fruiting trees which are vigorous. It is not susceptible to gummosis and scab, considered a good stock for lemons and limes and also for grape fruit tangerine.

Karna Khatta (*C. Karna,* L.) also known as large Karna lime is a commonly used rootstock for the propagation of mandarins and sweet oranges in U.P., Haryana and the Punjab.

Grapefruit (Citrus paradisi and Trifoliate orange) Poncitrus trifoliata) are also occasionally used as rootstock. Trifoliate orange is of special significance in

Table 7.1 Characteristics of citrus rootstocks

Characters	*C. auretium* L. (sour orange or serville orange)	*C. sinensis* (sweet orange, malta orange, mosambi, sathgudi)	*C. limon* (Rough lemon, jambhiri or jatti khatti)	*C. para disi* (Grape fruit)	*C. limett-ioides* (Sweet lime, mitha nimboo)	*C. limonia* (Rangapur lime, kagzi lime)	*C. reticu lata* (Mandarin or tangarine, santra)	*Poncirus trifoliata* (Trifoliate orange)	*Troyer citrange* (Hybrid)	*C. karna* (Karna khatta)
1	2	3	4	5	6	7	8	9	10	11
Production	Good, best for naval	Good	Good for 20 yrs.	Low average	Good for 10 yrs.	Good	Good during late yrs.	Average	Excellent	Good
Soil Preference	Sandy loam to clay	Light sandy, alkaline unsuitable	Sandy	Sandy to loam clay	Light	Medium to heavy	Sandy loam.	Sandy loam, absence of lime	Sandy loam to clay	Light
Growth in nursery	Vigorous upright	Less vigorous upright	Very vigorous	Slow bushy	Moderately vigorous upright	Vigorous upright	Less vigorous upright	Less vigorous upright	Rapid upright	Rapid upright
Longevity Cold hardiness	Average Hardy	Long Hardy	Short Tender	Average Hardy	Short Hardy	Average Hardy	Average Hardy	Average Very hardy to frost as well as water logging	Average Hardy	Average Very hardy
Suitable for	Lemon, sweet orange	Grape fruit	Limes, lemon, blood red malta, kumquat	Lemons and Lime	Orange	MandaRin Grape fruit	Manda-rins	Kumquat	Manda-rins	Grape fruit, musa- mbi, hill orange

Contd. on next page

	2	3	4	5	6	7	8	9	10	11
Most compatible scion	Lemons, oranges, grape fruits	Oranges, lemons, grape fruits	Navels blood red malta, sweet oranges	Lemons	Oranges, mandar-ins	Sweet oranges, mandari- ns, grape fruits	Orange, lemon, grapefruit	Kumquat	Mandarins	Grape fruit, musambi hill orange,
Fruit size	Above average	Below average	Average to above	Average	Good	Average	Average	Good	Good	Average to above
Fruit quality	Average	Average	Very poor coarse fruit	Good to excelle- nt	Good	Good	Average	Good	Ecellent	Av thick skinned, poor quality
Seeds/100 g	650–750	650–750	750–850	450–500	700–800	1000–2000	1000–1200	440–480	500–550	600–700
Seeds/fruit	15–20	15–20	15–20	30–40	3–5	8–15	10–15	30–50	3–5	15–20
% polyem-bryony	60–80	80–90	95–100	80–90	80–90	40–60	80–90	60–80	100	80–90
Gummosis	R	S	R	R/S	S/R	R	S	R	R	S
Quick decline (Tristeza)	S	R	R	S	?	R	R	R/T	R	S
Nematodes	?	S	–	S	?	?	–	R	R	?
Greening	S	?	S	–	–	–	–	R	R	S
Canker	–	–	–	–	–	S	–	–	–	–
Root rot	–	–	–	–	–	–	R	–	–	–

N.B. : R = resistant; S = susceptible ; ? = not known ; T = tender ;
R/S or S/R or R/T indicate that some species have opposing characteristics.

colder regions being the hardiest for withholding cold. It also tolerates waterlogging. Trees on this stock are dwarf and prolific. It is resistant to gummosis and tolerant to tristeza. It is compatible with most of the *Citrus* spp. and is commonly used in Japan. Grapefruit is suitable for rich, heavy, poorly drained soils, gives vigorous growth but poor yields. It is not used generally except for a 'wild' type for a budding lines. Besides these 'Cleopatra' mandarin, Palestine sweet lime citranges (*P. trifoliata* × *C. sinensis*) and Rangpur Lime are also known to be used as rootstock under specific conditions in restricted areas.

Budding

It can be done any time when the plant is in growth. It is generally done by the 'shield' method in which scion bud is inserted and then bound firmly with budding tape. Bud wood must be selected from well matured wood of current year growth obviously from high yielding trees with no off type bud mutations. Bud sticks are generally about 12-15 cm along with three or more unsprouted but well developed buds on them. Sprouted buds and buds with throns are unsuitable since the growing point is often damaged while budding. Bud wood can also be stored wrapped in *Sphagnum* or moist saw dust, if required. When unsuccessful budding operations may be redone after about showed green colour after about 3 weeks of the budding it may be unwrapped. When the bud is initiated and is about 3 cm long, the top of root stock is severed with a short narrow cut just above the bud. Stock suckers must be removed and the buddings topped at 60-90 cm height to force lateral buds. The lowest branches below the union are removed and upper 4 to 6 branches are allowed to develop. The time from budding to transplanting varies with local needs and conditions. Severing of the tap roots clearly and sharply at a reasonable distance from the stem at the time of transplanting helps to develop lateral roots that spread in the rich top layer of the soil. Lifting the plants from the nursery is done carefully so that the roots are injured to minimum. The plants are taken out with the ball of earth around the roots intact which may extend even to 15 cm in diameter. Practical defoliation, sometimes accompanied by pruning of a few off-shoots, is done at the time of lifting leading to a reduction of the water losses by the plant.

Propagation is possible by T-budding and inverted T-budding. Cirus can also be grafted or air layered. Some species may be propagated by layers, marcotts, cuttings or grafting. Lime, lemon, grape fruit and pummelo are propagated by air-layering, soft wood or leaf bud cuttings.

Bud mutations

Seed variations and the changes in somatic tissues (bud variation) of the citrus plant give rise to mutations or sprouts. Careful observation and isolation of desirable bud mutations help in selecting more productive trees which are uniformly standard. It is presumed by some that bud variation takes place by differential mitosis. Variations also occur even among the apogamic seedlings.

Care in Planting

Citrus when transplanted should be protected against sun and wind injury and must be planted immediately, after digging, in larger pits. Planting with bare roots permit inspection for diseases, nematodes and damages. Planting with a ball of earth permit transport over long distances. The distance recommended are 8, 7 and 6 m for graperfruit orange, lemon and limes respectively. Planting holes are generally 80 cm in depth and diameter. Each may be given 1/2 kg of mixed fertilizer containing bone meal : FYM : ash : bonemeal : : 1 : 20 : 2.

In regions of high rainfall a permanent cover crop such as Kudzu (*Pueraria phaseoloides*) is desirable. Manuring with N, P_2O_5, K_2O and MgO in the ratio of 4 : 2 : 4 : 1 is useful. Zinc and magnesium deficiencies may be controlled through foliar spray.

Pruning

It continues from the nursery to 2-3 years after the plant starts bearing. Suckers must be removed and dead wood cut out. Fruits should be picked when mature.

Selection of Site and Spacing

A suitable site is selected for orange cultivation, green manuring is done. Spacing is determined according to need of the cultivar, soil, irrigation facilities, climate, the rootstock used etc. The field is prepared well and kept clean of weeds. Most of the citrus cultivars do well, normally with a distance of 7-9 metres between plants.

Pits

The plants are put in well manured pits which are best made in the hot months and kept open until the following rains.

Planting

Healthy and vigorous budded plants are planted after the onset of rains when the atmosphere is a bit humid and not dry, in centre of the pits. The soil put over and around the roots is pressed well. Care is taken not to plant in way so as to bury a part of the stem in soil. In North India planting may also be done in February or March when the plants are dormant. In the south the planting season is from July to January. The pre-summer plantings are not advisable due to the approaching summer heat which is harmful for the establishment and early growth of the plants.

Irrigation

Immediately on planting copious irrigation is given followed by light watering, according to need, every four or five days. During the months of July to October when rains are expected watering is resorted to only when there is a long break of rains. Check basin, ring and furrow method of irrigation may be followed.

The basin system is perhaps the best method of irrigation where small basins are prepared with bunds, a little away from the drip of the young trees and these are connected with irrigation channels. The basins are widened gradually as the trees grow. The ring system is also recommended for young plantations. It consists in making a small raised mound of earth near the trunk of trees, and by the outer side of the mound and a ring about 6-150 cm. in width with an encircling bund is made to hold water. The furrow method is extensively employed in some citrus growing countries of the world. Several furrows are made at equal distances between the tree rows with a double mould board plough and irrigation water is allowed to flow in these. The entire soil surface is not wetted, yet an even distribution of water is ensured.

Manuring

Nutritional requirement cannot be prescribed for fruit trees specially in case of citrus owing to the rootstock factor operative in most cases. The rootstock influences the intake of nutrients from the soil and also its supply to the scion. A regular supply of adequate nutrients in proper amounts at the critical periods of requirement is needed for better production in contradistinction to ill balanced and faulty supply. The lemons, oranges, mandarins need macronutrients like calcium, nitrogen and phosphorus and potash according to the yielding capacity—the higher the yield the greater the need for macronutrients.

Nitrogenous fertilizers help the young plants to establish themselves well with rich green and healthy tops and an established root system. The supply of phosphates and potash is needed when the plants come to bearing, it encourages root development, stimulates and hastens maturity of fruits. Potash fertilizers give a balanced growth and maturity of shoots and fruits and also helps in formation of sugar and starch.

Recommended Manuring Schedule for one Citrus plant

Age (Years)	Nitrogen (gm)	P_2O_5 (gm)	K_2O (gm)
1	75	40	75
2	150	80	150
3	225	120	225
4	300	160	300
5	375	200	375
6	450	240	450
7	525	280	525
8	600	320	600
9	675	360	675
10 and onwards	750	400	750

Generally speaking calcium, nitrogen, potash and phosphorus can be graded in their requirement by citrus plants in decreasing order. Phosphorus increases the fruit size, influences the coarseness and thickness of the peel and ascorbic acid content. Potassium works as a limiting factor in yield under certain circumstances. Both the rootstock as well as temperature affect the uptake of potassium. Lower temperature is associated with less intake. Potassium retards degreening. Calcium acts both as a nutrient and as soil improver, its application reduces the nematode population and damage to the roats.

Application of fertilizers should be done in a manner that these are beyond 30 cm radius of the tree trunk and extending upto more than the drip area. It is safer to broadcast the nutrients in the recommended zone of each tree at a time when the soil is moist in 2-3 split doses. The liberal supply of calcium having been assured nitrogen (100 g), P_2O_5 (80 g) and K_2O (100 g) may be applied on three different occasions every year to each citrus plant which is full grown and has reached the bearing stage.

The citrus is also very responsive to micro nutrients like magnesium, zinc, copper, iron, manganese, boron and molybdenum which may be supplied through foliar sprays. The form in which these minerals should be supplied must necessarily be soluble in water and the pH should be more or less neutral. The concentration of the nutrients should be adjusted according to the age of the plant, form of nutrients, season of spray and degree of deficiency, if any. The need for Mn, Mg, Zn and Cu in single dose or split dose is likely to be more pronounced after fruit-set.

Soil aeration, freedom from weeds, adequate shallow tillage, avoidance of development of hard pan at some depth are important for citrus soils.

Deficiency Symptoms

Deficiency of some of the minerals gives characteristic symptoms for example copper causes 'die-back' or blisters on young growing branches, leaves etc. resulting in leaf and fruit-drop and disintegration of twings. N, Mn, Cu, Ca, Mg, Zn, Fe and B deficiency also give symptoms as below :

Phosphorus. Phosphorus is readily taken up by citrus and is seldom limiting factor. Reduction in foliage density, increase in premature fruit drop and increase in acidity of the fruit and common in acidic soils where the deficiency is easily pronounced. Excess phosphorus need not be applied in an effort to correct deficiency since the excess phosphorus may hinder in the absorption of zinc, iron and copper.

Potassium. Chlorosis in mature leaves, gumming of twigs, premature fruit shedding and reduction in fruit size appear, with deficiency. Fruit quality showed decline in some cultivars of orange. Potassium sulphate or potassium chloride may be used for correcting potassium deficiency. Any excess of potash increases fruit acidity. The requirement of potassium and its responses vary with cultivars.

Calcium. Symptoms of deficiency are not common. Leaves become smaller and thicker, premature drop of leaves and die back of twigs in case of deficiency. Application of domestic lime, gypsum or limestone or superphosphate can correct the deficiency. Its deficiency is rather uncommon but whenever it occurs the finer roots disintegrate to affect die back of young shoots and yellowing of leaves. 7-8 kg of lime per tree is recommended under these conditions.

Nitrogen. Its deficiency causes yellowing of leaves accompanied by sparseness of the foliage and reduced vigour of the tree. Foliar spray with urea or potassium nitrate or addition of organic manures to the soil in liberal doses, 2-3 times in a year is useful.

Leaves develop light green to yellowish green colour. Foliage sparse. Production affected adversely. NO_3^-N is preferred to NH_3^-N, both organic as well as inorganic forms of nitrogen are useful for the citrus plant. foliar spray of urea may be helpful to a limited extent. Addition of ammonium nitrogen checks the expansion of roots specially in soils with pH 4.3 – 4.6. Ammonium sulphate may increase sub-soil acidity.

Manganese. Leaves mottled. Leaf size unreduced (unlike zinc deficiency). Shade makes the symptoms more visible. In severe cases difoliation, loss of vigour and lower yield results. On acidic soils the deficiency symptoms are clear due to shortage of manganese but on alkaline soils the deficiency is due to the unavailability of mn having been fixed in an insoluble form. The veins form a dark green network with interveinal area assuming light green colour. A spray of $MnSO_4$ (2 kg) + quick-lime 1 kg (to neutralize the sulphate) dissolved in 900 litres of water is good for relieving the deficit on both sides of the mid rib which remain green.

Copper. Leaves become dark green, leaf fall, 'peach-leaf' appearance develops. Leaves develop dull brownish green colour, become narrow and elongated with rough margins. In case of acute deficiency the fruits show less colour. In its absence "exanthema" or die-back is seen to appear in citrus plants. Multiple buds develop in the axil of the leaves. Some gum exudes from the branch of the tree. Bordeaux mixture spray cures the ailment.

Magnesium. Leaves and maturing fruits affected more than other parts of the plant. Dispersed yellow areas on the sides of the mid-rib later developing to surround a triangular green area near the base of the leaf. Leaves become yellow all over including veins and may take a bronzed appearance in severe cases. A dose of epsom ($MgSO_4$, $7H_2O$) salt can cure it. 2-4% magnesium oxide as fertilizer or 0.1 or 0.2% solution of magnesium sulphate neutralized with lime can remove the deficiency when applied as spray.

Zinc. In acute cases bushy appearance of trees followed by die-back. Chlorosis between viens. Deficiency of iron increased symptom of zinc deficiency. Motling of leaves, reduction in size of developing leaves, younger leaves affected more than the older ones. Control : Spray of 1.3 kg $ZnSO_4$ + 670 gm slaked lime in

450 litres water is sufficient for 100 trees. It should be sprayed in February and again in October. In case the zinc deficiency is severe the quantity of $ZnSO_4$ and slaked lime should be doubled.

Iron. Leaves light green or yellow with network of green veins. Leaves become more yellow in cases of acute deficiency. Impaired growth and reduced yield. Die back of twigs visible. Top : root ratio affected. Iron chlorosis or lime induced chlorophyll deficiency sets in. Chlorosis of leaves. Veins remain green in the beginning but in advanced cases they loss colour. Use of large quantities of manures and spray with ferrous sulphate mixed with Bordeaux mixture removes the symptoms. The ratio in the ferrous sulphate : copper sulphate : quick lime is 5 : 4 : 8. For 17 kg of ingredients 900 litres of water is used for dilution.

Boron. Tendency to wilt. Leaves show buckling, yellow veins, vein splitting, thickening and puckering along the mid rib. Water soaked spotting of leaves and lumpy fruits appear. Older leaves affected more. Fruits have thick rind, gummosis and die-back are common. Boron application as spray of borax is recommended.

Molybdenum. Yellow spots in the leaves specially on acid soils. The spots are small in mandarins but larger on grapefruit. The deficiency is more pronounced on trees grafted on grape fruit rootstock. The deficiency can be corrected through foliar application of sodium or ammonium molybdate. In acid soils soil-applications are not as effective in correcting the deficiency as the sprays.

Sulphur. No clear cut symptoms of deficiency since the N, P, K fertilisers seem to supply enough sulphur as impurity.

Research workers of Horticultural Institute, Saharanpur, U.P. suggest a general nutritional spray containing 0.1% $CuSO_4$, 0.2% $ZnSO_4$, 0.1% $MnSO_4$, 0.1% ferrous sulphate, 0.5% borax and 0.3% slaked lime in 100 litre of water in Feb.-March before emergence of new growth and again in October as postmonsoon spray for maintaining good growth of *Citrus* trees. Whenever the attack of insect-pests is visible, insecticides like dimethoate may be mixed in the general nutrition spray stated above. Addtion of ordinary washing powder or sunlight soap to the mixture makes it more effective.

PRE-HARVEST FRUIT-DROP IN SWEET ORANGES

Pre-harvest fruit-drop in sweet oranges can be considerably controlled either by a single spray of 2.4-D (20 ppm) in August or two sprays 10 ppm each, one in August and the other in October.

Fruit Drop

The shedding of fully developed fruits, just prior to harvest known as 'pre-harvest fruit-drop'. Constitutes one of the major problems in the sweet orange orchards of north India. The extent of these losses varies greatly with different cultivars. Mosambi and Blood Red Malta were found to be more prone to this malady than others with total pre-harvest fruit-drop to the tune of 57.2 and

54.9% respectively during the 3 to 4 months period preceding harvest. The lowest fruit drop (34.1%) was encountered in Valencia.

Nature of Fruit Drop

Observations recorded at Regional Fruit Research Station, Abohar have revealed that dropped fruits can be grouped into the following two classes depending upon the causal factors of their shedding.

A. *Physiological fruit drop.* Included in this drop are fruits which look sound and normal and have no apparent signs of any disease or insect injury. The causes may be as given below :

(*i*) Fruit-drop due to the formation of abcission layer at the stem point,

(*ii*) Imbalance of growth regulators, such as auxins, cytokinins, gibberellins, etc.,

(*iii*) Excess or deficiency of certain essential elements,

(*iv*) Unfavourable weather conditions,

(*v*) Improper cultural practices

(*vi*) Improper cultural practices

B. *Pathological fruit-drop.* This type of fruit drop is due to the attack of certain pathogens. Two fungi namely *Alternaria citri and Colletorichum gleosporioides* are largely associated with pathological fruit drop.

Alternaria citri attacks on the styler-end of the fruit and disease caused by this fungus is called styler-end rot. This fungus infects the fruits at a very early stage of development but being a weak parasite, it remains quiscent. As the fruit reaches maturity, the fungus becomes more active and brings about shedding.

C. gleosporioides attacks on the stem-end of fruit and the disease, caused by this fungus is called 'stem-end rot'. Studies made at Abohar have shown that somewhat greater number of fruits shed due to stem-end rot than the styler-end rot.

Control of Pre-Harvest Fruit Drop

Sweet orange growers are advised to apply either a single spray of 2, 4-D (20 ppm) in August or sprays of 10 ppm strength one in August and the other in October. If the trees exhibit zinc deficiency symptoms, 0.6% $ZnSO_4$ may be added to the August spray. As infection of pathogens takes place soon after setting, it will be desirable to spray either with Bordeaux mixture @ 1 gm or Aureofungin @ 0.02 gm per lit. water. The application of 2, 4-D cuts down the loss of physiological fruit drop from one-third to one-fourth in Mosambi and Blood Red Malta.

Two sprays of NAA @ 30 ppm at pea and marble size stage of the fruit may also be given for minimising fruit drop. Citrus psylla is also one of the causes of fruit drop. Application of insecticides discourages the attack and reduces the fruit-drop.

Fruit Cracking

Furit cracking is common in lemon, it is sometimes noticed in Malta also. It is believed that the fruit cracking is related to fluctuations in relative humidity. Timely irrigation of the orchard in summer may be helpful in checking the cracking of fruits. Spray of potassium sulphate @ 4 gm/l water or GA @10 ppm three times in April, May and June is found to be deterrant to cracking.

Cultivars

The most important of the sweet oranges cultivated on a commercial scale in India are Malta Blood Red, Malta common, Pineapple, Mosambi, Satgudi etc.

Malta Blood Red. Dwarf trees, some strains are seedless and prolific, those with seed are less prolific and have 5-12 seeds per fruit. Rind thin and tight, pulp is blood red when fully ripe, juicy and sweet ; acidity 0.5%, pleasant flavour, excellent quality.

Mosambi. Small or medium fruits bear a circular ring (areole) around the apical end. The rind is marigold, thin and tight ; pulp less juicy, flavour not marked, rind green or yellowish green under tropical conditions, 18 to 20 seeds per fruit. 0.33% acidity.

Satgudi. Commonly known as Chini. Fruits medium sized, subglobose, a little longish than the diameter, aerole absent, juicy with fine flavour. Rind, medium thick, smooth, finely pitted and yellowish, pulp is straw-coloured. 12-20 seeds per fruit. This cultivar is grown extensively in the districts of Cuddapah, Chittoor and Kurnool in south India.

Pineapple. Medium sized, subglobose, orange coloured fruit on maturity. Rind thing, bright and glossy. Flesh primorse yellow, abundantly juicy, sweet and richly flavoured, with nearly 24 seeds per fruit.

Washington Navel, Jaffa, Valencia Late and Hamlin are well known cultivars of sweet oranges. Washington Navel the cultivar *par excellence* of California had excellent flavour and abundant juice. In India it could not be very successful.

Jaffa is the famous sweet orange from Palestine. Shamouti orange is known to be the world famous seedless orange grown extensively in Palestine. Hamlin is early maturing fruit of the Punjab. Some (Mandarin, *Santra*) cultivars cultivated on commercial scale are described below.

Nagpur Santra. Subglobose fruits of 100-150 gm are cadmium coloured with smooth surface and thin rind. Pulp of marigold colour. Excellent flavour, abundantly juicy, saffron yellow, sweetness and acidity are well blended. 4-11 seeds per fruit.

Khashi Orange (Mandarin). Fruits 125-200 gm in weight, orange yellowish in colour with smooth surface, thin rind. 10 segments, orange coloured pulp vesicles, juicy with good flavour and balanced sweetness and acidity 8-25 seeds per fruit.

Coorg Orange. The tree is large, columnar in shape with, compact foliage, sparingly spinose. Fruits are oblate to globose, pulp vesicles uniformly coloured, flavour good, juicy, 14-30 seeds per fruit.

Deshi Variety (Pathankot). Ovoid to subglobose fruits of cadmium colour, pitted surface and 7-10 segments. Pulp coloured, quite juicy, slightly acidic. Seedless or 30 to 7 seeds per fruit. Total solids 11% and acidity 0.72%. This cultivar is mainly grown in the Punjab hills but on a limited scale in the plains as well. It is grown by seedlings and not from seeds as in case of Coorg Orange.

Orther Cutlivars. Besides above cultivars Dancy and Kinnow (hybrid) cultivars are also introduced in India. Satsuma group of santra orange of Japanese origin has been successfully introduced in Saharanpur (U.P.) and Patiala (Punjab) regions. This group is commercially important because of the early ripening of the fruit and the hardiness or cold resistance of the trees. The fruit of the best strain is seedless and the trees are thornless. Another cultivar known as Darjeeling or Sikkim orange, a native of Sikkim is very popular in the Calcutta markets. It is similar to the Nepal and Assam (Sylhet) orange. Butwal and Emperor, other two cultivars introduced in Saharanpur are doing well on Karna Khatta stalk.

Kinnow

Kinnow is the first generation hybrid of the King sweet orange and the Willow leaf mandarins developed, and distributed by H.B. Frost at the California Experiment Station, U.S.A, in 1935. Its introduction into the Indian subcontinent seems to have taken place somewhere in the early forties of the present century. More recently, its cultivation has spread to greater parts of the Punjab and Uttar Pradesh besides some other citrus growing areas in the country.

Botany. Kinnow trees are fairly tall, vigorous, hardy with upright and spreading obovoid to round symmetrical top ; branches numerous, long, slender and smooth with a few scattered small thorns ; foliage dense and deep green, leaves numerous, medium sized, rather thin, broadly lanceolate to elliptical, base obtuse to normally cuneate ; apex acute to acuminate, margin entire to crenulate-serrulate, upper leaf surface dark green and glossy and the lower yellow-green, petiole short to medium (0.5 to 1.2 cm) distinctly articulated, wingless to narrowly winged.

Fruit Characters. Fruit colour : spanish orange to cadmium orange or marigold orange ; oil glands conspicuous on the surface ; surface very smooth and glossy, very shallowly pitted ; shape oblate, regular and uniform, size medium to large, base flattened, slightly depressed, apex rounded to flattened. It is a late and ripening between the first week of January to middle of February.

Kinnow, like other mandarins, can be successfully grown in all tropical and sub-tropical regions of the country, but seems to prefer a cool, less humid summer and comparatively cooler winter.

Soil. Although Kinnow can be grown successfully on a wide range of soils, yet it seems to prefer well drained, fertile, medium to light loams. Kinnow trees can tolerate more soil acidity (upto a pH of 4) though a pH range of 5.5 to 6.5 is ideal.

Prolific Bearer. Kinnow is prolific bearer and like of its parents, the King Mandarin, has a slight tendency towards alternate bearing.

Insect Pests of Citrus

Lemon Butterfly. Eight species of the lemon butterfly have been reported to feed on citrus in India of which *Papilio demoleus* is the major one. It lays eggs singly on the leaves. The caterpillar, on hatching, feeds on leaves specially on nursery plants of lemons and oranges in seedling stage during the rains. Spraying with malathion 50 EC @ 1 ml/l water or monocrotophos 40 EC @1 ml/l water twice at 10-day interval, and killing of the caterpillars by putting them in kerosene are recommended. The spray is necessary when the caterpillars are found on a large scale. Two sprayings are needed, the first when the green caterpillars are first seen and the second after 20 days. It is advisable to avoid use of broad spectrum insecticides to conserve natural enemies of lemon butterfly. A Commercial formulation known as Dipel of bacterium *Bacillus thuringieusis* Berl. is very effective at 0.05% concentration.

Chloridolum alcamene (Citrus Stem Borer). The stout grub of the bettle causes holes throwing out a fibrous powdery matter. It indicates its presence in the holes or galleries. Petrol or kerosene may be injected with a syringe to kill the pest. The holes may be plugged by wet or molten wax. Removing all the dead twigs, branches etc. from the orchard is a profitable sanitary measure.

Inderbela tetraonis Moore, and *I. quadrindata* (Citrus Shoot and Bark Borer). Laying of eggs by the moth under the losse bark and the emergence of the caterpillar marks the beginning of the malady. The caterpillars bores into the stem and the bark.

Prune the affected twigs and destroy. Insert an iron spike in the hole to kill the larvae when the infection is low. As soon as the caterpillar starts nibbling spray Dichlorophos, Endosulfan or chloropyriphos @ 0.05% on tree trunk and forking of branches. At tunneling stage insert cotton swab soaked with dichlorovos, carbon-bisulphide, petrol, kerosene oil or formalin in the holes and plug with mud. It is important to note that there is only one generation in a year and a regular watch from July onward to check the pest from damage is needed.

Diaphorina citri (Citrus Psylla). The pest is active from February to November and has 16 generations in a year. Adults are generally seen from November to January. It also transmits the citrus greening virus. Young leaves of oranges attacked just in the same manner as the young ones of *citrus* white fly. These orange coloured Psylla suck the sap of leaves and excrete a honey dew which develops the sooty mold fungus. In severe cases total loss of crop may be witnessed for several seasons resulting in the death of most branches or even the tree. Remove the vegetative growth by regulating irrigation and fertilizer management. Spray monocrotophos 40 EC @ 1 ml or carbaryl 50 WP @ 2 gm at bud burst stage and again after fruit set.

Phyllocnistis citrella (Citrus Leaf Miner). The pest is active in spring. It is quite demaging except during severe cold and also hot summer months. The mining may also lead to secondary infections by fungi and bacteria *e.g.* citrus canker. The greyish moth has a very small caterpillar. It mines into the surface of young *citrus* leaves which curl, and show irregular galleries on the surface in which the tiny caterpillars may be seen. The leaves ultimately dry up and fall. If the attack is mild the distorted leaves may be picked up and destroyed by burning. When the infection is severe spray consisting of 4% nicotine sulphate, 2 kg of soap, and 400 lit. water may be done three times at an interval of 3 weeks after the initiation of new growth. Spray of monocrotophos during flushing period every seven days is effective in control. Neem cake solution 60 gm/l water also helps in control.

Othreis fullonica and several species of *Othreis* cause devastating damage to *Citrus*. These may be conrolled by elimination of larval host, netting of the moths by light traps, baited traps with 0.05% malathion + sugar smudging and the disposal of fallen or decayed fruits. Checking the presence of ripening fruits in the autumn has been found to be a good method of avoiding or minimising damage to sweet orange. June-September flowering may be allowed to remain for fruit development.

Chaetodacus divursus. It seems to be a pest only in some parts of the country. Not much has been done to control these in India.

Chelidonium cinctum (Lime Tree Borer) is an important pest in Karnataka and Tamil Nadu. The young grub bores into the twigs and thence into the main stem. The damage is such that one borer is capable of killing of young tree. The treatment or control measures are the same as for *Chloridolum alcamene* (citrus stem borer).

Monohommus versteegi (Trunk Borer). A serious pest in Assam and Sikkim. The grub is active during May-August and bores the trunk near the ground level and if the grub is not killed in time the tree may die in the following winter.

Application of boiled coaltar to the truck before the attack, reduces the chances of incidence. Plugging of the hole after the bore has entered the trunk with any fumigant *viz.* carbon bi-sulphide is recommended to ward off the damage. Common control measure for borers are introduction of a small amount of petrol, carbon, bi-sulphide, charcoal or chloroform : creosote mixture : : 2 : 1 followed by plugging of the hole with clay. Painting the galleries and the surrounding bark may also be useful.

Rhinchocoris humeralis (Citrus Stink bug). This green bug is quite common in Assam, Kumaon regions and the eastern Himalayas. The nymphs and adults puncture the fruits of mandarin and orange specially when partly ripe and suck the juice.

Nezara viridula (Pumpking Bug). It also damages the *Citrus* when the alternative hosts are nearabout the *Citrus* orchard. *Cappoea taprebanica* and *Leptoglossus membranaceus* are also reported to attack the citrus. Brushing down the bugs in water with a kerosene layer has been suggested to kill the bugs.

Diseases of Citrus

Damping off which is caused by *Pythium, rhizoctonia* sp. appears in nurseries can be checked by rotation. *Greening* caused by mycoplasma can be controlled by the use of certified disease-free bud-wood. Greening has been reported from all the citrus growing tracts of the country. It can be spread through infected bud wood and the insect vector citrus psylla. The presence of the disease is marked by the development of chlorosis of leaves, thickening of leaf blade, shortening of twig internodes, off-season blooming, leaf shedding and die-back. Fruits are smaller and remian green longer, contain curved columellas, aborted seeds and fall prematurely. Removal of infected trees, control of insect vector through pesticide and use of certified disease-free seedlings besides injecting tetracycline are the measures which may be adopted to control the disease and its spread. *Drenching of soil with 1% Bordeaux mixture is very effective in control.*

Xyloporosis. This is more common in sweet limes, mandarins, kumquats and tangelos. Peg-like outgrowths appear on the inner side of the bark which get into the depressions formed in the wood. Gum pockets develop in the bark. This is bud-transmitted hence certified disease-free bud sticks and tolerant rootstocks can keep the disease under control. The tools in use should not have contacted the disease.

Psorosis. This affects the cultivars irrespective of the rootstock employed. Leaf flecking in young leaves and sometimes scaling of bark in the branches and trunk regions takes place. Destroying the plant, use of disease-free bud wood for propagation can save the crop from losses due to this virus.

Tristeza (Quick decline). A common virus infection in various parts of the world. In India it is transmitted by aphids. The symptoms include leaf fall; root decay, twig dieback and death. Control may be possible by the use of tolerant rootstock (rough lemon, Rangpur lime trifoliate orange, citrange, sweet orange, cleopatra mandarin). Sour orange and sweet lime rootstocks are susceptible and must be avoided.

Certain other viral diseases have also been reported *viz.* stubborn, bud union crease, woody gall, tatler leaf, satsuma dwarf, yellow shoat etc. from different citrus growing tracts but these are of little significance in India.

Citrus Canker (Xanthomonas citri). This appears in the form of tiny, circular, brown crater-like eruptions on leaves, stems and fruits chiefly or *Kagzi nimboo*, throughout India. The incidence is more in areas of high rainfall resulting in mild temperature and wet weather. This is caused by *Xanthomonas citri,* a bacteria, which spreads by contact. Severe attack causes falling of leaves, collapse of twigs and the disfiguring of fruits. The pest needs attention over long period. At the pre-monsoon stage pruning of infected twigs in the first week of June to be followed by a spray of copper oxychloride @ 0.3% or Bordeaux mixture @1% in the second week of June. At the onset of monsoon in the second week of July one more spray of copper oxychloride 0.3% or strepocycline (100 ppm.) Neem cake solution 60 gm/l water 2 to 3 times at 15-days interval induces resistance and increases the growth of the plant. In the first week of August and also September repeat the measures recommended for July (above). In the first week of February another spray of copper oxychloride may be repeated. Total eradication and sanitation has met with success in Florida.

Withertip (Anthracnose). The twigs wither and shrivel at the tips with pimple like water spots, the flowers and young fruits drop down. The damage is more in the drier zones. *Colletotrichum* species appear as saprophytes causes the disease. *Gleosporium limetticolumn* and *C. gloeosporioides* also cause withertip. Prune and burn the dead twigs, apply Bordeaux paint to the cut ends, spray the trees with copper-zinc-lime mixture.

Removal of dry twigs in February and their destruction, supply of good F.Y.M. to the plants and a combined spray with ferrous sulphate and Bordeaux mixture in the concentration of copper sulphate 2 kg, ferrous sulphate 2 kg, quicklime 4 kg and water 250 l is profitable.

Pink Disease. A fugoid diseas caused by *Corticium salmonicolor* on *Santra* and sweet oranges. The infected bark dry, become hard adhering closely to the wood and cracks vertically exuding gum thereafter a pinkish coloured fungus develops on the diseased surface, ultimately the bark dies out. Removal of diseased portions and spray with Bordeaux mixture (3 : 3 : 50) well before the rains help in control. Blitox (0.2%) may also be used to control the disease.

Gummosis, Brown Rot or Foot Rot. The basal part of the tree exudes gum, cracks, and the bark fall off exposing the wood. This is specially damaging if the stock-scion union is near the ground level. The rootstocks are known to differently susceptible *viz.* Jamburi is resistant, *Nagpur Santra* mildly susceptible, and Mosambi highly susceptible. Removal of diseased or infected bark completely and a paint with 20% creosote oil or with carbolic acid diluted with an equal amount of water checks the disease. It pays to avoid the contact of the bark with wet earth and apply Bordeaux paint to the trunk. Dust the pits before planting with zinc sulphate + copper sulphate + lime (5+1+4). When the rot has set in scrape the affected bark and apply a wash of zinc-copper-lime mixture.

Powdery mildew (Oidium tingtaninum). It is powdery growth on both sides of the leaf and twig. Leaves get distorted and smaller fall of leaves and fruits takes place in warm and humid season. Dusting of sulphur or spray of wettable sulphur (1 kg in 200 l H_2O) may be used to control the disease.

Scab (Elsinoe fawcetti). A major disease in the hills with low temperature and high humidity, in mandarins, citrons, lemons etc. Total destruction of infected plants. Bordeaux mixture or Blitox (0.3%) spray are effective in control.

Felt disease (Septobasidium pseudopedicillatum). Common in N.E. zone of India but not so in other parts of the country. Salt-felt-like leathery growth takes place in branches and twigs. Colonies of scale insects thrive just below the felt likes structure. The Khasi orange, pummelo, Assam lemon are commonly attacked. Bordeaux mixture may be used to control the disease.

Wilt (Fusarium sp. *Verticillium* sp.) Sooty mould (*capnodium citri)*, Leaf fall and fruit fall (*Phytophthora palmivora*), melanose (*Diaporthe citri*), Ganoderma root rot, mushroom root rot, dry root rot etc. have also been reported to be affecting citrus plantations. Sanitation, clean cultivation spray of Bordeaux mixture and other fungicides have been recommended for the control.

Phytophthora parasitica causes another type of gummosis on Santra. Scrapping of the affected tissues and application of Bordeaux paste is recommended for control.

Die-Back Disease. Also common in India. Deficiency of copper is considered to be the cause in the spread of die-back.

Mottle leaf (Zinc deficiency) may be controlled by a spray of zinc sulphate-lime mixture in water (10 : 5 : 100) and green manuring of the plot. *Exarthema* caused due to copper deficiency should be given spray of Bordeaux mixture or 200 gm. of copper sulphate may be added in the soil for every tree.

Chlorosis. Avoid digging near the root, add complete fertilizer regularly and green manure the soil.

□□□

CHAPTER

8

CUSTARD APPLE

(*Annona squamosa* L.)

Annonaceae

Origin and Distribution

The Annona genus comprises of more than 50 species of which *A. cherimola, A. diversifolia, A. reticulata* (bullock's heart or *Ramphal*), *A. squamosa (Sharifa* or *sitaphal*), sweet sop or sugar apple or Ate ; *A. muricata* (Sour sop) ; *A. glabra (A. palustris* or pond apple) have some importance as being edible. *A. cherimola* (Cherimoya) is considered to be the best fruit of the Annonaceae family. It is well recognised in India, Australia, California, Brazil and Mexico as fruit. Essentially a sub-tropical fruit has established in Andhra Pradesh with a tropical climate. On the basis of fruit form cherimoya has five recognised types of cultivars : (*i*) Finger-printed, (*ii*) Smooth, (*iii*) Tuberculate, (*iv*) Mammiliate and (*v*) Umbonate. All the members of the genus seem to have originated in tropical America. They can tolerate mild frost, prefer dry climate and do not ripen in cold weather. Hot dry winds at the time of flowering are injurious for the crop. Several genus of Annona were introduced in Egypt, India and other frost-free countries quite early. It grows wild in Andhra Pradesh, U.P., M.P., Assam, Tamil Nadu, Bihar, Karnataka, Maharashtra and Orissa. In India Andhra Pradesh has maximum area under custard apple followed by Bihar.

Annona with alternate, simple, entire, two-ranked leaves without stipules having super axillary, opposite of leaf, solitary or in clusters, bisexual flower produces large fruits the "apple" which are syncarpous and fleshy.

Annona squamosa is the most common species cultivated in India and in the tropics. *A reticulata* thrives in south India upto an elevation of 1200 m. It requires milder climate than '*A. squamosa* and is not cultivated on commercial scale in India. In south the fruits are ready in May. The seeds are small and brown in *A. glabra* and not black as in *squamosa* or *reticulata. A Muricata* is grown in Assam and Andhra Pradesh on limited scale. It is not popular due to the incompatibility of its scion with *squamosa* or *reticulata*. *A. diversifolia* (ilema) is common on Mexico and Central America. It can be successful in the subtropical regions of the Indian subcontinent.

Soil and Climate

The custard apple prefers sandy or light soils although heavy soils with proper drainage may also prove as good. The plant is not exacting in its climatic or soil requirements except that frost-free areas with rainless flowering season increase yield.

Propagation

The value of vegetative propagation is recognized as there is great variation among seedling trees in quality and yield. Seed propagation is common though it introduces many variations in quality and yield. Seeds are capable of germination when fresh. They may take 3-4 weeks time for germination. Scarification or soaking the seeds in hot water does not improve germination but soaking in gibbrellic acid solution (500 ppm) increases germination. The germinated seeds may be transplanted in pits made for the purpose. Seedlings growing wild from seeds may be topworked with commercial cultivars to the advantage of the orchardist at the end of the dormant period by cleft or bark grafting.

Inarching and layering have been followed for asexual propagation wherein the former has given almost 100% success. Budding has given 75% success but side grafting has been less successful. Commonly custard apple is propagated on commercial scale through grafting and budding. Clonal propagation through stooling and mist treatment has given encouraging results in maintaining uniformity and quality. Stem and root cuttings are not satisfactory for propagation of most species of *Annona*. Air layering is also unsuitable. Stools of *A. squamosa* treated with 20,000 ppm. IBA solution gives over 90% success.

Modified forkert, shield, patch or chip budding have given success when done in early spring on one-year old dormant stools. It is advisable to select one-year mature scion and an older and thicker rootstock for graft purposes. Grafting on a rootstock of different species of the genus *Annona* has proved better in imparting desirable character of the rootstock. *A. reticulata* for instance seems to impart greater longevity to the scion. Veneer, whip, cleft and splice grafting has given satisfactory results. Although custard apple may be grafted on its own seedling, graftage on stock of *Annona reticulata* proves better. Custard apple can also be propagated through cuttings taken from the dormant plant during winter in U.P. Mammoth and Red Custard apple are more or less true to type from seeds which remain viable for 7 years. Leaf-cuttings can also be rooted in a constant moist chamber. Planting should be done in dormant season to reduce mortality. Fertilizer application to the tune of 250 g N, 125 g P_2O_5 and 125 g K_2O in the beginning of rains gives consistently high yield. Distance between plant to plant may be kept at 5 metres for good performance specially with good manuring and other cultural operations.

Fruiting starts at the age of 3 years and continues to yield about 100-150 fruits weighing about 25-30 kg. per tree for about 12 years. Fruit drop can be reduced by the sprays of NOA, NAA, 2, 4-D, 2, 4, 5-T and GA_3. As a rule custard apple is not cross pollinated to any large extent although it is believed that artificial pollination may increase the yield. Pruning is done more for convenience of the farmer than anything else.

Fruits should be harvested when still firm as soon as they start changing the colour from green to yellowish green. The ripe fruits are rich in sugar. Ripening of fruits takes place in a few days after harvest. Ripe fruits can be stored at 5°C for 6 weeks. Mature fruits get chilled at a temperature below 15°C.

Cultivars

Barbados, British Guinea, Brandy, Islander, Red Sitaphal, Washington P_1 98797 are well known cultivars of *A. squamosa*. Promising genotypes (Rahuri 10/4 and 14/3 selections) have been evolved. Saswad regions of Maharashtra has given type Nos. 5, 6, 8, 10 and 15 which may proved better than local cultivars. Balanagar, Washington, Pi 107005, Red Custard and Mammoth are some of the cultivars of custard apple grown in the country. Fruits of Cherimoyer of Lakshmanphal (*Annona Cherimola*) are of good quality. *Annona atemoya* a hybrid between *A. squamosa* and *A. cherimola* is grown in the costal regions of Australia. African Pride (cherimoya × sugar apple) Pink's Mammoth and Neilson are also popular cultivars.

Recently some cultivars have been released with less number of seeds in the fruits. These important cultivars are Israeli selection, Israeli Hybrid (natural hybrid of custard apple and cherimoya) and Arka sahan.

Pests and Diseases

Mealy bug may be seen in between the fruit segments and on young shoot. 0.05% Dichlorovos controls the bug. Mealy bug infection may be suppressed with the help of its natural enemies *viz. Spalgis epeus, Scymnus coccivora* and *Croyptolaemus montrouzieri* when present in abundance.

Anthracnose (*Glomerella cingulata*) and leaf spot have been successfully controlled by sprays of 0.05% benlate or bavistin and 0.1% baycor thrice at 15-day intervals from the time symptoms appear.

Pink disease caused by *Pellicularia salomonicolor* responds to *Bordeaux* mixture spray after the removal and destruction of affected parts.

□□□

CHAPTER

DATE-PALM

(*Phoenix dactylifera* L.)

Palmae

Origin and Distribution

Date-plam, supposed to be a native of the region round about the Persian Gulf, has been grown in India, Africa and Spain since antiquity. Iraq is the biggest centre of date-palm cultivation. Dates are now being grown in deserts of South Africa, South America and Australia. Saudi Arabia, Algeria, Egypt, Libya, Sudan, USA, Spain and Pakistan are also date palm growing countries. In India, the chief producers are Uttar Pradesh, Gujarat and Madhya Pradesh, but there is a no commercial production of good dates.

Phoenix dactylifera (H. *Chuhara*), a monocotyledon, with unbranched stem may be seen growing even upto an altitude of 1500 metres in different parts of India, at random. No systematic plantation has been attempted. Dry districts of South India have some wild growth of *Phoenix sylvestris*, (H. *khajur*), *P. humilis* and *P. farinifera* that yield fruits of inferior quality. The net result is that production of high quality date is uncommon in India. Dried dates with 70% sugar ; 2% protein, 2.5% fats in addition to minerals are nourishing. Dates form a favourite food along with milk for the Arabs.

Soil and Climate

Non-alkali soils devoid of hard pan or canker in the top one metre, seem to be congenial for date-palm. Flowering and fruiting are enhanced in sandy soils provided there is no acute shortage of moisture or nutrients, well drained sandy soils seem to be ideal for the fruit. It can tolerate saline soils to a very great extent.

Date palm requires long summers with high temperature both during the day as well as night, a mild winter free from frost, absence of rains during flowering and fruiting, low relative humidity and full sun-shine. Optimum temperature for flowering and fruiting is considered to be over 22°C. Proper ripening of the fruit needs a temperature of over 26°C for no less a period than 30 days. Dusty winds lower the quality of the fruit.

The Dates

In general there are three types of dates : (*i*) soft with high moisture percentage and low sugar content ; (*ii*) semi-hard with high sugar and low water content; and (*iii*) dry with very high sugar and very low water content, these are hard.

Propagation

Date-palm may be propagated by seeds, offshoots or suckers. Seed propagation is not advisable since about five per cent of such plants give good quality fruit and it is difficult to get seeds. In addition seed propagation produces variability, the plant takes longer time to come to fruiting and the sex is not known for many years. It is desirable therefore to propagate through cutting and replanting the offshoots arising from the main fruiting plant near the soil level. The base of the offshoot should remain in contact with the moist soil for about a year before its separation to get better rooting. Four days in advance of the detachment of suckers all but innermost leaves are trimmed off to facilitate its removal from the main tree and also to reduce transpiration. The offset shoot is severed from the parent body in February-March or August-September, the former being more opportune time for field and the later for nursery plantations. Nursery plantation is *in vogue* under the impression that rooted shoots do better.

Tissue culture from shoot tips, lateral buds, stem portions also help in easy multiplication of date plants. Date palm plantlets have been initiated from clonal explants through embryogenic callus and transplanted to soil with success. Normal plants of the cultivar Khadrawy and Medjool were obtained by somatic embryogenesis technique in India.

Planting

Planting the freshly detached offshoot is done in March or September depending on the climatic conditions of place. Offshoots, if not planted immediately, may be kept in muddy water (not entering the crown), after detachment. Optimum offshoot size for planting is 3 kg or more (as in the Punjab, 6-8 kg) ; 1-4 kg offshoot planted in February give best results. The planting distance of about 6 metre is kept between two plants. The crowns must be about 15 cm about the ground level so that irrigation water does not harm.

Cultural Practices

Intercropping with legumes provides nitrogen and improves the fertility of the soil to the benefit of the young plantations. Green manuring with legumes may also be done to incorporate nitrogen rich organic matter in the soil needed so badly by the young trees. Grapefruits may be grown between two date plants to afford protection from sun.

The trees at the bearings stages benefit by application of 60 kg of FYM and

2 kg of sulphate of ammonia in equal halves one in January-February and the other in August-September depending on the needs of the plant. Maximum water requirement of the date-palm is estimated to be 8-9 acre-feet at the rate of 15-18 cm per irrigation every 15 days specially during the summers. Younger plantations need more frequent irrigations. Higly tolerant to salt-water irrigation.

Pruning

The maintenance of adequate leaf : bunch ratio is necessary for improving the quality of fruits in the season. Higher ratio brings high humidity round the fruit so that disorder like 'Blacknose' develops. Early pruning of leaves saves such situations. Better ventilation should be the chief purpose in such operations the extent of which depends on the stalk of the bunch and the amount of the leaves. Immature offshoots are removed in very young condition to improve fruiting.

Besides the pruning of leaves, removal of thorns in winters may also render pollination and subsequent handling of bunches easy.

Pollination

Pollination is affected in this dioecious plant by inserting the freshly opened male inflorescence on the female clusters as soon as these open to afford possibilities for the insect or the wind to pollinate the flowers. For hand pollination about 10 male trees are sufficient for 100 female trees. The male flower belonging to early flowering, thick and bushy inflorescence possessing large number of stamens containing viable pollens in good quantities should be taken. Viable pollen is confirmed by the appearance of yellowish dust when the inflorescence is shaken with water. If the pollination is successful and the fertilization does not take place all the three carpels develop to give worthless fruits but with successful fertilization only one carpel develops, while the two fall off while quite small.

Date palm being dioecious needs growing of male and female plants in equal numbers side by side. The selection of a good polliniser is of great value since the fruit size as well as fruit ripening are affected by the pollen. Deglet Noor's pollination with Fard 4 pollen shortened the ripening period of the fruits appreciably. Under Indian conditions shortening of the ripenning period is of significance as the damage due to rains may be avoided by early ripening of the fruits (before the onset of rains). Several species of Phoenix are good pollinisers.

Fruit Thinning

Fruit thinning, seldom followed in practice, may be done by removal of some bunches, or flowers from each bunch to ensures good flowering in the succeeding season, better sized quality fruits and early ripening. Bunch thinning may be done by shortening the length of strands, removing some of the central strand from each bunch and also reducing the number of fruits from each strand.

Fruit set is dependent on the temperature at the time of pollination and the viability of the pollen. 50-80% fruit set is considered to be a full crop. Fruit thinning is primarily done to ensure adequate flowering in the following season/year, improve the quality of fruit, prevent delayed ripening and reduction in the compactness of the bunches. In India 8-10 bunches are retained per tree, the extra or the defective ones are removed about 2 months after pollination. The thinning operation may be mechanical or through the spray of ethephon (100-400 ppm) 3 weeks after fruit set. Ethephon treatment hastened ripening, increased sugar content, developed larger fruits and reduced biennial bearing character of the palm.

Harvesting and Marketing

In date palm the development of the fruit takes about 24 weeks from fruit set. Five stages in the development of the fruit are recognised : (*i*) Hababouk (from pollination to 4 weeks) characterised by loss of two unfertilised carpels, (*ii*) Gandora or Chimri (9 weeks) green colour rapid period of growth followed by slow increase in fruit weight. (*iii*) Doka or Khalal (4-5 weeks) slow increase in fresh weight continues, (*iv*) Dang or Rutab (4 weeks) loss of fresh weight slowly and then rapidly and (*v*) Pind or Tamer (one week) fruit ripe, loss in weight, edible stage is reached. In India Gandora (hard green), Doka (full grown, hard yellow or red), Dang (softening) and Pind (fully ripe and dehydrated) are the four recognised stages.

GA (50 ppm) spray to unpollinated bunches gives a high percentage of seedless berries. 1-3 sprayings of 2, 4-D ; 2, 4, 5-TP ; 2, 4, 5-T ; IAA or GA_3 at 25-100 ppm after flowering produces seedless dates.

Dates may be picked at any stage of development of the fruits depending on the climatic conditions, local customs, the age, the availability of labour etc.

Curing the Dates

There is a practice of dipping the fruits is 1% boiling lye solution for a minute before curing. Curing or drying of the dates is done by spreading these on leaf matting in the sun. It may also be dried in a drier or by the application of external heat.

Yield

A palm tree starts fruiting, four years after plantation, under favourable conditions. The yield is 35 kg per tree or 100 quintal/ha for trees of 10 years of age.

Cultivars

There are probably 1000 cultivars of date palm in the world. In India, 36 commercial date cultivars were imported from Middle East countries and North Africa *viz*. California of which Barhee, Khalas, Medjool, Zagloul and Zahidi are of value. The cultivars with moderate heat requirement have succeeded in India ; those of commercial value are indicated below.

Scientists at the Rajasthan Agriculture University (RAU) have developed a date culture technology which may prove to be money spinner for cultivators in desert areas. Of the several datepalm varieties evaluated at the Beechwal Research Centre of the University, Halawy, Medjool, Khalas, Khuneji, Barhee and Sewi have been found highly suitable for the arid region.

The university has provided date suckers (shoots) to progressive farmers in the area, to popularise date culture in western Rajasthan. Each farmer has been provided with enough male and female suckers of Halawy variety to establish one acre of datepalm orchard and explore the possibility of its exports.

Barhee. A late heavy yielding, soft date cultivar from Iraq. The fruits are less damaged by rain, possess good taste and the yield is about 150 kg per tree, suitable for arid regions.

Deglet Noor. Algerian cultivar successfully grown in USA also. Rain and high humidity cause much damage to the ripening fruit. This semi-dry date also yields quite heavily, 100-150 kg per tree.

Haiyany. Soft cultivar commonly grown in Northern Egypt for home consumption. It has high moisture content and therefore does not cure easily but is good for fresh consumption. Humid weather is rather harmful and yield is about 100-125 kg per tree.

Halawy. A very popular early cultivar throughout the world due to its adaptability to high humidity and rains. The yield from a single tree ranges between 75-100 kg under favourable conditions. The fruits are small is size that shrivell during ripening. It is a cultivar between soft and semi-dry dates.

Khadrawy. It is slow growing short statured cultivar from Iraq which has proved second to Halawy is most parts of the world. The yielding capacity is about 50-70 kg per tree.

Khalas. It is a soft semi-dry mid-early date very famous in Iran. Khalas is resistant to rain and high humidity. The yielding capacity is the same that of Khadrawy.

Medjool. Again a soft date with flesh firmer than that of Khadrawy and Barhee. It is damaged by rain and high humidity. The yield is about 75-100 kg per tree. Fruits orange-yellow, seeds with ridges. *Khuneji* and *Sewi* have been found to be highly suitable for arid regions of Rajasthan (India).

Thoory. The leading dry-date all over the world. It is from Algeria and is late ripening. The yield recorded in USA is 100-125 kg per tree.

Zaidi or *Zahidi.* The cultivar has low moisture content and easy to be handled. It is good yielding culitvar from Iraq yielding about 100-150 kg per plant. Moderate humidity at the time of fruiting greatly improves the quality of fruit.

INSECTS, PESTS AND DISEASES

Not much damage has been reported by insect pests and diseases to the crop in India.

Palm leaf pustule (*Graphida phoenicus*)

This is also known as false smut or Graphiola leaf spot. Small, hard black, pustules scattered over the two surfaces of the leaflets occur. Removal of old and infected leaves and spraying with Bordeaux mixture may help in control. Spraywith 0.2% phytolan and 0.1% Bavistin or Bayleton checks the disease.

Indian palm weevil (*Rhynchophorus ferrugineus*)

The larvae of this red weevil penetrate through the tissue of the stem near the tip causing much damage. These are killed by the use of insecticides or by mechanical means. The attacks may be prevented by painting with tar at the cut ends, removal of old trees and leaves, plugging of all the holes in the trees, removal or seriously affected plants and clean cultivation. Timely application of Endosulfan 35 EC @ 0.1% (3 ml/lit) or Dichlorovas 70 EC @ 0.1% (1 ml/lit) or Carbaryl 50 WP @ 0.1% (2 gm/lit of water) prevents infection of weevil.

Rhinoceros beetle (*Oryctes rhinoceros*)

The black palm beetle damages *Phoenix sylvestris* by boring holes. Take out the adult from the holes during July-August and fill in the holes a mixture of mancozeb (3 gm) + sand (1 kg). The beetle can be effectively controlled by spraying the spores of *Metarhizium anisoplice.*

Scale insects (*Aspidiotus destructor*)

These hinder growth and ripening of fruit. The infection is more severe during summers. Application of Malathion 50 EC @ 2 ml, Monocratophos 40 EC @ 1 ml or Dimethoate @ 1 ml/lit water controls the pest. Cutting and burning of the affected parts helps in the control of the damage.

CHAPTER

FIG

(*Ficus carica* L.)

Moraceae

Origin and Distribution

The fig (H. *Anjeer*) native of Asia Minor, is one of the oldest crops, quite well known in Spain, Holland, France, Germany, Portugal, India, Greece, Turkey, Algeria, Morocco and Syria. *F. carica* is considered to be a native of southern parts of Arabian peninsula, Italy, Balkan peninsula and USSR. It is gynodioecious with two types of trees. The species *Ficus glomerata, F. hispida, F. auriculata,* and *F. palmata* are indigenous to India. *F. cuxia, F. pumila* and *F. rumphii* grow wild in India. In India fig is grown in Pune, Karnataka, Lucknow and Gujarat. The Indian fig has unlobed or faintly lobed leaves. Fig with lobed leaves did not do as well in the country. Foreign cultivars susceptible to nematodes have also not found favour with Indian orchardists.

Soil and Climate

The fig, a salt and drought resistant crop, is not very exacting in its demand of the soil and can be grown on a wide variety of soils, *viz.,* heavy clays, rich loams and light sandy soils. Nematodes and soils of low water holding capacity, more common in sandy soils, limit the crop. Save nematode infection frequent irrigation and adequate manuring can make light soils quite useful for fig cultivation. It does well in alkaline soils with frequent applications of lime.

Climate has a more powerful influence on the fig than the soil. The sprouting plant is not able to ward off damage due to early frost in autumn. Non-dormant grown up trees in full growth are damaged by temperatures between 1.5 to 4.5°C although when in dormant stage the plant can stand a temperature as low as –12°C. The optimum temperature for growth of figs is between 16-21°C. The size, shape, bloom, colour and quality of the skin and pulp as well as tendency towards parthenocarpy are influenced by climate. Although the crop can withstand temperatures between – 4°C and 45°C, temperatures above 38°C induce premature ripening, develop insipidness, tough skin and may spoil the fruit even before ripening. High summer temperatures cause sun burn of the skin. Fig cultivation in the Mediterranean subtropical climate with hot dry summers and cool moist winters is considered suitable. The winters are good for vegetative growth and the summers for good quality of the fruits.

Propagation

IBA treated hard wood cuttings stored in moist sawdust at room temperature for a month in February may be transferred to the nursery bed for rooting. Normally leafless, fully mature, dormant hard wood cuttings of about 25 cm length and 1.5 to 2 cm girth severed from the parent plant in December-February in North India or July to November in Central India are placed in bundles with the basal end upward in callusing pits for about 2 months before being planted in the fields in India at a distance of 5 metres between plant to plant (although more distance may also be given) at such a time that they establish themselves before the onset of the winters. Semi-hard cuttings may also be used for propagation in the usual way or in mist propagation beds. Air layering, ring budding or side grafting of the stock of *Ficus glomerata* resistant to root-knot nematodes gives vigour and yields good results. Sometimes transplanting is also done. Old established trees in orchards may be top worked by grafting for rejuvenation in which 1% NAA may be sprayed to check trunk and limb sprouts. Young trees are staked in the first year, trained to 3-4 well spaced branches in the next two years. Low headed trees have their own advantages.

When climatic conditions are such as to limit a well marked rest period the flowering and fruiting are badly affected. Under such conditions spray of Sandolin-A increases growth, earliness, blossoming and also yield.

Pruning is done to encourage new growth. The chief aim of pruning is to induce growth of flower bearing wood and increase the yield. In Pune, light pruning is done just after harvest but in U.P. severe winter pruning is done in December leaving only 3-4 buds on shoots of previous year's growth. Regular pruning during winters gives better yield. Generally, fruiting starts 3-4 years after planting. The one-year old branch (one that develped in the previous year) should be pruned leaving 4-5 nodes to encourage fruiting in all the branches. The fruits that get ready during summers are considered to be better. Figs are generally sold in the market after drying. Heading-back of the trees every year leaving only two buds on each one-year old shoot is practised in Karnataka. After heading back the first crop on the wood of previous year and the second crop on the wood of the second year is got. Painting is useful after pruning since the wound does not heal easily. The root must not be exposed. Intercropping with papaya and use as fillers in mango is not uncommon. Though fig is a crop which does not require irrigation two waterings during summer when the fruits are getting ready makes them sweet and large.

Manuring

Fig is capable of tolerating a high degree of sulphate or chloride in the soil or through irrigation water but are injured with even a small quantity of sodium carbonate. It responds to manuring. The plants must be regularly manured from the very first year. One year old plant does well when supplied with 9 kg of FYM + 0.54 kg cake or 14 gm nitrogen as $(NH_4)_2SO_4$.

The quantity of the manure is progressively increased, the increased being 6.8 kg FYM and 0.54 of cake or 14 gm nitrogen every year till a five-year old tree receives 36.2 kg of FYM and 3.42 kg of cake or 70 gm of N. This quantity is supplied every year. Very high yields have been reported with 600 gm of N and 400 gm of P_2O_5.

The manure should be applied just before the advent of spring in northern India, and in the beginning of the Monsoon for young plantations for the first three years in western India. Subsequently, when the trees are in full bearing, the manure is applied at the end of September when the dormancy of trees is over. In Northern India farmyard manure should preferably be applied before the growth starts in spring and the cake or inorganic fertilizer after the growth has started in March. In other areas, farmyard manure should be applied at the end of September while cake and inorganic nitrogen in early November.

In South India, at certain places where figs are raised with sewage water, manuring is not necessary.

Harvesting

Edible cultivars of fig generally bear two crops a year, the breba (spring crop) and the main or second crop. The main crop in North India ripening in May is borne on the current seasons growth while the breba on the previous year's shoots. The main crop is ready in July to September or February to May in Central India and South India respectively. Fig fruits have two periods of rapid growth separated by one of slow growth. Ethylene treatment or putting a drop of olive oil in the ostiole at the time of second growth accelerates the growth and also ripening of the fruit.

The stage of maturity, proper colour and ripeness is important for harvesting figs. Fresh fruits are delicate and must be marketed as soon as picked to avoid spoilage. Figs may be harvested when well ripe but no overripe if intended for drying which may be done in the sun or in a hot air oven. SO_2 may be used to prevent the growth of moulds, before drying in an electric oven at a temperature of 70 – 72°C.

There are four types of figs *viz.,* (1) Caprifig, (2) Symrnafig, (3) Common or Adriaticfig and (4) White San Pedrofig.

Caprifig. Also known as wild or primitive fig, it has both male and female flowers. The former are located near the opening of eye which is usually covered by bracts. The short styles are well adapted for laying of eggs by the fig wasp

plant but the extent of damage is not much. Beside these, *Alternaria fici, Cladosporium sicophilum* and *Botrytis cineria* also affect the fig plants at times. Spray 1% ferrous sulphate in late winter or 4% Bordeaux mixture in early spring for the control of these pathogens. In summer captan/zinab/ziram (0.1%) should be used. Surface rot of fruits can be controlled by chlorothalonil spray.

Fig scale infected by *Lepidosaphes fici* and the attack of Mediterranean fruitfly (*Ceratities capitata*) are also known to affect the fig plant. For control prune the affected shoats. If the infestation is serious spray chloropyriphos @ 0.5% (2ml/l water).

Gall makers. Two organisms are connected with this malady.

(*i*) *Udumberia nainiensis* is active during August-October and also March-May. It damages the flower buds. The larva feed on the infl. and causes discoloration, wrinkling and formation of small protiperances on it.

(*ii*) *Anjcerodiplosis peshawarensis* feed on the fruit pulp. It breeds throughout the year except the period April-July. Infected fruit becomes elongated, wrinkled and drops prematurely. Removal of the infected infl. and fruit destruction is recommended. Maggots can be controlled by malathion 50 EC (0.1%, 2 ml/) lit. of water).

Sunburn. Newly planted young trees and their branches are specially susceptible to sun-burn during winters. Affected part cracks, the bark peels off, giving access to borers. White-washing of exposed parts is useful.

Cracking of fruits. Sudden changes in atmospheric humidity are responsible for splitting of fruits, the pulp is exposed to insect and fungus attack. Excessive and prolonged rains during April-May damage the fig crop, considerably or even completely.

Mosaic (virus). Propagate from disease free material only on uninfected soil to control the disease. Resistant strains may also be used.

CHAPTER

GRAPE

(*Vitis vinifera* L.)

Vitaceae

Origin and Distribution

The area, in the neighbourhood of the Caspian Sea is considered to be the place of origin of grapes from where *Vitis vinifera,* the table grape of commerical value reached India. Over 100 species of *Vitis* grow wild in the country, some of which provide poor quality edible fruits, *viz. V. barbata* (Assam and Bengal), *V. rumicisperm* (Assam and Sikkim), *V. rugosa* (Kumaon Hills) and *V. parviflora,* (Kashmir).

Entire Europe, U.S.A., Africa, Queensland, South Wales, Palestine, Perisa, Afghanistan, China and Brazil are chief grape growing places. In India it is grown chiefly in Tamil Nadu, parts of Karnataka, some districts of Andhra Pradesh, Madurai, Nasik, Pune, Hyderabad, Aurangabad, the Punjab, Uttar Pradesh and Haryana. Chini (Himachal Pradesh) is being developed for growing raisin grapes with about 25% of sugar content. Grape vine sheds its leaves in winters of north India and the vegetative growth takes place only for a part of the year so that the yield is low, the cost of production and protection is also low. One hectare of a well established good quality vineyard may yield a profit of about Rs. 35,000 to 40,000 a year.

The grape (H. *Angur*) is considered to be one of the important subtropical fruits of the world. Grape vine yields about 18 tonnes of grapes per hectare in India. One of its varieties *Pacha-draksha* has yielded 70 tonnes per hectare in Madurai and 37 tonnes per hectare at Nasik. These production records are highest in the world. Anab-e-shahi another cultivar yields about 21 tonnes per hectare.

The cultivated types of the genus *Vitis* are classified into three groups.

(*i*) *Table grapes*. As the name indicates it is the group in which the grapes which are relished raw are included. These may include white, red and black (very dark blue or purple) cultivars. These are sweet with few seeds. The skin may be thin or thick. The important table grapes are Perlette, Beauty seedless, Thompson seedless, Anab-e-shahi, Delight, Himrod, Phakdi, Bhokri, Kali and Pandhri sahebi.

(*ii*) *Raisin grapes*. Raisin grapes are large, medium sized or small black or greenish white. These may or may not have seeds. High sugar percentage is their chief characteristic with thin or thick skin. These may be employed for making currants also. Sultana, Thompson seedless, White Corinth and Red Corinth are the more important cultivars grown in U.S.A. The raisins made from the cultivars of small seedless grapes grown mainly in Greece are known as currants.

(*iii*) *Juicy grapes*. Juicy grapes are principally used for making wines, bottling fresh juice etc. the important cultivars are Muscat and Black Prince. These may be either black or whitish with distinct strong aroma and sugar content. The juice is slightly acidic.

Climate and Soil

Grapes succeed commercially in a wide range of climatic variations. Absence of extremes of climate, well distributed rains broken by frequent spells of dry climate having short sharp winter and long dry summer are useful. Leaves are shed during winters and during this resting period frost is well tolerated by grape vines. In the south two crops a year are common, dry weathers helps in the continuous growth of vines throughout the year. Warm rainless summers during ripening and a dust, frost and hail storm free season throughout may be considered as good.

Light soils are ideal for grapes but any well drained soil should do. Even saline soils are tolerated by grapes to some extent provided these are well drained. Drainage is the first consideration though not the absolute for grapes. Light loams, black, medium black and shallow light soils with open textured sub-soils are supporting grape orchards in India. Granite soils rich in potash and felspar (or feldspar) produce grapes of execellent quality. Soils rich in lime are also well suited for commercial cultivation of grapes.

Propagation

Softwood cuttings may be used for propagation provided these are kept in mist but propagation by hardwood cutting is common. The cutting may be 20 cm to 30 cm in length taken from one season old mature canes after pruning. The lower end should be just below a node and it must be a square cut as against the upper one which should be slanting. These are planted in the month of December-January in slanting position in beds where these are allowed to take roots. In about there to four months roots are given out and the buds activated so that the plantlet is ready to be successfully transplanted by February-March. Seed propagation is possible but is followed only for breeding and selection of desirable types. The method of grafting vines has been introduced for improving the inferior varieties. Dormant one-year old wood cuttings are treated with IBA or NAA to promote rooting and kept in moist media to be planted later in nursery. Bud graft in the form of a side graft in which the scion has a single bud with a piece of wood attached to it is joined to the root stock from which an equal piece of wood is removed previously is followed in Spain, Italy and California. Budding, cleft, side whip and bench grafting have been tried, though only in cases where a good crop is not possible from own root plants due to attack of *Phylloxera* louse, nematodes, or any other cause. Layering may also be used in propagating vines specially for filling gaps in the orchards.

Cultivation

The field is properly tilled, humus added or green manuring done before the planting of vigorous cultivars in 3 m × 3 m × 90 cm pits 3 meters apart. Every pit is filled with well rotten F.Y.M. or with alternating layers of leaf mould and silt to the brim. The pits are watered occasionally to decompose the organic matter. When the pits are ready, at planting time rooted cuttings are planted in the centre of the pit. Triazines, substituted ureas and dinitros are safe herbicides for the control of weeds in grape orchards. The top is pruned to as many canes as desired. The cane is cut back to 2 or 3 buds soon after planting.

Manuring. Tissue analysis (specially of leaf petiole) is considered to give the most reliable index of the nutritional status of the grapevine. The ratio between potassium and magnesium gives the best indication of magnesium status. Higher the ratio more the deficiency of mg. Boron content was in some way related to bud failure ; 35.5 ppm content led to less failure of buds. Acute shortage (around 11 ppm) indicated high bud-failure. Nutritional needs of a cultivar may not remain same under different set of cultural and climatic conditions. The manuring schedule given here is based on the results of certain reports from the experiments conducted in Uttar Pradesh. The following manuring schedule should be adopted after planting of grapes. From 3rd year onwards fertilizer doses should be given in two instalments, first dose after pruning and second after the fruit set. In order to ward off deficiency of minor elements spray micron C just before flowering.

Manuring Schedule for Grapes

Age of Plant (Year of Planting)	Quantity (g/Vine)		
	Nitrogen (N)	Phosphorus (P_2O_5)	Potash (K_2O)
1	100	60	80
2	200	120	160
3	300	180	240
4	400	240	320
5th year onwards	500	300	400

Spacing of Vines

Spacing seems to play a vital part in the longevity of the vine. In recent years closer spacing have been found to be widely adopted by the growers as is evident from new planting experiments from a few states.

Spacing of Grape Vines

Culitvar	Relative Spacing of vines (metre)		
	Old	Recent	Latest
Anab-e-Shahi	3.0 × 6.0	3.7 × 3.7	2.5 × 3.5
Thompson Seedless	2.5 × 2.5	2.0 × 2.0	2.5 × 1.2
Tamil Nadu			
Gulabi	2.5 × 1.2	2.0 × 1.2	2.0 × 0.5
Maharashtra			
Anab-e-Shahi	3.0 × 3.0	3.0 × 3.0	2.5 × 2.5
Thompson Seedless	2.5 × 2.5	2.0 × 2.0	2.0 × 1.0
Karnataka			
Anab-e-Shahi	4.5 × 4.5	4.0 × 4.0	3.5 × 3.5
Thompson Seedless	2.5 × 2.5	2.5 × 2.5	2.5 × 1.2

Training

Training is considered to be essential for grapes. It provides desired shape to the vine besides facilitating cultivation, care and harvesting operations. The yield is also bettered both in quality and quantity. In short, proper training ensures economic returns. Better crop maintenance, leaf orientation as well as expansion, fruit distribution on the vine, and protection from sun-scald have been achieved through training.

The buds of newly planted one-year old cuttings sprout at the end of the dormant season giving rise to a number of sprouts depending on the number of viable buds present on the cuttings. Only one healthy shoot is retained (others are removed) and allowed to develop to a desired height. The apical bud of the stem is removed at this stage to force the lateral buds giving rise to primary arm. This process is later repeated till the secondary and the tertiary arms are obtained.

Now, is the time for training the grape vines as bower (pandal), Kniffin and trellis depending on factors like the economic position of the grower, vine-vigour, climatic conditions, intensity and angle of sun rays, facility of cultural operations, pruning etc. In any case the training should be done in such a way that the primary, secondary and tertiary arms are made to grow and develop slowly over a large period so that the arms do not become unproductive.

Bower System. This system is also called Arbour or Pergoal. It consists in planting vines at a distance of 3 × 3m and allowing these to branch only after attaining a height of 2 m or more. Two vigorous vines are trained horizontally in opposite directions over the arbour. Six bud spurs are retained on each primary arm during anuual pruning taking care that the branches are not exposed to direct sun. This system has been observed to suit Anab-e-Shahi cultivar and vines which make luxuriant growth. The three laterals on either side allowed to have 10-12 tertiaries which form the main fruiting canes after maturity.

Bower system of training suits the vines making vigorous growth all the year around. The operations like training, pruning and spraying become difficult making it an expensive system. The cost : benefit ratio in bower system has been worked out to be 1 : 2.09 against 1 : 0.96 for head, 1 : 1.42 for kniffin and 1 : 1.71 for telephone system of training at Coimbatore showing that the bower system was more economical with lowest cost ; benefit ratio. This system is specially suited to tropical climates where the vines show vigorous growth with pronounced apical dominance making pruning more meaningful and profitable. The system provides good protection to the crop against hot desiccating winds and ease in bird searing.

Head system. One single stem is allowed to grow to a height of 1.2 m, the distance between the vines kept at 1.8 × 2.4 m. The apex of the stem is removed at the desired height to initiate new shoots. Buds down below are nipped off regularly and only four laterals are allowed to grow 75 cm above the ground on four sides. These are cut to 2 buds at the first dormant pruning giving rise to secondary arms. At the times of second pruning 1-2 fruiting spurs are retained on the two arms. The vine takes the shape of a dwarf bush in the course of four years not needing any support.

The head system is considered to be least expensive and most simple but the demerit of the system is bud and flower drop leading to highest cost : benefit ratio. It is generally practiced in cultivar like Beauty seedless, Perlette, Gold and Delight.

Vines are planted in December-January about 2.5 metres apart in Western India. Side sprouts are rubbed off for about three-four months. As the vines attain a length of about 1.5 metres four to five buds near the tip are retained or allowed to develop and given a support. After about 15 months of growth all the new shoots are pruned back to a single bud and pruning procedure repeated again. Four bud spurs are retained and the one at the top bears fruits. After this the pruning is done every December-January and each time 4 buds are allowed to develop on the horizontal branches which are supported on trellis wires or wooden poles. The system suits the Bhokri.

Kniffin System

Originally Kiniffin system is a four-cane system. The vine is trained so that it has four canes, one along each wire and the bearing shoots hang freely without any other device. The modified form of the system which is now common is not as popular as the bower system. It is best suited for a moderately vigorous cultivar.

Two or three wirestrings may be streched horizontally at a height of 1 m, 1.5 m and 2.0 m from the ground level supported by vertical iron poles placed 5.0 m apart. Two wires are accommodated between two poles, the distance between rows is kept at 3 m. As the vines reach the height of the wire strings as single stem one arm is allowed to develop horizontally along each wire on either sides. So that each vine has 4 to 6 arms which are maintained permanently and the bearing units and the renewal wood are regulated on these arms while pruning. This system is suitable for Bhokari, Delight, Perlette etc. and cheaper than the Bower system. Lower arms become unproductive after some years. The system is common in the north with moderately vigorous cultivars. The cost : benefit ratio worked out at Coimbatore is reported to be 1: 1.42.

Overhead trellis. Popularly known as Telephone system, it is best suited to moderately vigorously growing cultivar with prominently greater apical dominance. Unlike the Kniffin system the lower arms do not become unproductive by streching. The lower arms at one height like the telephone wire. Overhead trellis provides better ventilation and more light in comparison to the bower system. This is more expensive than the Kniffin but less than the bower.

Growth of young vines is supported by bamboo sticks to a height of 1.5 m and then trained over head on a canopy of 3-4 wires 4-5 metre apart fixed at cross-angle arms supported by pillars. The tip of the single stem vine is removed when it has crossed the 1.5 m height. Secondary shoots are allowed to develop. Two vigorous and healthy shoots are selected as primary arms from which four laterals are allowed to develop as secondary arms. Each of these arms may carry 6-8 fruiting units.

The other systems now being developed are Duplex, Geneva, Double Curtain or other modifications control the yield but increase the quality of the fruit and also make pruning and harvesting more easy.

Pruning

Pruning is known to control or direct the functioning of the vines by concentrating the activity of the vine to the desired parts and extent. It is profitable when it is done with full knowledge of the bearing behaviour, growth habits or peculiarities of the plant to be pruned. Pruning should aim at limiting the number of buds in a particular branch so that the quantity of fruits produced or borne do not affect the plant adversely. It is known that overbearing delays ripening and reduces the capacity to yield in the next fruiting season besides lowering the price and increasing spoilage losses to the deteriment of the cultivator.

In grapes all the canes need heading back to a degree depending on the bearing habit of the cultivar. Severe heading back retaining only 1-2 buds which serve as renewal spurs is known as thinning out. Pruning also includes removal of damaged, diseased or over-crowded parts. Extent of thinning out and pruning depend on the inherent characteristic of the cultivar as well as its vigour, training system, soil productivity, climatic conditions and the need of the cultivator regarding the quality and ripening behaviour of fruit.

On the basis of the desirable extent of pruning cultivars may be two kinds (*i*) spur-pruned cultivars and (*ii*) cane pruned cultivars.

Spur pruned cultivar like Bangalore Blue, Beauty Seedless, Bhokari, Delight and Perlette produce fruits on the shoots originating from the basal buds on the cane and therefore the canes could be headed back to four or five buds. Cane pruned cultivars like Charnik, Gulabi, Kishmish and Pusa seedless are headed back to 8-12 buds.

IMPROVEMENT IN FRUIT QUALITY

Grape culitvars behave differently under different agroclimatic conditions. Anab-e-Shahi, Kali Sahebi, Chemma Sahebi, Pandhri Sahebi, Banglore Purple and Bhokari common in Western India; Pandhri Sahebi, Bangalore Purple and Bhokari in A.P. ; Anab-e- Shahi in Karnataka and Pachadraksha in Tamil Nadu do not attain that quality in north India where grapes are grown under marginal conditions and only early ripeners that escape rain damage can be popular. Perlette, Delight, Beauty Seedless, Himrod and Thompson seedless produce quality fruits under the agroclimatic conditions prevailing in north India.

Short spur bearing cultivars such as Perlette, Beauty Seedless can produce good quality fruits with optimum crop load on 'Kniffin system' or even on 'Head system'. On the other hand, a cultivar like Anab-e-Shahi which bears on long canes can only be trained on 'Bower system'. Such a cultivar can never attain acceptable quality on 'Kniffin' or 'Head system'.

A poorly trained vine can never produce good quality fruits. Pruning affects the fruit quality to a great extent. Pruning is profitable chiefly when done in accordance with the bearing habit of the cultivar. Perlette, for instance, bears well on first five nodes. So pruning is to be so adjusted that every bearing sour should have 4 to 5 buds. However, in cultivars like Anab-e-Shahi, long canes comprising 8-12 buds at the time of pruning are desirable. Pruning should neither be very severe nor very light unless there is a specific purpose behind it.

Irrigation during maturation and harvesting calls for a lot of care to obtain good quality fruit. Over irrigation during this period delays maturity and the fruits becomes insipid and watery whereas a severe shortage of readily available water during ripening period gives the fruit a dull colour and results in sun-burn.

The quality of fruit may also be controlled either by thinning, girdling or by the application of growth regulators.

Thinning

Thinning is employed for improving the berry size, quality of the fruit and hastening maturity. Bunches thus obtained are well filled. It is necessary in cultivars like Perlette to develop all the berries into normal ones. Thinning may be done by the following method :

1. *Flower Cluster Thinning.* Unwanted flower clusters should be removed between leafing out and blooming. It results in better berry set and larger percentage of normal berries.

2. *Cluster Thinning.* In cultivar like Perlette which bears very heavily, cluster thinning is of utmost utility. Undesirable and mishappen clusters should be removed just after the fruit set.

3. *Berry Thinning.* It includes the removal of parts of cluster or of some berries. The rachis of the cluster is cut far enough back from the apex and desired number of berriers retained.

4. *Girdling.* Removal of narrow ring of bark from the trunk is commonly done in individual fruiting spur or entire vine. Girdling of trunck affects the entire vine.

Girdling includes complete removal of a continuous ring of the bark, 5 mm wide. The best time for girdling is at the full bloom stage. The effect of girdling on fruit quality for improving sugar : acid ratio, berry colour and hastening the maturity is well known.

GA at low concentrations (10-25 ppm) improves berry set in Perlette and 'Pusa Seedless' grapes whereas a slightly higher concentration (50 ppm) increased cluster size and weight under Punjab conditions. The berries become elongated. When GA_3 is applied along with girdling, it results in improved size and fruit quality of 'Thompson seedless' cultivar.

Out of the various chemicals and growth regulators tried under the Punjab conditions, sevin (2000 ppm) applied at full bloom stage proved very effective.

Number of spurs per vine and buds in each spur

Grape Cultivars	Number of spurs per vine				No. of buds in each spur
	System of Training				
	Head	Kniffin	Telephone	Bower	
Anab-e-Shahi	–	–	–	80-100	5-6
Early-Muscat	8-10	20-25	35-40	45-50	2-3
Beauty Seedless	8-10	20-25	35-40	45-50	4-5
Delight	8-10	20-25	35-40	45-50	3-4
Gold	8-10	20-25	35-40	45-50	4-5
Kishmish charni	–	20-25	35-40	45-50	8-10
Kishmish beli	–	20-25	35-40	45-50	8-10
Khalili	–	–	40-45	40-45	4-5
Perlette	8-10	20-25	35-40	40-50	4-5
Pusa Seedless	–	20-25	35-40	20-25	2-4
Thompson Seedless	–	20-25	20-25	–	8-10

Harvesting and Marketing

The ripe fruits become soft and develop a transluscent bloom. A bunch of grapes is ready for picking after the berries bear the tip have changed colour and have become sweet. The picked fruit should not be exposed to the sun and should be packed with saw dust before transport to a distant place. The yield depends upon the cultivar, age of the vine, the cultivar and management. The fresh grape, though a perishable fruit, can be kept in good condition for about two months at 2 to 5°C in cold storage.

Cultivars

Anab-e-Shahi (Malta). It is white, oval, thin skinned sweet grape of excellent eating quality. Moderately compact large bunches, large barries, pale green, sweet. Susceptible to mildew, anthracnose and fruit rot. Intensively cultivated in Hyderabad. Heavy yielder.

Arkakanchan. Better known as "Anab-e-Shahi" or the "Queen of Vine Yard". It has big clusters and fruits of golden yellow colour having smell like musk. The T.S.S. 19-20%, good for fresh consumption and also for white wine. Less affected by insect pests and diseases.

Arkawati (Black Champa X Thompson seedless). Thin skinned, scented and yellow berries are sweet. T.S.S. 22-25%. Fruits are excellent for fresh use and also for processed product rasin.

Arkashyam (Bangalore Blue X Black Champa). Bluish black berries sweet to taste. T.S.S. 22-25%. It gives two crops in a year and is tolerant to diseases. The berries are used for making juice, red and dry and sweet wine.

Bangalore Purple or Bangalore Blue. It is widely grown in Karnataka. its colour is dark purple and has small to medium sized bunches with small to medium berries, very compact, uneven ripening, dark blackish purple with subacid taste. The berry is round and plump and the skin is thick and tough. Highly resistant to pests and diseases.

Beauty Seedless. Early grape cultivar recommended for northern India. No seed, ripen by mid May to early June. Deep purple Californian cultivar is noted for heavy cropping. Medium to large sized bunches heavy shouldered and very attractive. Berry contains 78% juice, 18% total soluble solids and 0.70% acidity. Promising for juice making industry.

Bhokari or Nasik Green. It is chiefly grown in Mumbai- Deccan. It is roundish, large-seeded, green grape with a thick skin and a watery, sweet juice. Poor quality.

Black Champa. A new promising, vigorous, medium-heavy yielder. Purple seeded grape of excellent quality. Susceptible to cracking and rotting during rains.

Dakh. It is popular cultivar of the Punjab, purple coloured round grapes with compact bunches. As a table variety it is considered inferior. The juice is coloured.

Delight. Early table cultivar of medium bunches which are conical, compact and attractive. Berry is green, small, round. Ripens in second week of June in Hissar, 68% juice, 18-19% total soluble solids and 0.68% acidity. Heavy cropper. Comparatively less susceptible to anthracnose disease.

Gulabi. It resembles Muscat Hamburg of Australia. A round, purple cultivar with a thick skin and sweet, soft pulp. It is an early yet good cropper. Medium sized bunches. Good keeping quality. Muscat flavour.

Kali Sahebi. It is purple, longish grape of superior quality extremely thin skin, sweet pulp and soft seeds. It is a poor bearer but the grapes keep well.

Kandhari. The berry is purple and oval of medium size with sweet pulp. The vine is shy bearer but the fruits stand well during transport.

Khalili. It is also grown in the Punjab. The berry is long, oval, yellowish green with white bottom. The flesh is soft and sweet. An early cultivar getting ready in June.

Kishmish Beli. Outstanding raisin grape cultivar recently introduced from USSR. Good sized cylindrical clusters of light golden berries having 94% juice, 2.8% solids and 0.66% acidity. Ripen in middle of June.

Kishmish Charni. Outstanding table grape and raisin introduced from USSR. Medium sized bunches well filled, conical, very attractive. Berry is deep coloured purple, uniform, round, slightly elongish, very sweet. Good keeping quality. 76% juice, 20-21% soluble solids, 0.71% acidity. Medium cropping. A promising variety for Haryana.

Muscat (Paneer). It is known for unfalling yield in Tamil Nadu. It has small to medium-sized bunches ; loose shouldered, uneven ripening, berries pink to blackish purple, light green pulp, soft, sweet, juicy, mildly scented.

Pacha-Draksha (resembles Bhokri). It is widely grown in South India near Madurai. High yielding, green grape.

Pandhri Sahebi. It is large, white grape of superior quality. It has very sweet and firm pulp but is a shy bearer. A selection from this is a vigorous and heavy yielder with large clusters of attractive medium sized berries.

Perlette. California cultivar, excellent for fresh market, famous for heavy cropping. Bunch medium to large, conical, heavily shouldered and attractive. Whitish green, medium in size, berry spherical ripens in early June, has 75% juice, 18% soluble solids and 0.82% acidity.

Phakdi. It is an oval, whitish green grape with an extremely thin skin and a soft, sweet pulp. Loose bunches heavily shouldered on sides and pendulous in the middle. Grapes of good quality.

Pusa Seedless. Released from IARI, New Delhi. Well filled large bunches, heavily shouldered, cylinderical and attractive. Berry is light golden in colour, oval, unifrom, medium size. Ready for harvest in June. Sweeter and high yielding than Thompson seedless. Berry contains 65% juice, 20-22% total soluble solids, 0.77% acidity. A table-cum-raisin grape.

Thompson Seedless. Excellent for raisin as well as of high table quality. Very sweet 69% juice, 20% total solids and 0.65% acidity. Ready by middle of June. Shy bearer in comparison to other seedless cultivars. White, round, small sized seedless grapes of excellent quality. T.S.S. 22-24%.

Sonaka. A new mutant developed from Thompson Seedless, very responsive to GA treatment. It was discovered in 1976 in Nanaj, Distt. Sholapur. It has higher TSS of 24-26% and good keeping quality. This cultivar is replacing Thompson Seedless in Sholapur (Maharashtra).

Arkhans. (Bangalore blue × Anab-e-Shahi). Sweet berries of greenish-yellow colour. TSS 18-21% tolerant to diseases with two harvests a year. Excellent for fresh use.

Pusa Urvashi. An early naturing cultivar resistant to diseases. Good for raisin and fresh consumption.

Pusa Navarang. Early maturing cultivar. Berries excellent for juice and coloured wine. It is resistant to Anthracnose.

The following are some of the promising hybrids of grapes developed at IARI, New Delhi.

(*i*) *Hur 'X' Pusa Seedless-67-37.* It matures by the end of the first week of June. The clusters are large (349 gm) and compact. Berries are attractive, large (2.5 gm), ovoid, yellowish green and seeded. An early maturing seeded hybrid with firm flesh.

(*ii*) *Hur 'X' Pusa Seedless-65-56.* It matures a week earlier than 'Pusa seedless'. Bunches are large (500 gm) and compact. Berries are attractive, medium sized (1.8 gm), round, green, tender, sweet, seedless (1 to 2 rudimentary seeds). It appears promising because of early maturity and increased berry size.

(*iii*) *Hur 'X' Pusa Seedless-62-80.* It matures 10 days earlier than the parents. Bunches are loose and medium sized (350 gm). Berries are large (2.6 gm) oval to oblong, greenish, yellow, tender, sweet with 1-2 seeds.

(*iv*) *Bharat Early 'X' Bhokari-62-20.* It matures during the first week of June. Bunches are medium (237.5 gm), compact with short berries. Berries are medium sized (2.3 gm), round yellowish green, muscat flavoured, sweet and seeded. It is early maturing cultivar.

(*v*) *Bhokri 'X' Pearl of Casaba-63-32.* It matures by the first week of June, Bunches are medium sized (230 gm) and loose to well filled. Berries are attractive, medium sized (2.3 gm), round, greenish yellow, tender, sweet.

(*vi*) *Hur 'X' Bharat Early-62-27.* The clusters are conical, medium sized (210 gm) and loose. Berries are medium sized (1.7 gm), yellowish green, tender, very sweet and seeded.

(*vii*) *Hur 'X' Bharat Early-65-21.* It mature during the second week of June. Bunches are medium sized (225 gm) and well filled. Berries are of medium size (2.20 gm), round yellowish green, firm, sweet and seeded.

Improved Cultivars. **Kishmish (Bedana), Muscat Hamburg, Black Hamburg, Gros Colman, Black Prince** and **Foster's Seedling** etc. have been introduced into North, West and South India. Kishmish mostly confined to Madurai district has medium bunches, spherical to oval, light green, soft, very sweet seedless and moderately juicy with good keeping quality.

Insect, Pests and Diseases

Flea beetle, mealy bugs, termite, vine girdling beetle, grape leaf hopper, grape leaf roller are considered to be of importance among pests of vine. Flea beetle (*Scelodonta strigicollis*) is a very destructive pest of the grape-vine in West India. The adult is shining coppery brown with six black patches on the elyters. The damage is caused by the adult when the vines put on fresh flush soon after pruning. It is advised to remove the loose bark after pruning and spray carbaryl 50 WP @ 3 gm/lit water.

The nymphs are yellowish white, whereas the adult is black and winged. Eggs are laid in sticks on the lower surface of the leaves. They remain active during the period March to October but the peak period of its activity is between March-April.

It car be controlled by giving two or three sprayings of Bordeaux mixture 3-3-50 or malathion (50 EC) 1 ml or Carbaryl (50 WP) 2 g in water.

Thrips (Rhipiphorothrips cruentatus). Spray vines with Endosulfan 30 EC @ 1 ml or carbaryl 50 WP @ 2 gm/lit water. Fenitrothion 50 EC @ 1 ml or malathion 50 EC @ 1 ml/lit water should also be sprayed soon after fruit setting. Sap is sucked from the lower surface of leaves. Silvery blotches are produced which turn yellow.

Chaffer beetle (Adoratus spp*).* The beetles appear after the break of monsoon. Feeds on leaves during nights to the extent that only vines are left. Hides during day time. Spray carbaryl 50 WP @ 2 gm or malathion 50 EC @ 2 ml/lit water.

Scales. Small insects harbouring below the bark. Suck up sap and the shoot dries up. Leaves turn yellow and the growth stops. Remove loose barks after pruning and spray with 0.1% fenthion 3-4 times for control.

Termites (Odontotermis sp.). Damage roots get into the trunk so that the vines dry up. Apply chloropyriphos or carbaryl @ 100 gm per pit before planting and apply 5 ml gamma B.H.C. 20 EC or chloropyriphoas 20 EC in basin of each vine 30 cm away from the trunk. Termites are serious enemy to vines and a regular schedule is recommended to check the damage through sprays of fenitrothion/malathion/carbaryl/ endosulphan at the time of fruit set (March), 3-4 weeks before ripening of fruits (May), during the months of July-August, and again during September-October. Care must be taken to avoid spray at the time of flower-bud initiation as the pest attack during the period is mild.

Downy Mildew. It is caused by the fungus *Plasmopara viticola* (B and C) Berl. It attacks leaves, canes, tendrils and bunches. The fungus makes the best growth at temperature of 25-28°C while it is seriously checked during periods of dry summer heat.

The fallen leaves and twigs should be collected and burnt. Fungicides should be sprayed at five stages beginning from the time of pruning, 20-30 days after pruning, before the sprouting of buds, soon after fruit setting and also during the growth of shoots to keep the disease under control. Use of Bordeaux mixture (4 : 4 : 50), copper oxychloride (0.3%), Zineb (0.2%) or captan (0.2 – 0.5%) have been widely used with success against the disease.

Powdery Mildew. It is caused by the fungus *Uncinula necator* (Schw) Burr. It affected all green parts of vine, leaves, shoots blossoms and fruits. Hot dry weather or heavy rain is unfavourable to the growth of the fungus. Since temperature and humidity have great influence on the development of the fungus, vines planted in open and high lying situations are less likely to be damaged by powdery mildew. Sulphur is the most effective fungicide for preventing the attack of powdery mildew provided all green parts of the vine are properly covered with a light coating of sulphur. Sulphur dusting also arrests further growth of powdery mildew.

In addition to sulphur dust (15 kg/ha), sulfex (0.25%), Karathane (0.05%), Calaxin (0.05%), Topsin M (0.1%) and Thiovit may also be used against powdery mildew. Baylaton (3 g) is most effective against powdery mildew of grapes. Use of sulphur is recommended at 7-10 day interval, for Karathane the interval is 10-14 days and for Baylaton 14-21 days for desired effectiveness.

Anthracnose. It is generally knwon as *Karpa* and caused by the fungus *Elsinoe ampelina.* Anthracnose affects all the green parts of vine. It is, however, more common on leaves and shoots. The fungus causes cankers (scars) on the petioles and veins of the leaves which become twisted and deformed. The diseased portions should be promptly pruned and destroyed. A spray of 4% Fycol or 0.4% Blitox or 3% Phytolan should be applied after prunning. This should be repeated 2-3 times during the rains at fortnightly intervals. On leaf emergence the vines should be sprayed with 0.2% benlate or bavistin once every 30 days.

Mineral Deficiency Symptoms

Boron. Young leaves show diffuse chlorosis, with deep lobes and malformation under conditions or boron deficiency. Old leaves are cupped and become necrotic at places. Short apical internodes are seen with brittle young growths. Fruit bunches consists of irrgularly spaced normal and dwarfed fruits, giving the so-called 'hen and chicken' condition.

Magnesium. Deficiency of magnesium is reflected in interveinal chlorosis of the older leaves in the green cultivars and strong purple or red tints in black cultivars with narrow green marginal bands. Later brown interveinal lesions develop. Early defoliation also takes place.

Physiological Disorders

Of the several physiological disorders in grapes the important significances are interveinal chlorosis, bud drop, flower drop, fruit drop, bud killing and blossom-end rot.

(1) *Blossom-end rot.* Black sunken spot develops at the blossom-end of the berry which spreads later. It is supposed to be due to faulty nutrition of calcium. Spray of 1% calcium nitrate is recommended as corrective measure.

(2) *Interveinal Chlorosis.* Nutritional disorder caused by iron, magnesium or zinc is responsible for the chlorosis. Measure to correct this disorder may be a spray of 0.2% solution of $FeSO_4$ + $MgSO_4$ + $ZnSO_4$. Grapes should not be grown on soils containing more than 0.3% salts. Over supply of potash also causes magnesium deficiency.

(3) *Drops of buds, flowers and fruits.* The conditions leading to such drops are supposed to be due to faulty nutrition, deficiency of auxin and defective plant protection operations. It has been suggested that the check may be brought about by (*i*) avoiding heavy irrigation at bloom, (*ii*) thinning, and (*iii*) spray of benzyladenine (200 ppm) and 4- CPA (20 ppm), (*iv*) spray of ethrel (500 ppm) at the stage of ripening, (*v*) Dip of bunches in microfix containing NAA (100 ppm) ten days before ripening to reduce berry drop, (*vi*) Maintaining proper C : N ratio by judicious application of fertilizers. Girdling (0.5% cm) and application of auxins have also been recommended to reduce the death of buds, which are caused by improper use of fertilisers.

CHAPTER

GUAVA

(*Psidium guajava* L.)

Myrtaceae

Origin and Distribution

Guava (H. *Amrood*) is a native of tropical America though widely grown in India, the tropics and the subtropics. A very common fruit, popular among the rich and the poor alike due to its moderated price, nourishing value, and good taste. The fruit is rich in vitamin C and pectin besides being a good source of thiamin and riboflavin. The fruits with white flesh have higher content of vitamin C, phosphorus, calcium and iron are present in good amounts. In India, Uttar Pradesh, the largest grower, produces best quality of guava. Bihar, Madhya Pradesh, Andhra Pradesh, Tamil Nadu, West Bengal, Punjab, Assam, Karnataka, Orissa, and Maharashtra are chief producers of quality guavas. Besides India, Cuba, Hawaiian Islands, southern Asian countries are also growers of guava.

Cultivation

It can be grown on all types of soils including the alkaline soils. While deep loam soil with free drainage is considered suitable for good growth of the crop, it also grown well on sandy loam soils. Alternation of summer and winter season in subtropical regions suits guava.

Ploughings check growth of weeds, allow good intake of water in the soil to lower depths. At no time the roots should be damaged by field operation. Irrigation is not given in the months of November and December. Water logged soils are unsuitable for guava orchards. It needs irrigation once every 15 days in periods of no rains or in summers. During the winters irrigation may be needed at an interval of 21 days. The irrigation is useful in the improvement of fruit quality and yield besides checking fruit drop.

The guava is very exacting in its demand of nitrogen, phosphorus and potassium. To the one-year old plant 50 gm nitrogen should be supplied in split dose, 1/2 in November and 1/2 in June-July. Phosphorus 25 gm and potash

50 gm should be supplied with the first dose of nitrogen in November. This quantity is increased with increase in the age of the plant. It is doubled in second year, trebled in the third year and so on till the sixth year when N : P : K quantity equals 300 gm : 150 gm : 300 gm. This quantity should be supplied to plant every year. When symptoms of zinc deficiency develop 2-3 sprays with 1/2 kg of zinc sulphate mixed with 0.35 kg of hydrated lime in about 70 litres water may be sprayed.

Planting

Plants may be planted in well manured pits during the months of July-October. The distance between the plants is kept at 8 × 10 m for seedling plants and 7 × 9 m for grafted plants. The distance between the plants may be varied according to the fertility of the soil, and the characteristics of the cultivar.

Propagation

Seeds must be sown soon after extraction from the ripe fruits. Seedlings may be raised in polythene bags and transplanted after two months. Guava seeds are viable for a short period and the plant can be propagated by seeds. Seed must be soaked in water for 48 hours before sowing. Guava seedling may develop from seed in about 15-20 days from the sowing time under favourable conditions. Vegetative propagation through cuttings, air layering, veneer grafting, budding, stooling have been employed successfully for getting good plants ensuring consistency in quality. IBA and NAA (50 ppm) induce rooting in cuttings. Vegetative propagation, however, is preferred since these bear early and give about 3-4 times more fruit each year. It may be done by inarching, shield or patch budding, air layering, or even root cutting.

Approach grafting or Inarching. Stocks of guava seedlings are used with about 80-90% success when done in July-August. This method is more prevalent and successful in comparison to other methods of vegetative propagation.

P. cujavillis, P. molle, P. cattleianum, P. guineense. P. friedrichsthalianum (Chinese guava) may be used as rootstock since these are compatible for the edible guava *P. guajava.* The Chinese guava stock is suitable for nematode infected localities, it is also resistant to guava wilt and may be used for dwarfing also. *P. pumilum* has also the dwarfing effect as root stock. *P. cujavillis* root stock gives the largest but non-uniform, thick skinned rough fruits. The rootstocks have profound effect on vigour, fruit quality, disease tolerance, resistance to cold and fruitfulness of the scion. Allahabad safeda cultivar grafted on *P. pumilum* produced fruits with higher sugar content while the graft on *P. cujavillis* had higher ascorbic acid content. The yield increased when Allahabad safeda was grafted on *P. cattleianum.*

Shield budding. Insertion of 2.5 to 4 cm long bud in winter or early spring in young root stock.

Air layering. Air layering is successful in about 4-8 weeks. Soil : sand : leaf mould mixed as 1 : 1 : 1 works as a suitable medium for transplanting the rooted cuttings. Forkert method of budding has proved better than other types of budding.

Though an expensive method, it is fairly common in India. Use of indolebutyric acid, indoleacetic acid and naphthalene acetic acid may be made, for better results, during the rains.

Root cutting. The plants, when 3-5 years old, are sawed off just below the soil surface in spring, shoots that appear are allowed to grow for three months, thinned and soil is banked round these. These are removed from the place in about 9-10 months time and planted.

Flowering and Fruiting. Guava bears fruits on new sprouts in north India, guava flowers twice a year. Two harvests are obtained, one in July-August and the other in November-January. The winter season crop fetches better price in comparison to the rainy season crop. It is essential to avoid rainy season crop, the flowers should be trained by stopping irrigation of trees from February till mid-May. In the end of May or June the soil of the orchard should be ploughed and harrowed followed by addition of cow-dung manure @ 45 kg for 5-year old tree. After mixing of manure, irrigation may be given at an interval of three days in the beginning and later, after every 10-15 days till the onset of rains. The tree puts up blossoms in June-July and the fruits grow to full size to give luscious and attractive guavas. Dry atmosphere at the time of flowering is advantageous though high temperature causes fruit drop.

Crop Regulation. Three crops of guava may be obtained in a year. Flowers of spring season (March-April) provide fruits in the rainy season (July-August) and are called *Amba Bahar*. Fruits resulting from flowers of rainy season mature in winter (Oct. Nov.) are known as *Mrig Bahar.* The winter flowers become mature fruits in spring and are called *Hast Bahar.* Fruits of *Mrig Bahar* are considered superior to others. *Amba Bahar* is more prone to attack by fruit fly and fruit borer. It is therefore desirable to regulate the crop for better and profitable yield.

The rainy season crop can be avoided or eliminated by the following cultural operations.

(*i*) *Withholding irrigation.* The flowers and fruitlets are made to drop by withholding irrigation during the period January to the onset of monsoon (June or July). The elimination of rainy season crop (*Amba Bahar*) results in higher production in winters (*Mrig Bahar*). This practice is specially useful in sandy and light soils which have low water retention capacity.

(*ii*) *Root pruning.* Roots are pruned upto the depth of 30 cm during March-April. Coupled with no-irrigation till the onset of monsoon it is useful in eliminating rainy season crop. The trenches may be filled with soil-manure mixture followed by irrigation at 7 day intervals. Thus this method has some adverse effect on the tree and its bearing capacity yet it serves the purpose of giving a fillip to the *Mrig Bahar.*

(*iii*) *Deflowering.* Removal of all flowers between the period April 20 to May 15 and again the flower buds after May 25 mechanically (or by hand) completely arrests the *Amba Bahar* and ensures good winter-crop.

(*iv*) *Use of bioregulators.* Two sprays of NAA @ 80-800 ppm in May at 15-day interval results in maxium flower drop encouraging winter season crop. The optimum concentration of NAA varies with the region, cultivar, age of the plant, the time of application etc. Once determined for an orchard it is an effective measure recommended for large orchards where time and labour are limiting factors.

(*v*) *Use of urea.* One spray of urea (10-20%) at peak flowering time in summer wards off the rainy season crop considerably. Optimum concentration of urea differs with cultivars *viz.* for Allahabad Safeda the optimum concentration is reported to be 10% while for Sardar guava or Lucknow 49 it is 20%.

(*vi*) *Pruning of new shoots.* Since guava bears fruits on new shoots their pruning at one-leaf pair stage during the last week of April to first week of May gives good winter crop.

Pruning

It may be done to remove dried and diseased twigs but it is not a practice in North India. It is one of the alternatives to ensure larger crop of good fruits. March to April pruning encourages new vegetative growth.

Some believe in severe pruning while other consider light or no pruning to be profitable. Severe pruning is essential for the formation of a strong framework and may be done in April-May to push up vegetative growth with proper spacing between branches. Formation of weak crotches is avoided to prevent plants from being injured due to the weight of the fruits and high wind blowing at the time of fruiting. For trees which have started bearing, annual pruning is desirable to stimulate growth and flowering.

Cultivars

Allahabad safeda. Fruit, medium with average weight of 167 gm and distinct flavour. Delicious, seedless, round, smooth skinned, white flesh and sweet. Very popular, cultivates in U.P., Bihar and being introduced in other parts of India. Medium keeping quality. May be stored for a month in cold storage (9-14°C).

Arka Mridula. A seedling selection of cultivar Allahabad safeda. Plants of medium vigour, high yield, medium-size fruits, excellent quality, high TSS with only few seeds in white pulp.

Allahabad Surkha. A selection from the village Sulem Sarai (about 4 km from the Allahabad Rly. station). Plants vigorous, domeshaped, compact, lamina large, dark green, elliptical, margin entire, flower borne singly or in a cyme, fruits large, slightly flattened at both ends, skin thin of uniform pink colour, flesh thick deep pink, sweat with a pleasant smell and aroma. An outstanding cultivar with good yield of about 120 kg fruits per year in the sixth year. Attractive, 112.0 gm average weight per fruit. Pleasant flavour. It is equally sweet as safeda but has apple colour. It is said that it can replace safeda. Tastes sweet, heavy bearer. Fruits spherical, dawn-pink in colour with deep minute dots on the surface.

Behat Coconut. Originated in village Behat in Saharanpur (U.P.), probably a selection from Allahabad Safeda. Fruit and keeping qualities both are good. Heavy bearer. Fruits round with rough surface having dots.

Chittidar. Originated from Allahabad, smaller than the Allahabad Safeda. Similar to safeda, with red spots on the skin, pulp white, somewhat sweeter. Average weight 111.6 gm, soft pulp acidic sweet in taste, good keeping quality and medium bearing.

Hafsi. Origin in Bihar State, medium weight of 104.90 gm. creamy white soft pulp with pleasant flavour. Fruits with smooth red flesh are not as sweet as the white ones. A distinct flavour.

Harijha. A cultivated cultivar of Bihar, fruits are round, white-fleshed and of good keeping quality. Profuse bearer.

Karela. Pear shaped, rough skinned and sweet, with white pulp.

Lucknow 49. A selection from Safeda made at Pune. Grown in south India and Maharashtra. Heavy bearing, large sized fruits (245 gm). Crisp pulp, soft and cream-white, acidic-sweet, dwarf, spreading, good fruiting. Good keeping quality. May be stored for a month at 10°C. It is also known as Sardar guava.

Mirzapur Seedling. Selection from Safeda picked up from Mirzapur, 167.6 gm fruit weight, pleasant flavour and taste acidic sweet. Fruit quality medium to good and keeping quality good.

Pear-shaped. Originated in U.P. (141.70 gm fruit weight). Attractive, pear shaped fruit, medium bearing good quality fruit. Smooth surface with large dots.

Banarasi. Sweet guava, no acidity, fruits round and yellowish of medium. Keeping quality.

Seedless. Originated from Allahabad. Shy bearer, a triploid with 33 chromosomes instead of normal number of 22. Few or no seeds. Vigorous trees bearing poor crops of rather small (71.8 gm), irregular fruits, not of much commercial importance, taste sweet. Excellent keeping quality.

Nasik, Dharwar, Nagpur Bedana, Saharanpur Bedana, Anakapalli etc. are also considered good cultivars. *Pant Prabhat.* and *Lafit* are the latest addition to the list of cultivars.

Pest and Diseases

(*i*) *Bark-eating caterpillar.* In U.P. *Indrebela tertaonis (Arbela tetraonis)* is a very serious pest which has been mentioned as a pest of Mango and Citrus fruits. It bores the bark and stem. Remove the dead shoots. Spray chloropyriphos (0.01%) or carbaryl (0.1%) on tree trunk after removing silken galleries. Inject Dichlorovos or Malathion (0.5 – 1.0%) @ 5 ml/hole. Plug the holes with mud.

(*ii*) *Leaf-eating insects. Myllocerus maculorus* is a serious pest at Allahabad, while *A. discolor* and *M. Sabulosis* have been reported from Pusa.

(*iii*) *Guava scale.* In Maharashtra *Pulvinaria psidii* is said to be only serious pest. It can be controlled by spray of Dimethoate (0.04%), Phosphomidon (0.04%) or Malathion (0.05%). Spray twice or thrice at 10-day interval. If the infection is severe the spray may be repeated at shorter intervals.

(*iv*) *Mealy bug (Cryptolemius* sp.). At Mysore it is reported to be a serious pest. 1-1.5 kg of fuel oil or 4-4-50. Bordeaux mixture is preferable as spray. Nicotine sulphate in water (1:600) may be sprayed for control 0.1% metacid spray is also effective. Soil treatment at the base with application of methyl parathion @ 250 gm/g tree may be useful. Besides alkaline banding is also necessary to cheek the nymphs from climbing up via tree trunk spray of phenthoate (0.05%) on the affected parts is also recommended.

Wilt. The wilt disease caused by *Fusarium* sp. is widely spread throughout U.P., M.P., Bihar, Rajasthan and West Bengal. Apart from *Fusarium* sp. guava trees are also affected by *Macrophomina phaseoli, Gloeo-sporium, psidii, Phasalospara psidii* (Cancker) and *Cercospora sawadae.* Wilt may be controlled by taking out of the plant with roots intact and burning these. The holes made due to the taking out of the plant, should be fully drenched with thiram solution (3 gm/litre water). As soon as the symptoms of wilt are noticed 0.1% Banlet or Bavistin should be sprayed in the vicinity of the plant till fully drenched. Treating the soil with lime, gypsum @ 1.8 kg/tree or margosa seed cake @ 10 kg/tree after exposing the roots to a depth of 15 cm has been found to be useful. A combination of *Trichoderma virdae* and compost (1 : 25) may be used. Two applications of Metasystox (0.05%) and zinc sulphate (0.3%) in March and September has also been used with success in the control of disease. The supply of nitrate nitrogen also helps the check of disease. Use of resistant root stock is the only solution to ward off the disease. Pruning the infected portions is also helpful.

Banarasi, Dholka, Sind, Nasik, Supreme are wilt resistant cultivars. Wilt is serious in alkaline soils. The infection can be minimised and the losses mitigated by drenching the soil with brassicol and spraying the plant with 0.1% bavistin.

Anthracnose. Common in western Uttar Pradesh. The tips become brown turning black after some time. It continues to extend towards the base of the plant and eventually die back sets in. The causal organism is *Gloeoeosporium psidii.* It can be controlled by the use of dithane Z-78, copper oxychloride or cuprous oxide, or difolatan.

Stem canker. The collapse of the tissue leads to wilting of the shoots as a result of cracks which appear on the stem. The casual organism is *Physalospora psidii.* Pruning of the infected parts and protection of the wounds with Blitox or Bordeaux mixture paste is recommended as control measure.

Leaf spot. Water soaked patches on the underside of leaves are caused due to the attack by *Cercospora sawadae.* Application of lime and sulphur (1 : 30) or copper oxychloride (0.3%) reduces the infection.

Yield

The yield of Allahabad safeda, Chittidar and Lucknow-49 was more than other cultivars grown under similar biotic and edaphic conditions. Accurate production figures are not available. The average yield per tree is estimated to be 90 kg for the seedling tree and 350 kg for the grafted ones under ideal conditions.

□□□

CHAPTER 13

JACKFRUIT

(*Artocarpus heterophyllus* Lamk)

Moracceae

Origin and Distribution

The jackfruit, (*Katahal*, H.) indigenous to India, giving the largest single fruit (40 kg) is heavy yielder producing more than many of the fruit trees. The fruits are borne on small leafless stalks arising from the trunks and the main branches of the tree. It is classed as commercial fruit but is seldom cultivated in regular plantations. It is a popular fruit in South India where it is specially grown as a shade tree in the Coffee, Areca, Cardamom plantation and has living support for *Piper nigrum* (pepper). The green unpripe fruit is taken after cooking although the ripe fruit is also tasty. Assam (8,000 ha), Bihar (4,000 ha) and South India (1,200 ha) are reported to be good cultivators of the fruit. U.P. also grows the crop but not as a commercial proposition. Malaya, Burma and Brazil grow the jackfruit to a considerable extent. The male and female flowers are separate in drooping catkins.

Climate and Soil

Flourishes best under humid and warm climate ; prefers rich, deep, well drained alluvial or textured loamy or lateritic soil. A tree of the tropical low land but capable of tolerating higher altitudes (2,400 metres) and cold, though the quality of fruit deteriorate above 1,200 metres. Warm and humid climate and good soil drainage seem to be the ideal. Any rise in water table or poor aeration of the soil is detrimental.

Propagation

Universally grown from seed *in situ.* Germination is improved by soaking the seeds in running water for 24 hours. Seed viability remains undiminished for two weeks after extraction from the ripe fruit, then it registers a notable decline. Seeds soaked in 25 ppm concentration of NAA gives maximum germination and good growth of seedlings. Seeds also give good germination when kept at 6°C within 3 days of their extraction from the ripe fruit. The seeds are not sensitive to light and may be kept in dark also. Generally 4-5 seeds may be planted *in situ* or seedlings in polythene bags. Transplanting in the orchard may be done when the seedlings are 24 month old.

Vegetative propagation may be done by budding (patch, chip, Forkert) or epicotyl grafting, inarching, air layering or cuttings. The success depends on the time of operation with respect to the age, cultivar, nutritional condition, vigour of the plant, the environment and the skill of the operator. Use of ethrel, IBA, NAA and some other growth regulators may facilitate the process when applied in proper dose and at proper time. Cuttings from forced etiolated shoots treated with IBA (5000 ppm) by the quick dip technique has given 100% rooting. Ringing and etiolation of shoots also helps in rooting. In addition to IBA, ferulic acid, Captan and IAA promote rooting. Air layering meets with greater success when IAA, IBA and NAA are used. Patch budding in the month of June in Uttar Pradesh has given best results. T-budding has been least successful.

Inarching or grafting meets with about 60.70% success, Transplanting of rooted seedlings is difficult. Propagation of selected clones is common by inarching or grafting to seedling root stocks of jackfruit *viz., A. champaden, A. hirsuta,* L. or *A. rigida.* In Java it is successfully budded by modified Forkert or flap method. Air layering with little use of Seradix A, 0.025% β indole butyric acid or napthalene acetic acid on to the exposed portion is very successful.

Planting

Planting is done in well prepared pits filled with compost or well rotten FYM previously exposed to the atmosphere for over 9 months in the Monsoon season at a distance of 12 metres. Phorate, Carbofuran and Lundave mixture in the proportion of 3 : 3 : 1.3 at the rate of 50 gm. per plant may be used to ward off the danger of white ants.

Culture

Where it is grown as a shade tree for the coffee, pepper etc. the jack fruit is given no cultural treatment or care and it seems to benefit by the care of the main crop. Excellent drainage and adequate watering make the jack tree give better performance. At no stage it should be exposed to drought or frost. In areas of severe drought or frost it is useful to provide some protection specially to the young trees.

Harvesting

The trees generally bear after eight years, the sole exception to late bearing is the Singapore cultivar which fruits at the age of 3 years only. The fruiting season lasts about 4 months. The expected yield from one tree is from a few to

250 fruits per annum each weighing about 3-5 kg although fruit as big as 55 kg have been obtained. The seeds may be stored at a temperature of 11-13°C for a period not exceeding six weeks.

Besides vitamins A and C, the edible pulp, which is about 30% of the fruit, contains water 73.1%, protein 0.6%, fat 0.6%, carbohydrate 23.4%, fibre 1.8% and ash 0.5%. The seeds contain water 51.6%, protein 6.6%, fat 0.4%, carbohydrate 38.4%, fibre 1.5%. Besides being used as vegetable, pickles, etc. the unripe fruit can also give a palatable beverage concentrate by cooking 5 kg of pulp with 11 kg of sugar, 35 g citric acid and 7.5 litres of water. The mixture is mashed, strained and may be preserved by adding a pinch of potassium metabisulphite.

Jackfruits differ in shape, size and quality of the fruits. The pulp of the fruits when crisp is more highly flavoured and relished while mushy or soft pulp is of varying quality ranging from very sweet to insipid. The jackfruit may be classified into two groups, (*i*) soft flesh, and (*ii*) firm flesh. In the soft flesh type finger can be easily thrust in, when the fruit is ripe but it is not so in the latter. The firm fleshy type is highly tasty, sweet and crisp. These two groups are further, divided depending on the taste, shape, size of fruit, odour of flesh, nature, shape and diversity of prickles on the rind for the maintenance of separate varieties.

Cultivars

There are some distinct types *viz., Padrauna, Rudrakshi, Singapore* or *Sri Lanka Jack, Khaja Allahabad. Gulabi, Champa, Hazari* are The Names denoting the quality of the cultivars *viz., Gulabi* (rose scented), *Champa* (scent like Michelia sp. and *Hazari* (bearing thousand fruits). Rudrakshi is a small fruited type, with a smooth and less spiny rind than the common jack and having a fleshy perianth. Singapore or Sri Lanka-jack is a recently introduced cultivar from Sri Lanka. Plant bears fruits in 3 years after planting and is extremely precocious in habit. The fruit are more or less of the same size as the common jackfruit. Besides the cultivars *Hadrhiyalva, Bhadonhi, Zarda, Bhusola* etc. are some of the local strains grown in Uttar Pradesh. Khaja Allahabad is a white fleshed heavy bearing cultivar of jackfruit commonly grown in the locality. A tree may yield as may as 250 fruits.

The Narendra Dev University of Agriculture and Technology, Faizabad, U.P. has made certain collections which may be classified into two categories.

NJT_1, NJT_2, NJT_3 and NJT_4 which have large bulbs with low fibre content with fruits of excellent quality recommended for Table purposes. Other selections for culinary purpose are NJC_1, NJC_2, NJC_3 and NJC_4 with smaller fruits of thin rind and soft flesh. Some strains have been reported from south India. More important of these are Muttam Varikka, Thean Varikka, Paladan Varikka, Palur-1, Pechippari-1, Hybrid Jack and Burliar-1.

Insect Pests and Diseases

Insect pests and diseases are of comparatively less or of minor importance. The more serious insect pests of jackfruit are given below.

Shoot borer caterpillar (Margaronia ceasalis). The larvae of this moth attack tender stems and developing fruits and may be controlled by hand picking or spraying 2% parathion. This pest has been reported from Andhra Pradesh, Uttar Pradesh, Assam and Bihar. Spray of carbaryl (0.1%) during flowering is advised.

Mealy bug (*Drasicha mangiferae*). It sucks sap from shoot and the affected parts. It can be controlled by spraying contact insecticides like carbaryl 50 WP @ 4 gm or monocrotophos 40 EC @ 1 ml/l water.

Jack scale. It covers the leaves and twigs and sucks the sap. It can also be controlled by contact insecticides.

Apart from these few others like Brown weevil (*Ochyromera artocarpi*) and moth (*Perina nuda*) have also been reported damaging the jackfruit.

The common diseases that attack this plant are (*i*) Pink disease, (*ii*) Stem rot, (*iii*) Fruit rot, and (*iv*) Male infloresence rot.

Pink disease (*Botryobasidium* sp.). Young woody branches loose their leaves and die-back. Thin pink incrustations appear on the lower side of leaves. Prune and burn the affected branches. Paste the cut end with Bordeaux mixture. Spray copper oxychloride @ 3 g/litre water after the rainy season.

Fruit rot (*Phytophthora palmivora*). Fruits develop soft rot.

The disease can be controlled by spray of Bordeaux mixture (4 : 4 : 50) or copper oxychloride @ 3 g/litre of water.

Premature drop of fruits can be minimised by spraying 0.04% benomyl at 15-day intervals. 2-3 sprays may be sufficient to control *Botrytis cineria* and *Rhizopus stolonifer* which are known to cause the premature drop.

CHAPTER

14

JAMUN

(*Syzygium cumini* L.)

Skeels ; Myrtaceae

Origin and Distribution

Indian blackberry, black plum, Java plum, *Jambul*, *Jam, kalajam, phalani* or *pharenda* and synonyms of *jamun*. It is considered to be indigenous to India or the East Indies, being cultivated in Laos, Philippines, West Indies, Africa, California, Florida (U.S.A), Israel and Medagascar. In India it grows in the Indo-Gangetic plains, lower ranges of the Himalaya and Tamil Nadu.

The genus *Syzygium* to which jamun belongs is a member of the family Myrataceae. *S. jambos* known as rose-apple or gulab jamun is a very popular fruit of south India and also West Bengal. The fruits are yellow. *S. fruiticosum* also yields edible fruits which are small and purple coloured and the plant can be used as wind-break. *S. zeylamica* is generally found on the western coast of the country. Other notable species of *Syzygium* are *S. densiflora* a compatible rootstock for *S. cumini* (jamun) with the property of being termite resistant. *S javanicum* (water-apple) is also seen growing in South and West Bengal. *S. malaccensis* and *S. uniflora* are also found in South India.

The trees are evergreen, attain medium to huge size. Fruits are attractive black or purple in colour, sweet in taste. The stone, bark, leaves etc. are used as an effective medicine against diabetes (passing of sugar in the urine) by the followers of Ayurvedic system.

Climate and Soil

Although a fruit of tropics and subtropics it is found growing in lower ranges of the Himalayas upto an altitude of about 1,500 metres. A little more susceptible to cold and drought than the seedling mango trees. The jamun requires dry weather at the time of flowering and fruit setting. Early rain is considered to be beneficial for ripening of fruit and proper development of its size, colour and taste. Loamy, deep and well drained soils are considered optimum for growth and fruiting. It is not exacting in its soil preferences. The plant benefits by dry atmospheric conditions during flowering and fruiting. Young plants are damaged by cold and drought.

Propagation

Besides seed propagation inarching, approach grafting, Forkert method of budding and air layering is also possible with Jamun. Jamun seeds give a high percentage of germination, if sown fresh. The seedlings are transplanted in their permanent places in the following spring (February-March) or Monsoon, *i.e.*, July-September.

The treatment of seeds with gibberellic acid improves the germination as well as the growth of seedings. Treatment of shoots with IBA and NAA in lanolin (0.1%) gives success in air-layering. Seeds from healthy jamun plants of desirable quality may be germinated and the seedlings used as rootstock, when two-year old, for inarching. During the period of union of rootstock and scion regular watering to the rootstock is considered useful. Venner grafting is done best when the seedling is one-year old. The best period for grafting and budding is July-August. Forkert method is considered best method for budding. Transplanting with intact soil ball, done in February-March or July-August helps the plants to establish itself with ease and greater certainty. Watering during the period of establishment is necessary.

Planting

Jamun tree may be planted in pits (1 × 1 × 1 m) made in May-June and filled with soil + F.Y.M. 10-12 m apart on road sides or in orchard ; it may alternate with thick growing but comparatively less tall trees (*Zizyphus mauritiana, Morus alba*). It gives a very effective windbreak.

Cultural Practices

Young trees need manuring, irrigation and protection from cattle. Almost no pruning is done. The branches are eaily broken under the stress of high wind and therefore windbreaks prove useful. Trenches one metre in width at a distance of about 3 metres from the stem may be opened each year and 2-3 quintals of well rotten compost be added, two weeks before flowering followed by irrigation.

Dry twigs and crossed branches are removed and care is taken to train the plant so as to provide a good framework of branches 0.6 to 1.2 m above the ground level.

The need of irrigation is more to young plant ; 8-10 irrigations per year but for the grown up and bearing trees 4-6 irrigations are considered sufficient.

Manuring with F.Y.M. or compost @ 20 kg per year in the pre-bearing stage and 80 kg per year to be bearing plant is profitable for better growth and higher yield. Fruit-bud formation, flowering and also fruiting are affected by judicious manuring and irrigation.

Harvesting and Marketing

New vegetative shoots appear twice a year, once in February-May and the other in August-October. 6-10 month old branches show bud formation in about 6 months between January to June. The February flush produces maximum growth and flowering. There seems to be a regular phenomenon of flower and fruit drop in jamun. In cases when the flowering period is between Jan. 21 and March 7 the first drop of more than 50% takes place by the end of April, the second drop of about 25% by May 15 and the third drop of 10% is over by July 15. About 15% flowers reach maturity.

The ripe fruits are picked singly by hand and marketed the same day. The average yield from a seedling tree may be 80-100 kg. The grafted trees may yield somewhat less but the fruits are of good quality fetching better returns.

Cultivars

The most commonly known cultivar is the Rajamun with a big sized oblong fruit, deep purple or bluish black in colour ripening in June-July. Its pulp is juicy, sweet and light pink or greyish in colour. The stone is comparatively small. A strain named Narendra Jamun 6 has been identified with desirable traits.

Besides the Rajamun several seedling strains are found in India. Late maturing strain of inferior quality fruits having bigger stones is quite common.

Physiological Disorders

Flowers and fruits drop at various stages in Jamun. This problem may be minimized by two sprays of GA_3 of 60 ppm concentration first at full bloom stage and the second 15 days after the first spray.

Insect Pests

There is no serious pest or disease known. Sometimes the fruits get wormy due to the attack of fruit fly. Leat eating caterpillar (*Carea subtilis*) and white fly (*Dialeurodes eugeniae*) generally attack the jamun tees in India, the former being active mostly in the South.

Use of rogor, Dimethoate or malathion 0.02% maintenance of sanitary conditions in the orchards, killing of the maggots and pupae are recommended as control measures. Two important diseases of leaf spot and fruit rot caused by *Glometella cingulata* have been reported. Regular spray Zineb (Dithane Z-78) @ 2 g/l of water.

CHAPTER

KARONDA

(***Carissa carandas*** **L.**)

Apocynceae

Origin and Distribution

The *karonda* is well known as a protective hardy hedge plant bearing berry-like sour edible fruits which are attractive in colour. Though widely grown throughout India, South Africa and Malaya, it is said to be native to India. This is also popular as *Caraunda* and Christ's Thorn. Plant is evergreen, shrub or small tree, generally not more than 5 to 9 metres in height. Its foliage is dark green with entire, ovate elliptic leaves.

Soil and Climate

It is grown on all types of soils in both tropical and subtropical climates. The karonda is very hardy in nature. This plant cannot be grown in the hills of Himalayas but thrives well at some places in the Nilgiris.

Propagation

It is commercially propagated from seed in the month of August-September. Though inarching is possible for vegetative propagation, cuttings may also succeed. Air layering and stooling with the help of IBA (5000 – 7000 ppm) have given encouraging results. The optimum time for the operation has been reported to be July-August. Rooted layers may be detatched three months after layering (Oct-Nov.). The cultivar pant Sudarshan may be propagated by soft wood cutting with the help of 8000 ppm IBA in lanolin paste planted in mist in polyhouse condition.

Planting

The karonda behaves as a xerophytic plant. A karonda hedge is impermeable to animals due to its thorny dense bushes that also yield fruits which can be sold in the market. So, it serves a dual purpose. Before planting, the trench is first dug out and soil is filled back with well rotten compost mixed in the ratio of 1 : 3 for the better growth of plants. It is necessary to irrigate every week in summer and fortnightly in winter for healthy plantation. Once the karonda hedge or solitary plant is established, it hardly requires any care. For hedge it should be planted 90-150 cm apart and for orchard 4 m apart.

Harvesting and Marketing

The karonda blossoms in February ; ripe fruits are ready by August and continue to be available upto Sepetmber. It bears heavily unless pruned severely. Unripe fruits can be available from May onwards and are used for chutney, tarts, and several other forms of pickles. Karonda contains Vitamin C. Fruits also make very good jelly.

Cultivars

Three selections from G.B. Pant University, Uttranchal are well known.

Pant Sudarshan. Fruits white with pink blush, av. wt. of fruit 3.46 gm. The plant is of medium height yielding about 29 kg of fruits/plant (bush) per year.

Pant Manohar. Plants are of medium height with dense foliage. Fruits are white with dark pink blush, average wt. of fruit is 3.49 gm and the average yield per plant per year is 27 kg.

Pant Suvarna. The plant grows upright with sprace branching. Fruits with deep purple blush over green background are bigger 3.62 gm (av. wt.). The yield per bush is about 22 kg.

Besides the strain knwon as *Maroon* type nos. 13, 16, 12 and 3 have been identified as promising ones from NDUAT, Faizabad, U.P.

The cultivars are classified according to fruit colour as (*i*) Green, (*ii*) Pink and (*iii*) White. There is quite a resemblance in the shape and size of the fruits. There are several species of the genus *Carissa*.

Carissa grandiflora, D.C. (Natal Plum). The Natal Plum has dark red fruits, papery skin and a few small circular seeds. The fruits ripen irrugularly throughout the year. It is only suitable for home consumption unless planted in large numbers. Fruit is a good source of ascorbic acid and makes a good jelly.

Carisa bispinosa, Desf (*C. arduina* Mill. *: Amatungula, Maritgula*). The fruits are of the size of berries and closely resemble those of Natal Plum but flowers are slightly smaller and the fruits are borne in clusters. Seeds are lanceolate.

Carissa edulis, Vahl. It is a straggling shrub. Flowers are white, pink and scented and come in axillary clusters of 10 to 25. Berries are oval, red but turn black at maturity.

Carissa ovata. It originated in Australia. Fruits are small and edible and are used in jam making.

Carissa spindrum, D.C. It is native of India and bears small edible fruits. This is the hardiest species of *Carissa* and is known to thrive in poor and rocky soils. May be helpful for afforestation on ravine and alkaline soils.

Insect Pests

Hairy caterpillar. It causes severe damage to the plant by feeding on the leaves such that the plant perishes. Endosulfan @ 1 ml or Dichlorovos @ 1 ml/l of water sprayed over the plant controls the activity of the pest.

Diseases

Anthracnose. Common and widespread disease of Karonda. The leaves show brown spots to begin with leading to the premature death of leaves and twigs. Spray of copper oxychloride @ 3 g/l water may be given to control the disease. The wound on the branches may be scrapped off and painted with Bordeaux mixture, Chaubatia paste or Blitox-linseed paste.

Fruit rot. Harvested fruits have this malady. Proper sanitation and care of the fruits avoiding injury may be helpful. Quick consumption of the fruits after harvest is also one way to ward off the fruit rot.

CHAPTER

16

LITCHI

(*Litchi chinensis* Sonn.)

Sapindaceae

Litchi grown in India for almost 200 years is indigenous to South East China where dried litchi nuts are considered a must for a dinner. It spread to the subtropical regions only recently. A monograph on litchi written by TS'ai Hsiang in 1059 A.D., perhaps the earliest publication on fruits, stresses its popularity and antiquity both.

Distribution

Litchi (H. *Lichi*) is very exacting in its climatic requirements. China and South Africa surpass in respect of area under litchi production. Countries where litchi has assumed importance are Africa, Australia, Brazil, Florida, Formosa, Hawaii, Indo-China, Japan, Mauritius, New Zealand, Indonesia, Thailand and West Indies.

In India, area under litchi was almost 11,500 hectares ; roughly 0.75% of that devoted to fruit cultivation. It is an important commercial fruit in Bihar where about 1,000 hectares are devoted to this fruit. North Bihar is the chief centre of litchi production. Submountaneous regions of Uttar Pradesh, Haryana, Hoogli (W. Bengal) and Gurdaspur in the Punjab are also of importance in this respect. Litchi can be successfully grown in the entire Sub-Himalayan region of India and made more popular and extensive.

Botanical Description

Litchi (*L. chinensis* Sonn. Sy. *Nephelium litchi* of the Sapindaceae) attains a height of 9 to 12 m. It has a broad round top consisting of glossy compound leaves with 2 to 4 pairs of elliptic oblong to lanceolate sharply acute glabrous leaflets. Long terminal panicles of small unattractive flowers appear in February in North India. Fruits, oval to ovate in shape, green to deep rose or brown colour depending on the stage of maturity are borne as loose clusters in groups of 3-20.

The external covering of the fruits is brittle and rough. The seeds in the grafted plants are small with more transluscent firm and juicy flesh or aril whereas seeds of the seedling trees are big. Fruits contain 20-30% total solids, 0.5% ash, 1.16% acid, 1.15% proteins and 15.3% total sugars. Besides, being a rich source of Vitamin C, fair amount of phosphorus, calcium and iron, it also contains small amounts of Vitamins A and B.

Soil and Climatic Requirement. Deep, well drained loams rich in organic matter with water table lower than 1.5 metres from the ground level (to escape water logging which causes decay of roots) are a welcome for litchi. Mildly acid soils are preferred to neutral or alkaline soils. Roots of plants grown in acid soils are associated with mycorrhiza. In Bihar it is grown in calcareous soils. It is essentially a fruit of subtropical regions with high atmospheric humidity, occasional rainfall, cool dry winters, freedom from frost and hot winds. In India moist summers and cool winters with spells of 1°C to 4.5°C temperature in winters favour economic production of litchi. Besides high humidity and abundant moisture litchi needs deep rich soil and freedom from pests.

Seasonal variations in temperature are necessary for proper fruiting. If seasonal fluctuations in temperature are limited, as in the south, it succeeds to hill slopes upto an elevation of 1,050 metres. Litchi growing tracts in India have temperature range of −4°C to 38°C during flowering and fruiting. Fruit splitting results due to alternate spells of dry heat and rainfall.

Propagation and Culture

Litchi seeds give a very high percentage of germination when sown fresh. Seeds stored in water for 24 hours do not show a decline in germination, but decline is registered to the tune of 25% after 7 days and 70% after 15 days of storage in water.

Sexual as well as vegetative propagation of litchi are prevalent. The plants raised from seeds are not true to the type and take 8-20 years for bearing. Litchi seeds lose their viability within 4-5 days after removal from the fruit and as such vegetative propagation has as good scope for this plant.

Air-layering is commonly employed for vegetative propagation of litchi. Non-flowering mature branches may be used for air-layering in spring or early summer under Indian conditions. Pretreatment of the stock plants with 200 ppm ethrel and treatment with IBA (5000 ppm) and p-hydroxybenzoic acid (200 ppm) yielded 100% rooting in Mumbai cultivar.

Rooting medium or soil is provided at the desired place above ground (air-layering) both in India and China. In fact, this method is so common in China as to be called Chinese layering but in India it is called Gootee. Air layering or gootee is practised when the atmosphere is humid (June-August). The young plantlets are detached from the mother plant in August-October, kept in nursery for better root growth, before planting in the field. *Sphagnum* and polythene aided gootee works better and needs no occasional wetting. Success in air-layering or gootee depends on temperature and relative humidity of the atmosphere besides the existence of favourable internal factors *viz.*, availability of carbohydrate, auxin and minerals. Ordinarily it takes 4-6 weeks for the roots to develop well enough to denote stage of separatation of rooted branch from the parent tree. Ringed stems when treated chemically with 50% aqueous solution of 'Rootone' or 100 ppm α-NAA in lanolin paste ensure success. Smearing the ringed stem with Seradex-3 or 2500 ppm IBA and tying *sphagnum* promotes rooting.

Euphorbia longan or Nephelium as a root stock gives good growth yet uncertain bearing in South India. Layering, however, has met with 100% success specially during the months of August and September.

Stem cuttings of litchi root with ease with the aid of auxins in sand, sand-leaf mould mixture or well rotten compost where aeration, moisture, sanitation and pH are favourable. Cuttings from two-year old young vigorous plants treated for 24 hours with 0.2% aqueous solutions of IAA or 0.05% to 0.09% solution of IBA have succeeded well.

For stooling 3-year old plants raised at close spacing are cut back to 25 cm stump in February, earthed up and allowed to root. These may be detached and planted in September. Apical grafting has also been successful in India, Israel, China and South Africa. Approach grafting and bud grafting have also been tried but to a small extent only.

Under favourable conditions the distance between trees should be 10 metres but where summers are hot with winds 7-8 metres is considered sufficient and profitable.

Manuring

Each litchi plant is supplied with about 250 kg of nightsoil each year in China. The bearing capacity and quality of fruits can be maintained only through proper supply of organic manure. A mixture of castor cake, neem cake, bone meal and wood ash in the ratio of 2 : 2 : 3 : 4 has been recommended with one unit equivalent to 1 kg. Two kg of zinc sulphate and a like quantity of hydrated lime in 400 litres of water fed through foliage shakes off the effects of short availability of zinc. The young plants do well with a supply of farm yard manure and bone-meal in proportion of 30 : 2. Under U.P. conditions each plant needs 100 gm (N), 50 gm (P_2O_5) and 100 gm of K_2O every year and for every rise in the age of the plant by one year this dose be added. Likewise 15 kg of compost for every one-year age of the plant is given per plant. Manuring may be done twice a year, *i.e.*, Dec.-Jan. and June-July. In order to ward off the deficiency diseases spray a solution of 30 gm slaked lime with 10 gms of boric acid, 10 gm copper sulphate, 25 gm manganese sulphate, 25 gms zinc sulphate and 1 gm ammonium molybdate in December and February may be sprayed.

It needs copious irrigation during summers and interculture operations for better production. Papaya, Phalsa and Pineapple may be used as filler trees. Evening irrigations and shade help to ward off cold or frost injury. Two irrigations a week is considered necessary during the rainless period of February to June. Irrigation may be given by flow or furrow method to the established trees and by basin method to the young plantations. Overhead irrigation is not harmful. Occassional shallow tillage to keep off weeds and the growing of cover or intercrops have helps litchi production.

Pruning

No major pruning is done, whenever done pruning aims at providing good frame to the young plant, increase bearing area during early years, reduce pest and disease incidence and damage caused by these improve penetration of sprays into the canopy besides facilitating access to routing cultural practices. Later snipping of old branches encourages new growth which alone bears fruits. When the fruits are picked up alongwith the shoots little snippings is necessary. Excessive growth of the roots or tops must be checked by pruning these. Heavy pruning becomes uneconomical and is not recommended unless the tree has become old and the fruit size reduced considerably. Thinning of branches is desirable to reduce the crown.

Harvesting

Plants developed by the layering method bear fruits within 3-5 years of age in the North and about 6-7 years in the South. Average yield of a full grown tree is estimated to be 80-150 kg per year. The fruit is harvested in bunches along with a portion of the branch with a few leaves. Picking of individual fruits is harmful since the skin at the stem end gets ruptured and initiates rotting. The mature fruit has flattened tubercles and smooth epicarp of bright reddish colour. Unless the fruits need storage or shipment to distant places the development of colour indicates the proper stage of maturity and harvest. In case transport is essential and local consumption be not possible, fruits are harvested as soon as the reddish hue begins to appear. Fruits must never be harvested soon after the rains when still wet. These should be kept in cool dry and well ventilated place. Non-metallic fungicides may be sprayed to cut down waste by 50%. Refrigeration at 1-4.5°C preserves litchi fruits for about three months.

Marketing and Yield

Fruits removed from the plant with shoots keep them in good condition for 2-3 weeks. Average yield 4,000 to 5,000 fruits per tree. It may be as high as 10,000. A litchi tree yields 100-300 kg under different growing conditions. Fruits when packed for transport must be kept compact with proper air circulation. China exports litchi nut which is the dried fruit throwing possibilities of yet another method of marketing in regions where the harvests are in excess of the seasonal demands.

Cultivars

In Bihar litchi cultivars grown are China, Deshi, Dehra Rose, Purbi, Bedana, Mclean and Muzaffarpur. The important cultivars of western U.P. including Dehradun are Early Seedless, Late Seedless, Early Large Red, Late Large Red, Calcuttia, Rose Scented, Khatti and Gulabi. Of these, Calcuttia and Seedless are of good quality and also profuse bearers.

Punjab and Haryana grow the cultivars : Saharanpur, Dehradun, Muzaffarpur, Rose scented, Calcuttia, Seedless late. In West Bengal the cultivars Bombai, Elachi, China, Bedana and Bengal are common.

Bengal. Probably a chance seedling of 'Purbi'. Trees vigorous and spreading with large clusters of 50 or more large bright red fruits (av. wt. 23 – 27 g) having distinct pointed tips. Flesh is soft, sweet and moderately juicy. Common in Australia.

Bombai. Most important commercial cultivar of West Bengal. Tree vigorous, 6-7 m height, 7-8 m spread, fruit in large clusters, heart shaped. Each fruit has another tiny underdeveloped fruit attached to the fruit stalk. Fresh sweet soft, white and juicy with pleasant flavour.

Brewster. Florida's exclusive cultivar. Trees small, foliage dense, lenticles on branches. Heart shaped fruits of bright pinkish red colour are borne in small loose clusters. Sweet when fully ripe, juicy and fragrant with small seeds.

Calcuttia. Large, round, pointed, glowing scarlet, heavy bearer, big bunches, very sweet, juicy, no splitting, suitable even for dry climates, commercially successful.

China. Trees semi-dwarf, small leaves, large globose fruits, very sweet, soft, juicy with pleasant flavour. Small shining smooth seeds. Less susceptible to sun-burn and cracking. Important cultivar of India.

Dehradun. Large, round, scarlet, big bunches, sweet, prolific, commercial cultivar.

Deshi. Early good cultivar common in West Bengal and Bihar. Fruits small, light red coloured with deep tubercles at ripening and juicy. Moderate but regular bearer.

Early Large Seedless. Large, red, sweet and prolific.

Early Seedless. Early Bedana : very sweet, fragrant, beautiful scarlet colour, very small seed.

Tai so. Trees vigorous and spreading. Fruits large with thin skin of bright red colour becoming dull red at maturity. Fruits are acidic-sweet in taste when immature. The sweetness increases with maturity. It is widely grown in Australia, known as Mauritius in South Africa.

Late Large Red. Large, round, red sweet.

Mclean. Bears very big bunches, large, round, glowing scarlet colour, very sweet, fragrant, excellent.

Muzaffarpur. Ripe fruits are available in 1st week of May in Eastern India and in middle of June in Northern India. Deep orange fruits of moderate size, sweet, moderately juicy with good flavour.

Rose Scented. Rosy flavour in aril. Trees vigorous, Fruits of varying shapes, deep rose pink in colour. Pulp greyish white, soft, very sweet suceptible to sun-burn and cracking. Seeds small. Important cultivar of India.

Seedless late. Very large, pointed, growing scarlet, very sweet, delicious, rose scented, big bunches. Very small seed, splendid cultivar for the market.

Shahi. Cultivar common in Bihar. Fruits large, oval, maturing in May. Heavy regular bearer. Suitable for canning.

Swarn Roopa. This non-cracking seedless strain has been evolved at Central Horticultural Testing Centre, Ranchi (Jharkhand). The fruits are dark-pink in colour with sweet and juicy pulp. It is a good yielder and is considered suitable for Chhotanagpur region.

Wai Chece. Trees of medium size, thick branches and compact foliage. Fruits small, rounded, deep red coloured. Flesh sweet soft and very juicy. Commonly grown in Australia and China.

Among the early cultivars Muzaffarpur, Early seedless, or Early Bedana, Early large red are common. The mid-early cultivars are Dehradun, Melean, Rose scented and Late large red. Calcuttia and late seedless are late cultivars.

Insect Pests and Diseases

Leaf damaging insect have been placed in three categories :

(*a*) *Leaf weavil.* These damage the leaves.

(*b*) *Leaf rolling caterpillar* (*Argyroploce leucaspis*). The larva rolls the leaves and feeds from inside and causes damage. The adult does not damage the crop. The pest is sporadic in Uttar Pradesh. Spray Quinolphos @ 0.03% or Carbaryl @ 0.2% at the time of emergence of new leaves.

(*c*) *Eriophyd mite (Aceria litchi* or Eriophy litchi). The nymph as well as the adult suck the sap from the under surface of the leaves and cause a formation of a brown velvety growth. The infected leaves first show white spots which later dry up ; the leaves become twisted. Curl and thickened. The severity of the attack is more during dry weather. The attack of the mite is during March-April and October-November. The adult mite is not visible with naked eye.

For control drenching of the stem and branches in dry weather, pruning and burning the infected shoots after the harvest of the fruits are mechanical measures. Spray Dechofol @ 0.02% in January and Dimethoate @ 0.05% in February according to the need.

Leaf miner (Acrocercops hierocosma). It mines the apical portion of the new flush and the midrib of the leaves. The toll is heavier with even a light shower during September-October. This insect is also known to cause leaf curl. Spray Endosulfan @ 0.07% or carbaryl @ 0.01%.

Fruit Borer. The fruits start rotting as the caterpillars bore into the litchi fruits. The damage estimated is within 20% in May and as much as 60-70% in June. It is necessary to collect all the rotten fruits falling on the ground and destroy preferably by burning. Two to three sprays of NSKE @ 40 gm/litre water during the months of May-June at 10-day interval according to the need. Alternatively, Phosalone, Fenthion or Fenvalerate may be used as spray (0.05%) at 10-day interval.

Litchi bug. The nymph and adult feed on the young leaves, twigs and fruits during March-April. When the attack is mild methyl-o-dimeton 0.05% may be sprayed in April to control the pest. At the time of the maximum emergence of new flush in July-August and the attack is severe monocrotophos 0.05% solution may be used to control.

Mealy bug (Droschica mangifereae). Bugs suck sap from the panicles. Summer ploughing, clean orchard, and alkaline banding of the tree trunk has been recommended for control. More details of control measures are available in the chapter on Mango.

Scale insects. These are also sucking insects. Phenthoate or chloropyriphos (0.05%) may be used for control.

Bark eating caterpillar (Inderbela quadrinoata Walk). These make holes at the joint of branches between May-September. Remove the dead shoots and the fallings on the ground and destroy. Spray chloropyriphos @ 0.05% in October. Dichlorovas (0.1%) may be injected during February-March at the rate of 5 ml in the holes made by the borer.

Diseases

Occasionally leaf spot and fruit rot may be noticed in litchi. The lower surface of the young leaves are found to be affected by leaf spot disease in July-December. Infection on the upper surface of the leaves is apparent only after injury later. Spray of copper oxychloride @ 3 gm/l water controls the disease.

Fruit rot (Aspergillus flavous, A. niger and A. nidulans). The ripe fruits when placed in unhygenic conditions show sign of rot. Captan @ 0.2% or copper oxychloride @ 0.1% spray before the harvest minimises the incidence of the disease.

Disorders. Sunburn and skin cracking of developing fruits are two serious problems when the humidity is low, temperature high and soil moisture limiting. It is held that a temperature of 38°C and atmospheric humidity lower than 60% favoured the cracking of fruits. The control measures suggested are (*i*) frequent irrigation during the critical period of aril growth (*ii*) spray of zinc sulphate (1.5% solution) every seven days from the pea-size of fruit upto harvest, and (*iii*) Spray Gibberellic acid (40 ppm) or ethephon 1 to 10%.

Physiological Disorders

Nutritional deficiencies cause several disorders, some of which are given below :

Essential elements	Disorders
Boron	Reduction in fruit size.
Copper	Die-back, small fruit, diminished flesh recovery.
Iron	Leaf yellowing, die-back.
Zinc	Browning of leaflets, small leaflets and small fruit.
Nitrogen	Yellowing of old leaves, stunted growth, poor flowering and small fruits.
Phosphorus	Tip and marginal necrosis of old leaves, leaf curl, leaf fall, poor fruit set, stunted growth.
Calcium	Death of growing points.
Magnesium	Leaves small, leaf necrosis, leaf drop, poor flowering.

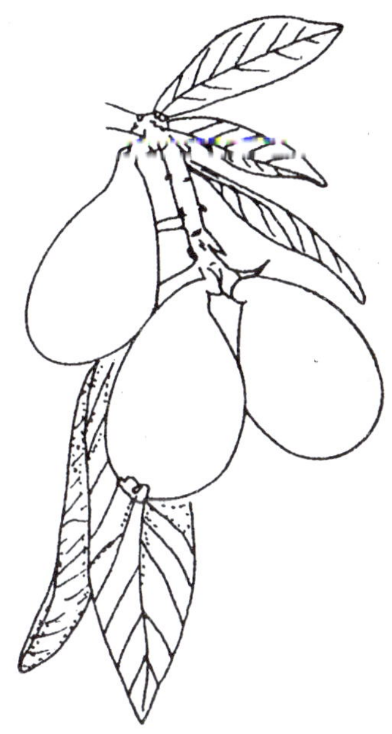

CHAPTER

17

LOQUAT

(*Eriobotrya Japonica* Lindl.)

Rosaceae

Origin and Distribution

The word Eriobotrya derived from two Greek words 'erion' and 'botrys' meaning wool and cluster respectively stands for wooly inflorescence. Loquat or Japanese plum, an evergreen sub-stropical tree orginated in Japan and China is cultivated in the subtropical regions of the world and is of commercial importance in Australia, South Africa, U.S.A., Japan, China and India. In India, loquat cultivation was first started in Saharanpur (U.P.) and now it is grown in Dehradun (Uttranchal Pradesh), Muzaffarpur, Meerut, Kanpur, Bareilly, Farukhabad, Punjab, Delhi, Assam, Gujarat, and hills of South India covering 1,500 hectares. It was introduced in India under the name of 'Japan plum' or 'Japanese medlar'.

Climate and Soil

A mild subtropcal climate with 75-100 cm of well distributed annual rainfall is the prime need of the loquat. It can be grown at an altitude of 2,000 metres and is well adapted to mediterranean climate. The trees can stand frost and tropical heat to considerable extent but cool and foggy weather during ripening robs off the sweetness and flavour of the fruit. Loquat, though not exacting on soil prefers a well drained light loam. Deep soil free of gravels but with copious water supply and good drainage is suitable for its cultivation. Excessive lime, hard pan and gravel in sub-soil do not suit loquat.

Propagation

Seeds should be sown immediately after extraction from the fruit since delays lower germination capacity of the seeds. Seed propagation though common in certain parts of India has the disadvantage of slow growth besides providing plants not true to type. Vegetatively loquat may be propagated by shield-budding, cutting, inarching or grafting. The buds should be taken from three month old branches and the best time for budding is January to February. Propagation by cutting is difficult but layering with 3% NAA is satisfactory. Inarching may also be done during the month of February or during the monsoon. Seedling loquat, apple, quince and pear may be used successfully for rootstock. Top working with promising or desirable cultivars gives economic returns.

Planting and Manuring

Loquat trees should be spaced 8 m on all sides and manured. Wind breaks are very useful to protect the plants from high winds. The best time for planting loquat is from September to October, when the possibility of damage due to heavy rains is minimum. Green manuring of the orchard should be done. A 60 kg manure-mixture of well rotten FYM, bonemeal and wood ash in the ratio of 100 : 5 : 10 should be applied to each grown up tree with early rains or in October with irrigation. The younger trees should be supplied half the quantity or even less according to the age of the plant. As the plant begins to bear these should be given more manure-mixture twice a year.

Recommended Level of Manures and Fertilizers for Loquat Plants

Age of grafts (years)	F.Y.M. (kg)	Fertilizers (gm)		
		N	P	K
1	5	25	25	50
2	10	50	50	100
3	15	75	75	150
4	20	100	100	200
5	25	125	125	250
6	30	150	150	300
7	35	175	175	350
8	40	200	200	400
9	45	225	225	450
10	50	250	250	500

Note. With advance in age the quantity recommended should be applied every year. Farm yard manure is generally mixed in the soil towards middle of April, after harvest, whereas fertilizers (N, P & K) applied in July-August near the drip of trees.

Culture

Inter-cropping with leguminous crops, vegetables, papaya etc. and mulching is practised. Cultural method must necessarily encourage mid-season flowering which when very energetic makes the plant lose much of its nutrients in a wasteful process. Pinching of the apical bud in June results in vegetative growth during the rains and there is vigorous flowering in the second flush leading to fruiting of good quality.

Pollination

While planting, care must be taken to plant varieties compatible with each other. Orchards cannot succeed if only self-incompatible cultivars are grown, *viz.,* Golden Yellow, Pale Yellow, Large Agra and Tanaka. Tanaka should be planted with California Advance and Golden Yellow also with California Advance which is a good pollinator.

Irrigation

The loquat needs copious watering but cannot stand water stagnation. The trees do not require irrigation during the rainy period but during long-breaks in the monsoon, in winters and summers, they should be irrigated at intervals of a week to a fortnight according to the severity of the weather.

Pruning

Loquat produces flowers on current year's growth. Annual pruning may be done to regulate bearing. The flowers appearing in October-November give the best crop and therefore the branches must be snipped off about 5 cm below the twigs towards the end of the May. Newly planted grafts need no pruning. Fruit thinning encourages development of large fruits, every season, and may be done early in the fruiting season.

Fruit

Round, oval or pyriform fruit with tough plum like skin of Pale Yellow or Golden Yellow according to the cultivar may appear in loose clusters. The flesh is firm, creamy yellow with abundance of juice. The fruits have mild, subacid and sweet taste, enriched by an apple-like flavour with 3-6 hard brownish, shining and oblong seeds. Like the pear, apple and quince it is also a pome.

Harvesting and Marketing

Plants start fruiting career from the third year and the fruit yield reaches peak value at the age of 15-20 years. It is advisable to start planting fresh trees when the orchard is about 25 years' old and remove the older plants that have lived for 30 years. Average yield is 45-50 kg per tree. Only the fully mature fruits should be picked up.

Cultivars

The cultivars have been grouped on the basis of their origin and time of maturity. The Chinese group has large fruits, pyriform in shape and of deep orange colour. The Japanese group has small fruits which are slender, and light coloured.

Loquat cultivars can be roughly placed in three categories : Early *viz.,* Golden yellow, Improved Gloden Yellow, Matchless ; Second early or midlate cultivars like Pale Yellow. Improved Pale Yellow and the late cultivars are represented by Thames Pride, California advance, Tanaka.

The commercially successful loquat cultivars adapted to Indian conditions of climate and soil are given below.

California advance. It is a late cultivar of good quality with large fruits. It is a shy bearer. It ripens from the middle of April. It is commercially grown in U.P. and Bihar and has been introduced from Australia. Fruits medium, pale yellow, thick pulp, excellent taste, sub-acid, few seeds.

Fire ball. The fruit is round, deep red in colour with fairly good size and quality. It is mid-season cultivar and is popular in U.P.

Golden Yellow. Fruits golden yellow, pulp pale orange, soft, mild taste sub-acid. Chief commercial early cultivar for Bihar, Jammu, Tamil Nadu and Uttar Pradesh. Fruits ripen in the second half of February.

Improved Gloden Yellow. It is a cultivar quite similar to the golden yellow and is grown in U.P. The fruit is double in size, orange yellow, thick pulp, mild sub-acid taste.

Thames pride. It is grown in U.P. and Tamil Nadu on a large scale. Fruits ripen in April. Fruits medium, deep yellow, pulp medium, pale orange, taste mild sub-acid, moderately seeded.

Improved pale yellow. This is a late cultivar and gets ready in the second fortnight of March. Popular in U.P. and Bihar and has been introduced from Japan.

Large Agra. It is a mild to late season cultivar of fairly good quality, ripening from 1st week of March. It is widely grown in U.P. and Bihar. Fruits medium, ovate globose, yellow.

Tanaka. A Japanese cultivar grown in Bihar, Jammu and U.P. The taste is good, flesh solid, but is a light bearing late cultivar and gets ready in middle of April.

Marchless. A good early variety.

Insect Pests and Diseases

Loquat, in India, is affected by the following pests and diseases :

Bark-Eating Caterpillar (Inderbela quadrinoata Walk*).* It attacks a very large number of trees including mango, litchi and guava. It damages the bark and destroys the conducting tissue. The tree trunks as well as the branches near the holes should be sprayed with 1% parathion emulsion once in November-December, plugging the holes with cotton wool soaked in carbon bisulphide and chloroform kills the caterpillar inside. Inject kerosene oil or creosote + chloroform or petrol into the holes and plug to kill these.

Trips. (*Heliothrips* sp.). These attack the flowers sometimes but the damage is not much. These can be controlled by spray of Dimethoate (0.05%) during the initiation of flowering.

Fruit fly (*Bactrocera dorsalis*). It causes deformation, rotting and fall of fruits. Spray Dimethoate 30 EC @ 1 ml/l water or mixture of malathion 50 EC @ 1 ml/l water and molasses @ 10 gm/l.

Scales (*Elecanium tiliai* and *Parlatonia oleae*). The insects suck the sap from the lower surface of leaves and shoots affecting the vitality of the plants and their fruiting capacity. First step to negate the challenge is to prune and burn the affected portions of all shoots. Spray of chlorpyriphos @ 0.05% to check the activity of the insects.

Die-back. It is caused by a species of *Macrophoma* and the infection is established presumably through surface of injured parts. A mixture of cold solution of potassium dichromate and 6% copper sulphate may be used to treat the branches to check die-back. Use of chaubattia paste in lanolin containing red lead 54 gm, copper carbonate 54 gm and lanolin 65 gm makes the plant free of die-back in a large number of cases.

Stem brown (Botryosphaeria ribis Gross and Dugg). This is not of much significance and the losses caused are small.

Collar rot and *root rot* of loquat is a disease caused by *Diplodia natelensis.* The affected portions should be scrapped with scalpel and a paste of copper oxychloride and lime with water applied. Smear a paste of phytolan 225 gm and lime 112 gm in 2.25 litres of water as the second alternative.

Composition of Fruits

Sugars 8-13%, acids 0.3-1.75% and pectin 0.3-0.5%. Jelly is made from unripe or ripe fruit juice. The seeds are removed and sugar added in ratio of 4 : 4 : : juice : sugar by volume with about 10 gm of citric acid for each cup of juice, and bottled.

□□□

CHAPTER

18

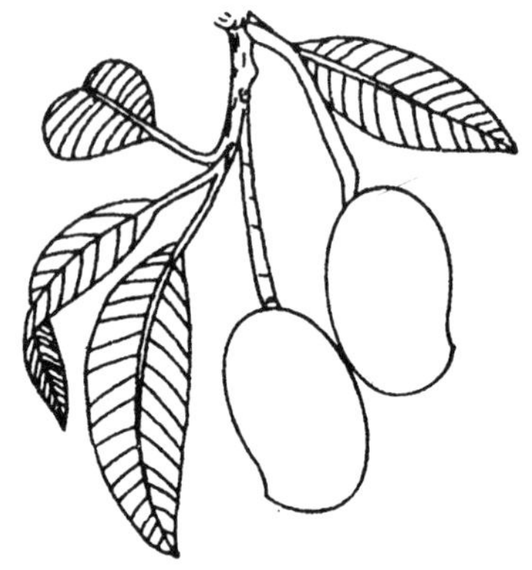

MANGO

(*Mangifera indica* L.)

Anacardiaceae

Mango (H. *aam*) is easily the most outstanding among the tropical fruits of India. It has great adaptability and thrives in a wide range of climatic and soil conditions. It has relatively hardy nature, low cost of culture and maintenance and above all, is the choicest of Indian table fruits having the premier place in the country. Besides the species *M. indica* which is found throughout the tropics of old world. *M. sylvatica* is also found in India. Mango has been cultivated in India since antiquity records show that Alexander the Great (327 B.C.), Hiuntsang (632-645 A.D.), Ibn Haukal (902-908 A.D.) and Ibn Batuta (1325-1349 A.D.) have testified to its cultivation during the time of their visit to India. Fruits are rich in Vitamin A and also C besides being a source of calcium and phosphorus. The area under cultivation in India is roughly 8,00,000 hectares.

Origin

Mango is believed to be indigenous to South Asia specially the base of the Malayan Archipelago although De Candolle, Popenoe, Vavilov and Mukherjee opine that mango originated in the Indo-Burman region. The introduction of mango into certain parts of the Asiatic Archipelago having taken place only in recent times goes in favour of the latter view. Mango is intimately connected with folklore and religious rites in India. Its cultivation for over four numerous wild and cultivated varieties, physiological archeological, and other literary evidences, the existence of several ancient names (including Sanskrit) phyto-geographical distribution, phylogenetic taxonomy of all species, relation to climatology and geology as well as abundance in the gardens of Bengal and Deccan, since antiquity, favour the Indo- Burman origin. India contributes 50-60% to the world's production of mango. U.P. has the largest area under mango among the states of India followed by Andhra Pradesh, Kerala, W. Bengal etc.

Soil and Climate

Mango trees grow well in a variety of soils from the well drained laterite soils to medium black soils. In fact it is least fastidious in its soil requirements. Soils which do not allow proper development of roots and growth give poor returns. Alkali, ill-drained soils with rocky subsoils are considered unsuitable for mango. A well drained fairly deep loamy soil is considered best for mango. The fruit produced on trees planted on sandy soils are of inferior quality. This essentially tropical fruit tree cannot stand severe cold and frost. Frost during flowering period (December or March) prove disastrous and warm to hot weather is considered good for the development of the fruit. Average range of temperature (between 24-27°C) is considered optimum for growth of the plant as well as fruit yield.

Rainfall during blossoming is highly detrimental, since it injures the unprotruded pistil, washes away the pollen, provides conditions that encourage powdery mildew and other diseases causing low yields and often crop failure. Excessive dew also washes away the pollen grain so that flowers remain unfertilized. Cloudy weather, during the periods of flowering and fruit setting, encourages fungus and jassid attack. A mango tree can give good performance in some arid regions with an annual rainfall of 60 to 80 cm. Late varieties cannot be successful in areas of high and early precipitation since it affects fruit drop and damages mature fruits. In fact mango trees can be successful if the annual rainfall ranges between 25 to 125 cm per annum provided there is no waterlogging and the rain does not take place at the time of flowering or fruit set. In any case the soils must have enough moisture in the root zone to satisfy the moisture needs of the tree. Commercial growing of mango is best done in areas where a humid rainy season with 35-70 cm of annual rainfall alternates with a well marked dry season during the blossoming period. Commercial growing of mango is possible upto an altitude of 600 metres provided there are no frosts although generally and latitude of about 150 metres is considered good. Within certain limits each increase of altitude by 30 metres retards flowering by one day. With each degree of latitude south or north of tropics a delay of four days is noticed.

Wind Velocity

Winds affect mango trees in several ways. Desiccating winds of high velocity deplete soils of necessary moisture and retard optimum plant growth.

Propagation

The mango may be propagated sexually by seeds which may be either mono or polyembryonic, and asexually by different types of graftage viz. (*i*) the scion is not severed from the mother plant until its union with stock is complete (inarching, tongue grafting, saddle grafting and root grafting), (*ii*) the scion is first severed from mother plant and is united with stock afterwards (rind or crown grafting, cleft or wedge grafting, whip grafting, side grafting, notch grafting) and budding of several types *i.e.* shield, Forkert, patch besides, (*iii*) rootage which includes propagation either by cutting or layering. Vineer splice, side and epicotyl grafting is also used for propagating commercial mango cultivars. Seed propagation, the nature's method has produced some of the choicest cultivars, most of which originated as chance seedlings, these need vegetative multiplication for large scale commercial plantations. Seeds were used for propagation and introduction of a variety in any country freely since remote antiquity but vegetative methods of propagation have advantages chiefly in respect of early fruiting, uniformity in bearing, size of fruits productiveness, faithful preservation of the characters of the plant. When propagated by seeds, monoembryonic varieties are never true to type, and the majority of them bear fruits of inferior quality. Old, diseased and inferior trees may be rejuvenated by asexual method of propagation (details of the types of grafting may be seen elsewhere in the books). Spacing (10 m × 10 m) : 100 grafts per ha.

Raising and Care of the Stock

Stocks obtained from fruits of good quality and flavour are used for raising the seedlings. Disease-free healthy stocks are planted in pots with drainage hole containing soil-manure mixture in the ratio of 1 : 1 or directly in the field under the share of a tree in a manner that the competition between the seedling and the shading tree for soil nutrients is the minimum possible. Seeds germinate in about 3 weeks. When the seedling attains the desired height and girth it may be transplanted in desired place directly or used as stock to support good variety scion for the production of desired quality of plants by artificial means. In the process of transplanting, care is taken not to injure the roots at any stage, this may be generally achieved by irrigating the soils well prior to the operation. Potted seedlings are preferably placed in shade and occasionally fed on liquid manure which may be prepared in any of the following ways :

(*i*) A handful of sulphate of ammonia dissolved in 9-10 litres of water.

(*ii*) One part of cattle urine diluted with 3 parts of water.

(*iii*) One part of cow dung with 4-8 parts of water.

About 1/2 litre of any one of above liquid manures is considered enough for each pot for one application. The application may be done a day after the light stirring of the soil. Copious watering is done once a week so that the seedlings in the soil get fully saturated with water to keep them free from water shortage for a span of 8-10 days depending on the climatic conditions. This would cut down waste of water considerably. Occasionally pruning of the stock seedlings have to be done to remove all side shoots arising in each flush. The pots with the seedlings should be lifted and rotated to prevent the roots from establishing contacts with the outside soil through the hole of the pots and getting sunshine on all sides.

Selection of Parent Tree and Scion

Young, vigorous trees of desirable quality should be selected for scion. In order to facilitate in arching the trees should be so trained as to help the process by ready availability of the shoots for the purpose as near the desired level as possible. The shoots to be used as scion should be of the same age and diameter as the stock, variations of over 1.8 cm in the diameter of the two make them uncongenial partners for satisfactory union and results.

Vigorous seedlings 45-60 cm in height selected to serve as stock for inarching or attached grafting are potted in 30 cm pots brought close to the tree which has to serve as scion source preferably at the beginning of rainy season. In Northern India inarching is mostly done during July-August and February-March. The age of the stock for Marching considered as best in between 6-15 months by western workers and also in Hawaii though in India 3-month old stocks have proved better. It seems to be a function of climatic conditions. Considering vigour of the plant and their mutual adaptability for the purpose some workers from India have even suggested 18-24 months as the best age of inarching.

A thin slice of bark of about 6-8 cm length and about 0.75 cm thickness is removed with a sharp knife from stock at a height of about 20 cm above the ground level. An identical cut is made on the scion. As it is done cambium layers of both the stock and scions are exposed. The two cuts so made on the stock and the scion are fixed together and no air is allowed to pass through union. The place of union is tied firmly with good materials like moist raffia fibre, waxed cloth, paraffined cloth or adhesive rubber tape and the bandage is made air tight.

As the cambiums of 'stock' and of 'scion', 'knit' the scion is separated from the parent tree by cutting it, in two instalments, below the point of union. It generally takes 45 to 60 days. As soon as the scion is severed from the parent plant, the grafted seedling may be directly planted in their permanent position or allowed to remain in the shade with regular watering etc. for about a year before transplanting in the desired places.

Any flower buds formed on such grafts in the nursery should always be nipped off. It is desirable that the grafts be topped about 60 cm above the union for training. The cleft or wedge methods of grafting and Forkert method of budding are also employed for vegetative propagation of mango in Sri Lanka. Occasionally seedlings of 3-4 years of age but generally 1 to 2 year old seedlings are used for the purpose. The potted seedlings are kept near the scion parent and topped at a height of 25 to 60 cm from soil level of the stock. The scion shoot is attached and tied securely with its exposed portion resting face to face on the exposed part of root stock. Crown and cleft graftings have also been practised for top working of old trees of inferior varieties. Although shield, Forkert and modified forkert methods of grafting have met with success their commercial possibilities yet remian to be established.

Budding with mango is difficult due to (*i*) inseparability of bark from the wood unless mature, (*ii*) easy coagulating of latex exudating from the cut end so that even a little delay in inserting the bud prevents attachment, at above all (*iii*) the sensitivity of the mango bud to seasonal variations. Budding though economical is only partly successful in areas with high humidity. Air layering, stooling and cutting have also been reported to be used for vegetative reproduction of mango.

Preparatory Tillage

Mango plantations are made on the selected site after cleaning the land of all wild growth, weeds etc. The seedlings are planted generally during the rains in pits made in well levelled areas. All growth near the pits is kept clean to provide open space for the growing tree. It is necessary that the land must be well drained.

Mango plantations may be (*i*) rectangular, (*ii*) square, (*iii*) quincunx (*iv*) hexagonal and (*v*) contour. The square system of planting is more common. The mango plants are spaced 12-14 metres in rich soils and about 10 metres in poorer soils according to the spreading nature of the cultivar. The 1 × 1 × 1 m pits are made during the hot weather. It is filled with a mixture of well rotten F.Y.M., compost, superphosphate, ash etc. to half of its depths followed by top soil.

Mango planting can be done during July-September or February-March. Monsoon period gives better results. As much soil is taken as necessary to accommodate the earth balls round the roots of the young plants and then the ball with the plant is put in the pit. The remaining space is then filled with moist, surface soil and trampled lightly but firmly taking care not to break the ball. After planting watering is a must. At planting the graft union should be about 20-25 cm above the ground level. Young plants need protection from desiccating winds in summer and frost in winters which is provided by thatch-screen made of grasses, leaves etc.

The depth and frequency of irrigation varies with the cultivar, cultural treatments and environmental conditions so that rapid and vigorous growth of the tree is encouraged. Generally no irrigation is necessary during the monsoon months except when long spells of drought occur. For the remaining part of the year young plants of less than two years need irrigation every third or fourth day. Plants of 2-5 years need irrigation every 10-12 days in the winters and once a week during the summers.

Trees that attain the age of bearing need not be given profuse irrigation. Adequate irrigation during the period of fruit set is considered helpful in checking fruit crop. Some workers have reported vigorous bloom by irrigating with the tree with the pre-bloom fertilization. Irrigation should not only be timed with the needs of the crop but also the variations of the season. While the regions with heavy rainfall and high humidity may not be given any irrigation those that are comparatively dry necessarily benefit by timely irrigation chiefly through the basin and the ring method.

Manuring of mango trees by bone meal, F.Y.M., slaughter house-waste, fish manure etc., to fulfil the needs of nitrogen, phosphorus and potassium is not uncommon. As the plants attain bearing stage their need for nitrogen and phosphoric acid becomes more acute and may be supplied in the form of sulphate of ammonia, bone meal, F.Y.M., niciphos etc.

For the sowing of stones a mixture of 8 parts of leaf mould, 8 parts of good soil and 1 part of sand is considered best. After 2-3 months fortnightly application of liquid manure may be given. Liquid manure may be prepared by dissolving a handful or sulphate of ammonia in 12 litres of water or by mixing urine or fresh dung of cattle with water in the proportion of 1 : 3 and 1 : 4 respectively. The nutrient solution may also consist of KNO_3 (275 gm), K_2SO_4 (120 gm), acid phosphate (350 gm) in 100 litres of water and supplied at an interval of 10-15 days. The manuring of pits in which the plantings are done is considered more advantageous than the after-planting application. The pre-plantation manuring schedule for the pits has been recommended as 50 kg of F.Y.M. + 2 kg of bone meal + 50 kg of wood ash. In areas where the white ant menace be a problem the lower portion of the pit should be filled with a mixure of 15 kg of F.Y.M., 6-15 kg of bone meal and 1-2 kg of neem oil cake and 50 gm of 10% BHC. Use of aldarin dust @ 150 g/plant is recommended for the control of termite menace.

One year old plant is recommended an annual dose of NPK to the tune of 100, 50, 100 gm. This quantity is increased as many times as the no. of years (age of the plant) up to the 10th year when the dose works out to be 1 kg, 0.5 kg and 1 kg for N, P and K respectively. Mango plants do well when given neem cake 2 kg along with 100 kg of well rotten compost manure. This quantity is added every year to maintain the plant's health and fruit yield.

The bearing trees have been recommended supply of N (725 g), P_2O_5 (180 g) and K_2O (670 g)/tree per annum. In the 'on' years the application of 90.8 kg (FYM), 1.82 kg (castor cake), 4.54 kg (bone meal), 2.5 kg (sulphate of ammonia) and 13.60 kg of wood ash has proved profitable. Foliar spray of 0.3% urea per year after pruning, besides 1 kg (N), 0.5 kg (P_2O_5) and 1 kg (K_2O) increased the yield. Manuring schedule must be determined with regard to the soil of the orchard and the time of application according to the stage of development of the plant.

Intercropping with seaonal vegetables and small short-term crops may be done, care being taken that the crops are not very close to the young mango plants. Some of the crops which are generally used are sannhemp, guar, *kulthi*, groundnut, pea, bean, brinjal, bhindi, cabbage, knol-khol, turnip, onion, etc. Cover crops may be grown for green manuring purposes only in bearing mango orchards. Proper tillage is given on levelled ground where intercropping is possible. Clean cultivation is seldom followed.

High density planting. This technology developed at Pantnagar University for getting high yield is suitable for alternate as well as regular bearing types. Amrapali and Dashehari cultivars have been recommended high density planting, in north India. The spacing for Amrapali and Dashehari planting distance recommended are 2.5 × 2.5 m (1600 plants/ha) and 2.5 × 3.0 m (1333 plants/ha) respectively. It is necessary that annual pruning of fruited shoots be done in the last week of June or early July. This be followed by the addition of 50 kg nitrogen/ha to the soil and a spray of 1-2% urea and 0.2% copper oxychloride. Soil drenching with paclobutrazol @ 2 ml/tree in November should be followed by irrigation of the orchard. High density mango orchard are given 100 kg N, 75 kg P_2O_5 and 75 kg K_2O per hactare every year. Total quantity of P and K are applied in December and the nitrogen half dose (50 kg) in January and the remaining nitrogen (50 kg) may be given after harvest.

Top Working

Trees of inferior varieties are top worked with better varieties by cutting off the top of the tree to force new branches. The sap drainers (2-3) are retained till the new grafts have made some growth. The tree is headed back and the tree is covered with gunny bags or white washed with 2.5 kg of lime and 250 gm of $ZnSO_4$ in 8 litres of water. The new shoot may be grafted by venier-grafting or side-grafting. Cleft grafting or crown grafting is also practised.

Top working an old tree in a declining state will serve no purpose unless steps are taken to invigorate the tree. Old mango trees in weak health can also be invigourated by infusing fresh sap of younger seedlings into them. For this many stocks are planted near the old tree and when these have attained the height of 1 m or more from ground level they are headed off, the cut ends are then shaped to wedge and inserted into barks of the old tree and tied up.

Rejuvenation of Old Mango Trees

It is a well known fact that mango trees live for a hundred years or more and grow to huge dimensions. Old trees have a tendency to develop hollows in their trunks and in main branches starting from crotch or the point of bifurcation. Rain water is deposited here and this causes the bark and wood inside it to rot, and as the process continues year after year, large cavities are formed. Similarly when a branch is broken by wind or by other means the untrimmed stub left behind usually rots and dies. Gradually a cavity may be formed at this point too. Old trees in neglected condition with unhealthy branches full of *Loranthus* and stem borers need to be rejuvenated by removing all borers and *Loranthus*. Addition of 10 kg of bone meal, 100 kg of well rotten F.Y.M. and 2.5 kg of oil cakes is recommended for one plant.

Fruit Bud Differentiation and Blossoming

Grafted mango generally bears fruits from the fourth or fifth year of planting. The time of flowering differs in different regions. The mango produces blossoms and bears fruits mostly from terminal buds of its shoots and in the absence of terminal buds from the axillary buds. Dry weather stimulates flowering while cloudy weather or winter rains tend to retard. The maturity and the age of its seasonal vegetative growth (leafy, almost) influence and control flowering in the mango. The number and period when there is flush in the mango trees varies with the region as given below :

Regions	Periods of flush		
	I	II	III
Western India	Feb.-March	March-April or beginning of monsoon	October-November
Bihar	Early spring	April-May	July-August
U.P.	March-April	July-August	–
South India	February-June	October-Nov.	–
Punjab (Full flushes between April and August ; April-May, heaviest)			

Generally eight to ten month-old mature shoots produced in the spring and the early summer that cease growing at least four months prior to blossoming season bear flowers. The spring and summer shoots get sufficient time to grow, rest and mature for the purpose but the later formed shoots, generally, do not bear fruits.

Biennial Bearing

The older trees naturally settle down to a more or less rhythmic habit of producing (1) profuse blossoms and sparse vegetative growth in spring and summer of one year, (2) profuse vegetative growth and spars blossoming in the spring and summer of next year leding to biennial bearing in the mango so that an 'on' year follows an 'off' year. For producing blossoms high C/N ratio is needed and the reverse favour vegetative growth.

Poor flowering in the 'off' year is the result of poor vegetative growth in the 'on' year and the profuse flowering in the 'on' year is the result of profuse vegetative growth in the 'off' year. Thus the tree develops 'on' and 'off' year rhythm which goes on regularly until 'on' year is transformed into and 'off' year. Not all trees of the same cultivar may have the 'off' and 'on' years. In an 'on' year the flowers may be destroyed by natural enemies such as pests and diseases so that the year is 'off' year for all practical purposes. Not all cultivars of mango are generally rigid in their habit of bearing biennially.

Better mango crop can be assured by avoiding crowding of trees, use of superior clones, use of commercial cultivars *i.e.*, Dashehari, Langra, Samar Bahisht, Chausa and Bombay Green in northern India : Himsagar, Langra, Zardalu, Krishnabhog. Gulabkhas and Fazli in estern India ; Alphonso, Rajapuri, Sardar, Pairi, Malkurad, Jamadar, and Kesar in western India : Neelam, Bangalora, Rumani, Boneshan, Swaran Rekha, Mulgoa, Neelum, Bangampalli, Badami and Pairi in the south. Better cultural and handling practice *viz.* training at young age, watering at appropriate stages, supply of nutrition, clean cultivation, fungicidal and insecticidal sprays whenever needed, careful post-harvest handling, including harvesting the fruit with some stalk, delicate handling, cleaning from dust, wrapping in tissue paper and keeping in aerated crates, one layer to be separated from the other by paperscraps, acid test for maturity (sinking of the fruits in water after harvest, immature fruits do not sink in water), no transportation of ripe fruits, addition of N : P_2O_5 : K_2O @ 0.75 : 0.2 : 0.7 kg to bearing trees (organic in October and inorganic later in parts after January) and control of mango hopper by 0.25% DDT or 0.08% malathion and also mango mealy bug by 10% BHC (in soil) in January help considerably in larger production and profit.

Inducing sufficient vegetative growth in the 'on' year by improved cultural practices *viz.* :

1. A moderate supply of nitrogen as sulphate of ammonia but in no case excess of the element is desirable. Regular ploughing in autumn, winter and beginning of monsoon also helps in regular bearing. 8-10 cartloads of compost to an acre and 4-5 kg of sulphate of ammonia per tree is considered desirable. Immediate irrigation after harvest. Although root pruning has been recommended it is hardly practicable in giving results. Deblossoming early in the season induces early growth of the shoots in the same season so that these bear fruits in the next season.

2. Most cultivars give a high yield but fail to bear regularly every year after the age of about 10 years. A year of good crop may follow that of no crop or poor crop. Sometimes there is only one good crop in 3 years (Maharashtra) and at times in 6 years (Bihar).

Physiological factors affecting the nutrition and the metabolic conditions may sometimes be responsible for alternate bearing in mangoes. Several 'off' years following one 'on' year, *i.e.*, normal bearing year may be related to nitrogenous fertilizers which produced a fair amount of vegetative growth in spring and early summer even in the 'on' years to enable fruiting in the otherwise 'off' year. It is not a purely nutritional problem as believed in the past. Shoots arising after September do not mature in time to bear flowers in the following year except in cultivars like Bangalora, Neelum, Suvarnarekha and Baramasi. An early vegetative growth in the 'on' year helps to ward off alternate bearing. The cultivars that produce mostly terminal inflorescence and only a few axillary types of flower cluster are more markedly biennial bearers whereas those cultivars that produce a greater percentage of axillary inflorescence are moderately regular, annual bearing. Unfortunately most of our good desert cultivars are biennial croppers. Rumani, Totapari Red small, Mallika, Ratna, Amrapali, Bangalora and Neelam are notable examples of regular bearing habits. Attempts are being made to evolve cultivars, which possess good and regular yielding capacity, through breeding.

3. Deblossoming suggested as a means to conserve the nutrition in the 'on' year to bear a substantial crop in the otherwise 'off' year has not succeeded.

Ringing of about 15 cm thick branches at least four months before flowering time has also been suggested to increase C/N ratio to force better flowering in terminal shoots. This is not advisable on account of the general health and sanitation of the trees and therefore it cannot be followed by orchardists as a regular practice. Many of the regular bearing cultivars are poor yielders. Culivars like Alampur, Baneshan, Himayuddin, Jahangir, Ratul possess high fruit quality but low yield.

4. The flowering panicle with 100 to 4,000 flowers may develop only one fruit due to the presence of a large number of flowers, in the panicle, with male functional parts only. The percentage of perfect flowers is 1-25 in *Jahangir,* 6-11 in *Alphonso*, 40 in *Dashehari* and 60 in *Langra*; about 90% of these flowers fail to set due to lack of cross pollination. Low fruit set is also a problem in regular bearing cultivars with mixed panicles and bearing more than one crop a year. Heavy fruit drop damage is another cause. This needs a year. Heavy fruit drop damage is another cause. This needs a solution which is possible by breeding superior cultivar.

5. Over-vigorous trees with active vegetative growth may be induced to flower and fruit by exposing to sunlight, withholding irrigation in the dry season and stopping nitrogen supply. Application of common salt is also helpful (2.5 kg/tree) in checking further leafy growth so that the existing leafy growth accumulates carbohydrates for inducing flower bud formation.

Factors generally responsible for alternate or irregular bearing may be (*i*) *Internal factors* : Physiological, nutritional, C/N ratio, sex-ratio hormonal disbalance and hereditary make-up regarding bearing and flowering habit and (*ii*) *External factors :* Light, temperature, moisture, wind, pests and diseases.

Alternate bearing, in all probability, is an inherently controlled phenomenon. Higher level of auxin-like substances and an inhibitor similar to ABA coupled with lower level of gibberellin- like substances are vital for floweing in mango.

The practical measures which may help to overcome alternate bearing may be listed below :

1. Cultural operations to force the shoots to bear flowers in the 'off' year also. Adjusting the fruit-set in 'on' year in such a way that the plant's nutritional status as well as the C : N ratio remains capable of supporting a crop in the following year also to eliminate 'off' year.

2. Deblossoming with 3-chloroisopropyl N-phenyl carbonate @ 250-300 ppm.

3. Ringing, use of Naphthylphalonic acid etc. Training and pruning operations. Rejuvenation of trees. Use of common salt to over-vigorous trees and maintaining proper C : N ratio.

Fruit Drop. Some varieties of mango, Langra, for example, show tendency of fruit drop before these are ripe. It is generally true of younger plants. This wasteful malady can be controlled by encouraging pollination from amenable cultivar by planting such plants in the orchard. When the fruiting has taken place and fruits are held in position fortnightly irrigation is beneficial. Application of adequate amount of proper additives to the plants and two sprays of 10-20 ppm naphthalene acetic acid during April-May is useful.

Harvesting and Ripening of Fruits

Grafted mangoes are usually harvested before these are ripe on trees. Generally the fruits are supposed to be ready for harvesting when a few naturally ripe mangoes fall from the tree. There appears to be two reasons for the harvesting of mango fruits while still green. Storage for a few days after harvest improves sweetness and general quality. **The stage of picking varies considerably in different** cultivars depending upon factors like slight yellowing of rind as in the case of *Alphonso.* In green cultivars the colour changes to oil green while in Pairi and Cowasji Patel the surface of the fruit becomes somewhat uneven.

After the harvest the process of ripening is hastened. The fruit is kept in the room for a day after harvest and thereafter it is ripened between layers of straw, grass, hay etc. Leaves of *Albizzia lebbeck, Butea frondosa,* mango, banana, sugarcane, sawdust and treatment with ethephon quicken the process of ripening. While making use of ethephon the concentration, the temperature and the period of treatment are of importance. The temperature may be kept between 24-28°C and duration should be 5 minutes. The strength of the chemical ethephon should be kept at 500 μl/l or 1000 μl/l for accelerated fruit ripening and improved surface colour. Ripening generally takes about 7 days. The temperature of the fruit rises by about 9-10°C in the stacked state. If the temperature gets more than 35°C spoilage sets in. As soon as the temperature gets to 32°C for few days the fruits should be taken out and kept on top of straw layer to complete the process of ripening. Fruits ripened at 20°C are better than those ripened at 32°C. Calcium carbide kept in a heap of mangoes quickens the ripening process.

Yield

The yield, within genetic limitations, depends on climate, vigour, age as well as 'on' and 'off' year habit. The young trees, in the beginning, bear 15 to 20 fruits per annum which increases to an average of 400 to 600 fruits in the 10th year and to 2,500 fruits per tree after twentieth year. The average of 1,500 fruits per tree is common but 2,000 to 5,000 fruits may be obtained in well looked after trees of 20-year age. A decline in yield is often noticed after the tree has given 40 flushes.

Cultivars

Olour, Goa, Salem, Chandrakaran, Bellary, Bappakai, Kurkkan, Mazagaon, Nileshwar Dwarf and Mylepalium are polyembryonic varieties reported from India where the seed may give rise to several seedlings with independent root systems. Of these only one is the outcome of sexual union of male and female counterparts of the flower. The other plants are true to the parent trees. The monoembryonic plants do need vegetative propagation to have plants true to type. Indian climate seems to restrict polyembryony *viz.* Mulgoa behaves and polyembryonic in USA but is monoembryonic in India. Other polyembryonic cultivars are Binoboy, Carabao, Corazon, Higgins, Cambodiana, Paho and Pahutan which have been reported from Philippines; Apple, Cecil, Florigon, Peach, Saigon and Simond have been reported from Florida while Sabre and Warburg from Israel.

Seed propagation of mangoes, in the past, led to a large number of variations and in many cases the names do not denote any particular characteristic. Some of the common cultivars grown in the country are listed below :

Andhra Pradesh	:	Swarnrekha, Banganapalli, Chinna, Rasam, Cheru, Kurasam, Mulgoa, Borsha.
Bengal	:	Maldah (Bombay Green), Shadwala, Himsagar, Kala Pahar.
Bihar	:	Langra, Zardalu, Gulabkhas, Khasulkhas, Fazli, Mithua, Bathua, Bharat bhog, Sepia, Bombaya, Krishnabhog, Himsagar, Sukul, Hathijhool, Kanchan.
Bombay and Goa	:	Alphonso, Bishop, Malkurad, Fernandin, Cowasji Patel, Pairi or Mulgoa.
Karnataka	:	Badami, Raspari, Totapari.
Punjab	:	Langra and seedling types.
Saurashtra	:	Jamadar, Kesar, Alphonso, Vanraj, Dalimbya, Langra, Rajpuri, Slebhai-Amdi.
South India	:	Neelam, Bangalora (Totapari), Rumani, Mulgoa, Pairi, Alphonso, Himayuddin, Alampur, Baneshan, Jahangir, Raspari.
Uttar Pradesh	:	Fazli, Langra, Dashehari, Khasulkhas, Samar Bahisht, Fajri, Jafrani, Rataul, Taimuriya, Chausa, Raspoonia, Prince, Baramasi, Kishanbhog, Sonatol, Gopal-bhog, Gulabjamun, Gulabkhas, Sepia, Shah Pasand, Zafran, Bombay Green, Bombay Yellow, Safeda Malihabad, Safeda Lucknow, Mithua Ghazipur, Gaurjeet, Himsagar.

Early cultivars like Gourjeet, Bombay green, Bombay yellow, Himsagar and Gopalbhog get ready before mid June. The second earlies, like Dashehari, Langra, Rataul, Zardalu, Krishnabhog get ready between mid June to mid July whereas the late cultivars Samar Bahisht, Chausa, Fazli, Lucknow Safeda, Neelam and Amrapali ripen between July and August.

Alphonso. (Syn. Hafus, Appus, Badami, Gundu, Khader and Patnam Jathi). Originated in Goa, medium ovate oblique, yellow, very good cultivar for the humid coastal climate, excellent fruit and keeping quality, suitable for canning and cold storage. Medium bearing habit.

Bangalora. (Syn. Totapari, Collector, Sundersha, Gilli Mukku and Vadiyamuthi). Greenish, extremely oblique fruit with bright red blush. Quality of the fruit is poor. A heavy and regular bearer with an exceptional keeping quality. Fruit is prone to mango-stone weevil.

Banganapalli. (Syn. Baneshan, Chapta, Chaptai, and Safeda in North India). Large obliquely oval, yellow colour, good quality fruit keeping well. A mid-season heavy bearer, suited to dry areas.

Bombay Green. (Syn. *Maldah* in U.P. and *Saroli* in Delhi). Obliquely elongated fruit of green colour, very sweet with strong flavour. An early cultivar susceptible to malformation. Good bearing and good keeping quality.

Bombay Yellow. Early cultivar known as Maldah is quite successful in U.P. and West Bengal. Good in fruit quality, good bearer, poor keeping quality.

Chausa. Sweetest mango, late, maturing in late July or August. Good bearing after 10-15 years of planting in the orchard. High susceptibility to mango malformation.

Cowasji Patel. Also known as *Mulgoa*. Large, roundish-oblique, yellow fruits of good quality, very suitable for making preserves, good keeping quality, poor yielder. Late, taste sweet, pleasant flavour.

Dashehari. Mid season cultivar of fibreless flesh and delicate taste. The fruit is elongated, dull yellow in colour. Moderate, somewhat regular bearing. Good for canning and cold storage. Good keeping quality. Maturing in June.

Fazli. Late cultivar with large, oblique green fruits, fibrous pulp, medium quality and moderate keeping quality.

Fernandin. Medium large oval yellow fruits, origin in Goa, oldest cultivar in Konkan area. Latest west coast cultivar. Heavy bearer and good quality.

Gulabkhas. Elongated, with red blush and excellent keeping quality. Shrivells in storage (does not rot). Fruit good, bearing habit good, somewhat regular.

Himsagar. Mid season cultivar, ovate fruits of yellow colour and good quality. Fairly heavy bearing, good keeping quality.

Krishnabhog. Mid season, origin in Bengal, roundish-oblique, medium sized yellowish fruits of good keeping quality, heavy bearing.

Langra. (Syn. Banarasi Langra. Hajipur Langra and Hardoi Langra). Greenish, roundish-oval, medium sized, yellow, sweet pulp and thin stone. Thin skin and fruit of good keeping quality. Takes long to come to full bearing but the yield is very high after about 30 years. Alternate bearer, widely favoured in U.P., Punjab, Bihar and on the banks of the river Narbada in Gujarat.

Neelam. (Syn. Kaja Loddu), Small, ovate-oblique, yellow fruits, fair quality, good in storage, heavy yielder, late, regular in bearing, two crops a year. Good taste, acidic blend.

Pairi, (Syn. Peter, Yerra, Goa, Nadusalai, Grape, Peter Pasand, Gohabunder and Raspuri). Attractive medium ovate with prominent red blush, thrives in humid south. Commercially important for Western India. Fruit quality much inferior to Alphonso. Heavy yet alternate bearer with fair keeping quality.

Ratna. Regular, released by hybridising *Neelam* × *Alphonso.* Moderately vigorous, starts bearing in fourth year. Flowering period December to January somet... even February. Fruit matures in May-June.

Rataul. Early cultivar, small but very sweet fruits, low yielder, Gopalbhog and Zardalu are early while Taimuriya is late.

Rani Pasand and Safdar Pasand are better in vitamin C content than Bombay yellow, Mallika, Ratna, Fazli, Chausa and Langra.

Rumani. Apple shaped, yellow colour with red blush on shoulders, heavy yielder, good keeping quality. Moderately good quality when harvested slightly early, remains sour. May give a second small crop.

Samar Bahisht. (Syn. Chausa, Khajri). Late cultivar of soft yellow colour. Large sweet fruit, moderate and alternate bearing, fibrous flesh.

Sukul. Largish, oblong oblique. Yellow fruits. Late cultivar, good keeping quality, the fruit quality is ordinary.

Swarnrekha. (Syn. Chinnaswarnrekha, Sundari or Sindhuri). Medium sized, ovate-oblong green fruit with a prominent red blush on the shoulders, quality fruit of good keeping quality. Early cultivar of coastal Andhra Pradesh. Moderate biennial bearing. Fruits fibrous but taste good.

Mallika. It is a hybrid of the parents *Neelam* × *Dashehari,* evolved at IARI, New Delhi. After 11 years of experimentation it was released in 1971. It is a regular bearer and fruits are of good size, bigger than *Dashehari* with good taste. This hybrid between a regular bearing cultivar and a biennial bearing cultivar overcomes the problem of alternate bearing in mangoes.

Amrapali. It is another promising cultivar, a hybrid of the cross between *Dashehari* × *Neelam* evolved at IARI, New Delhi. Its performance was tested for 14 years before being finally released to the orchardists for planting in 1980. *Amrapali* has very sweet fruits. It is dwarf like *Neelam* attaining only 1.5 to 2 meters height in 10 years. It is high yielding regular bearer. One hectare of area can be sufficient for 1600 plants as against 100 plants of *Mallika* or *Dashehari.* Both Mallika and Amrapali excel the parents in quality of fruit.

Sindhu (Ratna × Alfonso). Plants high yielding, regular bearer, dwarf and compact, infl. coppery red, almost 35% flowers are bisexual. Fruits may be parthenocarpic, stone very thin, non-viable, free from spongy tissue, borne in clusters. It was developed from RFRS, Vengurla, Sindhudurg, Maharashtra.

Arka Puneet (Alfonso × Bangarapalli). This was developed from IIHR, Bangalore. It is a regular bearer. Fruits are of attractive colour, medium sized, free from spongy tissue. Good keeping quality and aroma.

Amba (Amrapali × Janardan Pasand). A late maturing strain having fruits of medium size with thin yellow skin and fibreless flesh.

Manjeera. (Rumani × Neelam). A regular bearer with high yield. Fruits of medium size are fibreless. Yellow skinned and sweet.

Niranjan. Off season, fruiting starts regularly in June-July and maturing in October.

Dashehari clone 51. CISH Lucknow selection. Regular bearer.

Basti-5. A hybdrid between Langra (Varanasi) and Rumani is also expected to be a promising strain.

Mango Malformation

It is a complex malady and although active work on this problem is being done since 1945 in India we are still in the dark about the mode of spread of this disease and also its transmission under field conditions.

It is common in north India, negligible in West Bengal, Bihar and Orissa and virtually unknown in the South where incidence of the disease is low possibly due to relatively high temprature during most part of the year. Temperature below 20°C during the initiation of panicle increases the severity of the disease. Incidence of the disease is higher if the temperature ranges between 0–8°C and the relative humidity is high. The floral malformation reduces the yield. Chausa and Bombay green suffer to the extent of 90%. Seasonal variations also influence malformation. Healthy panicles free from the infection have higher level of RNA, DNA, soluble protein and total nitrogen than the panicles with malformation. Higher amounts of carbohydrates induced greater malformation while higher nitrogen content has less infection of malformation.

It has been found that mango malformation is due to imbalance between growth promoters and growth inhibitors. Hence different concentrations of NAA were tried to overcome the effect of antiauxin like substances. NAA also increases the proportion of perfect flowers in the panicle. The spray of NAA (200 ppm) during October reduced the number of malformed panicles considerably in 'Chausa', 'Bombay Green' and 'Dashehari'.

The amount of RNA, DNA and protein are always higher in the healthy panicles than in the malformed ones in all the cultivars under investigation, except 'Bombay Geen' where DNA content did not show much variation. This suggests that if normal synthesis of organic metabolites like RNA, DNA, and protein can be brought about by biochemical manipulation, the disease can be controlled.

It is a serious malady with symptoms both in seedling as well as in the tree stage. It is quite common with cultivars like Bombay Green and Fajri jafarni which are susceptible. The chief symptoms of the disease in the seedling stage are the production of numerous thick vegetative shoots at the growing point. The growth of the seedling is checked and the whole seedling gets stunted. The bunchy shoots which are green in the initial stages dry up later. This condition is known as malformation of vegetative shoots.

In the case of adult tree malformation of both vegetative shoots as well as inflorescence takes place. Compact masses of sterile flowers appear in the form of a bunch. After some time part of inflorescence becomes vegetative giving out small leafy structures. Following two causual organisms are associated with mango malformation.

Fusarium monoliniforme. This is a fungus by which the malformation takes place. It is seen experimentally that the fungus may not show any symptom of the disease until and unless there is any injury on the top of the plants. It also needs spreading agent for causing infection from one plant to other plant.

Aceria mangiferae saged. It has been shown by several experiments at IARI, New Delhi that the mites and any fungus on the body causing injury to the plants provide a means for entry of pathogen into the host plant.

Control Measures

If the disease is to be controlled chemically then it may be effective only when the combination of the following chemicals are used : fungicide and insecticide. The fungicide will kill the fungus and insecticide will finish the mites. Besides these, some sticking substances should also be used to provide a better residual effect of the spray.

All the infected shoots should be prunned off. Then a mixture of fungicide (0.1% captan), miticide (AKAR 33.8, 0.1%) the sticker (Tenac) should be thoroughly sprayed. The spraying should be repeated every 10-12 days. Management of the disease needs systematic care and caution for several months.

Soon after harvest the affected parts should be pruned about 15-20 cm below the point of incidence. A spray of mangiferin zinc chelate ($CuSO_4$ 4 gm + lime 2 gm + mangiferin 40 gm + one lit. water) in July-August. This should be followed by another spray of mangiferin zinc chelate ($ZnSO_4$ 4 gm + lime 2 gm + mangiferin 40 gm + one lit water) in September. In October spray NAA @ 200 ppm, (2 gm NAA + 75 ml ethyl alcohol + 125 ml water + soap. Cobalt sulphate 1 g/l water also reduces the incidence. In order to maintain the C/N ratio 1% solution of urea may be sprayed in mid-November. Thereafter four sprays of carbendazim 50 WP @ 1 gm/lit water every 15 day. In January the twigs and inflorescence should be pruned 5-7 cm down the top followed by deflowering through the use of ethrel @ 500 ppm at the initiation of panicle. Besides these measures fumigation with fungicides *viz.* Vapam, planting of resistant types like "Ilaichi" identified by CISH, Lucknow. Injury to roots during planting should be avoided.

Diseases and their Control

Anthracnose. It is the most common and widespread fungus disease of mango. It is prevalent in almost all the mango growing regions of the world. Black spots are caused on shoots which dry up. Dry spots appear on leaves, the fruits fall or remain stunted and the mature fruits develops cracked black spots, specially towards the stem-end. The blossoms are also blighted. Humid weather helps the *Colletotrichum gloeosporioides* to cause anthracnose of ripening fruits. Bordeaux mixture (6 : 6 : 100) is the most reliable and effective fungicide so far tested for the control of anthracnose. Use of Bordeaux mixture (3 : 3 : 50) with clacium casinate as spreader or Burgundy mixture has been recommended. Several applications are needed, the first three for the protection of the flower cluster, and the other five to protect the fruits. Bavistin 0.1% (1 gm/lit water) or oxychloride or Dithiocarbamate 10.2% ; (2 gm/lit water) may be sprayed in nursery in February, April, September, October and December. Fruits may be dipped in the solution of benomyl (100 ppm) before storage. These fungicides are also useful against Blossom blight. Spray in Feb., April and December.

Application of carbendozine 50 WP @ 1 ml/l water during rainy season and fruit dip before storage has greatly reduced the damage. One spray of captan @ 3 g/l water followed by two sprays of zineb @ 2 g/l water at 14 and 21 days interval has also proved effective.

Black spot. The disease makes its appearance on the leaves as numerous small, angular, water soaked areas, usually crowded towards the tip and varying from 1 to 4 mm in diameter. The spots are dark brown or black slightly riased and are definitely more angular than those of anthracnose. The lesion spots have a rough surface due to development of cracks from which there is gummy exudation. The casual organism of the disease in *Bacillus mangiferae.* Satisfactory control of the disease is obtained by five applications of Bordeaux mixture (4 : 4 : 50 with addition to sticker between August to December.

Mango scab. Symptoms produced by mango scab are very much like those of anthracnose. When the infection is heavy, crinkling and distortion of leaf takes place followed by premature setting. The blotches on the stem bark are greyish and irregular in shape. The disease is caused by the fungus *Elsinoe mangiferae.* It can be controlled by Bordeaux mixture.

Blossom blight. Inflorescence is badly affected, so as to limit mango production seriously, during rainy or humid seasons. The early symptoms of flower infection are minute black spots, which later enlarge and may coaleasce to cause flower shedding and drying. It is caused by the fungus *Colletorichum gloeosporioides.* Control measure for blossom-blight are eight applications of 3 : 3 : 50. Bordeaux mixture and calcium caseinate, commencing from the time flower buds swell and extending at intervals of 4 to 21 days over a period of about three months.

Powdery mildew. Oidium mangiferae attacks fruits, flower scales, buds of tender flower-heads, axils and stalks. The whole surface of leaf and twig is covered with a powdery substance. Wettable sulphur 0.2% (2 gm sulfex in 1 lit. water) or 1 gm of Bavistin or 1 ml Karathane per lit. of water are employed to check powdery mildew. First spray sulphur just after the emergence of panicle, second of Bavistin and the third of Karathane may be given at 15 day interval. Benlate spray (0.1%) also controls the disease.

Sooty mould. Sooty mould on mango is caused by the fungus *Meliola mangiferae.* Prevention and control of the disease lie in eliminating the honey dew insects. Spray Diazinon 20 EC @ 1.5 ml/l water with wettable sulphur or starch @ 2 gm/l water. Starch removes the black scoop from the leaves. If the sooty blotch is on the furits, they should be dipped for two minutes in a bleaching solution of 1/4 kg each of chloride of lime and boric acid per 4.5 litres of water and afterwards washed with clean water.

Red rust. Red rust or red spot is a common disease on the mango in the tarai and other humid regions of India. It affects the foliage, bark etc. of host plant and is both parasitic and epiphytic in nature. Initially the spots are greenish grey in colour and rough in texture but later on the surface bears reddish brown hair like structures which give it the characteristic red rust appearance. The organism causing red rust in mango has been identified as *Cephaleuros virescens.* Copper sulphate, copper oxychloride (0.3%) or bordeaux mixture (1%) sprays have been reported to be quite effective in controlling it.

Brown Rot. Brown rot is a systemic disease and is caused by *Physalospora perseae.* The characteristic symptoms of the disease is formation of longitudinal cracks on the inflorescence stalk. It is suggested that at least one dusting with 50 : 50 : : sulphur : copper mixture should be given at the maximum blossoming period. When the fruits reach nearly marble size, foliar spray of Bordeaux mixture 6 : 6 : 100 with addition of sticker should be given. Spray of Blitox 3 gm/l water is also effective in control.

Sclerotium Rot. The disease is caused by fungus *Sclerotium dephini.* The control of the disease is secured by adopting suitable cultural practices and strict sanitation. Diseased plants should be uprooted and burnt.

Recife sickness. The outward symptom, the first to be visible, is the gummy exudation is the stem from the perforations made by beetle. Destruction of xylem vessels causes wilting of the tree. The disease has been found to be difficult to control. The method suggested to arrest its further spread in field is to cut and remove the diseased trees from the orchards. Spray Blitox 4 gm/l water in July and September each year.

Trunk infections. A few mushrooms are known to attack mango trunks but the damage caused by them is not of serious nature. They are mostly saprophytes or weak parasites which attack the dead bark of the trunk. Lichens can be removed by spraying the stem with a solution of 1.5 kg commercial caustic soda in 90 litres of water.

Cercospora mangifera is reported on mango leaves in Japan. Mango twigs from Bengal have shown the presence of *Dimerosporium mangiferum* which turns the wood ash grey. Damping off of mango seedlings due to *Rhizcotonia solani* is liable to appear in nurseries as in Indonesia.

Dieback (*Botryodiplodia thobromae*). Disease appears in October-November. Prune the diseased twigs and apply carbendazine or theophomate methyl @ 1 gm/l water solution. When the young plants are pruned Bordeaux paste is applied. The use of copper oxychloride or phytolan @ 4 gm/l water also serves the purpose. Streptocycline 30 g/l water has also pruned effective.

Withertip of young branches due to *Colletotricum gloeosporioides* has been recorded. Die-back due to *Botryosphaeria ribis* has been observed on the mango in Hawaii. In India the twigs of young and fully grown trees are found to wither and die from the tips downward. A gummy substance, brown or reddish brown in colour but turning brownish black after drying has been noticed to accompany the disease.

Phoma Blight (Macrophoma mangiferae). Incidence of the disease is visible in November with light brown patches on the leaves which later turn black or dark brown. Apply 2-3 sprays of copper oxychloride @ 3 g/l water every 15 days in Sept.-October to ward off the disease attack.

Gummosis. Imbalance of minerals *viz.* Zn, Cu, B and calcium causes cracking on the bark of the twigs from which a little drop of gum exudes. As the infection gets heavier the exudation of gum increases. This may be seen on branches, twigs or even trunk of the affected tree. For control apply a solution mixture of $CuSO_4$ (250 gm) + $ZnSO_4$ (250 gm) + slacked lime (100 gm) in l of water to each tree during April-May. Irrigate the trees soon after the application of above mixture. Besides, a spray of copper oxychloride @ 3 gm/l water in July and September is also recommended.

Bacterial Leaf spot or canker (Xanthomonas campestris p.v. *mangiferae indicae)*. Small water soaked lesions toward apex of the leaves spread during the rains when the relative humidity is above 30% and the temperature ranges between 25–30°C sprays of streptocycline @ 300 gm/l and copper oxychloride @ 3 gm/l every ten days in the month of May controls the disease successfully. Agricmycin @ 100 mg and Bovistin @ 1 gm/l water may also be helpful in reducing the incidence of disease. Commercial product of *Bacillus Coagulans* is used as biological control.

Insect Pests

Mango stem borer. Old mango trees are generally infected with the borer. If the bark is tapped with a piece of wood the affected part gives a hollow sound. The tunnel in which the borer has penetrated should be plugged with cotton dipped in kerosene oil, or Guesarol. 0.3% Endosulfan may be used as spray to check borer menace. Occasional spray of emulsion of 0.05% Monocrotophos and sealing the hole is considered best for controlling stem borer as well as bark easting caterpillar. 0.05% Eldrin wash is also known to be effective.

Mango hopper (Indioscopus clypealis, Amritodus etkinsoni)

Lays eggs inside the flower buds, wingless nymphs appear in 4-6 days these suck sap from the flower and fruits and develop into winged adults. These insects also seek large amounts of honey dew which cover the flower and interfere with their fertilisation. Black sooty mould develops on the secretions.

Three sprays, first of cypermethion @ 0.01% and second of carbaryl @ 0.01% and a third of NSKE @ 4% are considered enough to get rid of the damange causing happers. The first spray should be at the appearance of blossom, the second when the fruits have attained the size of a pea seed and the third 15 days after the second spray. Avoid spraying when the flowers are in full bloom. Biological mode of control by *Verticillium leconii* has also been recommended. When the hopper and powdery mildew problem exist at the same time a mixture of insecticide and fungicide may be sprayed.

Mango mealy bug (Drosicha mangiferae). The insect having a mealy coating on its body causes maximum damage to fruit yield and encourages growth of sooty mould. The nymphs crawl up to the terminal shoot and remain these throughout the development period. They may be seen adhering to the shoots, leaves, panicles etc. It is active during December-January. The critical stage of the crop for control of the pest is a month before the bud sprout.

Several steps have been recommended to save the crop from the damage by the mealy bug. Keeping the orchard clean of weeds like Bhatter (*Clerodendrum inflortunatum*) and other grasses, post-harvest summer ploughing, spreading sand around the tree trunk. In the month of December adhesive band of polythene may be used to cover the tree trunk about 50 cm wide from the ground level. The lower portion of the banding must be kept slippery by applying rosin + castor oil in the ratio of 4 : 5 or grease + Coaltar mix (1 : 2). The main purpose is to check the nymphs from crawling on of trunk. Greasing once a week of the banding is necessary till February.

A spray of monocrotophos 40 EC at 1.5 ml or Methyl parathion 2% dust @ 200-300 g around tree canopy to destroy the nymph in November is very effective control. Spray NSKE (4%) or crude garlic oil (1%) on tree trunk before banding to kill those already present on the trunk. Whenever possible the natural enemy of the mealy bug (*cocconallids* and spiders) be allowed to remain active.

Mango fruit fly (Bacterocera dorsalis). It is active during April-May and is considered to be a pest to be largely associated with the ripening mango fruit. It lays eggs under the rind of the fruit when it is about to ripen. The maggots eat the pulp, the affected part drops and the maggots pupate in the soil and emerge as fly to continue the cycle. It is necessary that all the fruits falling from the tree must be destroyed.

The control measures should be taken a month before harvest. Generally two sprays are considered necessary to combat the damage. First spray (also known as bait-spray) of mixture of carbaryl 0.2% or malathion 0.05% with molasses @ 10 gm/lit water one month before harvest. Repeat the spray 10-12 days after the first spray. Raking the soil during summers is also helpful for the next season crop. The annihilation of the male is also useful along with the bait.

Gall Midge. It is also known as inflorescence midge. The infestation is best visible in January. The bud-sprout stage is the best stage to control the damage. First spray of the insecticide may be done at the bud burst stage and subsequent sprayings may be done to ward off reinfection of midge. The insecticides commonly used are Fenitrothion 50 EC @ 1 ml, Dimethoate 35 EC. @ 1.5 ml or Methyl dimeton @ 1.75 ml/lit. water. When the mango hopper is also active along with midge the control measure for mango hopper is sufficient to check the midge-damage.

Mango leaf webber (Orthago euadrusalis). It needs control measures during harvest and also after harvest. The twigs with web should be cut off and burnt before harvest. Avoid application of chemical spray. After pruning spray the trees with carbaryl 0.1% or quinolphos 0.05% every 10-15 days.

Mango shoot gall psylla (Absylla cistellota). Remove the gall-bearing twigs from the tree and burn them. Spray quinolphos @ 0.5% in the beginning of August followed by two sprays of monocrotophos @ 0.05 and Dimethoate @ 0.05% at interval of 3 weeks in sequence.

Shoot borer (Chlumetia transversa). At the time of the initiation of new flush remove the infected shoots and burn. Spray carbaryl 0.1% followed by Dimethoate @ 0.05% and cypermethion @ 0.01%

Bark eating caterpillar (Inderbela quatrinotata). This pest is active throughout the year. Clean the infected portion and insert cotton soaked in kerosene oil or petrol and seal the hole with clayey mud.

Mango scale (Pulvinaria sp.*)* This is also active during April-May when the fruit is quite small. The spray of Quinolphos 25 EC @ 2 ml, monocrotophos 40 EC @ 1.25 ml or Dimethoate 30 EC @ 1.5 ml/lit. water at 15 days interval controls the pest and its activity.

Mites. Minute insects not visible to the naked eye. These attack the leaves and flowers which become silvery white and dry up and drop. Monocrotophos may be sprayed when the mite menace increases.

Red ants. Live in leaf nests on the tree. They distribute injurious scale insects all over the tree. 0.05% monocrotophos with sulphur in the proportion of 2 : 1 may be used to destory them. Dust calcium cyanide into nests : spray 0.05% monocrotophos.

White ants. These destroy the young grafts specially during the dry weather in summer. Apply 0.2% DDT around each graft soon after the monsoon for control against white ants. Good nitrogenous manures (cakes etc.) may also be supplied. Soils, in the vicinity of the tree, should not be allowed to dry. BHC (5%), aldrin, chlorodane or heptachlor may be mixed with the soil in the basin.

Jhumka (Clustering of fruits). Clustering of fruitlets at the tip of panicles is caused due to several causes. The more important ones are as (*a*) Dense planting leading to heavy infection of pests and disease, (*b*) monoculture of single variety or strain resulting in non pollination, (*c*) withholding irrigation over long periods, (*d*) deficiency of Boron , Zinc and copper, (*e*) reduction in the number and activity of pallinators due to large scale use of insecticides, (*f*) adverse climatic conditions during the months of Feb.-March.

The control measures may include effective pruning, planting distance to be not less them 10 m × 10 m, coplanting with compatible variety, adequate fertilization after soil test and timely interculture.

Lichen. Old trees are generally infected by it specially in the wet regions. Light, erased and white spots are noticed on the stem. The lichen should be removed with the help of gunny bag and a spray of caustic soda @ 15 g/lit water be given.

Loranthus (Dendrophthoe folcota). A partial stem parasite, found on the trees with small leaves of unhealthy green colour. The infected branch must be cut off at least 20 cm down from the point of attachment of loranthus. It is necessary that the loranthus be removed before it flowers while attached to the mango tree. Infection may be eradicated by a spray of 3-4% emulsion of diesel oil, washing soap and water. Injection of $CuSO_4$ and 2, 4-D into the affected branches also helps in the eradication of the parasite.

Physiological Disorders

Black tip or Mango necrosis. The first symptom of the disease is the development of small etiolated area at the distal end of fruit, which gradually spreads, turns nearly black and covers the tip completely. The fumes of the brick kilns are supposed to cause black tip disease of mango. Formerly it was considered to be the effect of the fumes like ethylene, sulphur dioxide etc. emanating from the kilns. Later it was shown that a deficient supply of boron to the scion may cause of the disease. The incidence of the disease has been checked by the sprays of boron in the form of borax. 4 sprays each of 3 kg borax in 400 litres of water at preflowering, flowering, fruit set and once between the time the fruit is set and attains the size of 3-4 cm in diameter are recommended for control.

Mango-Leaf Scorch, a Nutritional Malady. This invovles collapse of tissues in clearly defined marginal areas of brick red colour towards the tip. Both young and old plants are susceptible. It has been found the IARI that this malady is due to chloride ion toxicity.

At present the following recommendations can be made to growers for minimising the extent of damage caused by this malady :

(*a*) use of potassium sulphate fertilizer in place of potassium chloride ;

(*b*) removal and destruction of dried leaves falling near the base of tree ;

(*c*) use of irrigation water having low level of chloride ions ; and

(*d*) 4-5 foliar applications of potassium sulphate (5%), on newly emerged flushes showing acute conditions.

Spongy tissue. It develops in the mesocarp of the fruit in a small area or the entire fruit without any external sign of the disease. The fruits become unpalatable and give offensive smell. Harvesting of the fruits at 3/4 maturity stage reduces the incidence to negligible degree. This is a physiological disorder to which the cultivar *Alphonso* is highly susceptible. Involvement of calcium in the incidence of the disorder is also indicated.

□□□

CHAPTER

19

MANGOSTEEN

(*Garcinia mangostana* L.)

Guttiferae

Mangosteen (*Mungustaan* H.) is delicious, sweet, soft, parthenogenetically produced tropical fruit with a very pleasant flavour used as a dessert or as an astringent. It is industrially important because the rind contains 7 to 13% tannin. The evergreen trees cannot stand cold weather, dry winds, and strong sun and prefer a moist climate with plenty of irrigation throughout the year.

Origin and Distribution

Generally cultivated in the lower elevations of Nilgiri Hills in India, it is supposed to have originated in Malaya archipelago and the Molucca and Sunda Islands. It is quite well known in Burma, Sri Lanka and the southern part of Philippine islands ; some cultivation is also done in Kerala and West Bengal. It is cultivated throughout the tropics.

Propagation

Seeds lose viability soon after extraction from the fruit. Seeds of large size are planted immediately after harvest for better germination. The plants raised from the seeds are exactly like the mother plant because the female parts of the flower produce seeds without fusion with male parts. The seeds are formed from nucellar tissue of carpel-wall in the fruit and are not zygotic. The original roots of the seedling die off and new roots are given out. As such the seedling needs good care during the period and the growth is also slow. The roots are devoid of root hairs. An equal proportion of soil and compost may be regarded as a good medium of growth of seedling. Shade from sun during hot periods is helpful to the young seedling in establishing itself.

Vegetative propagation is done by rooting of hardwood cuttings and grafting (inarching) on some of the wild allies of the mangosteen such as *Platonia insignis, Garcinia tinctoria, G. morella, G. livingstonei* and *G. xanthochymus*. Success with budding has also been reported from Manila. IBA treatment promotes rooting in air-layering.

Planting

2 or 3-year old plants with a height of 0.5 – 0.75 m are fit for setting out. Transplanting must not be delayed lest the roots get deep and are damaged during the act of lifting from the original place. The young trees do require shade. The mangosteen trees are best planted in pits (1 × 1 × 1m) with a spacing of at least 10 metres between plants.

Cultural Practices

Regular watering during the drought periods and occassional mulching preceded by stirring of the surface soil for weeding is needed. About 40 to 75 kg of cattle manure and 5 to 7.5 kg of bonemeal or oil cake per bearing tree per year gives satisfactory results. The trees do not require pruning as a rule.

Fertilizers are applied thrice a year : at preflowering, after fruit set and after harvest. Magnesium deficiency and Fe/Zn imbalance should be corrected by proper and timely application of the required additive N (0.25 kg), FYM (25 kg) and P_2O_5 (0.5 kg) are needed for a plant per annum. Nitrogen should be supplied at the three stages mentioned above while FYM and P_2O_5 should be supplied after harvest. Mg deficiency needs spray of $MgSO_4$.

Harvesting and Marketing

Mangosteen bears two crops a year; the main fruiting season is August-September and then there is a second crop in March-April. On an average a tree yields 300-500 fruits in India which is roughly less than half of the yield reported from other countries. Harvesting of the fruits should be done only when fruits become mature as prematurely harvested fruits do not develop flavour.

Jolo, a variety with large fruits, big seeds and delicious pulp is known to be from Philippines. In the south the three edible species of mangosteen (*Garcinia indica, G. tinctoria, G. cambogia*) have poor quality fruits as compared to *G. mangostana*.

Insect Pests and Diseases

Gamboge is a serious disease or physiological disorder of the mangosteen in which the pulp becomes yellowish, sticky and unpleasant. There is yellow exudation of gum on fruits and branches. The disease is possibly caused by injury or due to extremes of temperature and humidity prevailing during fruit-set or even prior to that. Summer crop is affected more than the monsoon crop. The fruits which are gamboge affected are of poor quality with unattractive appearance.

Little leaf disease also occurs sometimes to a considerable extent. Removal of affected parts helps to check the spread of the disease. This condition was also associated with Fe/Zn imbalance. Mg deficiency if detected may be corrected by spraying $MgSO_4$.

Leaf eating caterpillar. The mites attack the foliage and the fruit. May be controlled through the use of 0.05% Malathion.

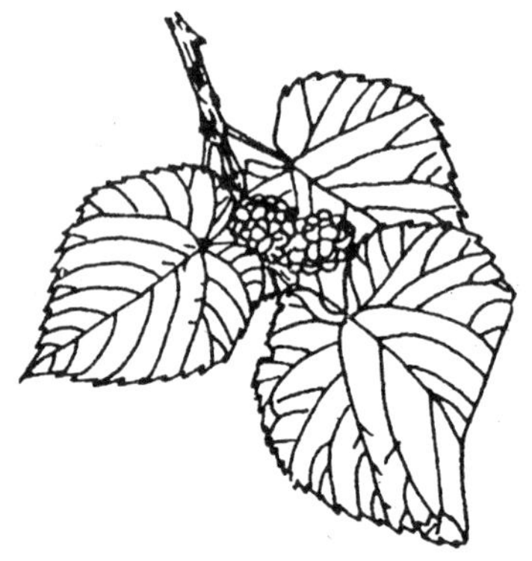

CHAPTER 20

MULBERRY

(*Morus* Species)

Moraceae

Origin and Distribution

Though mulberry is considered to be a plant of temperate regions, its trees are found all over the Indian subcontinent. Several species of mulberry (*Morus alba, M. nigra, M. rubra*) are grown in India. Common mulberry (*Tut*) is generally found in the sub-Himalayan tracts. In Kashmir and Karnataka it is used to rear silkworms. The mulberry tree, dioecious in habit, grows to a height of 4 to 18 m. *Morus alba* L. is moderate sized short lived deciduous tree, native of China, grown in the Punjab, U.P. and Kashmir. Mulberry speads rapidly to different parts of the world where its bark is used for paper making, the sweet and edible leaves and cattle fodder and the heartwood for sports goods, furniture, boat oars. Its root is used as an astringent. Fruit juice cures soar throat and fever and cases of dyspepsia and melancholia. The fruits contain about 8% of sugar and 1% of acid. Karnataka has the largest acreage of about 28,000 hectares. Kashmir is the close competitive grower.

Soil and Climate

Mulberries have no special demand on soil though a clay loam suits better than clay or sand. Ample sun proves useful, it can stand drought and frost considerably well.

Propagation

Though seed propagation is quite possible, vegetative methods of propagation are prefered to ensure plants true to type. Cutting, shield, ring or flute budding and inarching is common in mulberry. Mulberry propagation from cuttings is easy. Hardwood cuttings of about 20-30 cm length from mature stem with about

4-5 buds are planted in nursery buds 10-15 cm apart slanting, and 2/3 embedded in the medium of growth in the month of December or January. These sprout in about 3-4 months time and may be transplanted as the Monsoon sets in. Optimum time for budding is the end of March or early April, the buds are initiated by May or June.

Planting

Planting is done in pits (45 × 45 × 45 cm) made in well levelled and ploughed land at a distance of 6-7.5 metres between them. The pits are supplied with 2-3 baskets of well decomposed compost or F.Y.M. raked well and then watered extensively at least 10 days before planting.

Pruning

It helps better production of fruits if done properly. The stem is kept clear of all lateral shoots upto a height of about one metre. This operation should best be done after leaves have been shed off. The plants needed to rear the silk worm do better by more severe pruning which should be done nearer the ground surface to encourage more foliage.

Cultural Practices

Plants used for sericulture are manured heavily to encourage more leafy growth which is made possible by heading-back to ground level. The shoots that come up are allowed for as many years as found profitable. After about 18-20 years new plantations are needed.

Harvesting and Marketing

The flowering takes place in February and March and the fruit is ready by June. Fruits should be hand picked as soon as near ripe, pressed and dried. No other method succeeds for storage. The fruits are also collected as they fall on the ground. The fruit yield from a single tree may be 10-20 kg. About 3,000 to 6,000 kg of leaves are picked in 6-7 instalments from a hectare during the period March to May every year for sericulture.

Cultivars

Morus alba. A species suitable for feeding the silk worms. It is the Chinese species of the same family at the fig and the jack fruit. It has several cultivars *viz. Laciniata* Beissn, *Pendula* Dipp, *Pyramidalis* Ser and *Tatarica* London. *Laciniata* the skeleton leaved mulberry has uniformly divided leaves into narrow lobes. *Pendula* has slender and drooping branches. *Pyramidalis* derives its name from the pyramid like shape of trees. The *Tatarica* cultivar (Russian mulberry), a dwarf cultivar with its bushy top is generally used as root stock or wind break. It bears black insipid fruits of low quality.

Besides these, cultivars of *Morus alba* var. *S. Vinosa* Del, *Japonica* Audib, *Muticaulis* Perr, and *Celtidifolia* Hbk. are commonly grown. *Vinosa* is ornamental, *Japonica* has oval fruits of red colour, *Multicaulis* has black coloured sweet fruits and is of commerical importance in China to silk industry ; *Celtidifolia* produces inferior fruits.

Morus nigra L. With large dull coloured entire leaves produce black, juicy and dark coloured fruits. It is supposed to have been introduced from Persia and is known as black mulberry possibly due to the colour of the fruit.

Morus rubra L. It is known as Red or Virginia mulberry with plants of 18 metres height and an average diameter of 1 metre. The unripe fruit which is deep red becomes almost black tasty when ripe. A tree may yield as much as 20-40 kg of fruits.

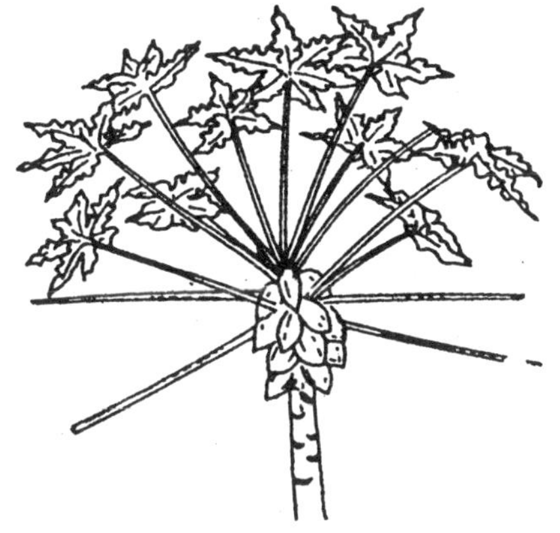

CHAPTER

PAPAYA

(*Carica papaya* L.)

Caricaceae

Origin and Distribution

The papaya (Melon tree or *papita*), a native of tropical America, is widely grown all over India in kitchen gardens as well as in commercial orchards. It is a herbaceous, quick growing plant with usually unbranched stem and a crown of palmately lobed leaves with long hollow petioles.

The fruit has 0.5% protein, 9.5% carbohydrate, 0.1% fat and good amount of minerals (Fe, Ca and P), 2,500 international units of Vitamin A and 70 mg of Vitamin C per 100 gm fruit. It also has Vitamin B_1, B_2, nicotinic acid and riboflavin. Unripe fruit is a rich source of papain, an enzyme helpful in digestion of proteins, removing skin blemishes, useful in the treatment of stomach ulcers, diphtheria and even cancer. Among its industrial use clarification of beer, tanning and manufacture of chewing gums are common. The ripe fruit helps in the proper functioning of the liver, spleen and the digestive tract.

Papaya is grown in central and south Africa, Australia, Hawaii, (USA), Pakistan, India and Bangladesh besides many other tropical and subtropical regions. The area under papaya in India is estimated to be roughly 10,000 ha of which Bihar alone has 4,000 ha followed by Assam, U.P., Gujarat, Andhra Pradesh, Karnataka, Tamilnadu, Maharashtra, Madhya Pradesh etc.

Climate and Soil

Papaya enjoys sun, proper irrigation and frostless conditions. Well drained, adequately manured light soils are considered best for the growth of papaya plant. Alluvial and medium black soils are also good. Papaya cultivation cannot be a success in sticky and calcarcous soils or where water accumulates even for a few hours.

Papaya plant is normally dioecious. The female flowers produce fruits which are variable in shape, size and quality. Unripe fruits have smooth and green skin, it becomes yellow when ripe. The flesh is yellow to orange in colour. A number of dark brown, small seeds, each with a transluscent mucilagenous covering are attached to the walls inside the fruit.

Propagation

There being no variability in seedling trees the papaya plant is raised from fresh seed though it is also possible to propagate from cuttings and graftings ; the cuttings root well after treatment with 3-5% of indolebutyric acid. Seeds should be rubbed with ash from cattle dung to remove the mucilaginous coating before these are dried in shade to facilitate good germination. The seed is treated with Captan or Cerasan and sown about 1 cm deep on well-prepared raised beds about 15-20 cm above the ground level. The seed should be planted 2-3 cm apart, in rows 7-8 cm apart. The seed-bed should be watered daily if there is no rain. The seed-bed needs protection from hot sun and heavy rains until the germination is complete.

The young seedlings are susceptible to damping-off. The soil of the seed-bed may be sterilized ten days before sowing by steam or by the application of a solution of formalin (1/2 kg formalin in 28 litres of water) and drenching the soil with copper fungicide. To avoid the loss of formaldehyde fumes, the soil surface should be covered with several folds of newspaper for a day. No sowing be done for 10 days immediately following formalin treatment. Micropropagation methods may be employed for seedless cultivars.

Planting

The seedlings are planted, two per pit in July-August or February-March where frosting does not occur. The size of pits should be 50 × 50 × 50 cm spaced 2.5 to 3 metres apart and well manured with 25 kg of compost and 1.5 kg of bone-meal and 1 kg of neem cake as basal dressing.

There are two methods for papaya plantation (*i*) raising the plants from seeds at the places these are destined to remain for whole life time, (*ii*) raising the plants in nursery or seed bed and then transplanting. Seedlings may have male and female plants in equal proportion and thus a minimum of two seeds be placed at one place to increase the chances of the presence of plant bearing female flowers. Male plants of deciduous cultivars are retained in the proportion of 1 : 20 female plants. The extra being removed. The papaya plantations are mostly done in the immediate vicinity of cities and towns for an easy, quick and profitable disposal of the crop. Use of sewage water which is usually available in some areas obviates the need for frequent and heavy applications of organic manures and copious irrigation.

Manuring and Care

The basal dressing is followed by the addition of 0.5 quintal of well rotten compost or F.Y.M. four months after planting. One-year old plant is given nitrogen, phosphorus and potash through chemical fertilizers to the tune of 200 gm (N), 100 gm (P_2O_5) and 250 gm (K_2O) in two split dose, one in February-March and the other in September-October to ensure better yield. Magnesium and calcium are also removed by the papaya plants in quite good amounts and these should also be supplied in split doses at the time of growth, blossom and fruit set. The manuring schedule may be repeated every year. Adequate watering is necessary to ward off drought and frost, weekly in summer and biweekly in winter. The summer irrigation should be followed by earthing. In any case water-logging should not be allowed, in fact the plants must not come in contact with water by placing 20-30 cm of earth all round of young plants. Hoeing and weeding must be done regularly. Young plants also need protection from high winds, hot or cold, by raising quick growing plants in rows to serve as shields three months ahead of the impending dangers of desiccation.

Regular weed control is desirable, it is always paying, paraquat @ 0.25-0.5 kg/acre and Karmix (diuron) @ 1.5-2.5 kg/acre controls weeds effectively. Annual weeds may be controlled by the use of glyphosphate at 1.1 kg/ha and double concentration prove effective in control of perennial weeds. The use of herbicides may be done after three months of plantation and repeated every 60 days. Some of the chemicals like diuron caused necrosis chlorosis, and stunted growth, and their use should be done with caution.

Papaya itself is used as a filler tree between newly planted mango and orange groves. Chillies, onion, cabbage, radish, knol-khol and tomatoes may be grown for about six months with papaya and the main crop. The land should be ploughed and harrowed cross-wise at least twice during the year. Frequent light cultivation and stirring up of the soil are essential for improving the quality and quantity of the crop. Deep cultivation must be avoided.

Sometimes papaya bears several fruit set in a cluster instead of a single fruit at each node. Unless the fruit is of small size, there is likely to be some damage resulting from overcrowding and some fruits fall off. Some of the fruits may also be deformed. Under such conditions, it pays to thin the fruit clusters, leaving not more than two fruits per node immediately after fruit-set for best results.

Harvesting and Marketing

Papaya trees bear their first fruit in 10 to 14 months from the time the plants are set in the orchard. Fruits mature throughout the spring and summer in North India. The mature papaya fruits are harvested when they are still hard and green but turning yellow. It is best picked by giving it a careful twist with the hand, fruits which are not within the easy reach of hands, may be picked by means of a ladder, taking care to avoid the bruising or splitting of the skin. The harvested fruits require bamboo baskets, and wooden crates for packing. The largest yield of fruit is obtained from the second or third harvest. The yield varies from 350 to 750 quintals/ha in different parts of the country. On an average a healthy plant may yield 30-50 kg fruits every harvest.

Jam, marmalade with shreds of oranges, candy, pickle, canned fruit, salads, squash and jelly are prepared from papaya fruit.

Cultivars

In western India *Washington, Coorg Honey* Dew, and Gujarati ; in south, *Coimbatore 1, Coorg Honey* Dew, *Washington*, Long, and *Round* ; in east *Ceylonese* Round, *Coorg Honey* Dew, *Ranchi, South African*, and *Washington* ; and in the northern region *Coorg Honey* Dew, *Saharanpur Selection, Washington* are commonly grown.

Washington. It is distinguished by purple rings at the nodes and dark purple petioles. Fruit is medium to large sized, round to ovate each over 20 cm long with a circumference of over 50 cm at it maximum bulge, an individual fruit weighs about 2.5 kg. Fruit is sweet and good flavoured.

Co. 1. Dwarf, selection from the progenies of Ranchi. Fruit-size medium, shape spherical, smooth skin with golden yellow colour, nipple slight and ridged at the apex, flesh orange in colour, soft firm and moderately juicy.

Co. 2. Plants of medium height, fruits large and ovate of yellowish-green colour. Flesh is orange coloured and soft. Papain content high.

Co. 3. A hybrid between *Co. 2* × *sunrise solo*. Tall, vigorous plants, fruit of medium size. Sweet with good keeping quality.

Co. 4. A hybrid between *Co. 1* × *Washington*. Plants of medium size, fruits large sweet and keeping quality good.

Co. 5. A selection from '*Washington*' cultivar. It is excellent in the production of Papain. The plants start bearing at a height of 50 cm from the ground level. Average yield of papain/plant may be 14-15 gm. The plants give 75-80 fruits in a span of two years.

Co. 6. Recommended for the production of papain and also for desert purposes. They are dwarf, yielding 80-100 fruits in 30 months. It is a selection from Pusa majestic.

Co. 7. A dwarf and gynodiocious variety of dwarf stature yielding about 80 fruits in $2\frac{1}{2}$ years. The papaya fruit yields about 7-8 gm of papain. This is suitable for dual purposes.

Coorg Honey Dew. A dwarf heavy bearer Madhubindu, the fruits are almost oblong, nearly of the same size as that of Washington but only less elongated slightly less sweet but with equally good flavour. It is believed that the cultivar is also fairly true to its type and has only a small proportion of male flowers. Popular in south India with high degree of uniformity in seed progeny.

Honey Dew. Heavy bearer, less seeds, good taste and flavour. Suited for north India.

Peradeniya. It is a good cultivar for extracting papain. Fruits medium to small, shape obovate, yellowish, green skin, flesh orange in colour, soft, firm, juicy, keeping quality good.

Ranchi. Fruits are oval, of medium size, with bright yellow flesh, which is sweet and almost free of the characteristic papaya odour. Popular in south India with high degree of uniformity in seed progeny.

Solo. It is probably the only well established cultivar in the world. It produces plants with only female flowers and perfect flowers (no pure male flower). The name was given due to the small fruits (av. wt. 300-500 g only). The pear shaped fruits are of good quality, some of the improved *solo* are line 5 solo, line 8 solo, line 10 solo and Kapoho solo.

Selection No. 7. Fruits medium sized. Flesh red in colour, soft, firm, moderately juicy, keeping quality good, good papain yielder.

Philippines. Medium sized fruits, shape oval to oblong, yellowish green skin, good papain producer.

Pusa Dwarf. Plants dwarf, first fruits borne at a height of 25-30 cm from the ground level. Fruits are mildly sweet weighing 1.5 to 2.5 kg. Suitable for areas where the winds are common. Dioecious.

Pusa Giant. Average weight of fruit 2-3 kg, quality moderate. Plant dioecious. First fruits borne at a height of 90-95 cm from ground level.

Pusa Majestic. Plants may bear female flowers or may be hermaphrodite. Plants bear fruits of medium size (1.25 to 1.50 kg) and quality. Fruits have good storage quality. Tolerant to virus.

Pusa Delicious. Plants of moderate size, fruit weight 1.5-2 kg. Flesh deep orange coloured. Sweet with good flavour. Possibly the best quality fruits with less number of seeds. The plants are hermaphrodite or bear only female flowers so that all the plants bear fruit.

Pusa Nanha. Extremely dwarf variety well suited for kitchen garden or pot cultivation or high density orcharding.

Pant Papaya – 1. Dwarf, round fruits with smooth skin. Fruit bearing starts at 40-45 cm height stage. Average fruit but may be 1 kg to 1.5 kg. A plant gives 40-50 fruits in a season. This has been found suitable for high density orcharding.

Pant Papaya – 2. Plants of medium height. As it approaches 50-60 cm height it starts bearing. The fruits are longish in shape with an average weight of about 1.5 kg each. A plant yields 30-35 fruits.

Taiwan 1, 2, 3, 15, 16 and *super disco* are also notable cultivar of papaya.

Insect Pests and Diseases

The Red Spider. (*Tetranychus* sp. and *Brovipolus phoenicis*). The insect affects ripe fruits which develop rough and brownish patches. Affected leaves turn yellow. Dusting with sulphur or spraying with lime-sulphur or 1% solution of Morocide controls the menace. Affected leaves may be clipped off and destroyed. Dimethoate 30 EC @ 1.5 ml/l water may be used for control.

Anthracnose (Colletotrichum gloeosporioides). It causes depression on the affected areas of the leaves and fruits followed by a premature drop. Spraying with Carbendazime or Thiophenate methyl @ 1 gm/l water or mancozeb @ 2 g/l water two weeks before harvest prevents the damage due to the disease which is checked.

Collar Rot. The soil borne fungus, *Pythium aphanidermatum* causes rotting of the tree trunk at the point where it leaves the soil due to stagnation of water. Subsequently the leaves wilt. Prevent water stagnation at the roots, to stop the onset of the disease. Bordeaux mixture 5 : 6 : 50 or Blitox or Captan 0.3% conc. may be applied near the base of the plants to the soil to control the disease. Copper oxychloride @ 2 g/l water checks the spread of the disease. Seed treatment with organo-murcurials or Thiram @ 2 gm/kg seed or Difoltan @ 2.5 gm/kg seed improves the emergence.

Papaya mosaic (Ring spot). The disease is caused by Papaya ring spot virus (PRSV) which is transmitted by a number of aphids of which the most important vector is *Myzus persical*. The infestation of insect vector should be checked by DDVP 70 EC or Dimethoate 30 EC @ 1 ml/l water or Methyl demeton 25 EC @ 1.75 ml per litre of water every 10-15 day interval.

Aphids (Aphis goosypi and *Myzus persicae*). Nymphs and adults suck sap from the leaves and transmit mosaic virus disease. Spray of Demothate @ 0.3% or methyl demeton @ 0.05% is effective in checking.

Root-knot nematode (Meloidogyne sp.). Chief symptoms are the yellowing and then shedding of leaves, premature fruit drop, root galling and finally death of the plant in severe eases of infestation. Application of 2 tons/ha of neem cake tends to reduce the incidence. Addition of Carbafuran 3 G @ 40 kg/ha in the field controls the nematodes. Commercial product of *Paecilamyces lilacinus* (a fungus) @ 8 gm/plant reduces the nematode activity. VAM fungus has also been shown to reduce the number of nematodes, hence the damage also.

Papaya leaf curl. The disease is caused by the infestation of tobacco leaf curl virus. Reduction in the size of leaves follows severe curling, crumbling and distortion of the leaves accompanied by vein clearing. Inverted cups are formed due to the rolling of leaf margin, flowering is highly suppressed and fruit produced may be only few. The virus is transmitted by white fly. *(Bemisia tobaci)* Avoid growing of tobacco, tomato and zinia near the papaya orchard. The measures for the control of white fly is the one recommended for the Ring spot of papaya.

CHAPTER 22

PASSION FRUIT

(***Passiflora edulis*** **Sims**)

Passifloraceae

Origin and Distribution

The passion fruit (*Passiflora edulis*) a native of the Amazon region of Brazil also found growing wild in New Guinea, Java, Sumatra, Malaya, Mexico, Hawaii, Sri Lanka and some parts of India is essentially subtropical. It is widely cultivated in South Africa, Kenya, Fiji, Australia, New Zealand, Hawaii, Indonesia, Florida, Sri Lanka and South California. In India it grows wild in the Nilgiri hills, Kodaikanal, Coorg, Malabar, Kerala and Himachal Pradesh. The perennial vines cling to support, the flowers are ornamental with prominent crown and the fruit with a hard shell is acidic in taste. It is not used as a table fruit and is a good source of vitamins A and C. The fruit is used in the preparation of passion fruit squash.

Climate and Soil

The passion fruit (H. *Jhumkalata*) needs high rainfall of the tropical and sub-tropical areas. A mild climate free from extremes of heat and cold is necessary for the passion fruit. The optimum temperature range is between 20 – 30°C. The sub-montane areas, foot-hills and low hills are preferred for successful production of this fruit. The yellow or golden fruited cultivar *Passiflora edulis* f. *flavicarpa* does well in the plains whereas *Passiflora edulis* (the cultivar with purple fruit) is cultivated at an altitude of 1,000 to 2,000 metres in South India. Heavy rainfall is not considered suitable for the cultivation of passion fruit. A soil having sufficient moisture, humus and lime is considered suitable for its cultivation.

Propagation

The passion fruit plant is mostly raised by seeds. The seeds are removed from the pulp after fermentation for 4-6 days, washed thoroughly, dried in shade and kept in a sealed container in a refrigerator. These seeds give better germination (more than 95%) after a period of 3 months. Seeds do well in sterilised seed beds with well drained medium and maintained in heated glasshouse. As the seedlings develop 2 or 3 leaves these may be transferred in pots. Seedling are healthy when both temperature and humidity are high. The seeds are sown in beds with a spacing of 12 to 15 cm between rows and 22 cm apart in February or June-July. The seeds germinate in a fortnight. After 2 to 3 months the seedlings are fit for transplanted in the field after about 6 weeks. Cuttings of suitable size from well matured wood and with 3-5 well developed buds may be planted in February or August in the usual way for vegetative propagation. The cuttings readily provide rooted plants fit for transplanting in about three months time. The purple fruited cultivar may be grafted to the yellow fruited to ward off the damage from fusarium wilt.

Planting

The field in which passion fruit is to be planted should have well pulverised non-saline soil prepared in the same way as is done for wheat crop. Sandy soils specially need a supply of high amounts of organic manures *i.e.*, F.Y.M. compost or green manuring. The pH of the soil should be 5.5–6.5. Well drained heavy soils are also suitable. The seedlings or the rooted cuttings are transplanted in the spring or the monsoon in pits 3-4 metres apart. Planting distance may be kept 2-4 m between rows and 3-4 m between plants. The plants need support which may be either vertical or horizontal trellises ensuring maximum and also even exposure to sunlight. It is necessary to remove all the side-shoots below the graft and after a month or two a vigorous leader is selected and trained to the top wire. With advance in the growth of the plant other branches which come up are also trained giving a complete support to the vine.

Cultural Practices

The vines are irrigated after suitable intervals according to season and the needs of the plants regularly after planting. Plants need irrigation which must be provided according to the needs of the crop depending on the nature of the soil, atmospheric conditions and the extent of growth of vines etc. Water-stress at any stage in the life of the vine diminishes the yield by reducing vegetative growth, flowering and inducing premature fruit crop. Over-watering as well as water logging are detrimental and should be avoided. Short of 1200 mm well distributed annual rainfall must be supplemented by irrigation. The vines require protection against frost. Adequate watering and protection against frost seems to be the only demand of the vines. A little before flowering one dose of 10 kg of F.Y.M. supplemented with complete fertilizer (NPK in the ratio of 10 : 6 : 10) @ 0.5 kg per grown up vine does well.

The phosphate fertilizers may be given in split dose in spring and late summer. Potassic nutrition is needed in maximum amounts during the maturity of the fruits. The vines benefit by the adequate supply of Ca, Mg, Cu, Fe, Mn, B and Fe. In the case of micronutrients foliar spray of these is the best way to remove the deficiency. Replenishment of the nutrients taken up by the vines is necessary but care should be taken to avoid excess of nitrogen supply which encourages vegetative growth at the expense of flowers and fruits.

Pruning and Training

A few years old canes give new shoots which bear fruits. Regular pruning encouraging new wood production should be done during winter when the vines are dormant. Pruning may be done to reduce disease and pest and encourage growth. Pruning is not beneficial from the yield point of view since unpruned vines outyield those pruned. Pruning should be done only when the vine is in dormant stage. Summer pruning of mature trees increases fruit size and colour. Use of growth retardants like paclobutrazol helps in widening croch angles.

Two systems of training are *in vogue.*

(*i*) *Trellis*, more or less identical to the kniffen system practised in the case of grapes, is easy to develop and maintain. The vines thus trained yield heavily.

(*ii*) *Arbour* or *Pergola*. The vines trained by this system bear even heavier crops than those trained in trellis system, but the cost of putting up an arbour is high and the pruning becomes comparatively more difficult.

Harvesting and Marketing

The fruit may be round or oval, greenish yellow or purple when ripe. It has a thin, shell-like rind enclosing many black or brown seeds which are covered by a yellow or orange aromatic juicy flesh with acidic but pleasant flavour.

The vines bear fruits from May to November which becomes ready for harvest when these attain characteristic colour. The fruit is either picked from the vine or allowed to drop on the ground and then collected. A yield of 7.5 kg or 100 to 250 fruits per vine is generally obtained every year.

For storage a temperature of 5°C and humidity (80-90%) is beneficial. The fruits may be dipped in some fungicide to extend the storage life of the fruit by reducing the chances of rotting. A dip in wax helps in reducing loss of moisture from the fruit.

Cultivars

Of the several cultivars of *Passiflora*, six bear edible fruits and the rest are wild species. In India only two types are commonly grown.

Passiflora edulis, the (PURPLE fruited) is common in India. Fruit round to oval, purple with well marked and pleasant flavour and a prominent white bloom. The fruit is acidic. The vine starts bearing after 2 years. The annual yield per vine is 200 to 250 fruits of 10-50 gm each.

Passiflora edulis var. *flavicarpa* (YELLOW fruited). The vine grows vigorously in the plains of South India. The fruit has a more pronounced and pleasant flavour. The flowers being protoandrous cross pollination must be resorted to for better bearing. Fully ripe fruits yellow in colour and spherical or ovoid in shape are available in May-June and September-October. It starts bearing after a year and the yield may be 12 tons per hectare. It is resistant to Fusarium wilt and Phytopthora blight and tolerant to leaf spot and nematodes.

Passiflora quadrangularis. The name is possibly derived from the four sided elongated large fruits that it bears. It is grown on the lower elevations of the Nilgiris. The fruits are of inferior quality as compared to the above two strains. This is a woody vine, native to tropical America where it is grown for its edible fruit (giant granadilla).

Besides these yellow granadilla, *P. laurifolia* L.- Waterlemon, *P. maliformis* L., the sweet calabash *P. mollissima* (banana passion fruit), *F. caerulea* (blue passion fruit), *P. lingularis* (sweet grandilla) also produce edible fruits.

Passiflora holosericea Linn. It is a woody climber with fragrant flowers. The plant flowers in flushes from April to November. It is native of Vera Cruz in South America. This silky leaved passion-flower bears the specific name holosericea because of its pubescent stems and leaves.

Passiflora lunata J.E. Sw (Syn. *P. biflora* Lamk). It is native of Jamaica and Mexico. Ornamental flowers are produced from March to November in flushes.

Passiflora incarnata Linn. (Wild passion fruit). The plant is well suited for screening and ornamental planting. It is native of South-East United States of America. The flowers are purplish and sweet smelling. It tolerates cold.

Kaveri (Purple × yellow). It is a high yielding hybrid tolerant to most of the diseases and also nematodes.

Neel's special. It bears fruits quite early at the age of one year. This is tolerant to the disease Brown spot caused by *Alternaria passiflorae* and is self incompatible.

Insect Pests and Diseases

Several pests have been reported to attack passion fruit vines causing losses due to damaged fruits or lowering of yield. The pests are fruit fly puncturing and damaging the fruit and downgrading the quality. Dimethoate provides effective control when applied once every fortnight from October to April. It also controls mealy bug. Natural enemies of the mealy bug *viz.*, wasp parasites, lacewing larvae and ladybirds keep the activity of the pest under control. Ants which interfere with natural control of mealy bugs must be removed first to help natural control. Passion vine mite affects both young and old vines during summer and winter. Effective control is possible by sprays of wettable sulphur every 2-3 weeks.

Among the important diseases reported are wilt disease (caused by a species of *Fusarium*), woodiness virus, Rhizoctonia root disease, Grease spot (*Pseudomonas passiflorae*), chlorotic spot and brown spot caused by *Alternaria passiflora.* Brown spot, septoria spot and alternata spot infect the fruit. Golden passion fruit and hybrid Purple Gold are less susceptible. Phytophthora blight causes serious damage and the fruits are rendered unmarketable. Brown spot may be controlled through spray of Dithane Z-78 @ 2 gm/l water. Copper oxychloride may be used to control leaf blight @ 3 gm/l water.

CHAPTER

23

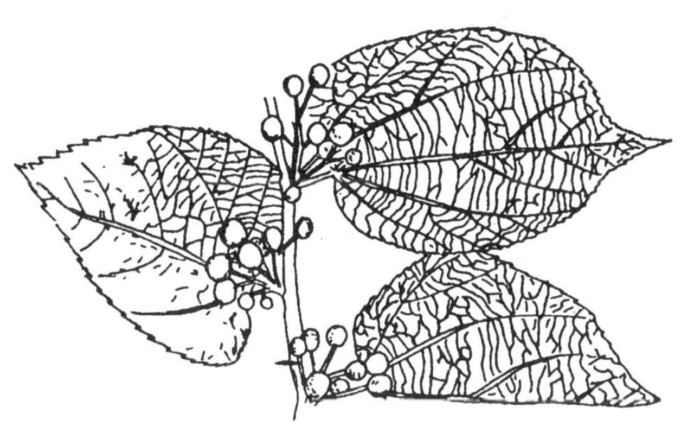

PHALSA

(*Grewia asiatica* L.)

Tiliaceae

Origin and Distribution

Indigenous to India *Phalsa* is highly esteemed as fresh fruit, juice and syrup. Phalsa a typical berry of summer, grown for a very short period in a year, is high in vitamin C content (natural ascorbic acid) besides being a good source of calcium (129 mg), phosphorus (39 mg), iron (3.1 mg), 72 kcal energy, low carbohydrate (14.7 gm), fat (0.9 gm) and carotene (419 microgram) which makes it slightly alkaline with a cooling effect, when taken as sharbat/drink. The fully ripe fruit (berry) does not keep long. The small hardy bush bearing small berry like fruits are marketed in March to June in different parts of India. It is grown in U.P., Punjab, Rajasthan, M.P., Maharashtra, West Bengal, Bihar and South India.

Climate and Soil

Capable of growing and fruiting in the hot, dry environments phalsa trees can also thrive in the humid tracts of coastal India. Phalsa production responds to distinct alternation of winters and summers. The plants remain evergreen in regions devoid of winters and flowers more than once every year but the produce is less and of inferior quality. Rich loamy soil is considered to be good for the deep root system of phalsa. Non-alkali soils free from layers of 'Kankar' are considered suitable for phalsa although certain cultivars can be cultivated in soil with pH 7.5-8.5. Any type of soil of loamy texture, free from pan formation is suitable. It is adapted to do well in drought conditions.

Propagation

Seed propagation of phalsa is common but cuttings may also be used for the purpose with the aid of growth regulators. The seeds are sown in the monsoon in nursery beds at a distance of 5-8 cm from one another and 22 to 30 cm between rows. Plants may be put in pits at a distance of one metre in

December-January or July-October. The phalsa can be planted, in February by lifting the seedlings without ball of earth. Layering may also be done in the rainy season. IBA (10,000 ppm), NAA (10,000 ppm), 2,4-D (1000 ppm) and boron (100 ppm) help in rooting. Rooting of hardwood cuttings is facilitated by IBA and ethrel treatments.

There are two kinds of Phalsa grown by cultivators who have chosen to call them as "local" and "Sharbati". The plants of some phalsa are tall and others dwarf as has been reported by the Haryana Agricultural University. The yield of the dwarf type is more than the tall type.

Cultural Practices

First irrigation is needed immediately after the application of fertilisers in second week of February thereafter every 20 days till needed. Irrigations are supplied whenever required, at regular intervals during flowering and fruiting. It is a hardy fruit tree resistant to drought and not exacting in its requirements. The fruit is borne in clusters in the axil of the leaves only on the growing shoots. Severe annual pruning encourages new vigorous shoots and ensures regular and heavy fruiting. Pruning should be done when the plants shed their leaves and winters (North India). The plants should be cut off at a height of about 0.5 to 1 metre from the ground between December and January when these are dormant. In the south the trees are placed at greater distance and allowed to grow and pruning is not done. Intercropping may be done with profit. The roots are exposed to the sun for 15-20 days followed by a supply of ample amount of F.Y.M. + 250 gm superphosphate + 100 gm Urea + 100 gm Potash and covered by soil. Light irrigation after manuring is good for the growth of the plant as well as the yield.

Harvesting and Marketing

A good crop is obtained during the third year after planting. As the fruits ripe, in the hot season the colour changes to deep red brown or purple and pulp tastes sweeter indicating the picking time. A bush may yield 5-10 kg of ripe fruit. Fruits must be disposed off within 24 hours to ward off spoilage. At the time of flowering spray of Gibberellic acid (60 gm/l) +5 ppm 2, 4, S-T encourages fruit production and improves its quality. Two sprays are recommended.

Insect Pests and Diseases

This hardy plant is resistant to the attack of insect pests and diseases. However, bark eating caterpillar has been observed to cause some damage at certain places. It can be controlled by inserting cotton swab dipped in 0.05% solution of Dichlorovos in the holes caused by the caterpillar. The holes should be closed by mud. Besides this the mealy bug may also be found when phalsa is grown in mango groves *Cercospora grewiae* causes leaf spot disease, which may be controlled by dithane Z-78 spray (0.3%) or Bordeaux mixture (4 : 4 : 50) or 0.3% copper oxychloride.

CHAPTER

PINEAPPLE

(*Ananas comosus* Merr.)

Bromeliaceae

Origin and Distribution

Ananas comosus syn. *A. sativus* Schult., the pineapple, is one of the most important commercial fruits of the world. The area between 15°-30° south latitude and 40°-60° west longitude is considered to be the place of origin of pineapple. The wild Brazilian pineapple (*Ananas microstachys* Lindl.) is considered to be its ancestor. The pineapple, a native of Brazil, is known as Pina in Spanish and Ananas in French. It is grown in Cuba, Hawaii, Africa, Philippines, Malaysia, Australia, Kenya, Mexico, Brazil and Bangladesh. In India, Tamil Nadu, Karnataka, Andhra Pradesh, Maharashtra, Tripura, Goa, Pondicherry, Assam, West Bengal are the chief growers.

Pineapple provides a range of health promoting plant chemicals. Modern scientists believe that the enzyme bromelain, present in the fruit, which breaks down protein has healing property. Bromelain assists in digestion of protein-rich foods besides being anti-inflammatory and helping disperse blood clots. Pineapple also protects against stomach cancer through the plant nutrients P-commaric and chlorogenic acids. Pineapple should be consumed raw, when cooked it has no theropetic effects.

Climate and Soil

Although the belt between 25° north and south provides climate suitable to the growth of pineapple in Australia and South Africa the growing region extends upto, 34° south. Temperatures between 15°C and 35°C may be considered tolerable while 0°C – 20°C is considered favourable. Pineapple is sensitive to frost and also intense insolation. It may be grown successfully upto altitudes of 1,100 m above sea level. Annual rainfall of 100-150 cm is considered optimum for pineapple. Mulching, shade and drainage wherever needed help in increasing the yield.

Mild tropical climate with partial shading gives good crop. A temperature range of 15°C to 32°C is considered suitable and a cool winter improves the quality of the fruit. Light soils are considered suitable for the fruit. Sandy, loamy and the slopy laterites with good drainage seems to be ideal. High manganese/ferrous ratio is considerably injurious and spray of 1% $FeSO_4$ is recommended where the soils are high in Mn content. Alkaline soils are not considered suitable. Heavy loams must be avoided. The optimum pH range reported for the successful growing of pineapple is between 5.5 to 6.0.

Propagation

Mature seeds are obtained from the ripe fruit, dried and kept in carbon bi-sulphide (a fumigant) for 48 hours and then treated with concentrated sulphuric acid for 40 seconds before being kept in the sterile sand for germination. The seeds are allowed to germinate and attain a height of 10 cm and then transferred to new places for a period of about 18 months before being planted in the fields. Plants raised from suckers need about 17 months to fruit while those from slip take 20 months and the ones raised from crown need almost two years. Aerial suckers, basal slips, crown slips, the underground parts of stem or the crown may be used for vegetative propagation. Suckers from main stem bear early while the slips (arising immediately below the frond) give large fruits. The suckers should be stored, left to dry, before planting. Seed treatment with Bordeaux mixture or dipping of the basal end in a mixture of 'Blitox' or 'Bordeaux' mixture prior to propagation is recommended to ward off rotting and mealy bug infestation.

Plantings are done during rains on ridges. Plants that send out basal slips are considered unsuitable as source of propagation material. Planting material should be collected from high yielding well maintained plantations. Planting may be done in 'single row' or 'double row' system. The distance between rows and plants in the former system varies between 150-170 cm and 90-150 cm respectively. In the double row system the distance between rows is 45-60 cm and between plants 30-45 cm. Ten to fifteen thousand plants may be put per hectare in India which is half as much in other countries. The usual distance between plants is 45 cm and that between rows 90 cm. Trials at the Indian Institute of Horticultural Research, Bangalore, have proved conclusively that dense planting in pineapple resulted in high yield without affecting the quality and size of the fruit.

Natural Mulch

Close planting results in the overlapping of the basal leaves forming a sort of natural covering over the soil, preventing the evaporational losses and thereby reducing the frequency of irrigation.

Merits of Close Planting

Close planting has the following merits over traditional method :

(*i*) It helped in reducing the frequency of weeding operations and in reducing the cost of cultivation.

(*ii*) Increase in planting density resulted in a single, more acidic fruits.

(*iii*) By increasing the plant population upto 63,758 plants per hectare (by planting at a distance of 22.5 × 60 × 75 cm in two row beds) the yields could be increased nearly six times against that obtained by the conventional methods, without adversely affecting the fruit size and shape.

Light and frequent irrigation during the off monsoon season is also important. Manuring and weeding give good returns. Flowering may be regulated by NAA spray and calcium carbide treatment. Green yellow fruits are harvested and the yield is reported to be 8-25 tons per hectare. The planting material required depends on the cultivar and the local conditions. Generally ratoons or aerial suckers (or "Suckers") are used on large scale. Ratoons give earliest crop. The crown when planted takes 2 years. Generally these are used in the smooth Cayenne cultivar. Stumps may also be planted on furrows and the shoots that develop on these, removed and planted.

The health and vigour of the plant from which the material is taken are of great importance. Poor growth is due to lack of fertility; suckers or slips from undernourished plants may produce vigorous plants under favourable conditions. As the plants develop fruits, slips, and suckers also grow some of which get rooted and these give rise to still more slips and suckers so that the second ratoon crop comes mostly from suckers of the main crop which did not fruit during the first ratoon season. This indefinite cycle continues for the benefit of the grower on productive soils for over two decades.

Weeds may be removed either by mechanical means or by the use of weedicide like pentachlorophenol, tapazine, tafapon, 2, 4-D must not be used as it affects pineapple. Use of Dalapon before planting eradicates weeds completely, perennials like *Cyperus* and *Oxalis* may be controlled by phenoxyacetic acid, phenylacetate, benzoic and sulphuric acids. Monouron @ 2 kg/ha in 90 litre of water has been used successfully in the control of weeds in pineapple fields. Bromacil and Diuron @ 2 kg/ha and Gramoxone @ 3 kg/ha are being used freely in pineapple growing tracts every 6 months with success.

Fertilizers in dilute form and with high inorganic content are admirably suited for the purpose. On an hectare basis 60-75 cart loads of F.Y.M., 680 kg ammonium sulphate, 340 kg of P_2O_5 and 680 kg of K_2O in 4 split doses is recommended. Ammonium sulphate is preferred to sodium nitrate as the source of nitrogen since the latter leaves an alkaline residue. 5-10 gm phosphate per plant is considered sufficient, high doses are not needed.

An application of 15-20 kg (per 1,000 plants) of a 10 : 6 : 10 NPK mixture soon after planting with two or three more applications of 25 kg each during the year is reported from Assam. Top dressing of 1,750 kg of fish manure is practised in two doses, three and five months after planting in Maharashtra.

All suckers should be removed except two per plant for the ratoon crop.

Flowering

Pineapple flowers when the ripeness-to-flower stage is attained. The plant takes more than a year and has about 40 leaves at that stage. Pineapple may be induced to flower, if the stems are kept in horizontal position, due to geotropism. The change in tropism initiates accumulation of auxin which induces bud initiation. Gravity leads to the displacement of statocysts causing disturbance of the cell membrane of lower surface giving rise to altered electrical potential and modified translocation pattern resulting in accumulation of more auxins on the lower surface causing negative geotropic curvature.

Growth regulators can also induce uniform flowering in pineapple and can thus considerably reduce irregular fruiting. Ethrel, calcium carbide with water, auxin-like substances (NAA, 2, 4-D) and ethylene have been used to induce flowering. NAA has proved very effective.

Fruiting

Pineapple cultivars are associated with the problem of irregular fruiting. Pineapple plants normally produce fruits during June to August. Occasionally a few fruits may be found during September and October and practically no fruiting occurs during the rest of the year.

Researches conducted in Bangladesh Agricultural University on the control of irregular fruiting has led to the following conclusions.

The pineapple plants treated with 5, 10 and 15 ppm of NAA required 45, 42 and 40 days respectively for flower initiation. Plants treated with 10 ppm of 2, 4-D required 52 days for flower initiation. A positive correlation was found to exist between leaf number and fruit weight in pineapple. All the treatments of NAA and IBA had significant effect on plant height. A gradual decrease in plant height was observed with the increasing concentrations of 2, 4-D upto 50 ppm.

Harvesting and Marketing

Special problems with harvesting are that pineapple should not be harvested unripe since it undermines the quality of the fruit. The whole crop gets ready in about 2-3 months so that the problem of spoilage sets in the harvested fruits. Spray of parachlorophenoxyacetic acid 10 days before harvest has been practised to reduce physiological break-down during transport and storage. The fruit gradually changes from dark green to light green yellow, sometimes to a deep orange with the onset of maturity. As it matures the pineapple fruit also becomes less angular about the eyes, and the ends of the bracts protecting the 'eyes' dry up. When at least two-three rows of eyes at the base turn yellow pineapple may be ready for harvest. The eyes flatten as the fruit matures. Harvesting is done by cutting the fruit stalk about 3 cm below its base. The cut end is dipped in 10% solution of benzoic acid in alcohol to check the attack of fungus so that the storage life is prolonged. Successful marketing depends on the proper maturity of fruits at harvest, development of uniform colour and shape of the cultivar. Fault-free fruits without injury with proper wrapping of individual fruits in transport, careful handling with no drenching in rains, helps in getting more profits.

Ripening and Storage

Treatment with ethrel enhanced ripening of pineapple giving attractive appearance and imparting a little acidic taste. The stage of maturity at harvest, the temperature of the atmosphere, as well as the characteristics of the cultivar determined the storage life of pineapple. A treatment with NAA (100 ppm) just after picking prolonged the storage life of the fruit. Gibberellic acid spray of 9-90 ppm strength at 28°C retained the green colour for a fortnight or more.

Yield and Quality

Yield vary to a large extent with the soil and the climatic conditions, culture, density of plants and the cultivar. The yield may be 60-70 tons of fruit per hectare in Hawaii, about six times that in India. High density planting increases pineapple yields.

The quality of pineapple rests on the amount of sugar and acid present in the fruit. On an average the sugar content is 10 to 12%, protein 0.6% and the acid content 0.6% to 0.8%. Besides, it has got vitamins (A, riboflavin and C) and minerals like calcium, phosphorus, iron etc.

When the juice is extracted for canning dried waste (pineapple bran) a valuable feed is obtained. Other by-products are alcohol, citric acid vinegar. The leaves yield good fibre for making cloth.

Cultivars

Indigenous types. These are inferior in fruit quality. In Kerala *Kallara local, Trichur local* and *Thaliparamba local* named after the places where these grow as hedge plants. *Jaldhup* and *Lakhat* (sour) are grown in Assam. In Andhra Pradesh *Simhachalan* is largely grown. Fruits are small and sour in taste.

Some of the common cultivars (Queen, Ripley Queen, Mc-Gregor and Alexandrica) belong to the 'Queen' group. Charlotte Rothschild, red and yellow Mauritius represent the Spanish type. Kew and Giant Kew represent the Cayenne type. Besides these three, indigenous types are also found throughout the country. As with all other long established fruit crops, innumerable cultivars of pineapple are grown all over the world for home consumption.

Queen. Plants spiny, dwarf and compact with serrate leaves, very early, best dessert cultivar. Ready in June-July. Fruits of medium size (Av. wt. 1.25 kg) with prominent deep eyes, deep yellow on maturity. The flesh is firm, crisp, sweet and golden yellow in colour. It suckers freely, necessitating vigorous thinning of the suckers in ratoon to maintain size of the fruit. Mac Gregor, a selection from Queen is also very popular.

Ripley Queen. Earlier than the common Queen with more conical fruit. Another cultivar *Alexandrica* is a selection from Ripley Queen producing larger fruits.

Charlotte Rothschild. A spiny cultivar of the Spanish group which compares very well with *Kew* in taste. The fruit is smaller in size.

Kew. Smooth leaves and large fruits having broad and shallow eyes (suitable for canning), ripe fruit is yellow. The flesh is firm, juicy and pale yellow in colour. Shy suckering habit.

Mauritius. Mid-season cultivar ripening in July-August. Medium sized fruits (2-3 kg) may be yellow or red skinned with flesh of the same colour. The fruits are oblong, sloping slightly towards the apex which are later rounded at the base and sloping abruptly towards crown. The fruits of yellow types are green in colour when mature, these gradually turn deep yellow when fully ripe. The yellow fruits are less sweet than the red ones.

Red Spanish. A leading hardy, vigorous, disease resistant cultivar of Cuba, Puerto Rico and Malaya. The fruit size averages 0.9 kg to 1.4 kg. Second-rate quality, not good for canning. Flesh pale-yellow, fibrous with pleasant aroma.

Sugar Leaf. Widely grown in America, medium sized (0.7- 0.9 kg) sweetest and best flavoured fruit of the white fleshed cultivar. Difficult to handle.

Abakka. Like red Spanish with rich, sweet flavour, weight average, 1.8 to 2.7 kg, needs care in handling. Basal slips on fruits are common.

Natal Queen. The leading cultivar in Australia and South Africa. Average of 0.5 to 1.0 kg sweet fruits having distinctive aroma and delicious flavour but less juice than other cultivars. Good for home consumption. Many suckers are found on the plant. There are many strains of the cultivar.

Eleuthera. Very much similar to Abakka with smaller fruits. Tender slipper with absence of basal slips. Double or multiple crowns are common. White flesh with a yellowish tinge, sweet rich fragrance and juicy character.

Lakhat. Small fruited wild acidic type of Assam.

Simhachalan. Small fruited cultivar of coastal Andhra Pradesh.

Jaldhoop. A small fruited cultivar of Assam. The fruits are sweet and the sweetness is well blended with acidity. It has a characteristic alcoholic flavour.

Cabezona. Large sized fruits (upto 7 kg). Grown in Puerto Rico.

Congo Red. Medium sized fruits, yellow flesh and good flavour are its characteristic features.

Giant Kew. (Smooth Cayenne). It is the most important cultivar in the world, fruiting all the year round in Hawaiian climate (sometimes acetylene treatment is given for regulating the ripening of the fruits). Leaves are spineless, ash grey with smooth margins and the fruits are large oblong, slightly tapering towards the crown, suitable for canning, flesh yellow when ripe, excellent flavour, almost fibreless, very juicy with aggreable taste. Shy suckering habit, Av. weight of fruit is 2 to 4 kg.

Cayenne. Very heterozygous. A number of clones have arisen as somatic mutation. The height may be 30-45 cm with about 80 reddish leaves mottling above and silver-grey beneath. Fruits large (av. wt. 2.5 kg), dark blackish green, when unripe. Juice has acidic taste. Sugar content about 12%. Not cultivated in India.

Pest and Diseases

There is no serious disease or pest in India although it is affected by various pests and diseases.

Nematodes (Heterodera radicola) threatens the crop very seriously and can be controlled to a considerable extent by treating the soil with chloropicrin gas before planting. Fumigation with Dowfume W-85 may also prove effective in controlling nematodes.

Root Knot Nematode (Meloidogyne spp.). The nematode is a serious threat to the crop. The galling of root is visible as a symptom soon after its attack. The plants become unhealthy. Growing of sugarcane in the fields where the symptoms of this nematode appear in pineapple is recommended as crop rotation.

Mealy bug (Dysmicoccus brevipus). The bug may be seen in colonies. These suck the sap of the plant from almost all the parts and considered to be a major pest of pineapple in India. The first change visible is the turning of leaves to red or pink colour followed by the drying of the leaf apex and the margins turn inward. Planting of healthy suckers dipped in 0.05% solution of fentrothion or phenthoate for 10-15 minutes, three months after planting phorate 10 G @ 15-20 kg/ha should be applied and the field must be kept clean of all weeds and grasses for control.

Treating the planting material with paraffin, cyanogas or methyl bromide also serve as preventive. *Pseudococcus bromeliae* B are reddish mealy bug which infect fruits and shoots of pineapple. In the field plants may be sprayed with parathion. The cultivar Cayenne is very susceptible to mealy bug. Two natural parasites *Hambletonia pseudococcina* and *Anagrus cocciduorus* are known to control mealy bug.

Fruit eating beetle (Carpophilus dimidiatus). The beetles feed on ripe fruits therefore the fruits should not be allowed to remain in the fields to ripen. Damaged fruits should be promptly collected and destroyed.

Termites (Odontotermes spp.*).* Termite infestation is noticed in sandy soils. The dried suckers should not be allowed to remain in the fields. Whenever the fields receive timely and regular irrigation the termite infestation is reduced.

Drenching the field with Lindane 20 EC @ 3.75 l/ha or chloropyriphos 20 EC @ 2-3 l/ha conrols the termites effectively.

Soft rot, storage rot and fruit rot caused by Ceratostomella paradoxa. A serious menace in Java and Gorakhpur district of Uttar Pradesh. Paint the cut end of stems within five hours of harvest with 10% solution of Benzoic acid in alcohol. Fruit rot may be controlled by dipping planting materials in 1 : 1 : 3 Bordeaux mixture or Phytolan before planting. Fruit rot generally takes place in storage.

Heart or stem rot (Phytophthora parasitica). It is not common in India but occurs in other countries. It causes slight wilting and twisting of the leaves accompanied by discolouration and sometimes the disintegration of the stem.

A dip before planting in fosetyl-Al prevents the onset of the disease to considerable extent. The treatment is seen to be effective for a period of one and a half years. The fungus develops resistance to the fungicide after some time and just to overcome the capability of the fungus to develop resistance it is advised to mix mancozeb with fosetyl-Al both in the dip treatment and also spray over the plants.

Good drainage, careful handling to avoid wounding the planting material and preplanting dip of the material in 1 : 1 : 3 Bordeaux mixture, also helps in control of the disease.

Mealy bug wilt. The bug transmits the disease and the loss caused to the crop is about 10-15%. The symptoms include the wilting of leaves from the tip downward, colour of the leaves change to bronze to red. The leaf apex curl tightly back, the roots decay, wilting sits in and finally death. A dip of the planting material in 0.02% methyl parathion and application of Phorate 10 G @ 16 kg/ha to the affected plantation every 3 months is advised for control.

CHAPTER

POMEGRANATE

(*Punica granatum* L.)

Punicaceae

Origin and Distribution

Pomegranate migrated from Iran and is cultivated extensively in Spain, Morocco, Egypt, Afghanistan, Arabia, Baluchistan, Burma, China, Japan, Azerbaijan and California. In India it is grown in Maharashtra, Rajashthan, Haryana, Gujarat, Andhra Pradesh, Uttar Pradesh, Tamil Nadu and Karnataka.

The pomegranate (H. *Anaar)* is very much liked for its cool, refreshing juice with sweet acidic taste and also for its medicinal properties. The juice is useful for patients suffering from leprosy. Citric acid and sodium citrate are manufactured from the juice of pomegranate in Azerbaijan. The juice is also fermented to produce wine. Its barks and rinds are commonly used in dysentery and diarrhoea. The rind is also used as dyeing material for cloth.

Climate and Soil

Cool winters and hot dry summers produce the best quality fruit. It can tolerate frost and may be seen at an altitude of 1,600 metres. The fruit quality deteriorates in humid climates. It thrives in hot dry regions with irrigation facilities. Although the tree can withstand frost it is injured by temperatures below -11°C. The deep loams or alluvial soils are considered ideal; even medium or deep light black soils are useful. Alkaline and wet soils may also be tolerated by pomegranate plants. The cold winters make the trees deciduous which are otherwise evergreen.

Propagation

Seed propagation is possible though not very desirable. The seed may be sown in October and seedlings are ready to be transplanted during the following monsoon. Fruits of uniformly high quality may be obtained or propagated by hardwood cuttings from one-year old, fully mature wood or by layering or by root sucker.

Planting

The rooted cuttings are transplanted with the start of spring. The transplanting may also be done during monsoon. The plants are dug from the nursery with a ball of earth. Planting of the rooted cuttings is done 4 m apart in pits 1 × 1 × 1 m just before the onset of monsoon. As the plants start growing some pruning may be done to remove most of the side shoots and leaves. Use of growth promoting substances *viz.* IAA and IBA greatly improve rooting of cutting. Increase in the moisture level also increases rooting.

Cultural Practice

The plants need watering daily for about a week after planting. After the plants are well established, in about 6 months, they can stand considerable amount of water stress and irrigation may be given at intervals of 2 to 4 weeks as per need. The pomegranate is inter-cropped with vegetables. Pomegranates may be (1) double flowering ornamental type and (2) white flowered type. One-year old plant is manured with about 11 kg of F.Y.M. and 1/2 to 1 kg of cake and 125 gm nitrogen + 62.5 gm phosphorous + 62.5 gm potash each year for five years thereafter the dose may be determined as per need of the plant. The fertilizers as well as manure should be applied to the bearer orchard in the basin, one metre away from the trunk during the 'bahar' in split doses.

Pomegranate has been known to improve the fruit production by the spray of $ZnSO_4$ + $CuSO_4$ + Borax mixture in the respective concentrations of 0.6%, 0.4% and 0.2%. Two sprays may be given during flowering and fruiting period. The response is favourable since the micronutrients (Zinc, Copper and Boron) play a vital role.

The plants should be properly trained to form a single stem with a number of well distributed scaffold limbs. It is essential that the suckers are removed as soon as they arise. Flowering is regulated by withholding irrigation 2 months before the normal flowering time. At the end of this imposed drought period the fertilizers may be applied followed by light watering. After 4-5 days heavy irrigation is given to be followed by the normal watering schedule. By exposing the roots in April followed by irrigation and manuring early flowering may be induced and the fruits are ready by November-December. On a small scale the removal of flowers may be done by hand or any other mechanical means. This is necessary to regulate the cropping behaviour. In orchands chemicals like Ethophan (2000 ppm) or Alar (500-3000 ppm) may be sprayed during the flowering time to achieve the desired outcome. The following seasons are recognised in Deccan (*i*) Mrig-bahar in June-July or break of monsoon, (*ii*) Ambia-bahar in Feb.-March and (*iii*) Hasta-bahar in Sept.-October. The Mrig-bahar fruits mature in October-November. Fruiting should be allowed only once a year by regulating the flowering.

limits for the economic production of apple. Optimum temperature range for growing apples is between 13-21°C. High temperatures and lower C/N ratio affect the crop adversely.

Site and Soil

Sloping sites, to allow free drainage, are considered more suitable than level tops. The slopes considered suitable are between – 4 to – 1°C temperature zones with altitudes no more than 2,500 meters.

The ideal soil *viz.*, loam, sandy loam and silt loam with open, porous and well drained sub-soil free of weeds suits apples. Deep organic loams (Kashmir), light loams (Kulu valley), brown or reddish brown soils (Kumaun) find favour for apples in different parts of India.

Propagation and Culture

Apples are ordinarily propagated by T-budding or grafting on seedlings of Crab apple, Yellow Newton (Rus Pippin) or Golden Delicious. Winters are best for grafting though shield budding is done in June with the season's bud. Both whip grafting and shield budding are widely practised in India. Apples do not prove profitable on their own roots and these are therefore grafted on root stocks of different apples. Seeds obtained from fruits of well mature commercial apple trees are used to raise rootstock besides commercial varieties of Crab apple also serve as root stock. One species of *Malus baccata* var *himalacia (pareo)* is also good for the purpose.

Rootstocks released as M, Merton and M M have been found to be suitable in different situations. Only very prominent characteristics of some of these is given here in the inset below :

Characteristics of some clonally propagated rootstocks

Rootstock	Characteristics
M_2	Withstands drought, suitable for sandy soils and for weak cultivars. Susceptible to woolly aphid.
M_7	Easily propagated, vigorous growth for five years, profuse branching.
M_9	Very dwarfing, very early bearing, weak stock needs staking. Resistant to woolly aphid. Poor average roots.
M_{16}	Suitable for bushy trees and half standards on very poor soils.
M_{25}	Suitable for half standards specially in light soils.
M_{26}	Easily propagated, better than M_9.
M_{27}	Ultra dwarfing.
M_2, M_{12}, MM_{111}	Invigorating.

Harvesting and Marketing

The trees may bear fruits from the fourth year and continue to do so for about 25 to 30 years. Cracking of fruits caused due to hardening of skin of the fruit during droughts followed by rains may be avoided by picking premature fruits or by regular irrigation. Cracking is more common in Mrig-bahar crop. The regularly irrigated crop Ambia-bahar does not show cracks. Premature picking lowers the quality of the fruits. The moisture retaining capacity of the soil should be increased and boron deficiency warded off to avoid cracking or splitting of fruits.

Application of GA_3 (15 ppm) and boron (0.2%) has been found to reduce cracking and improve the fruit colour. Another physiological disorder known as internal breakdown leading to the blockening of the arils. This may be greatly minimised by harvesting of the fruits 120-135 days after fruit set.

Cultivars

Two types of pomegranate are recognized (1) Evergreen and (2) Deciduous. There is no leaf fall in the Evergreen type *viz*. Bassein Seedless, Ganesh Kabuli, Musket red, Dholka and Jodhpuri cultivars which can succeed in areas with less cold. Deciduous types are not suitable for the plains of Uttar Pradesh. These need low temperature during winters for good flowering and fruiting. In the absence of low-temperature in winters the yield of the deciduous type dwindles.

In northern India, there is no well recognized commercial cultivar. *Paper shell*, popular in South India, is a cultivar with soft seeds and pinkish flesh. *Daru* a wild cultivar with non-edible fruits may be seen growing in the Himalayas. It yields the well known condiment "anar dana". In western India, two types locally known as **Alandi** and **Dholka** are commercially propagated by seed. The **Dholka**, an important cultivar of Gujarat, bears fruis of large size, the colour of the rind is greenish-white and the flesh pinkish-white or whitish, the seeds are rather soft and the juice acidic sweet. Alandi or Vadki a cultivar common in Pune area has medium size fruits deep pink flesh sweet acidic taste and hard seeds. Ganesh Khind No. 1, a selction from Alandi, has pinkish flesh, sweet juice, medium sized fruits, high yield and soft seeds, Bedana and Kandhari with large fruits, deep red rind, fleshy tasta, juice sweet, acidic, and hard seeds have been known to be successful in Himachal Pradesh. G.B.G.I. and **Spanish Ruby** with small fruits, thick rind and soft seeds are prolific bearers while **Musket Red** with small fruits, thick rind, moderately sweet juice and seeds not very hard and **Vellodu** are low yielding cultivars of pomegranate of Andhra. *Jyothi*, highly attractive, yellowish red fruit of medium size, red aril and soft seeds. *Yercaud-1* is a cultivar released by Tamil Nadu Agricultural University for commercial cultivation. *Bedana, Bassein seedless, Ganesh, Paper Shell, Kandhari, Spanish Ruby* and *Dholka* are considered superior due to soft seeds.

Some commonly grown cultivars are given below :

G 137. A clonal selection from *Ganesh* is a prolific bearer. The tree has spreading habit, the fruit surface is reddish yellow and colour is pink. The fruit is sweet in taste.

P 23. Seedling selection from *Muscat*; fruit surface greenish yellow with red tinge aril pinkish white, heavy bearer of sweet taste.

P 26. This is also a seedling selection from *Muscat* :fruits large, rind yellowish green with red tinge, aril pinkish white, heavy bearer, sweet to taste.

Mridula. A hybrid (*Ganesh* × *Gul-e-Shah Red*) fruits of medium size, rind smooth dark red with blood-red aril, juicy and sweet to taste considered superior to other commercially grown strains of pomegranate.

Besides these S 242, 114 h selection and Co 1 are also known to be promising.

Insect Pests and Diseases

The Pomegranate butterfly or fruit borer. It causes maximum damage to the crop (20-80%) in India. The female lays eggs on the calyx in July. During monsoon the newly hatched caterpillars bore into the fruits and feed on the developing seeds leading to rotting and premature fruit drop. Bagging of the fruit with polythene or butter paper reduces the infection considerably. Economically beneficial alternative to bagging is giving two sprays of Demothoate (0.045%) or Deltamethion (0.003%) and Carbaryl 50 WP (0.2%) every 20 days interval in rotation after fruit set.

Bark eating caterpillar. Remove silken galleries and spray with Chloropyriphos (0.07%) or Cranbaryl (0.1%) on affected parts. Inject Dichlorovos (0.1%) into the holes made by the caterpillars and plug with clayey mud.

Stem-Borer (*Aleurodes* sp.). A caterpillar which eats through the bark of the tree is mostly found in old and neglected plantations. The pest can be controlled by clearing the spot especially the hole and killing the insect by inserting a small cotton plug dipped in petrol, chloroform, carbon bisulphide or kerosene oil and closing the hole with mud. Painting with coal tar also helps in checking the attack and in the healing of the wound caused by the borer.

Leaf spot and Fruit rot. The disease is caused by *Cercospora punicae* and *Colletotrichum gloeosporiodes*. Light brown to dark brown spots develop on the leaves and fruits. Humid condition make the disease severe and monsoon rains cause maximum damage to the *Mrig Bahar* crop. Mancozeb (2 gm/l) should be sprayed during the monsoon to protect the leaf from the infestation. In order to keep the fruit set free from the disease Carbedazine or Theophorate methyl (1 gm/l) should be sprayed during September-October.

Bacterial leaf and Fruit spot caused by *Xanthomonas campestris* pv. *puniacae*. Spots, deep red in colour and small in size appear on the leaf blade. The infection on fruits are raised, dark brown. The disease spreads rapidly in March-July when the temperature is high and humidity low. Three sprays of Pausomycin (500 ppm) + copper oxychloride (0.2%) at 15 days interval starting from the initiation of the disease significantly reduces the incidence.

TEMPERATE FRUITS

CH/

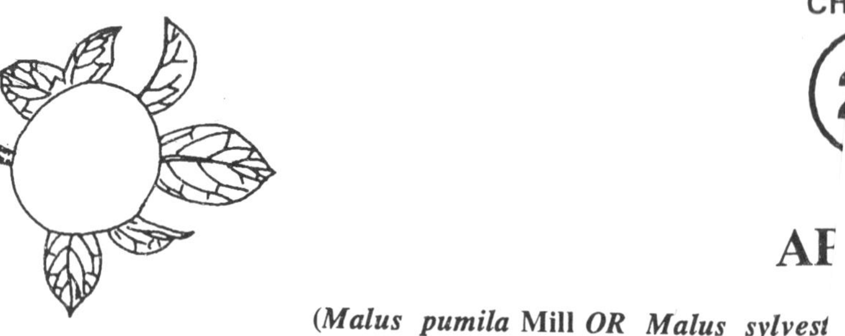

AI

(***Malus pumila*** **Mill** ***OR*** ***Malus sylvest***

Origin and Distribution

Apple (*Malus pumila, Pyrus Malus* L. or *Malus sylvestris*) an temperate fruit, a native of S.W. Aisa, occupies nearly 2.6% of the fruits in India. *Ambri*, a dessert cultivar indigenous to Kashmir and the region is well known for its table qualities. *Pyrus* or *Malus ba* apple) may be seen growing wild in the Himalayas. Other wild apple (*Pyrus coronaria, P. angustifolia, P. rivularis, P. isensis*) are forests of north America. The production was estimated to be per capita by Thapar in 1967.

Apple (H. *Seb*) cultivation is one of the major industries of t Pradesh in India yielding a gross income of about Rs. 50,000 per hec important apple growing areas are Jammu and Kashmir, Kumaon hil areas of West Bengal, Assam, Nilgiris, Manipur, Tripura, NEFA a Generally apples are grown at high altitude 1,600- 2,400 meters al but a few plantations are also reported from Bangalore and Nilgi an altitude of 1,000 and 1,300 meters. Firminger reported attemp to grow apples in various parts of India and good fruits were Bihar. In fact some cultivars require chilling below 4-5°C for ab for the buds to open in spring, *i.e.*, to break the rest period. below 3°C inactivates normal metabolic activities of the trees.

Leaf buds generally need a little more cold than the flower l where the leaves do not develop simultaneously with the flow lower altitudes) fruits do not set in due to lack of proper C/N r intensity and duration of cold is in a position to set the northe

Rootstock	Characteristics
M_7, MM_{106}	Semi-dwarfing.
M_8, M_{26}, MM_{104}	Dwarfing.
M_9, Merton 778]	Resistant to woolly aphid.
MM_{101}, MM_{104}, MM_{106}]	
M_{13}, Robusta no. 5	Vigorous.
M_{16}, M_{25}, MM_{109}	Very vigorous.
MM_{106}, MM_{111}, MM_{109}, M_2	Suitable for sandy soils with no rains for large part of the year.
MM_{104}	Very good anchorage, useful in place of staked bush trees.
MM_{106}	Suitable for dry sandy soils and heavy cropping.
MM_{111}	Suitable for warm climates and sandy soil.
Merton 793	Vigorous, colar-rot and woolly aphid resistant, early-fruiting, recommended for Kumaon hills of Uttranchal.
M	Malling.
MM	Malling Merton.
Apple seedling	Very invigorating.

M Stands for rootstock classified by East Malling Research Station.

MM Denotes those produced by East Malling Research Station in collaboration with John Innes Institution.

Merton is the one produced by John Innes Institution.

Suckers arising from roots of mature plants may also be used for vegetative propagation. Soak the roots in water for an hour or two if the plants have remained detached too long from the soil during transferring before planting. The planting holes should be one meter across and 20 cm deep so that all the roots may be accommodated in well spread condition. Remove all damaged roots. Stake the trees efficiently with the lower 60 cm treated with wood preservative (Cuprinol) for 24 hours. Speed out the roots before planting the sapling, making it firm if necessary. Supply well rotten compost or F.Y.M. Apples do not need liming. Desert cultivars prefer slightly acid soils and usually need more potash than nitrogen. Supply of excess nitrogen should be avoided. Apple trees are fed in advance of the spring season when they actually utilize the added inorganics.

Planting is done late in winter to avoid frost. For adequate root development a temperature of 7°C is considered useful. The distance between trees depends on cultivar, fertility of land, water holding capacity of soils, root stock type etc.

The dwarf trees are planted in the south side so that Sun's rays reach other trees easily. Planting of suitable pollinisers at suitable distance must be done preferably in alternating rows or one such row for every row of the desired cultivar.

The pollinisers can be put in three categories depending on the time of their flowering as early, mid-early and late. Among the earlies, Fanny and Early Shanbury can pollinate each other. Red and Royal Delicious, Jonathan and Golden Delicious are most suitable pollinisers among the mid-season or mid earlies. In this class are McIntosh, Winter Banana and Red Gold with abundant fertile pollen. The triploid varieties *viz.* Baldwin Tomkins, King and Blenheim Orange need pollinisers like Golden Delicious, McIntosh and Winter Banana. Of the late varieties Maharaji (Rymer) and Winter Delicious (Buckingham) set good crop with Red Delicious.

Fig. 26.1. Staking apple tree, Diagonal stake is considered more useful for short stemmed trees. Drive in the top of the state towards the prevailing winds to give a support to the tree against stiff breeze.

Apples are planted in ground free of weeds, regularly irrigated for about two years in the beginning, insurance against drought is a must for the crop and mulching with leaves etc. is recommended. Dalapon may be used to control weed in apple and pear orchard, when the trees are in dormant stage. Mature apple trees thrive better under sod, but the younger ones need protection from weeds etc. Intercropping with potato, pea, bean, cabbage, strawberry, etc. for the initial 8 years of planting has given good results. Orchards located on steep slopes need mulching. 5 quintals of bonemeal and 10 quintals of wood ash per hectare are given annually besides the manure given at preplanting stage. Ample supply of organic matter and superphosphate is good for the growth of the trees and also for bearing. Liming specially on acidic soils, is needed every 5 years.

It is desirable to add 100-150 gm of nitrogen as sulphate of ammonia or the like in the first year of planting specially after the plant has set itself. The fertilizer should be mixed with the soil in a radius of 1 m away from the plant. As the years pass by the amount of nitrogen may be increased by 150 gm every year. Foliar feeding is also useful.

Pruning and Thinning

Pruning may be done at (*i*) First dormant (*ii*) Second summer (*iii*) Second dormant (*iv*) Third summer and (*v*) Third and fourth dormant stages. Open centre system admits light prunings and has been practised owing to need.

Modified leader system with clear trunk of 1.5 meters and 3-5 main branches is generally practised in South India in respect of apple trees. When main scaffold branches are cut back to about 0.5 meter during first dormant pruning, some secondary branches emerge. Some of these are allowed to develop. Second dormant pruning consists in the removal of unwanted crowded and diseased branches. The process is repeated year in and year out for 5 years and only 10 branches are allowed to develop. Open centre type heavy pruning becomes necessary to keep the centre open. Pruning of bearing trees is necessary only to maintain a suitable balance between vegetative growth and fruit production. The short slow growing branches (the spurs) bear fruits in case of apples for several years. Less pruning is needed in these varieties. Pruning limits the extension of branches in a year upto 25-20 cm.

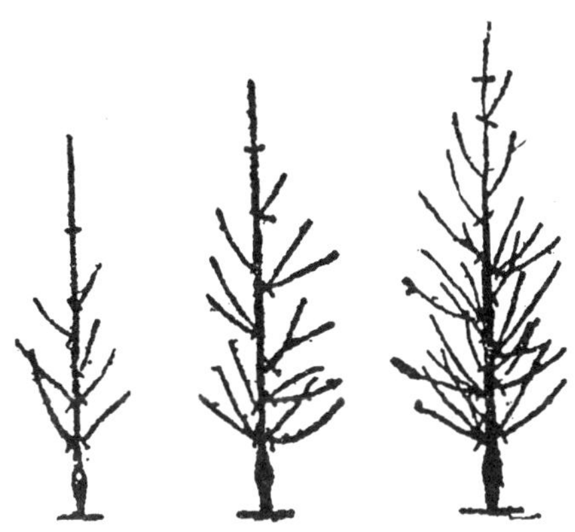

Fig. 26.2. Training a pillar for apple trees in which a central stem produces large number of laterals easy to prune and harvest.
One year old. **Prune back the stem to 70 cm high and cut back all lateals within 1.5 cm of the main stem.**
Two year-old. **Cut back the leader by half to just above a bud. Remove half of the laterals and leave the rest for the flower bud to be formed.**
Three-year old. **Prune back laterals that have fruited ; 2-yr-old laterals will fruit in the following year.**

In the regions of hail storm, pruning is reduced to be minimum to allow crowding and hanging down of long branches and afford protection to the fruits. Apple trees need training from the second year for cordons, bush trees dwarf pyramids, pillars or spindles and for half standard and full standard from third year, for espalier trees from fourth year. The following systems of training are generally followed for the apple trees.

1. *The full standard.* The trunks are about 1.8 meters with strong root stocks, planting distance 9 meters.

2. *The half standard.* The trunks are 1.35 meters, grafted on strong stocks, tolerate high nitrogen content of the poultry manure, planting distance 7.2 meters on all sides.

3. *The open centre bush tree.* Much of the success in apple production depends on proper picking, storage and disposal. When fruit separates easily from the spur, desirable firmness of flesh as well as taste are obtained and it may be picked up. The fruits may be graded for profit. 0.6 to 0.75 meter trunk, sends out branches from goblet shaped trees so that both branches and fruits get plenty of light and air. Bush trees may be grafted to semi-dwarfing or developing stock keeping the planting distance 3.5 m.

Fig. 26.3. Pruning for bush apple, the pruning its manner and extent is shown. *First year.* Prune the single stem to just above 70 cm from the ground. *Second and third years.* The growth that develops after the first pruning should be pruned to half in the following winter. Next season again prune in a like manner.

Fig. 26.4. Half standard usually grown on grass.

4. *The delayed open centre bush tree.* Quite popular ; blossom buds are readily found on usually distributed branches. Trees are strong. Main branch grows on (for a year or so before cutting back unlike the open centre bush tree). The trunk should be about 1.6 meters. The side growth is pruned in two years. After 5-6 years one or two of the older branches are removed to encourage new growth. Branches which are bent down should be trimmed in a manner to encourage branching upwards.

5. *The cordon.* A tree on single stem is known as cordon. These are generally trained at an angle of 45°. Cordon on M IX are planted 2.7 meters apart in rows 3 meters apart. The scion is 8 cm above the ground level. Wires are situated at a height of 0.65 meter from the ground and then every 0.6 meter. To ensure strength growth of cordon below stock are tied and the leaders of one-year old growth develop on the cordons. For 3 years the leaders are cut back by half to a point just above a bud. The system continues for 10-12 years. Leaders may be pruned whenever found useful.

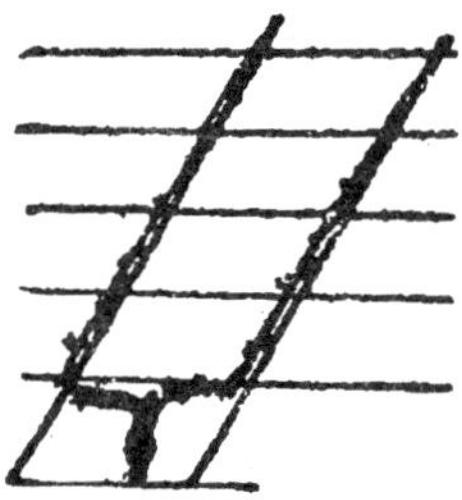

Fig. 26.5. Double cordon system.

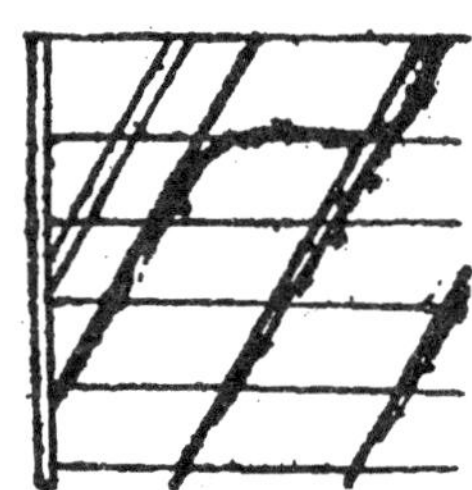

Fig. 26.6. End cordon system.

6. *Pillar.* Pillar system is convenient for plants grown at a distance of 1.5 meters in rows 3-6 meters apart. They succeed on the root stock MM 104. The tree may grow to a height of 3.6 meters. The central system has a large number of lateral growth. Train the one year, two-year and three-year tree as indicated in the Fig. 26.2.

Other methods followed are Larcure, Spindle bush and Espalier. Espalier trees have their branches trained to run parallel to the ground. The mode of pruning bush trees is indicated (Fig. 26.3). Each branch is treated as an individual cordon. Trees must be planted 7.2 meters apart.

Fig. 26.7. Pruning of apple trees.

Thinning becomes essential in the event of branches cracking or injury to the plant due to heavy bearing. Thinning of fruits gives good results. In the centre of the cluster is the king apple which is often spoiled. It must be removed first and then the deformed ones. Early varicties need thinning before the natural drop, a month after full bloom but the late ones may be thinned after about 45 days. One fruit every 20-25 leaves is considered satisfactory.

Fig. 26.8. Thinning of fruits. The centre of each cluster of fruit has the king apple which is often misshapen and should therefore be removed first. Other deformed fruits should be removed when these are 3 cm in size.

Preparation of Rootstock

Generally, in U.P. seed plants were used as root stock but of late, vegetative root stocks have become popular.

A. *Seed rootstock.* Well washed and dried viable seeds obtained from mature healthy fruits of varieties which are not triploid may be placed in fields with rich well decomposed farm yard manure and superphosphate 25 cm apart at a depth of 1 cm and covered with grass in November. As the seeds germinate in March, the mulch is removed.

It is useful to place the seed, at 4.5°C for 60 days to break dormancy, before sowing.

B. *Vegetative rootstock.* Rootstocks are planted in soil 1.5 m × 75 cm apart. During dormancy in the first year 10 cm portion of the root stocks are cut off so that new branches are formed in March lower portion of these new branches are covered with soil upto 20 cm. This increases rooting during the rainy season and the new branches are separated out in winter.

For better stooling, ringing at lower end of branches along with exogenous application of IBA solution (0.25%) is recommended. Rooted branches should be removed in December.

Propagation through Hardwood Cuttings

Apple cultivars commercially propagated on seedlings root-stock exhibit a lot of variability in vigour and degree of precosity. The increase in the period of juvenile phase adds susceptibility to root-rot, collar-root and woolly aphid attack, which can be mitigated to a considerable degree by the use of clonally propagated Malling and Malling-Merton root-stocks. Standardisation of the method to use hardwood cuttings is desirable for convenience and less expensive propagation of apples.

Cuttings are collected from well established 3-yr old hedge of clonal root-stock with easy rooting potentiality during the dormant period by removing shoots that include the swollen shoot base. November-December or early. February cuttings root quickly. The cuttings should be made to 50-60 cm length with a straight cut at the basal end slanting cut at the distal end before a 5 second dip in 2,500 ppm IBA for increased rooting. The excess liquid is shaken off and the base allowed to dry for nearly 30 minutes before planting.

Warm Storage

Where the ground temperature is too low, as under Chaubattia conditions, during December to March, provision of bottom heat may help root initiation. Use of specially constructed thermostatically controlled bed with electrical installation accommodating 8,000 to 10,000 dormant cuttings have given good results where winter temperature is low. Before removing cuttings from the bins, the temperature should be lowered for a few days to allow the new root system to harden off.

During February collection, cuttings should be taken out only after a week's storage. The rooted or callused cuttings should be lined out for establishment and for further propagational purposes.

Harvesting and Marketing

Much of the success in apple production depends on proper picking, storage and disposal. When a fruit separates easily from the spur, retains firmness of flesh and taste which are desirable it may be picked up. The fruits may be graded for profit. A damp clean shed is prepared so that there is sufficient humidity in the atmosphere. Chill the fruits by keeping in the open for the night. The harvested fruits may be stored for 120-150 days at 4.5°C in the temperate climate provided there is good circulation of air. Temperature between 0-1°C is very good for the storage of apples for about 8 months specially for Ambri and Red Delicious cultivars. The storage life of early cultivars is short. The per tree yield is 10-20 kg in South which is considerably lower than that in the North where the average is about 40-100 kg. The fruits of the late cultivars sometime need chilling for about a week to 'sweeten' them. The fruits may be wrapped in oiled paper so that CO_2 given out during its respiration does not escape out and O_2 does not take its place.

Cultivars

Cultivars of apple suitable for some of the more important apple growing tracts in India

Season	Himachal Pradesh	Jammu & Kashmir	Uttaranchal
Early season	Tydeman's Early (P), Michael molies Delicious, Schlomit Starkrimson, Red June, Worcestor	Irish Peach, Benoni	Early shanburry (P), Fenny, Benoni, haubatia Princess, Chaubatia Anupam,
Mid-season	Starking Delicious, Red Delicious, Richard, Vance Delicious, Top red, Lord Lambourne, (P), Red cheif, Orgeon spur, Red spur, Red Gold (P)	Cox's Orange Pippin (P), American Mother, Razakwar, Jonathan (P), Red Gold (P), Queen's Apple, Rome beauty, Scarlet Siberian	Red Delicious, Starking Delicious, MC Intosh (P), Cortland, Golden Delicious (P)
Late-season	Rus pippin, Grany Smith (P), Golden Delicious (P), Yellow Newton (P), Winter Banana.	King Pippin, American Apirouge, Kerry Pippin, Lal, Ambri, Sunhari chamure, Golden Delicious (P), Red Delicious, Ambri Baldwin, Yellow Newton (P), White Dotted Red.	Rymer, Buckingham (P),

P – Pollinizer

High yielding varieties for Himachal and Uttaranchal. Gala and Red Fuzi.

Scab-resistant varieties

(1) Himachal Pradesh. Prima, Priscilla, Sir Prize, Jonafree, Florina, Macfree, Nova Easy Grow, Coop 12, Coop 13, Nova Mac, Liberty, Freedom, Florina.

(2) Jammu & Kashmir. Firdous and shireen.

Low-chilling varieties

Varieties of this group perform well where winter chilling is less than 800 hrs below 7°C. Apple varieties suitable for :

(*i*) **Table purpose.** Michal, Schlomit, Anna, Tamma, Vered and Neomi.

(*ii*) **Processing purpose.** Tropical beauty and Parlin's beauty.

The new semi-dwarf spur type bud sport strains of Golden and Red Delicious in U.S.A. need a fair trial in different apple growing tracts of India.

Of hundreds of cultivars of apple in cultivation only a few are recommended for India. The cultivar Ambri that used to be most extensively grown in Kashmir is considered to have been introduced from Central Asia. It is a regular bearing cultivar with a high yield. Its fruits are oblong to ovate with well developed red stripes. The fruit is sweet with firm flesh. It bears an attractive aroma and is good for transport. Excellent dessert and aroma, good keeping quality. Maturity—September.

Baldwin. It is grown in the Kulu valley very successfully. Fruits large-sized and of deep bright red colour. Flesh yellowish white, crisp, juicy, tender and mildly sub-acid. Fruits harvested in October keep till March. Self unfruitful. Introduced from U.S.A.

Beauty of Bath. It is a prominent and good dessert cultivar of H.P. Fruits are small to medium sized. Flesh white, fine grained and of excellent quality. Fruits are generally borne on the tips of shoots. Self unfruitful, shy cropper.

Ben Davis. It is grown in Kulu valley. Picked in October. Fruits are medium to large, round, oblong. Flesh yellowish, white, firm, juicy, aromatic, mildly subacid. Skin tough, yellowish green, waxy, bright and red streaks and splashes mingled with carmine. Self-fruitful cultivar.

Blenheim Orange. It is grown in Kulu valley. Fruits large and roundish, skin yellow; flesh yellow, crisp, sweet, juicy and aromatic. Fruits are picked in September and keeps till February. A good cropper. Self unfruitful.

Blood Red. It is a well known cultivar of Kashmir valley. A dessert fruit of good quality. Picked in October. Fruit medium sized, slightly flattened, oval. Flesh pale yellow, crisp, tender, juicy, aromatic. Skin slightly ribbed and entirely covered with deep scarlet red colour and a bluish bloom. Alternate bearer.

Cox's Orange Pippin (Kesri). It is extensively grown in Himachal Pradesh.

Fruits medium-sized, round, oblate, symmetrical. Flesh yellow, firm, crisp, tender, very juicy, sub-acidic taste, aromatic. Skin orange red, deepening to bright red Picked in early August and keeps well till November. A regular cropper and a heavy bearer. Self unfruitful.

Early Shanbury. It is widely cultivated in H.P. Fruits medium-sized, roundish, but tapering towards the eye. Flesh white, tender, subacid and fine in taste, Skin smooth, yellow with deep red colour.

James Grieve. It resembles Cox's Orange Pippin and grows in Kumaon Hills. Fruit medium-sized. Flesh juicy and sweet. Skin yellow, striped with deep crimson. Harvested early in July and keeps till late August. Self-fruitful.

Jonathan. H.P. and U.P. hills are the widely growing regions of Jonathan. Fruits medium-sized, oblong to conic. Flesh light yellow, tender, juicy and sprightly sub-acid. Skin smooth, covered with rich red yellow dots. A good keeping cultivar. Picked in September and keeps till March.

King of Pippins. It is grown in Kulu Valley. A good dessert and cooking cultivar. Picked in July and keeps till October. Fruit medium-sized, inclined to be oblong. Flesh yellowish white, firm, crisp, juicy, lightly flavoured and very sweet.

Lady Sudelev. It is a cultivar of H.P. Fruit medium-sized. Flesh fine-grained, juicy and with rich aroma. Skin striped or flushed with crimson. Does not keep for long.

Newton Wonder. It is grown in Himachal Pradesh. Fruit large, roundish and slightly flattened. Skin yellow with brownish red check. Flesh firm, crisp, juicy and aromatic. A good cropper and self fruitful.

Red Astrachan. A Russian cultivar, early bearing. Fruit medium to large, roundish. Skin thin, tender, smooth, pale yellow with deep red cheek. Flesh white. Fruits ready in late June. Poor keeping quality.

Red Delicious. It is widely grown in Himachal Pradesh and Kashmir. Fruits medium to big with dull dark red colour tapering in shape with characteristic five lobes at the apex. Fruits are ready by September which may keep for a month. Flesh white, sweet and of excellent quality. A regular high cropper. Self fruitful. Maturity—September end. Good dessert cultivar.

Rome Beauty. It is grown in South India. Fruit large, round. Of fair quality, may be ready by September. Bears on the tips of shoots. Self-fruitful. Fruit keep long.

Stayman Winesap. It is a cultivar of H.P. Fruits large, round, conic of yellow flesh, sub-acid flavour and fine quality.

Winter Banana. It is only grown in H.P. Fruit large, skin pale yellow with red cheek. Quality very good. Fruit does not have good keeping quality and is unfit for commercial purposes. These are ready by September and may keep for about three months.

White Dotted Red (Syn. *Imperial*). It is grown in Kashmir. Fruit with white dots medium to large, roundish, flattened with smooth, glossy skin. Flesh yellowish white, juicy. Ready in November, may be used for cooking or as dessert.

Worcester Pearmain. It is mostly grown in Himachal Pradesh. Fruit medium to small, skin orange with scarlet flush. Quality good. It is regular and heavy bearer. Self-fruitful. Fruits are ready in September and may be kept for 2-3 months.

Yellow Newton. It is grown in Kulu Valley and Kangra regions. Fruit large, roundish to slightly flat. Skin greenish yellow with brownish, red cheek. Flesh cream, juicy and of excellent quality. Fruits keep well.

Other cultivars like Rymer (Maharaji), Granny Smith and Irish Peach are also grown in India. Richard and Red June, (low chilling cultivars like Tropical Beauty from Australia and Beverly Hills from California) are also of interest for cultivation in India.

Insect Pests and Diseases

Apple root borer (Dorysthenus hugelii Redt). Very few of the attacked trees bear normal fruit, and many, specially the young trees, die. Adults emerge with the monsoon, pair and lay eggs in soil. Grubs feed on thick roots and other organic matters. The attacked plants become shaky, weak and die. Sandy soil is preferred by the pest. Remove and destroy from the roots of affected trees during October-November. Drench the basin (1.25 m) of affected trees with chlorpyriphos 20 EC @ 3 ml/litre in March and again in July. In case of severe infestation apply Lindane 1.3 G @ 300 gm/basin in July. Avoid dry sandy soil for apple cultivation.

Codling Moth (Laspeyresia pomonella). It has not been reported in India but has long been a serious pest in Afghanistan. Care should be taken to prevent this or to eliminate the pest immediately if it is introduced.

San Jose Scale (Quadraspidiotus perniciosus). It is very serious in Kulu Valley. It is also reported all over India. This polyphagous pest attacks more than 700 plant species. The damage is caused by nymphs and female scales by sucking plant juice from all the aerial parts of the plant. The overwintering nymphs become active in mid-March. The females reproduce in mid-May producing 200-400 nymphs in a month. The infested leaves, twigs, fruits and stems are covered with ash grey scale. In winter (November- March) spray 2% miscible tree spray oil. In case miscible oil is not easily available, spray diesel oil emulsion @ 4-6%. *Aphytis* sp. and *Encarsia perniciosi* @ 2000/tree may also proves effective. Use of *Chilocorus bijugus* @ 20 beetles/tree may control the scale insect effectively. If infestation occurs during April to Oct-November spray Methyl-o-dimeton @ 0.04 % or Monocrotophos @ 0.04%. Oil spray should not be mixed with insecticides and other fungicides.

Woolly aphid (Eriosoma lanigerum Hausm). Aphid attacks the roots and also shoots. This pest is active from April to December but found in large numbers on aerial parts after rains (late August-October). Appearance of whitish cottony patches on stems and branches are the marks of indentification of the pest infestation. Treat nursery with Carbafuran 3 G @ 60 kg/ha. In non-bearing trees place phorate 10 G @ 10-30 g or Carbofuran 3 G @ 30-50 g per tree at 5 cm depth in the root zone of 1-4 years old trees. In bearing trees, apply Dimethoate @ 0.03% or Demeton methyl @ 0.025% in March, April and October. Use resistant rootstocks *viz.* M 21, MM 111, MM 114, MM115, and M 25. Woolly aphids are controlled by using natural parasites like *Aphelinus mali* which attacks the 4th and 5th stage of nymphs. Insecticidal sprays should be avoided in areas where predators (*coccinellids*) and parasitoids (*Aphelinus mali*) are present. Summer application of systemic insecticides should also be avoided.

Stem Black (Coniothecium chomatosporum). Painting pruned wounds with a paste containing 50 gm each of red lead and copper carbonate and 100 ml of raw linsed oil is helpful in preventing infection. Pre-bloom spray of lime-sulphur (1 : 60 dilution), starting at the dormant stage and continued every 15 days throughout the blooming period.

Stem-Brown (Bortryosphaeria ribis). It also enters through pruning wounds, and can be effectively controlled by protecting wounds with the above mentioned paste. Control as Stem Black. Destroy the diseased branches Apply *Chaubatia* paste or Bordeaux mixture paste to the cut surface.

Pink disease (Pellicularia salmonicolor) and *collar-rot (Roselinia)* are serious in Kumaun. These can be controlled by one spray of 2 : 10 : 40 Bordeaux at the open cluster or petal-free stage. The affected parts may be treated with paste containing 57 gm of red lead and of copper carbonate in 100 ml of raw linseed oil to control the infection.

Brown rot (Sclerotinia fructigena). It causes heavy loss. If necessary the tree may be sprayed with 4 : 4 : 50. Bordeaux at the pink bud and calyx stage and again two weeks later.

Mildew (Oidium sp.*). Pink disease* and *collar rot (Sclerotium rolfsii), Powdery mildew (Sphaerotheca bucotricha)* have been reported as very serious in South India, can be controlled by dusting with sulphur in spring. Powdery mildew and other disease associated with apple may be controlled by removing the affected fruits and cracked branches before spraying Bordeaux mixture (4 : 4 : 50).

Sooty blotch and *fly speck (Leptothyrium pomi).* Only the appearance of fruits is damaged. It can be checked by spraying lime-sulphur or colloidal sulphur. Washing the fruits in 5% solution of bleaching powder or 3% solution of sodium chlorate removed the blotch are prevented increase of specks during shortage.

Soft rot or blue mold (Penicillium expansum), rot *(Phytophthora cactorum), bitter rot (Glomerella cingulata)* are also serious diseases in many parts of the country. They can be controlled by regular, and frequent spraying with 4 : 4 : 50 Bordeaux.

Venturia inaequalis Wint. This is one of the serious diseases of apple. The characteristic symptoms of the disease are the development of light-brown or olive-green, irregular spots on the lower surface of leaves. Spray of urea (5%) on trees once before general leaf fall and again just before bud break (2%) followed by Difoltan @ 3.3 g/litre and Dithane M-45 @ 2 g/litre of water at 10-15 days interval after petal fall will effectively control the disease.

Rubbery wood recognised by extremely flexible branches, drooping behaviour of trees with cheesy texture of soft wod is caused by Mycoplasma.

Star Crack. It is a viral disease in which star shaped cracks are noticed in the skin of fruit. Mosaic, also due to virus, forms yellowish flocking and mottling of the leaves. In these cases use of certified virus-free bud-wood. Hot water treatment of affected bud-wood at 50°C for 15 minutes is recommended.

Mineral Deficiency Symptoms

Several mineral deficiencies have been noted of which those of phosphorus, calcium, magnesium, potassium, iron, manganese and boron are common. We may consider these in short.

Calcium. Dieback of tips of shoots. Margin of leaves rugged. Scorched and turned towards upper surfaces. Scorching develops progressively from young to old leaves.

Magnesium. Shoot growth may or may not be affected *i.e.*, the growth may be normal or nil. Leaves show severe interveinal necrosis towards end of season. Defoliation of terminal shoots takes place, beginning at bases. The defoliation may be total with lapse of time. Fruits may not mature and ripen.

Phosphorus. Short and thin shoots, sparse foliage, leaves small, dull green with purple or bronzed tints, early defoliation. Sparse fruit buds and blossoms, smaller fruits.

Potassium. Branch shows dieback, stunted growth of shoots, leaves bluish green, slightly chlorotic near margins. Leaf margins scorched (gray or brown). Small fruits. Immature appearance.

Iron. Leaves on tips of shoots chlorotic. Leaves may be bleached, margins show brown patches, chlorosis descreases down shoots, branches may dieback. Fruits of pale green colour and highly flushed.

Manganese. Interveinal chlorosis beginning near margins and progressing inwards. Small veins, not visible in chlorotic leaves. Chlorosis may be general, all over tree, though young leaves on shoots may be green.

Boron. Shoots defoliated, terminal leaves in rosette form. Bark rough, split. Fruits malformed. Skin cracks and flesh may show brown patching or browning in the core region.

Copper. Both apple and pear show symptoms of deficiency of copper and zinc. In the deficiency of copper young shoots have their normal growth early in the season but by summer the young tip leaves show distortion and browning and are shed which continues down the shoots and is accompanied by dieback of the shoot. The bark of the tree may become rough and separate from the tree. Gradually the larger shoots also die. Sucker growth develop from bases which also die from the tips downwards. Fruiting almost depends on the severity and stage of the deficiency.

Zinc. Zinc deficiency is marked by dwarfing of the shoot growth and the internodes become progressively shorter until leaves only carried bushy rosettes on terminal growths. Leaves become small and narrow and may be accompanied by interveinal chlorosis. Fruits are several dwarfed. This holds good for pears also.

CHAPTER

APRICOT

(***Prunus armeniaca*** **L.)**

Rosaceae

Origin and Distribution

Apricot, a drupe native of Armenia, though known to the Chinese since 2,000 to 3,000 B.C., is also grown in the warmer regions of the Himalayas. Apricot, *Khubani* (H.) is grown in Kashmir (India) to an extent of about 400 ha. The Himalayan region is full of apricot trees of inferior quality growing in semi wild state. These are of value for the seeds while the Kernels are eaten like almonds. The trees may be medium to large.

Apricots

Zardalu, the wild apricot is indigenous to India being cultivated in Shimla Hills for its oils and fruits. The common apricot plant is 6-9 metres high with reddish bark, glabrous twigs, leaf petioles have conspicuous glands and the flowers pinkish to white. The fruits are valvety when young, peach-like in colour and shape with yellowish orange flesh and smooth flattened stone.

Bailey classified the two forms of common apricot *viz*., the Russian apricot (*Prunus armeniaca* var. *sibirica* L.) and the Manchurian apricot (*Prunus armeniaca* var. *mandshurica* koehne) as varieties of the same species. Cross between apricot and Japanese plum (*P. salicina*) have given rise to plum cot.

Since apricot is very much susceptible to frost due to early flowering, late blooming hardier type is required for better fruit production. Apricots are thus not easy to grow. Blossoms are liable to suffer from frosts or cold winds during the period early February or March. In order to make them grow frost-free sites are also necessary because low temperature, often limits its culture. Temperature limit for blossom lies near about $-1°C$ and for fruit setting it is 0°C. The trees need protection from frost at the flowering time which may be achieved by muslin or polythene covers.

Propagation

Apricots are propagated vegetatively by shield budding of desired cultivars on root stocks of plum, peach or wild apricot, the latter two giving better results. Autumn is considered to be optimum for budding though it may be done in spring and summer also if the environment is suitable. Plum stock is useful for heavy soils. Bromyton, a vigorous plum stock is preferred for large trees while St. Julian (East Malling) plum stock is recommended for small trees. The alternative to plum stock is the Common Mussel. Apricots may also be propagated on Myrobalan plum stocks which is specially useful due to nematode resistance. The apricot bushes may be given a diagonal staking (Fig. 27.1). Light summer pruning may also be given.

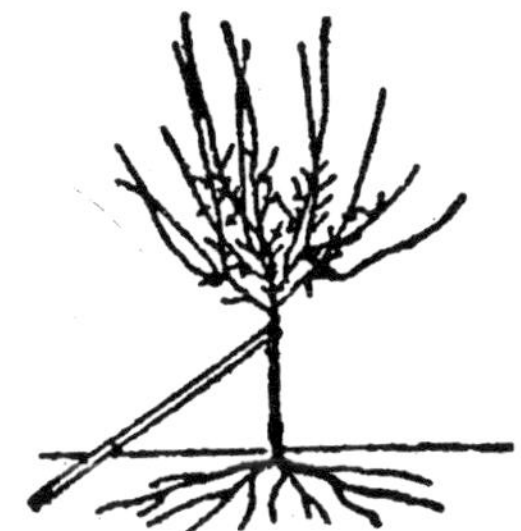

Fig. 27.1. Diagonal staking.

Planting

Apricots may be planted at a distance of 6-8 metres between plants. The distance between the plants is regulated with regard to the fertility of the soil. The fertility level requirement of apricot is lower than that of peach. It needs well drained loam, rich in humus. Early planting should be done to allow the plants establish well before the onset of flowering. The trees need protection from frost. The hydrogen ion concentration of the soil for apricot should be near about neutral, acid soils should be corrected by adding lime to bring the pH to 7.

Cultivation

Cultivation of apricot is similar to that of plum and peach. Clean cultivation, freedom from weeds and incorporation of manures and fertilizers is considered useful for apricot. Apricot soil needs good soaking upto the root-depth (2 m) when the fruits have developed to about the size of hazel nut. Flow-method of irrigation is practised. Mulching may be restored to help the soil to retain moisture.

Bushy forms do not require much watering except in dry seasons. Watering becomes specially imminent with the setting in of drought conditions at the fruit swelling stage. Adequate irrigation, freedom from weeds and maintenance of productivity are the key notes of successful cultivation of apricot.

Pruning

Year-old apricot plants possess numerous lateral branches. Proper spacing of these branches is essential to form the head. In case the plant from the nursery does not have the desirable shape, the main stem should be severed at a height of 60-75 cm to force laterals. Growth of undesirable laterals must be checked by pinching off during the first year of planting. Adequate pruning should be done to develop a well balanced head with 3 to 5 strong scaffold branches by the end of first growing season. The scaffold branches are later headed back breast-high to force 5-7 secondary scaffold branches. Care must be taken to have the upper main scaffold branches longer than lower ones to compensate for the fast growth of the lower branches. Later pruning is confined to the removal of branches that crowd or cross each other to ensure healthy development of the desirable branches allowing adequate aeration and good sun. Since spur bears fruit, for about three years severe pruning is not desirable. Pruning may be practised profitably to secure new place for those infused or broken during picking. Honestly speaking the extent and type of pruning depend on the bearing habit of the variety. Pruning may be light if the new growth in the year is 30 to 60 cm and heavier if this growth is less. Heavy pruning must be avoided leaving old wood in good proportion on the trees to avoid delayed fruiting and yield reductions. The laterals are tipped and branches or laterals growing away from the walls (in case of trees growing near walls) during summer. Training of the trees to a modified leader system increases the yield. During winter the branches of fan-trained plants (Fig. 27.2) by the side of the walls or fence, may be fastened to nails, by the help of strings and the laterals may be cut back.

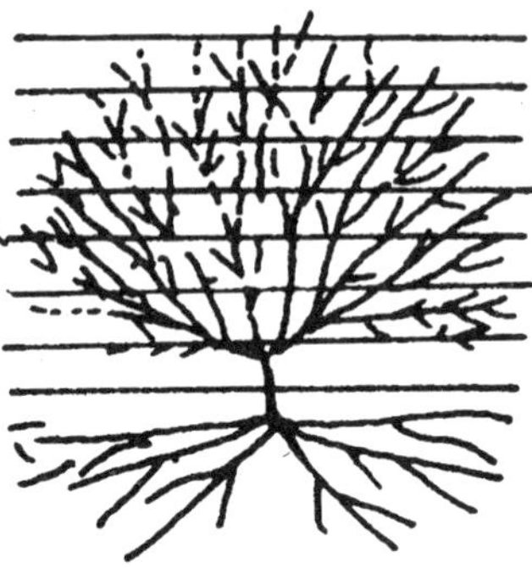

Fig. 27.2. Fan-trained Apricot

Pollination

Apricots are generally self fruitful but at the flowering time bees and insects are less active and therefore fully open flowers may be dusted with a camel hair brush to ensure pollination and setting of fruits.

Thinning

Heavy fruit set needs thinning which is done by removing one fruit from each pair after some of the fruitlets have had natural drop. More thinning may be done, if necessary.

Harvesting and Marketing

Apricots must be picked by hand individually and carefully when ripe or a little earlier when still hard for being in attractive condition at the time of retailing. Wasp damage is warded off by picking hard fruits and ripening them in a warm window.

Cultivars

There are a number of apricot cultivars of importance to India. These are mostly exotic and belong to the species *Prunus armeniaca* L. The chief characteristics of those successfully grown at varying elevations in Himalayan region, are portrayed below. The recommended cultivars *viz*, Royal, Kaisha, Shipley early, St. Ambroise and New Castle are self fruitful and may be grown in compact blocks or units.

Cultivars of Apricot may be grouped according to their suitability for cultivation in different areas.

I. Himachal Pradesh

Mid hills. New Castle, Early Shipley, and Shakarpara.

High hills. Suffaida, Charmagz, Mari, Kaisha, Nugget, Royal.

Dry temperate. Charmagz, Suffaida, Kaisha and Shakarpara.

Early maturing. Early Shipley, New large early, Charmagz, Baite, Beladi.

Late maturing. Turkey, Moorpark, Kaisha, Royal, St. Ambrioise, Farmingdale, Alferd.

II. Uttaranchal

Chaubati Madhu, Chaubatia Alankar, Chaubati Kesari and Bebeco.

III. Jammu and Kashmir

Halman, Rakchakarpa, Tokpopa, Margulam, Narmu, Khante, Turkey, Australian, Rogan and Shakarpara.

Hemskerk. Fruits of orange-yellow appearance with blotched red flesh and good flavour. Ready as early as July-August.

Kaisha. The fruits are ready in June. These are medium-sized and round in shape. The fruit is yellow in colour with reddish cheek while the flesh is good flavoured orange-yellow.

Moorpark. A late maturing, widely grown cultivar with a liking for clayey soils. Fruits large and orange-yellow.

New Castle. Characterized by having a nearly spherical pit with roundish medium-sized orange-yellow coloured fruit. The fruits are ready early in June with juicy and sweet flesh. The production is steady and regular.

New Large Early. An early bolting cultivar of fine flavour with orange-yellow skin and red coloured flesh. The kernel is an important source of cooking oil and the oil cake is used as cattle feed. Oil of bitter kernel is used for lighting.

Royal. The fruits have dull yellow skin with reddish tinge and yellowish flesh which is firm, juicy, possessing high flavour. The stone of this large roundish oval fruit may be free or partly clinging. This cultivar is synonymous with Blenheim and ripens in-mid June. A good cultivar for desert.

Shipley Early. The fruit of this cultivar is also round and of medium-size as those of Kaisha. This is the earliest cultivar getting ready in May. The fruit is creamy-white with red flesh on the sunny side. The fruit is of average quality.

St. Ambroise. A mid-season to late cultivar, increase in fruting with rise of altitude upto 1,800 metres. The fruits are symmetrically oval, large, firm and tender with orange-yellow skin and free stone.

Gilati sweet. Fruit medium sized, oblong to ovate, cavity rather deep, skin yellow, flesh turns brown yellow later. Juicy, sweet, moderately flavoured. Ready in last week of June in Kashmir.

Insect-pests and Diseases

Bacterial gummosis–*Pseudomonas syringae.* Gumming lesions are produced on bark or outer sapwood of trunk. Clean the weaping wounds just before dormancy break and apply Mashobra paste. Repeat the process on new lesions before the rainy season. Spray streptocycline (0.05%) before the onset of rains. Spray copper oxychloride (0.3%) after leaf falls.

Ingredients and their Relative Proportions in Mashobra Paste :

Streptocycline	–	25 g
Marphorin	–	150 g
Steric acid	–	425 g
Lanolin 2	–	2.25 g
Water	–	5.5 lit

Fruit composition

The apricot fruits contain 1.94, 0.29 and 3.01 gms of nitrogen, phosphorus and potassium respectively for every 1,000 gms of flesh.

CHAPTER

28

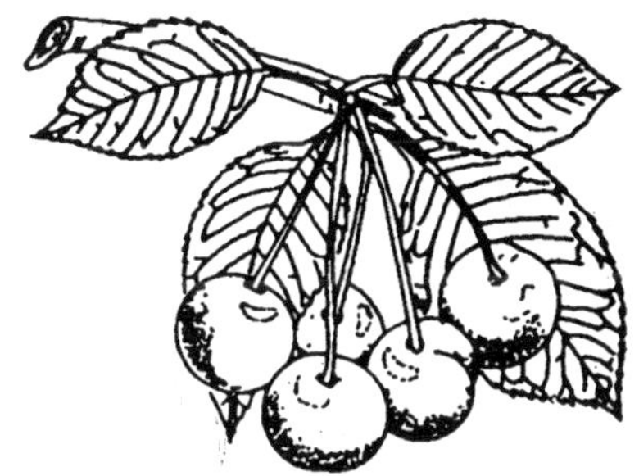

CHERRY

(*Prunus avium* L. and *P. cerasus* L.)

Rosaceae

There are two main types of cherries (H. *Gila*) the sweet (dessert) and the sour. The cultivation of sweet cherries (*Prunus avium* L.) has been taken up on a commercial scale in the Kashmir Valley of India. The cultivation of sour cherries (*Prunus cerosus* L.) better known as cooking cherries has not been taken up in India on any significant scale. The cultivation of these are quite different, it is easier to grow sour cherries than the sweet cherries since the former presents fewer problems of pollination and the birds do not attack them. Sour cherries also bear earlier than the sweet ones. It is a temperate fruit.

Sweet cherry tree is tall with fruits of different shapes and colour *viz*., globular, somewhat depressed or heart shaped, yellow or red.

Soil and Climate

Sweet cherries are more exacting about soil than any other fruit. Well drained chalky or deep sandy soils suit invariably. Manures applied as top dressing prove better than the basal dose. These are well adapted to grown to an altitude of 1,500 metres.

Propagation

Seed propagation is possible only after stratification for 4 months at 2-4°C when germination can occur under favourable conditions. Sweet cherry is grafted on seedlings of the wild cherry *Prunus paddum* (Paja), *P. avium* (Mazzard) and *P. mahaleb* seedlings. Grafting is done in January by the whip and tongue method and the grafts become ready for transplanting in two years. Sometimes the stocks are planted in permanent positions in the orchard and grafted *in situ*. Suckers from wild cherries can also be used as stocks. Sour or cooking cherries are usually used for making jam or cherry brandy. They grow well when cultivated as fan-shaped trees on a north wall, especially the cultivar Morello. The sour cherry is usually budded on Malling F. 12/1 a Mazzard stock. The Yellow Morello does well when grown on its own root. T-budding on seedling rootstocks has also been reported.

Planting

Type and age of trees. 3 to 4-year old standard or half standard trees should be planted and allowed to grow undisturbed without any pruning for the first few years, so that these then bear well. Two reciprocal varieties be planted to pollinate each other. Sweet cherries do not crop well as fan-shaped trees; the sour cultivars give a better performance. In sour cultivars a 2-year old bush tree or 3-4 year old fan-trained tree is considered best for planting. It does well in cultivated soils. Planting is best done just before the onset of winters when the soil is still warm. The recommended distance for fan-trained sweet cherry trees is 6 metres and for standard trees 12 metres.

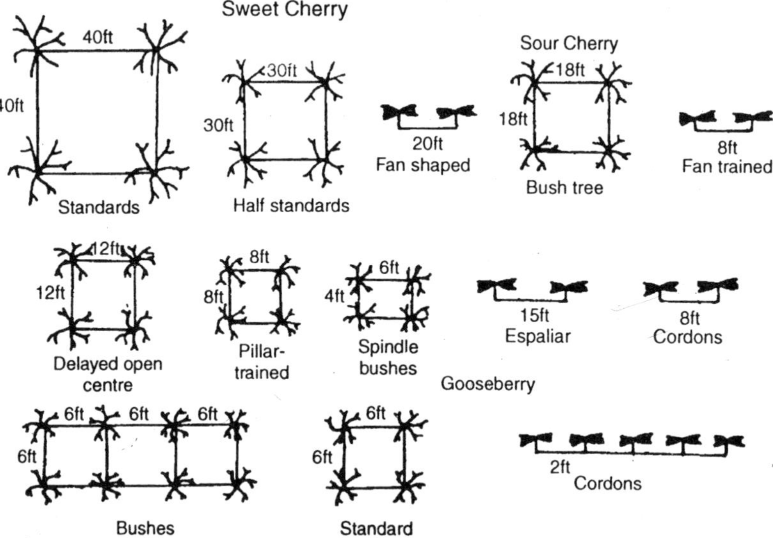

Fig. 28.1. Showing planting distances for sweet cherry, sour cherry, pear and gooseberry recommended under different methods (1 feet = 30 cm).

Planting and staking. Cultivars of sweet cherry may be planted late though it is better to plant in November. Early plantings help the roots to settle before the winter. Cherry trees grow to a considerable size and remain in a profitable condition for over 60 years. The distance between the trees should vary from 6 to 12 metres according to the needs of the cultivar, the fertility of the soil and the type of growth pattern. The trees should be staked after planting.

Cultural Practices

Manuring. The cherry orchards are best put under grass. The grass is kept in check by sheep that supply organic manure to the vegetation or maintain a lawn in which the cuttings are left on the ground and not removed. Besides sheep manure, phosphates may be applied to obtain good growth. In the initial stages a distance of 1 metre diameter round the young plant is kept free from grass so that the young roots are not robbed of moisture and nourishment.

If the soil is acidic, supply carbonate of lime in the autumn round the trees over the surface or mulch at the rate of 0.25 kg. per sq. m. every four years and let it be washed in gradually. Some straw is put all round the tree covering an area of about 1 metre wide and 3 cm deep sprinkled with fish fertilizer at the rate of 100 gm per sq. metre. This mulching does away with the necessity of very frequent watering during summers. Each watering needs about 45 litres of water.

Pruning

Cherry trees grow into shape without much pruning. Crowding of branches should be checked by thinning out. The dead wood should be removed in the dormant season and the cut ends be painted with tar to prevent the growth of fungus. Summer pruning is necessary specially when the fan-shaped tree is grown against a wall. "Prune, when dormant", does not always apply to cherries as also with certain stone fruits *viz.*, maples, walnuts etc.

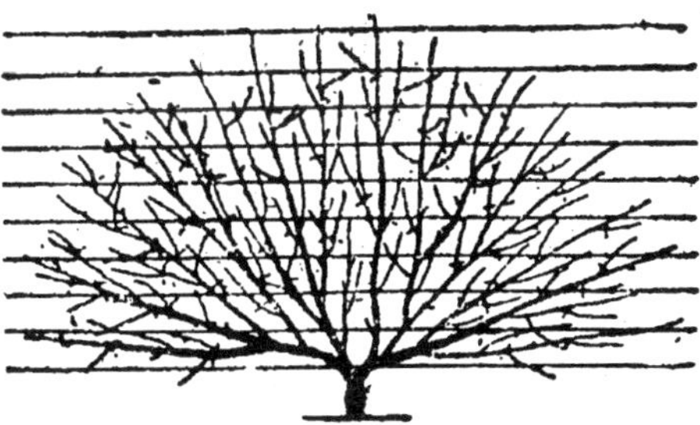

Fig. 28.2. Summer pruning of cherry.

Pollination

A large number of cherries are self-unfruitful and thus do not set fruits with their own pollen. They are also cross-incompatible with some commercial cultivars. Thus, only those compatible cultivars be grown together whose period of flowering sufficiently overlaps to affect cross-pollination.

Harvesting and Marketing

The cooking cherries can be left on trees until fully ripe. The fruit is picked up by hand when dry so that it may be transported well to the consuming markets. It should be securely packed in specially made baskets to avoid damage in transit. Covering of branches with fish netting is necessary to protect sweet cherries from birds during the period fruits are ripening. The trees may do well when sprayed with 5% solution of tar distillate wash. The withered tip and dead twigs may be removed and burnt.

Cultivars

A large number of cultivars of sweet cherries are of importance of which some are mentioned here.

Amber Heart. Mid-July, pale-yellow with red cheek, pollinated by Napolean or Florence.

Bigarreau de Mezel. Early July, large deep reddish-black, excellent flavour, pollinated by Florence

Bigarreau Napolean (Gilas double). Large and handsome, skin creamy red, flesh firm, sweet, juicy, slightly acidic, good keeper, ripens at the end of June, late cultivar, flowering late, pollinated by Roundel Heart and Florence.

Black Eagle. Mid July, purplish-black, sweet juicy, pollinated by Bigarreau de Schrecken. Other promising sweet cherry cultivars are as follows :

Gilas Awal Number (Guigne Pourpera Pecoce). Medium size, light red, fleshy, juicy and sweet with acidic taste. Marketed in May in Kashmir.

Gilas Misri (Biggarreau Noir Grossa). Large sized, red coloured, firm skin and sweet juicy flesh. Cultivated in Kashmir.

Bigarreau de Schrecken, Early Rivers, Elton Heart, Emperor Prancis, Florence Early, Frogmore Early, Governorwood, Knight's Early Black, Merton Bigarreau, Merton Bountry, Merton Favourite, Merten Glory, Merton Heart, Merton Premier, Noir de Guben, Peggy Rivers, Roundel Heart, Water Loo, Morello. Commercially successful cherry cultivars in India have a light colour, firm flesh and belong to the Bigarreau group.

Sam, Summit, Sue, Sunbrust, Lapins, Compact Stella and hybrid (13-17-40) are new promising varieties.

Insect-pests and Diseases

Insect-pests :

Leopard moth *(Zeuzera multistrigata)*. Pinkish-white young caterpillars bore into branches and stems during July-August and feed. Kill caterpillars in tunnel by inserting a pointed wire or Aluminium phosphide or Dichlorovos after cleaning the tunnels and later plaster with mud.

Cherry black fly *(Myzus cerasi)*. It is the serious pest of cherry. The leaves curl and drop and shoots may be killed. It can be controlled by spraying Dimethoate (0.15%) or Dimeton methyl (0.1%) at 15-days interval.

Cherry blossom wilt *(Circospora rubrolinata)*. It damages the flower, browning of flower tissues and leaves on spurs. *Control :* Remove affected spurs, spray Bordeaux mixture.

Diseases

Bacterial gummosis *(Pseudomonas* spp.*)*. Control measures should be adopted as for Aprocot.

Canker and silver leaf *(Chondrostereum purpureum)*. Foliage shows peculiar silvery lustre. The papery brown canker affects the limb of tree. To maintain the vigour of tree, proper fertilization is essential. Protect the fresh wounds with Chaubatia paste or Copper oxychloride paint.

Composition of Chaubatia paste. 1 part $CuCO_3$ + 1 part red lead oxide + $1\frac{1}{4}$ part linseed oil.

Minerals Deficiency Symptoms

Copper. Copper deficiency is reflected by the die-back of shoot as in the case of apples and pears. In the case of cherry and plum, this may be associated with much distortion of growth and some gumming along with multiple bud formation.

Magnesium. Older leaves necrotic. Dull purple tint may precede necrosis. High and orange tints develop later. Early defoliation.

Potassium. Bluish-green leaves, chlorosis near margins and brown marginal scorch.

Iron. Leaves of shoot tips become chlorotic. Detailed vein pattern visible in chlorotic areas.

Manganese. Interveinal chlorosis spreading from margins inwards. Chlorosis generally extending to whole of the tree.

CHAPTER

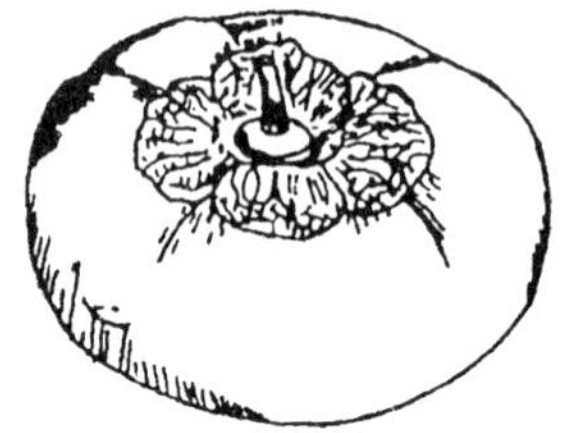

JAPANESE PERSIMMON

(*Diospyros Kaki* L.f.)

Ebenaceae

Origin and Distribution

Japanese persimmon (H. *Kaki vhondo*) or Chinese date plum sometimes considered to be a subtropical fruit is a temperate nut like the chest-nut, hazel nut and strawberry and is native of China. It is a monoecious tree growing to a height of 15-16 meters or near about. In India it is mostly cultivated in Kulu Valley where it was introduced. It grows successfully at elevations varying from 900 to 1,500 meters in the Himalayas. The cultivars recommended for growing are Fuyu, Hachiya, Hyakume. Indian persimmon (*D. embryopteris,* Pers, gaub) trees are found in Madhya Pradesh, western Peninsula (India) with edible fruits; dye obtained from fruits is used for colour cotton and timber.

Propagation

Seeds are taken out of a fruit which is still firm and mature since these show better viability. Seeds are washed with water and the floating ones are discarded. Seeds need stratification for about 3 months for improvement in germination of certain species. Good germination may be obtained by keeping the seeds at 28°C. Seeds take 2-3 weeks for germination. The viability of the seeds may be improved if the seeds are dried at 45% humidity and stored at 0°C.

Seedlings may take a year to attain a size suitable for grafting. The seedlings are very susceptible to transplant shock which should be eliminated through proper manipulation. Leafy summer cuttings taken from forced *D. kaki* seedlings can be used for rooting provided the base of the cuttings is treated with 1000 ppm IBA is 50% ethanol and kept in a mist chamber maintained at 23°C. Root cuttings may also succeed but it is tired scarcely.

Seedlings of *Diospyros lotus* are used as root-stock. The seedlings are collar grafted by the whip and tongue method and are kept covered with soil till sprouting. Crown grafting on seedlings of *Diospyros lotus* (Amlok) is also done.

Autumn budding may be read for transplanting earlier. T-budding or chip budding is suitable if buds are taken from mature wood.

Planting

The persimmon plant is headed back before planting, in autumn, in well prepared pits. The planting distance should be 7 to 8 meters from plant to plant, The plants should be planted about 8-10 cm deeper in the pits than they stand in the nursery. Weed control has been recommended by the use of paraquat. Irrigation is essential and must be supplied according to the need of the plant. Soils with good water holding capacity minimise the frequency and depth of irrigation.

Pruning

The tree should be trained to form a low head. To achieve this, trees are headed back about a meter from the ground at planting time to develop a framework of 4-5 strong branches. The adult trees need no pruning but the dead, broken or interferring branches may be removed preferably once a year. Fruit thinning gives good results by way of helping regular bearing and large sized fruits.

Harvesting and Marketing

The fruit is harvested when it has attained the colour characteristic of the variety, but is still hard. It is clipped from the tree, keeping intact the calyx and a short piece of stem. The fruit is wrapped in tissue paper and then packed in two layers for despatch to the market.

Cultivars

Most cultivars are astringent until fully ripe. The berry becomes soft when ripe the develop proper sweetness and flavour. The astringency of the fruits can be done away be storing the fruits with tomato, mangosteen, passion fruit, gooseberry and banana etc. The commercial trees of persimmon generally bear only female flower and bear seedless fruits. Trees with male or both male and female flowers are found.

Dai Dai Maru. A cultivar successful in the Nilgiris. Fruits can be ripened by keeping in closed containers with pears for three days.

Fuyu. Tomato shaped fruits, size medium to large. Skin smooth, glossy and red in colour. Flesh light-orange, non-astringent, firm and of good quality. The fruit can be eaten even before it softens. Good storage quality, very late.

Hachiya. Fruit oblong, conical, very large and pointed. Skin glossy, deep orange red and attractive. Flesh deep yellow, firm, astringent until ripe, practically seedless, very good quality.

Hyakume. Fruit oblong to roundish, ovate, large, skin buff yellow to orange. Flesh dark crimson, firm sometimes fibrous but of a very pleasing quality.

Izu. Very early, fruits orange red, attractive, 200 g av. wt. Prone to severe fruit fly attack.

Hana Fuyu. Late, large sized fruit, 250-300 g av. wt.

Fuyu Hana. Mid early, slight dome-shaped appearance, good quality fruit, 220 g av. wt. A Brazilian cultivar. Also known as Pomelo 152-7.

Suruga. Late, small-medium size of fruits (200 g av. wt.) dome shaped with definite wrinkles at calyx end. Quality of fruit-very good. Susceptible to leaf spot disease.

Maekawa Jiro, Ichikikei Jiro, Jiro and Matsumoto Wase Fuyu are mid early culitvars commonly grown in Japan. Mackawa Jiro has biennial bearing habit, fruits of excellent quality. Crops may be obtained even without cross pollination. Ichikikei Jiro has good quality orange red fruits. Needs adequate pollination for uniform production. Matsumoto Wase fuyu fruit has a short shelf-life. Fruits are of medium size and susceptible to dehiscence.

Non-astringent cultivars of persimmon are Izu, Maekawa Jiro, Ichikikei Jiro, Jiro, Hana Fuyu, Fuyu, Suruga and Twentieth century.

Insect-pests

Leaf-roller *(Hyplocala rostrata)*. Insects active throughout the rainy season. Caterpillars fold leaves and feed inside. This can be controlled by spraying Endosulfan (0.05%) or malathion (0.05%) in mid-July and again in mid-August.

CHAPTER

PEACH

(*Prunus persica* Stokes)

Rosaceae

China is considered to be the original home of peach though some hold it to be the native of Persia from where it spread to Europe. Peaches (*Aru*) may be found growing in the Himalayas wild. Submountain regions of U.P. and Punjab have the cultivars '*Chakli*' and '*Noki*'. Their fruits are of poor quality. Peach need the warmest climate of all the temperate fruits and is grown in lowest hills except for varieties of the Honey group. The cultivation of peaches is being tried at different elevations in the Himalayan hills. The peach, *Prunus persica* stokes (*Amygdalus persica* L., *Persica vulgaris* Mill) bear glabrous twigs ; leaves oblong, lanceolate to broad lanceolate, serrate margin, flowers solitary sessile, fruits vary in their size and shape, flesh white or yellow, stone very hard and deeply pitted. Varieties which need more cold for breaking the dormancy are commonly cultivated in hills of Uttar Pradesh, Jammu and Kashmir, Himachal Pradesh and the hilly tracts of the Punajb. The cultivars which need a subtropical climate are grown in Saharanpur, Meerut, Bulandshahar and Muzaffarnagar in U.P., Dehradun in Uttaranchal besides Haryana, Punjab, Delhi etc.

Climate and Soil

Peaches (H. *Aru*) do well in wet and humid climate with cold winter and dry summer. According to estimates made under Indian conditions 45 cm of annual rainfall is sufficient for peach. Some cultivars like 'Floridasun' require less chilling. Spring frost is more harmful. In its soil requirement, light sandy soil is considered best. Deep fertile loam or sandy loam with good drainage also favours peaches.

Propagation

Propagation by seeds as well as vegetative means is favoured in peaches. Peach is propagated by T-budding on peach seedling root-stocks, using seeds of "Lovell" or 'Elberta' cultivars. Almond and wild apricot also serve as root stock for peach. Peach root stock-succeeds best on light, deep and well-drained soils. In heavy and water logged soils plum does well as a root-stock although not all plums are compatible with peach. Shield budding may be done on one-year old seedling of wild peach or subtropical plum as root-stock. One year old grafts or buddings are planted in the orchard in pits 90 × 90 × 90 cm, 6 m apart.

The older trees are planted at a distance of 10 meters each way. The trees are white-washed to protect the bark from sunburn after planting.

Cultural Practice

Cultivation is necessary in winter and continued till April-May for profitable as well as superior quality of crop of peach. A suitable cover crop may be sown in the rainy season after the fruit is picked and the same ploughed in during winter. Besides green manuring, nitrogenous and potassic fertilizers 30 kg each and 60 kg of phosphate be applied before spring to one hectare of the orchard. Non-bearing trees need application of 300 gm of nitrogen and tree per annum for each one-year age of the plant.

Buds are borne laterally by the one-year old wood and on short spurlike twigs. Under normal conditions the trees develop two fruit-buds and a leaf-bud at one node. Peach varieties are mostly self-fruitful and set good crops without cross-pollination. J.H. Hale, June Elberta, Harlberta, Chinese Cling, Mokodo, Kandoka, Giant are self-unfruitful and require to be pollinated by other cultivars.

Pruning

Young trees are trained to open centre system. At the time of planting, the stem is cut at a height of about 60-80 cm from the ground and only 3 to 4 branches allowed to develop on it. These main branches should be well-spaced and distributed on all sides, but not lower than 1/2 meter height from the ground. Shoots, other than the three selected ones, should be rubbed off during the first summer. This process is repeated twice or thrice to allow the selected shoots to provide wide laterals and make proper growth. During the second summer forked branches are cut unequally to make strong crotches. In pruning cut is made always in a manner to encourage a spreading shape. Fan training of peach trees is not an uncommon practice.

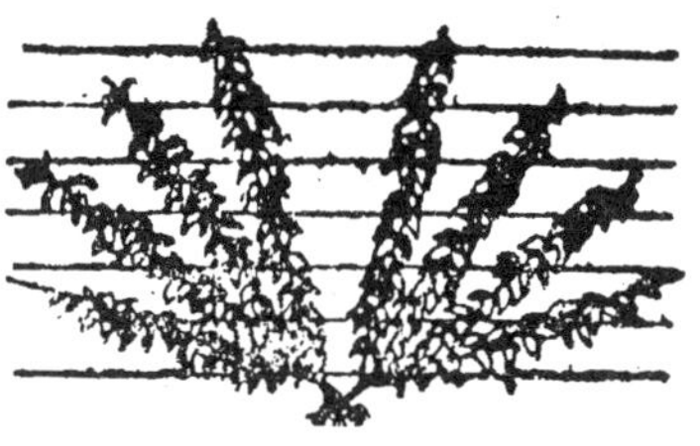

Fig. 30.1. Fan-training of peach trees.

Annual pruning of bearing trees is necessary to maintain open centres with view to admitting light which is essential for the satisfactory development of the colour of fruit and to prevent fungus disease and stimulating strong new wood for better production. Peaches bear fruits mostly on one year old wood and its pruning is of importance. The best rule for thinning is that only 3-4 main limbs each having secondary branches be allowed to develop. Heading back is done annually. The bearing trees are headed back ; the one year old shoots are chopped off 2/3 and also some are thinned out. Cut is made to the outward-pointing side-branches to encourage a spreading growth. Heavy pruning causes excessive vegetative growth and poor cropping.

Thinning is applied after the natural drop has taken place in May and June for the development of proper size and attainment of good quality.

Harvesting and Marketing

Peach fruits should be handled quickly when ready for harvest. Peaches are picked when they are still hard, as they can ripen well in storage or in transit. The plants bear for about 20 years in India and the yield per young tree is roughly 20 kg per year. The yielding capacity increases with the age of the plant and well looked after plant yields 1-1.5 quintals at the age of 8-10 years.

Cultivars

Several cultivars of peach have been recommended.

1. *For Hills.* These cultivars which do well in the hills need low temperature (below 7°C) during the period of leaf fall and flowering. These have been further grouped as :

   (*a*) Early – Alexander, Crawfords Early (Totapari)

   (*b*) Mid season – Peregrine, Early Rivers, Harles Early, Elberta

   (*c*) Late – Chaubatia Red, Red Nectarine, Red Wing, Princess of Winter.

2. *For Plains.* About two dozen subtropical cultivars which need less cool atmosphere are grown in the plains. Some of these are listed here. Sharbati, Chakaiya, Saharanpur No. 6, Safeda (L.R. Brothers), Saharanpur Hybrid 3, China Flat, Double Flowering and Dr. Hogg, (ornamental), Floridasun, Florida's own, Bidul's Early, Early Amber, Sun Red, Tinston.

Alexander. Fruit large, round. Skin greenish white nearly covered with brilliant red. Flesh greenish white, juicy and sweet, early cultivar.

Duke of York. Early cultivar, fruit medium-sized, skin crimson. Flesh juicy, sweet and excellent flavour.

Early Beatrice. Early cultivar. Fruit small-sized, skin brilliant coloured with red cheek, flesh delicious, melting and very juicy, clingstone.

Early Rivers. Fruit large-sized, round, skin pale straw-coloured with delicate pink cheek. Flesh juicy and highly-flavoured, ripens in early July.

Elberta. Fruit large, attractive, free stone, quality medium.

J.H. Hale. Self-unfruitful (male sterile) culitvar, fruit very large, round, yellowish, blushed with red. Skin a little fuzzy, flesh of good quality ; not fit for canning.

Late Devonian. Late cultivar, fruit, very large sized, solid, handsome ; flesh juicy, fine flavoured ; ripens by the end of August.

Noblesse. Fruit good-sized; skin pale yellow green with dark red streaks ; flesh melting, juicy, sweet and highly flavoured, ready by mid-August.

Peregrine. Fruit large-sized, skin brilliant crimson, flesh juicy, sweet and fine-flavoured.

Triumph. Fruit yellow, partially clingstone, quality excellent.

The subtropical peaches need 60-100 hours of chilling temperature. Important ones are mentioned below :

Chakaiya. The tree is of medium vigour with stout branches. The bearing intensity is very high. Fruits are roundish and slightly depressed both at the base and apex. The flesh is juicy and sweet in taste.

Floridasun. It is less chilling cultivar developed at the Florida Agricultural Experimental Station (U.S.A.) and now grown in the state of Punjab, Haryana, Delhi, U.P., Rajasthan, Kashmir, submountainous Himachal Pradesh etc. Fruits are attractive and ripen at the end of April. It is free stone and yellow fleshed. Fruits are round having uniform red blush.

Early Amber. It has been introduced from Florida. It is a free stone medium-sized cultivar which ripens after *Floridasun* in mid-May. The pulp is yellow, soft, juicy, sweet with good aroma. It has succeeded as a good commercial cultivar. Needs less chilling.

Tinston. It is an European cultivar. The tree-size is medium, the fruits are small roundish in shape. The pulp is yellow, soft, sweet, free-stone and heavy fruiting.

Sun Red. It is similar to Floidasun. It ripens in the month of May. The bearing is so profuse that it would require thinning. The fruit shape of Sun Red is round without a suture, bulge or a beak. It is nectarine.

Sharbati. It bears delicious fruits in the month of June. The skin is greenish yellow with vermillion tinge on the base of fruits. The flesh is soft, cling stone, juicy and sweet. Sharbati does well in the submountainous areas of U.P., a honey type needing less chilling.

Quetta. Popular in Kashmir. Fruits medium to above medium, pointed, halves accuminate. Skin, thick, downy, yellow base with scattered red patches. Flesh firm, creamy yellow, moderately juicy, sweet with acidic blend when fully ripe. Free stone. Marketed in third week of August.

Dr. Hogg and *Double flowering* are two ornamental cultivars of subtropical peach, which are used for decorating lawns and park corners. Both are free stone cultivar and bear double coloured flowers. The Double flowering culitvar has flowers of rosy colour while the flowers of Dr. Hogg (cultivar) are deep red.

Elberta, and World Earliest are also grown in India. Of these Elberta has the widest climatic adaptability. Chakli belonging to *Prunus platycarpa* is grown in the plains but its fruit is of poor quality and is flat.

Recommendations. Stark Earlyglo, stark Early, white Giant, starking delicious and candor new promising peaches for mid hills of Himachal Pradesh ; Flordasun, Dawn Rambler and Dawn Rose for subtropical regions of north India ; Safeda Early Cream and SRE 6 for *tarai* region of Uttar Pradesh ; and Praire Dawn, Praire Rambler and Praire Rose for cold areas of dry temperate regions.

Pest and Diseases

Leaf curl aphid *(Brachycaudus helichrysi)*. A serious pest of peach, generally attacks in November-May. Nymph and adult suck the sap of new leaves and curl them. Spray the infested plants 7-10 days before flowering with any of the insecticides like methyl demeton (0.025%), Dimethoate (0.03%), monocrotophos (0.04%), Fenitrothion (0.05%) or Dichlorovos (0.03%). If needed repeat the spray at 15 days interval.

Defoliating beetle *(Anomala lineatipernis)*. Serious pest during rainy season, persists almost throughout, pest is nocturnal in habit, eats leaves and fruits. Spray 0.1% carbaryl at 7 days interval after first appearance of beetles.

Leaf Curl. A disease caused by *Taphrina deformans.* The fungus grows on the epidermis and cuticle. Prune off the affected twigs. Spray copper oxychloride (0.3%), Carbendazim (0.05%) or Dithane M-45 (0.3%)at bud swell stage.

Gummosis. Due to excess moisture or due to attack of bark-borer etc. Affected parts are pasted with lime 1% mixed with 0.1% Metasystox in March or September for control.

CHAPTER

31

PEAR

(***Pyrus species*** **L.)**

Rosaceae

Origin and Distribution

Pear has been grown in Europe since antiquity and wild pear (*Pyrus pashia*, **mehal**) is found growing in the Himalayan region in large numbers. The cultivated pear, *Naakh* (H.) (*Pyrus communis* L., common pear) and *Nashpati* (H.) (*Pyrus pyrifolia* var. **Culta nashpati**) are distinct in their susceptibility to fire blight disease. Pears are temperate fruit grown extensively in Kashmir as well as in the South. In India it is grown in Himachal Pradesh, Kulu Valley, Dehradun, Meerut, Muzaffarnagar, Nainital, Saharanpur, Tarai areas of U.P. and the Nilgiri Hills. It resembles apples in many aspects yet differs in certain details *viz.* it is grown on a lower altitude than apple, deterioration of trees is more rampant yet self unfruitfulness is common to both necessitating the planting of suitable pollinators.

The European pear (*Pyrus communis*) with a persistent calyx fleshy pedicels and pyriform shape is different from *Pyrus pyrifolia* or *P. serotina* (oriental pear) having diciduous calyx, non-fleshy pedicels and shape like an apple.

Cultural Requirements

Besides being a temperate crop it needs sun, water-retaining medium of growth and shuns cold winds. Dry hot summer winds are harmful. A deep open sandy loam soil is considered suitable for the crop. Good moisture supply during summers must be ensured. The climatic requirements are a little warmer than that of apple. Soil pH between 5.5-6.5 suits well.

Propagation

Wild Himalayan pears *Pyrus pashia, P. pyrifolia* are generally used as root stock on which the cultivars selected for growing in any area should be budded or grafted. In orchards one may come across "double worked" trees, this is a well known method to grow suitable cultivar even on incompatible stock through the agency of a mutually compatible yet substandard cultivar. Whenever dwarfing is required quince (*Cydonia oblonga*) may be used a rootstock with or without the help of interstock according to need. Bartlett pear is double worked on quince stock on one side and Old Home cultivar of pear on the other. *Pyrus communis* (Naakh) may be propagated by T-budding on wild pear (*P. pashia*) in the beginning of rains. Quince (*Cydonia oblanga)* may be used as rootstock for some cultivars.

Planting

In pears, dwarf pyramids are usually desired and training is done. Winter planting of pear tree may be done in one meter square 20 cm deep pits. Support the fleshly planted young trees keeping the roots well spread with stakes driven about 60 cm deep in the soil. Since all the commerical cultivars of pear are self incompatible, disploid cultivars like Bartlett, Beurre hardy, Deyenne du Comice, Conference and Winter Nellis set good crop if planted in groups of 2-3 cultivars. Jargonella, a triploid, should be planted with diploid cultivar to set good crop.

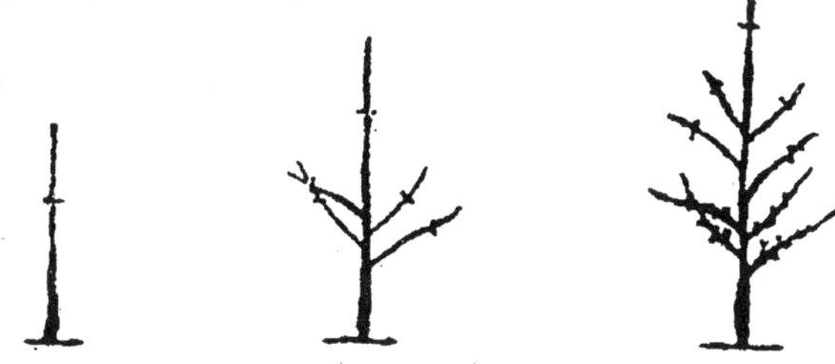

Fig. 31.1. Training a dwarf pyramid in pears.

***First year* : Cut back the maiden to a height of 40 cm.**
***Second year* : Prune the new growth on the main stem about 40 cm. and side growths to 20 cm. just above downward pointing buds.**
***Third year* : Fruiting spurs are formed on branches. The leader and upper branches are cut back again.**

Pear trees, planted at a distance of 35 meters can be trained as bushes with open centres, as bushes with delayed open centres, or as pyramids ; trees trained in any of these ways should be maintained at a distance of 2.5 meters all round. The spindle trees are planted 1.2 meters apart with 2.8 meters space between rows. Oblique single cordon needs only 60 cm space between two plants while the espaliar trained trees against walls or fences do well when put at a distance 3.6 meters from one another. The distance for some varieties is kept 6 × 6 m. Planting distances of some patterns of pear *viz.* delayed open centre bushes, pillar-trained trees, spindle bushes, espaliar, cordon are depicted on page 444 (Fig. 28.1).

Manuring

Good growth and profitable yields can be achieved by the application of nitrogen, phosphorus and potash as sulphate of ammonia, superphosphate and potassium sulphate in the ratio of 1.5 : 6 : 1; average requirement being 30 gms of sulphate of ammonia, 120 gm of superphosphate and 20 gm of potassium sulphate per square meter. With advance in age of the trees the need of superphosphate increases.

Planting time. Roughly, the application of manures and fertilizers should be in 2-3 instalments each year at planting, pre-bearing and bearing stages. Apply 30-40 kg compost + 250 gms superphosphate + 50-75 gm urea in one meter wide ring around each tree in spring when growth begins.

Pre-bearing stage. 30 gm superphosphate + 5 gm potash should be applied for 10 years every December along with 10-15 kg of compost and 10 gm nitrogen in spring to every tree.

Bearing stage. Every tree should be supplied with 40 kg compost + 250 gm phosphorus + 50 gm potash in the month of December-January. Apply 100 gm of nitrogen in spring.

Pruning

Pears with upright growing habit can be trained to form, pyramids, bushes or cordons if the stock is quince. Pear stocks make vigorous growth and needs heavier pruning. Pear usually needs harder pruning than apples, although the method is the same. Pears produce fruit buds very easily, so that after ten years they usually need thinning out by cutting back the spur bearing the fruit buds to half their size. The main consideration in its pruning is to thin out central branches and encourage lateral growth.

Harvesting

Pears being highly perishable need picking while still green. When all the pears appear to have attained full size, lift up carefully by the palm to see if the stalk comes away easily from the spur. If it is so the fruit is ready for picking. The late cultivars must not be picked up too early as they shrivel and do not give good appearance. The optimum temperature for storing pears is between 1-2°C. 40-50 kg. fruits are reported from a tree each year.

Cultivar

Some of the important cultivars of Kashmir (Gosh Baghu and Baghu Gosha also known as Bartlett) are well known for their quality and yield. Kieffer, cultivated in South India is a cross between *Pyrus communis* and *Pyrus pyrifolia*. Beurre Hardy and Louise Bonne of Jersey, good quality pears are also grown at an altitude of over 700 meters in South India. Cultivars recommended in the Punjab are *Pathar Naakh, Leconte, Smith* and Kieffer. The plains can grow successfully the cultivars which need less cool environment *viz. Kashmir pear, Leconte, Pathar Naakh, China Gola etc.* Nashpati which is cultivated in the Punjab and U.P. (submontane areas) bears a heavy crop of poor quality fruit. It needs no cross pollination. Thumb pear or Chusni and China Nashpati are early pears from the Kumaon and Himachal Pradesh. These are marketed in June.

Thumb pear is early season while 'Beurred' Amanlis, Victoria, Beurre Hardy and Gola are mid-season cultivars. Among the late cultivars *Doyenne du comice* and *williams* Bon Chritien are suitable for canning and drying.

Baghu Gosha (Citron Des Carmes). Fruits are of medium size greenish yellow skin with russet patches. Flesh greenish white, firm and juicy with moderately sweet flavour. Fruits are harvested in mid-July.

Conference. Fruit, medium-sized, pyriform. Skin green with russet patches. Flesh salmon-coloured, sweet, juicy.

Clapp's Favourite. Fruit large sized, pyriform ; skin pale lemon yellow. Flesh granular, tender, rich and melting. It is picked up before July.

Doyenne du Comice. Fruits obtuse pyriform, skin yellow with russet patches. Flesh : melting, declicious and aromatic.

Dr. Jules Guyot. Flesh juicy, melting and fine flavoured. It is harvested in early July.

Easter Beurre. Fruit, large sized, roundish obovate, pyriform with a short stout stalk. Skin yellowish green. Flesh, white, fine-grained, delicious, richly flavoured. It is picked in early September and keeps till March.

Emile d' Heyst. This cultivar of pear bears large, oblong but slightly tapering fruits towards the stem end. Fruit flesh melting, juicy and fine flavoured. Fruit is harvested after mid-July.

Marie louise d' Uccle. Fruit resembles that of Emile d' Heyst.

Nashpati. It bears fruit of inferior quality. The flesh is gritty, very prolific, successful in the Punjab.

Thompson's. Fruit is of medium size, skin yellow. Flesh fine grained, melting and of good aroma.

William's Bon (Chritien (Bartlett). It is an English cultivar. Fruit is like Clapp's Favourite ; skin yellow with faint blush. Flesh buttery, fine grained, juicy, sweet. It is excellent for canning.

Winter Nellis. It is a Belgian cultivar. Fruit medium- sized, roundish obovate, narrowed near the stalk. Skin yellowish. Flesh delicious, juicy, aromatic. Fruit is picked at the end of July and keeps till December-January.

Nakh Kashmiri (Chinese sandy pear). Conical in shape, small to medium size, crisp, white and juicy flesh. Thick green skin, turns yellow on ripening. good storage quality. Excellent dessert cultivar.

The pear varieties belong to three groups namely European, Asian and hybrids. The varieties recommended for different states are stated below :

Jammu and Kashmir

Early. China pear, Beurre-de-Amanlis.

Mid-early. William Barlett, China sand pear, Genta Drauord Fertility, Clapp's favourite, Doyenne Bussoch.

Late. Viear of Winkfield, Hardy.

Himachal Pradesh. Patharnakh (Sand pear), Keiffer (hybrid), China pear.

Uttaranchal (*i*) *High hills*—Max-Red Barlett, William Barlett, Conference, Hardy, Winter Mellis, Clapp's Favourite, Flemish Beauty, commice. (*ii*) *Lower hills and plains*. Patharnakh, Gola and Le Conte (hybrid).

Minerals Deficiency Symptoms

Magnesium. Older leaves of terminal shoot dark brown or blackish. Interveinal necrosis near leaf centres. Margins remain green. Defoliation starting from base in severe cases.

Potassium. Leaves dark brown, marginal scorch. Poor shoot growth. Fruits appear immature.

Iron. Leaves of the tips of shoots smaller, strongly chlorotic or totally bleached. Branches die back. Fruits highly coloured.

Manganese. Leaves slightly faded green with interveinal chlorosis progressing from leaf margins. Symptoms may be fairly general over the tree.

Zink and boron deficiencies may be offset by foliar spray of 0.5% $ZnSO_4$ and 0.3% borax to the affected plants.

Insect-pests and Diseases

Insect-pests

Pear psylla. To control this pest spray Endosulphan @ 0.2% or Methyl Demeton @ 0.25% about three weeks after petal fall. Heavy infestation may require two sprays, one at petal fall and another three weeks later.

Disease

Fire blight *(Erwinia amylovora)*. Infection occurs on flowers, twigs, fruits, branches and the trunk Usually the first symptoms appear on flowers. The most obvious symptom is the rapid wilt and browning or blackening of blossoms or leaves. Spray of Bordeux mixture (2 : 6 : 100) at silver to green tip stage of flower bud may reduce the disease incidence when disease infection is not high. When conditions are highly conducive spray Streptomycin at 3-5 days inteval.

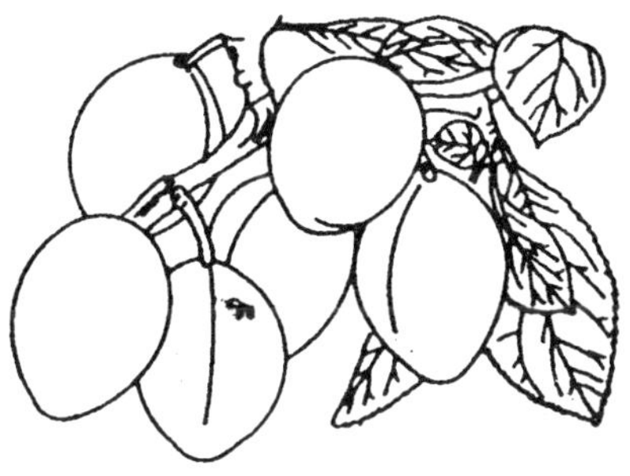

CHAPTER

32

PLUM

(*Prunus domestica* L.)

Rosaceae

Origin and Distribution

Plum (H. *Alucha*), a native of North America and Japan is extensively grown in Canada. The chief plum producing areas are Oregon, Weshington, Missouri and Missicipi Valleys. The Japanese and hybrid varieties are grown in Maryland and West Texas. Of the sources of cultivated varieties the Japanese (*Prunus salicina,* cultivated at lower heights) and the European (*Prunus domestica,* cultivated at higher elevation) are important. Bokhara plum was considered by Naik under *salicina* though it is held to be a variety or type of *domestica.* There is yet another view putting it under cerasifera group. Myrobalan is important as root stock because of its general graft compatibility and adaptability to many soils. The *P. domestica* L. subsp. *insititia* assumed popularity for jams, jellies etc.

The plum (*alu bukhara*) tree of the *domestica* species is robust about 8-10 meters high with large ovate and thin, irregularly serrate leaves. The flowers are white or cream in colour. The Japanese plum is a small tree with glabrous parts showing reddish or brownish twigs and closely blunt serrate leaves. Flowers are white. Plum should be cross pollinated for good crop. Beauty and Santa Rosa are self-fruitful. The domestic species is cold resistant just like apple but the Japanese plum is comparable in hardiness to peach in wood although the bloom being early are damaged by frost.

Soil and Climate

Well drained soils of the clayey loam type with good subsoil in rooting zone work better for plums as in the case of apple and pear which need soils of slightly poor quality. Well manured soils having enough of peat of leaf mould to cover about 5 cms of the surface with annual dressings of fish guano or bonemeal at the rate of (150-200 gm) per square meter prove useful. Well rotten farm yard manure or 30 cm deep straw mulch is a good substitute. Acid soils must be given a dressing of carbonate of lime before planting the plums. The temperature requirement during the growing season of plums is specific, for the type. Plums are also grown commercially in Dehradun, Saharanpur, Muzaffarpur and Tarai areas of north-west U.P.

Propagation

Plums are very variable in rootstock requirement and fruitfulness. It is mostly propagated on wild apricot (*Zardalu*) rootstock by shield budding in India. *Myrobalan* B, and *Myrobalan* 29C (*P. cerasifera*) as rootstock is hardy, deep rooted and long lived though less vigorous. Other rootstock used for plum may be peach (*P. perscia*). Vegetative propagation is commonly practised plum. Cuttings of cultivars like Garden Green are planted in November or February to serve as rootstock. These may be used for shield budding in April-May or for tongue grafting in February. In alkaline soils with nematode menace apricot rootstocks prove better. In dry soils almond may be used for the purpose. Japanese cultivars may do well on rootstock of European cultivars but reverse is not true. *Pershore* as well as *St. Julian* bring about early bearing, the latter is compatible with large cultivars of plums. *St. Julian* probably works better than common Mussel, the latter being used for plums grown as bushes or fan-trained trees. Common Mussel rootstock has the disadvantage of sending up suckers. The common plum rootstock is highly compatible with *Victoria* and helps to resist the silver leaf disease. The trees need a space of 4-5 meters between themselves. The cultural practices are essentially those followed for peach. Hardwood, cuttings of Myrobalan, serve as rootstock, when rooted.

Planting and Growing

Plum trees may be grown as bush, half standard, standard or fan-trained depending on the variety and the growth environment. The bush tree shoud be purchased when two year old, fan-trained when 4-year old, while rest when 3-year old. Bush plums are discouraging due to the drooping character of the branches ; the seedlings to be planted must necessarily have at least one meter clear high stem. The fan shaped trees are usually trained against walls or fences but standards and half standards are preferably planted in field.

Planting must be done firmly with roots spread evenly in one meter square about 15 cm deep pits. Before planting the damaged roots should be removed clean. Planting of one-year nursery trees gives best results because the interruption in growth of young trees is much less although two-year old trees may also be transplanted. The space needed for plums varies with the types of the growth of trees. Bush plums may ordinarily be given 5 m sq., half standards 6 m square and standards 8 m square space although the distance may vary with cultivar and the fertility of the soil. The period of planting may be November to February according to environmental conditions and cultivar requirements of the cultivar.

The trees should be staked while planting to check 'wind rocking death' due to winter winds. While the European plums (*P. domestica*) are self-fruitful the Japanese plums (*P. salicina*) often requires cross pollination.

Pruning

Plum trees may be trained with profit, the 3 year scheme is generally followed. Variations in the type of growth of plum call for different pruning systems to develop a modified leader type of tree. European cultivars respond to light pruning against Japanese where pruning encourages good vegetative growth resulting in increased fruit size.

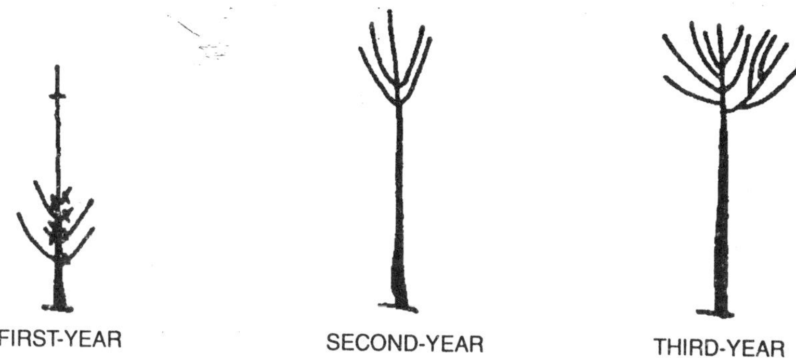

Fig. 32.1. Pruning of plum trees for standard (suitable for growing in grass)
***First year* : The main stem is cleared of all-side growth lower down upto a height of 1.8 meters from the ground.**
***Second year* : The branches in the main stem are cut back.**
***Third year* : The side branches are also pruned sufficiently.**

One-year old leaders may be pruned back by about half their length in February following planting. Pruning of this type for three years with the strong growing cultivars on vigorous stocks, and for four years in the case of heavy cropping varieties on weak stock is desirable. Cuts made in the winter should be smoothed and coated with thick white lead paint to check the onset of silver leaf disease by stopping the entry of spores. If pyramid shape is needed the trees are given 3 meters square space, their laterals cut back late in July to 20 cm and the central leader allowed to grow upwards. But for cutting back one-year old growth a few months after planting, no winter pruning is done in this case.

Pruning also helps at striking of proper C : N ratio of the trees which have reached the bearing stage. Thinning may be done occasionally to allow full light and air preferably during the hotter months. When gumming takes place, quickly seal the wounds. For spreading cultivars, the leaders needs be pruned to an upward pointing bud but the upright cultivars *viz. President* and *Czar* must be cut back to just an outward-pointing growth to encourage spreading. Japanese cultivars need more fruit thinning than the European cultivars. Most cultivars require more sever thinning than Santa Rosa which needs only slight thinning.

Pollination

Plums are known to be generally self-fertile, incompatible as cross and for possessing different periods of bloom, characteristics that make pollination a problem. The climatic conditions influence the fruitfulness of a cultivar. Plum growing can be made successful by planting, one pollinizing cultivar blooming at the period when the main cultivar blooms, at least every second tree in every second row. It may, however, be noted that not much reliance can be placed on cross fertilization between species. Bees are necessary for pollination and must be encouraged besides providing ample pollen grains in orchard.

Thinning

Proper spacing of fruits encourage attractive colour, large size and good flavour. Thinning may be done in April or just after the natural fruit drop. As the plums attain the size of the fruit of oak hand thinning is recommended to a spacing of about 5-8 cm. Japanese varieties being heavy bearers need more thinning. Thinning reduces chances of branch breakage, increases uniformity in size and colour of the fruits besides ensuring regular cropping. A heavy crop after thinning needs support to the branches. If the fruits are close enough the spread of brown rot disease is encouraged in humid regions and therefore thinning is profitable. Heavily cropped branches may need support by maypole or forked stakes.

Harvesting

Unless long distance transport of the plums are desired those of superior quality and high sugar content must be harvested mature as firmness and colour change. Rough handling of the fruits so as to cause bruising or removal or waxy coating must be avoided. Fruits are gathered in 3-4 pickings as they mature. Plums are generally harvested at a more ripe stage than peaches and may often be knocked down to the ground by shake to the branches or may be allowed to fall when fully ripe. Precooling and transit under refrigeration give longer life to harvested fruits.

Plum types. Plums are of several types. Closely allied to plums are Gages requiring almost similar treatment culture etc. Damsons are small fruits full of stones, less popular than other types of plum. Japanese Plums are ornamental, heavy cropper with decorative and juicy fruits. There are also less common types of plums *viz.* Cape plums, Cherry plums, Shiros, Bullaces etc. Use of the word plums is so common that we have coco-plum, date-plum, hog-plum, madagaskar-plum, marmalade-plum, sapodilla-plum, tamarind-plum, plum-tree etc. indicating several genera.

Cultivars

Bryanston Gage, Cambridge Gage, Coe's Golden Drop, Early Transparent Gage, Early Laxton, Green Gage, and Jefferson President are some of the recommended dessert cultivars which need Greengage, Victoria, Early Laxton,

Majorie's seedling and Denniston's Superb respectively as pollinators. Denniston's Suporb, Giant prune, Early Transparent Gage, Laxton's Gage, Majorie's seedling, Qullin's Golden Gage, Purple Pershore, Victoria etc. are self fertile. The important Japanese cultivars Santa Rosa and Beauty are self fruitful and may be planted alone. Other important cultivars are Grand Duke, Manipora, Kelsey, Wickson and Bright Red Burbank, Alucha red, Alucha yellow and Titron are prolific bearing cultivars of good quality grown in the north-western plains. *Titron* or *Jamuni* is a commercial and paying cultivar of Saharanpur. The fruits are of good quality yielding good profits.

Cultivars Recommended for the Himalayan Region

Beauty. The delicious fruits have deep red colour. These are medium sized and as early as Early Transparent Gage. The stone is not free. The cultivar is a regular and prolific bearer. A Japanese cultivar, self fruitful like the Santa Rosa.

Early Transparent Gage. A regular and heavy cropper, prolific bearer with delicious fruits which are rather small and roundish. The colour of the fruit is pale yellow with red dots. Fruits are ripe in July and ready to be picked up by the third week. Self fertile.

Grand Duke. A dark purple skinned fruit. It is good for cooking and jam making. The fruits are very large, oval in shape with unequal halves flattened apart. A prolific bearer ready by August.

Kelsey. The fruits are heart shaped greenish yellow in colour. These are picked up in August. The flesh is yellow, juicy, firm and of good quality.

Santa Rosa. The fruits are large, dark purple in colour with whitish bloom. The flesh is very juicy and of gloden colour. It is an early cultivar which is ready in July. Plants are regular and prolific bearer. Self fruitful.

Victoria. Late cultivar, carmine with pale and blue bloom. Large, oval shaped fruits with dark skin. Regular cropping habit. Fruits are good for canning, cooking and dessert purposes. Susceptible to silver leaf disease. Self fertile, very popular.

Wickson. Fruits with cherry red coloured skin are heart shaped and large. The flesh is very juicy, sweet, aromatic and of amber colour. The picking is done in August. The cultivar is a regular and good bearer.

In the plains of U.P. following cultivar of plum are grown. Early round and Calcutta are good pollinisers for these.

Jamuni. The fruits are ready in the second week of May. Fruits are round, brownish in colour yellow flesh, juicy and sweet. The stone is small which separates easily from the flesh. The cultivar is known as *Teetar* or *Titron.*

Zardalu chittidar. The fruits ripen between May 25 to June 7. Round in shape with red coloured dots, yellow and very sweet to taste. The stone does not cling to the stone and is large.

Zardalu. Fruits ripe in June. The shape may be roundish, dark yellow with yellow flesh, sourish but tasty.

Cultivar use for cooking. Czar, Giant Prune, Yellow Egg, Purple pershore, Majorie's Seedling, Qullin's Golden Gage, River's Early Prolific.

Mineral Deficiency Symptoms

Magnesium. Older leaves have chlorosis beginning from the margins or between veins. Interveinal necrosis may also develop.

Potassium. Leaves bluish green or chlorotic. Margins rolled forward parallel with mid-rib, margins scorched.

Iron. Leaves at shoot tips are chlorotic, vein pattern may or may not be visible. Branches show die back.

Manganese. Interveinal chlorosis of leaves beginning at margins. Leaves on terminal shoots tend to be green.

Pest. When aphid attack is visible two sprays of 0.03% Diazinon may be given at an interval of 15 days as in Peach.

No specific insect or disease attacks the plum trees. Spray materials as suggested for peach and apples can cantrol the insects-pests and diseases of plum if any.

CHAPTER 33

STRAWBERRY

(*Fragaria* species)

Rosaceae

Origin and Distribution

The Himalayan region abounds wild strawberry (H. *Hisaalu*) a small herb with shallow roots which produces small fruits of inferior quality. Cultivars commercially grown in India are not indigenous. Culitvated garden strawberry of India is said to be related to *Fragaria vesca* L. (everbearing Alpine strawberry) by some although F. *vesca*, a deploid, has obviously little in common with the evolution of the strawberries which are mostly octaploids. Cultivated strawberries are either 'June-bearing' or 'ever-bearing'. The fruit is an enlarged receptacle bearing numerous achenes the so-called seeds. Strawberry is grown mostly in Dehradun, Nainital, Mahabaleshwar and Meerut. The soil is kept covered by straw during fruiting or cold winters giving it the name strawberry.

The anti-bacterial juices at strawberry have been used for decades to protect against typhoid epidemies. This property is due to the presence of a variety of plant nutrients, bioflavonoids.

Soil

Soils of high fertility are prepared thoroughly one year in advance of planting. Soil management should aim at giving rapid growth in first year. Strawberries may be grown on a large variety of soils. Soils rich in organic matter with capacity to retain moisture yet drain off any surplus give best results. Slightly acid (pH 5.5 to 6.5) light soil, good drainage, clean shallow cultivation, adequate irrigation and enough organic matter are the prime needs of successful strawberry cultivation. Protection of the soil from erosion should be done by growing cover crops. For clean cultivation use of geese has been done in U.S.A. About a dozen geese per hectare do the jobs well. They are removed from the orchard just before fruiting, till then they are harmless to strawberry.

Propagation

Seeds need stratification at 0°C for 90 days before sowing to give good germination. One year old runners may be employed for propagation. These are transferred to soil and as the roots develop, and planted. The cultivated strawberry is supposed to be derived as a result of a cross between *Fragaria chiloensis*. Duchesne, a native of pacific Coast of North and South America and *Fragara verginiana* Duchesne, wild species of the eastern coast of north America, having as many number of chromosomes (30 pairs) besides being fully fertile with them.

Planting

Healthy plants obtained from a virus-free mother stock raised on soils rich in organic matter may be planted in well prepared holes, one in each, roots spread carefully and the soil put back and tread down had followed by light hoeing. The area for one plan may be maintained at 75 cm × 45 cm. The season for transplanting varies with the elevation. At higher altitude transplanting may be done in March-April. In the plains September-October plantings prove most profitable. August planting yield heavily for about four years. Late plantings do not fare well since the plants are not able to establish well and winter losses occur. With spring planting no harvest can be expected in the first year.

Cultural Practices

Cultural practice that increases the size and number of leaves is practised for better yield of large fruits. Early planting, absolute freedom from weeds and pests, adequate irrigiation and fertile soils are needed. Strawberry roots do not penetrate deep and therefore supply of ample moisture and shallow hoeing is recommended. As the plants blossom plantation is bedded with straw to keep the fruits off the soil. The straw is removed after fruiting. Removal of runners is also helpful in increasing yields.

The soils is worked in succeeding months. Strawberry is often grown in rotation with vegetable crops which are looked after well so that both the land and the strawberry benefit besides getting extra income from the yield of vegetables.

The plant needs a top dressing of organic manure in the winters. Application of 2.5 quintals of NPK mixture per acre in the ratio of 1 : 2 : 2, a week before planting and a similar does when suckers begin to develop is recommended. Side dressing with 50 kg. of nitrate of soda twice a year *i.e.* at pre-blossoming and blossoming increases fruit yield. The addition of about 12 quintals/has fish-fertilizer in April has been found to be a profitable practice in India. Similar dose of bone meal may be given immediately after fruiting. Strawberry needs protection from frost and birds.

Training

For strawberry three training systems have been *in vogue viz.* the hill system, the spaced row system and matted row system.

(*i*) **Hill System.** Runners are removed so that the available nutrients may be utilized by the mother plant. In the hill system, though the cost of production is more, the yield is less, the fruits are bigger.

(*ii*) **Spaced-row System.** Runners are not removed and the number and spacing of new plant is regulated. It helps in the increase of the size of the plants and also yield. Three to four plants may be allowed to develop from each runner.

(*iii*) **Matted-row System.** The mother plants may be spaced 30 cm × 90 cm apart in rows 120 cm to 150 cm apart. Runners are allowed to develop so that the plants occupy a space 90-124 cm. wide. The yield depends on the number of plants and the spacing.

Picking of Fruit

The fruit is a soft, highly perishable berry (an aggregate, botanically). The ripe fruits are picked by picking of the stalk by hand every 2-3 days after a period of about a month from the date of blossoming ; generally the barries take only 15-30 days to be ripe.

Harvesting

Harvesting in the early or late hours of the day with low temperature and storage under proper refrigeration prolongs the life of the fruits which are otherwise highly perishable. Too much handling must be avoided. 15,000 to 25,000 quarts of fruit yield are possible from a hectare.

Cultivars

The following cultivars are recommended,

Cambridge Favourite. Mid-season, large, salmon-scarlet, heavy cropper, compact grower.

Cambridge Late Pine. Late, conical, crimson, sweet flavoured, resistant to frost and mildew.

Cambridge Prize winner. Early, large, conical, crimson, good cultivar for sandy soil.

Cambridge Rearguard. Very late, rounded, dark crimson very susceptible to red core disease and mildew, loves sandy soils.

Cambridge Sentry. Mid-season, glossy crimson, resistant to disease, good for bottling and jam making.

Cambridge Vigour. Early, glossy crimson, resistant to disease and frost.

Merton Princes. Mid-early, large, crimson, susceptible to red core.

Pusa Early Dwarf. Successful in plains, plants are small, they give a storage in September-October. Plant to plant distance should be 22.5 cm to 30 cm. 50,000 plants are required for one hectare.

Red Gauntlet. Mid-season, large, scarlet, vigorous compact grower, large trusses.

Royal Sovereign. Mid-season, scarlet large, delicious, does well on heavier soils.

Talisman. Late, large, crimson, excellent for jam, resistant to mildew, soft in rainy season.

Chandler. Plants are resistant to physical damages and tolerant to viruses. Fruits large, high dessert quality with outstanding colour and flavour. Average weight of berry is 15-18 g.

Jioga. An early maturing, tolerant to viruses, fruits very large, dessert and processing quality good. Average weight of berry is about 9 g.

Belva. It produces fruits in off-season also. A day-neutral variety having frutis large, flesh and skin firm with red colour. Its keeping quality is fairly good. Average berry weight 15-18 g.

Pajaro. It is very successful under summer system. Plants tolerant to virus. Fruit has good dessert and processing quality, fruit is quite susceptible to physical damage caused by rains. Average weight of fruit is 7-6 gm.

Chaubatia abhichal. It is a hybrid between Premier Speece. Fruit is red and large. It produces 295-300 g furits per plant.

Insects-Pests and Diseases

1. White grubs and cutworms. These insects damage the root and stems of young plant. Deep ploughing and treatment of soil should be done with chlorpyriphos.

2. Spider mite. This small red colour mite infests the upper surface of the leaves. Sucks the sap of the leaves causes failure of the crop. It can be controlled by spraying malathion (0.05%) twice.

Diseases

1. Anthracnose (*Colletotrichum fragariae C. acutatum, Glomerula Cingulata*). Anthacnose is one of serious disease of strawberry. Diseased caused by *Colletotrichum* is termed as "strawberry anthracnose." These include anthracnose, crown rot, bud rot, flower blight, fruit rot, black leaf spots and irregular leaf spots. To control the anthracnose use disease-free planting stock, removal and destruction of mummies, fumigation of nursery and main field with a mixture of methyl bromide and chloropicrin, regular fungicide (copper oxychloride, 0.2%) sprays all these integrated approach should be adoted.

2. Mycosphaerella Leaf Spot (*Mycosphaerella fragariae*). Small, purplish, circular spots are appear on the upper surface of young leaflets. This disease may also occur on petioles, fruit pedicels, suckers and calyx. Destroy the crop refuse, proper spacing and drainage may reduce the disease severity. Spray one of these fungicides *viz.* carbendazine (0.1%), benomyl (0.1%), captafol (0.2%) and Ziram (0.2%).

3. Red Stele of Strawberry (*Phytophthora fragariae*). Older leaves show the symptoms of yellowish, reddish or brown tinged while new leaves remain small sized. Fruits are undersized and dry before repening. The pathogen first destroys the feeder roots and rootlets. Good drainage and the planting stock should be from disease free field are the prerequisite for healthy crop. Dipping of planting stock and foliar spraying with metaexyl, fooetyl-al, ethazol and propamocarb should be done.

Mineral Deficiency Symptoms

Phosphorus. Early defoliation of older leaves. Poor growth. Few crowns. Leaves dull green with a little purple tinge which increases as defoliation approaches.

Magnesium. Older leaves bright yellow with red tints in centre and near the margins.

Potassium. Dark green leaves with scorched margins.

Iron. Leaves chlorotic with visible vein pattern.

Manganese. Faint leaves with interveinal chlorosis beginning at margins.

THE NUTS

CHAPTER

ALMOND

(*Prunus amygdalus* Stokes Syn. *Amygdalus communis* Linn.)

Rosaceae

Origin and Distribution

Almond, (H. *Badaam*) considered to be a native of Mediterranean region is a dual purpose tree *i.e.* for nut as well as beauty. Spain, Italy, USA, Morocco, Portugal, Turkey, France, Algeria, Afghanistan and Persia may be taken as the almond growing countries. India is not on the exporting list. It is most favoured nut with Indians who regard it as highly nourishing and of great medicinal value. A very valuable oil called *badam roghan* is extracted from it. Today the major almond growing in India is the Kashmir Valley. The cool and dry areas of Himachal Pradesh especially Chini area bordering Tibet so also Lahaul and Kinnaur have been found to be suitable for the crop.

Climate and Soil

The almond grows successfully between 30° – 40°N. It requires a farily warm dry weather during ripening of the fruit, cool and foggy summers do not suit it. The temperature should not go beyond – 9 to 36°C. Deep, fertile and well drained or light loam soils are considered suitable for almonds, very heavy soils are not desirable. It draws heavily on soil resources. Acid soils need liming for successful almond production. Due to early flowering habit it is very susceptible to spring frosts. Wind breaks are considered useful. Blossoms, with petals exposed but not yet opened, have been known to withstand cold upto – 1°C but those with fallen petals are killed at 0°C.

Propagation

Almond trees grown in India are of a seedling origin. Planting of seedling is considered to solve the pollination problem. Its propagation by T-budding on almond or peach seedling rootstock is becoming popular.

Raising of Seedlings

The almond nuts harvested between July-September, are stored in a cool and dry place till the following December. For breaking the rest period, the nuts must be chilled at a low temperature for 40-60 days. Seeds are sown 3 cm to 5 cm deep on well cultivated and manured beds in rows at a distance of 15 cm from seed to seed keeping 30 cm distance between the rows. Germination starts in 2 to 3 weeks.

Budding as well as grafting entail the use of rootsocks such as those of appricot, plum, peach and almond ; almond rootstock proves best. Apricot seems to be ill compatible as rootstock. Plum (Myrobalan) and peach root-stocks succeed but almond rootstock gives vigorous and heavy yielding trees.

When seedlings are two to three-month old and have attained a height of 20-30 cm ring budding may be done preferably in April or May. Seedlings unsuitable for ring budding or "T" budding due to less growth are cleft grafted in the following January or February. Tongue or whip grafting and bark or crown grafting is done according to convenience in spring after the sap dries and the stock is headed back at a height of 15 to 22 cm. from the ground.

Planting

The trees are planted in 1 meter × 1 meter pits at a distance of 6-7 meters from each other in autumn. It is advisable to plant good pollinisers (more than one cultivar) for this self-unfruitful nut.

Generally two cultivars, though not all, with overlapping flowering period intercross and set good fruits under favourable climatic conditions. Long flowering period, abundant sunshine and low wind help to solve pollination problems.

Non Pareil, a cultivar well known for quality, productivity succeeds commercially with Ne-Plus-Ultra, Peerless, Larguedoc, and Eureka which prove good pollinisers. Naturally these may be kept together. The best method to plant the different cultivars together is to put 3 rows of the main cultivar and one row of the polliniser cultivar. It is also useful to have 3-4 colonies of bees per hectare of orchard.

Cultural Practices

Almond tree has a very short period of rest. Irrigation and manuring should, therefore, aim at delaying the onset of dormancy in almond trees so that the rest period concludes after the winters to ward off the damage due to frost leading to poor yield.

Irrigation and Manuring

Frequent dry summer irrigation is good for proper development of nuts and prevention of fruit drop besides laying foundation for the next year's crop. In

Kashmir it is grown under rainfed conditons. About 60-75 cm of rain per annum is regarded as minimum for proper growth.

Manuring

In addition to bulky nitrogenous manures ensuring good humus supply, 2 to 3 kg of ammonium sulphate and 1 to 2 kg of superphosphate/tree/annum is considered desirable.

Pruning

The fruit is borne largely on short-spur and on one year old wood. Since the spurs remain productive upto 5 years it may be a good practice for bearing of almond trees to prune one-fifth of old bearing wood every year. If the bearing trees make an extension growth of 20 cm to 30 cm per year if it profitable. Thinning of branches rather than heading back is generally recommended.

Harvesting

The harvesting should be started when the hulls covering the hard shells begin to split. The nuts are cleaned of any remaining dried skin and sun dried for a short time. Nuts especially of hard shelled type may be bleached by sulphur fumes which give them attractive colour and increase their market value. Almond trees begin to bear in 3 to 4 years and they reach full bearing in 8 to 10 years. The yield of 4-8 quintals per hectare is common in India.

Cultivars

The trade in India recognises only three types of almond *viz.* (*i*) Sweet Hard Shelled, (*ii*) Bitter Hard Shelled, and (*iii*) Paper Shelled or "Romali". The following cultivars are popular in India and abroad.

Drake. The tree bears medium sized nuts of good shape and colour. The kernel is brown, elongated and well developed. It is sweet and delicious. The shell is soft whitish or light brown in colour. Belongs to "Romali" type.

Eureka. It tolerates adverse conditions such as shortages of water and poor soils better than other cultivars. The nut is of small size.

IXL. Generally it is grown in the foot hills and valley areas. The kernels are often subject to gummosis. Inter-sterile with the Non-Pareil.

Ne-Plus-Ultra. It is valued as a good pollinizer for Non-Pareil. The nut is big and its shells have attraction. The nuts are frequently gummy when the soil becomes slightly water-logged.

Non-Pareil. A thin shelled nut cultivar with high percentage of kernels, good quality, high price and regularity in bearing, it is reckoned as the best commercial cultivar. The nut is flattened, elongated and pointed at the apical end. The shell is white, papery and soft. The kernel is flat, light brown, fairly plump, sweet, and good flavour. Belongs to Romali type.

Peerless. It has got a thick heavy shell. It is also good polliniser for Non-Pareil. It is highly susceptible to frost.

Texas. A late ripening cultivar with nuts small and poor in appearance. The kernel's flavour is slightly bitter. Blooms late and escapes frost.

Thin Shelled. A regular bearing cultivar that flowers in February-March and the nuts ripen by the middle of August. The husk separates easily from the shell at maturity. Precocious cultivar with large nuts. The shell of the 'Romali' is very thin type. The kernel is sweet and delicious.

California Paper Shell, Pathic's Wonder, Stout Paper Shell and H.P. Selection 5 are some promising cultivars of almond which are under trial in the Punjab, Haryana and U.P.

(*i*) *Varieties for Jammu and Kashmir.* Makhdoom, Parbat, Waris, Shalimar, Afganistan seedling, IXL, Jordanolo, Merced, Non-Pareil.

(*ii*) *Varieties for dry temperate zone.* Neplus-Ultra, Texas.

(*iii*) *High and midhills.* Merced, Non-Pareil, IXL.

(*iv*) *Himachal Pradesh.* Drake, Katha, Peerless, Neplus-Ultra.

(*v*) *Varieties for the plains (low-chilling).* California Paper shell, Hybrid 15, Pethicks Wonder, JK 55.

Insect-pests and Diseases

Insect pests. The insect pests attacking almond are the same as for other temperate fruits (Apple, Pear, Peach etc.).

Diseases. Almond suffers from diseases like blossom blight (*Monillinia* spp.), shoot hole (*Cercospora* spp.), collar rot (*Phytophthora* spp. and *Phythium* spp.) and bacterial gummosis (*Pseudomonas* spp.). Fungal diseases may be controled by spraying copper oxychloride @ 3 g/litre or Dithane Z-78 @ 2.5 g/litre water at bud burst, petal fall and fruitlet stages. Bacterial gummosis can be controlled by applying Mashobra paste after cleaning the weeping wounds at the time of dormancy break and again before rainy season. Spraying the tree with Streptocycline @ 0.03 g/litre water before onset of rains is also recommended.

CHAPTER

CASHEW-NUT

(*Anacardium occidentale* L.)

Anacardiaceae

Origin and Distribution

The cashew, (H. *Kaaju*) a tropical dollar earning fruit, considered to be a native of Brazil was introduced in India from Brazil some 400 years ago by the Portuguese settlers. It is called *Parangi andi* or *Portuguese nut* in Kerala. In India the chief producing areas are Andhra Pradesh, Goa, Karnataka, Kerala, Maharashtra, Orissa, Tamil Nadu, West Bengal, Tripura and Pondicherry. It is estimated that cashew is grown on 600,000 hectares and the production is around 250,000 MT. It is distributed up to 27°N and 28°S.

Uses

The seeds are the source of cashew nuts which are obtained by shelling of the roasted fruits. They yield an edible oil. The shells (pericarp) yield cashew-shell oil called cardanol an astringent and a vesicant used for water proofing, as preservative, insulating, varnishes and in the manufacture of typewriter-rolls, oil and acid proof cement and tiles, break-lining, inks etc. The sap from the bark provides an indelible ink. The bark also yields a yellow or reddish gum like arabic slightly soluble in water that repels insects. It is used for tanning of fishing nets.

Cashew nuts yield kernels to the extent of 25%. The milk of cashew nut yields 7% butter and 5.5% ghee. Its most important products are the nut and shell oil which are important items of export. Its juice is reported to be a fairly good source of vitamin C. The nuts contain about 5% water, 20% protein, 45% fat, 26% carbohydrate, 1.5% fibre, 2.5% mineral matter (Ca, P, Fe), vitamin B_1 and riboflavin. The shell contains approximately 50% of cashew-shell oil while the cashew apple contains water 88%, protein 0.2%, fat 0.1%, carbohydrate 11.6%.

It is also grown in other regions *i.e.* Indonesia, the Malayan states, Sri Lanka, Goa, Daman and Diu, Madagaskar, Portugal and the British East Africa. It is very important commerically in India. India, with an average of about 4,18,000 ha under the fruit, has an annual production of about 1,84,000 tones of cashew nuts. Its contribution is roughly 26% of the total world production (Mozambique 37% and Tanzania 24%). Nearly 90% of international trade in cashew kernels and cashew shell oil is monopolized by India. Kerala with an annual production of 24,000 tons of nuts is the leading centre of cashew production in India. In South India it thrives on the coastal areas and on hills upto 1,000 meters. It can be grown in regions of high rainfall

Climate and soil

It does well under hot and humid climate of the west coast. The cashew strictly a tropical plant, is hardy and drought resistant though is less resistant to frost and chilly weather. It cannot stand dry hot summers. It is able to blossom and develop its fruits during a long dry period that delays the blossomings of the trees until the commencement of rains. It is said to be rather highly resistant to strong winds. Temperature is the limiting factor in its cultivation and a mild tropical climate is most suitable. If the temperature falls below the freezing point, the trees are injured. It thrives under a variety of climatic and soil conditions and may be grown upto 750 meters above sea level receiving a rainfall of 80-375 cm. It seems to thrive on sandy belts along the east coast with about 15 cm. of rain per year and with 300-375 cm in the west coast. It is a sun-loving tree, cloudy weather and heavy rains reduce the yield.

It is very tolerant and is able to grow and fruit well in shallow and impervious soils like the sand hills of west coast, where few other species would perhaps produce any economic crop. Nevertheless, it is more productive and profitable on better soils and does best on deep loam. It is reported to be less tolerant to saline soils. Loams and sandy loams, pH 6.3-7.3, well drained, rainfall 125-250 cm/yr, nearness to sea, Temp. range 15°C in winter and 32-38°C in summers, humidity 80% frostless winters are the situations considered favourable for cashew cultivation.

Propagation

Cashew is usually propagated through seeds collected from the trees bearing large crops of high quality nuts and planted *in situ* with 3 seeds per hole. Vegetative propagation by inarching, grafting or air-layering is also possible. The best time for layering is between February and April.

(*a*) *Raising of seedlings.* Medium-sized nuts (about 120 to a kilo) selected from heavy and regular bearing trees of known parentage stored in a cool dry place are sown on carefully prepared seed-beds in December. Seeds may be sown in polythene bags (15 cm × 20 cm) with stalk end pointing upwards and the suture slightly inclined. Protection against birds, rodents and other animals pays.

Germination. Rather poor and slow, generally the earliest seedlings may appear 15 days after planting though many nuts take much longer. Nuts with high specific gravity germinate more quickly and give more vigorous seedlings, have a higher absolute viabiliy than the seeds with low specific gravity. Very large nuts with low specific density do not germinate well.

(*b*) *Planting.* Pits 60 × 60 × 60 cm are made 8 M apart well in advance of the monsoon (June) and allowed to weather for a month or two. Ash, if available is applied to the pits. Germination is usually complete in 30 days and seedlings 12 cm in length are transplanted in the pits in orchard with the first showers of the monsoon in June or with summer showers in May, and thinning is done as the plants grow in size.

Budding, air-layering or approach grafting are generally employed for the vegetative propagation of cashew. Vegetative propagation of cashew is valuable for getting plants true to type. The shoots of the cashew tree exhibit a natural tendency to 'root' where they come in contact with the ground. This tendency of natural rooting as well as the ability of the shoots dipped in 100 ppm IBA for 24 hours to root has given impetus to nursery men to resort to vegetative propagation by air layering. Air layering can be done in February to April so that the plantlets are ready for planting in June. Air layers from selected trees five high and consistent yields. Selected clones may be easily multiplied by side veneer grafting as in Venezuela. Softwood grafting is suitable for vegetative propagation. Epicotyl grafting did not give more than 30% success. In less humid region veneer grafting gives greater success.

Cultural Practices

After cultivation. There is hardly any field operation, except prevention of cattle trespass for the first 2 or 3 years in the orchard and a gradual thinning out of the plants. Manuring and irrigation is uncommon. 15 kg FYM, 1 kg SP, 150 kg p.s., 1.5 kg cake per plant before planting. A full grown tree needs 250 g N. 125 g P_2O_5 ; and 125 g K_2O per year, for high yield, in 2 split doses.

Sometimes weeding and also intercropping is practised in the south. Where intercrops are grown the cashew trees benefit by the manure and irrigation given to the other crop. Sweet potato, ginger, black gram and even paddy can be grown as intercrop in between the cashew trees. Intercropping is not desirable, after the cashew trees commence bearing.

Pruning

It is not done. As a rule the trees are shaped by removing the water shoots and lower branches during the first 3 years of growth. Dead and diseased twigs and dried branches are removed periodically to reduce losses through diseases like 'Die-back' and also minimise chance of serious losses due to forest fires.

Cropping

The tree commences economic bearing in 3rd year, attains full bearing by the 10th year and continues to yield ordinarily for 30 to 40 years and in deep and well drained soils upto 70 years. Flowering commences in the month of November and extends upto February and the fruit ripens from March to May. It takes about two months after pollination for the fruits to be ready. The shell of the nut attains full size (in 3-4 weeks after flowering) after which the pedicel swells to produce the cashew apple. The fruits are harvested when fully ripe. They may be allowed to fall on the ground and collected in dry weather but in wet weather they must be collected daily and dried in the sun.

Cultivars

Some important cultivars from India.

BPP_1. Hybrid Tree no. 1 (♀) X tree no. 273 (♂) Av. yield is around 17 kg/tree. Av. wt. of nut 5 g.

BPP_2. Hybrid as above, 19 kg/tree yield. Av. wt. of nut 4 g.

BPP_3. Selection from germplasm from Vishakhapatnam, Av. yield 16 kg/tree, Av. wt. of nut 6 g.

BPP_4. Selection from germplasm accession of 9/8 Eprupalam, av. yield 12.5 kg/tree/yr., Av. wt. of nut 6 g.

BPP_6. Selection from AP yield 42 kg in 50th yr. Nut wt. 5.2 g.

Vengurla-1. Selection, yellow apple colour, yield 23 kg tree, Av. wt. of nut 6 g.

Vengurla-2. Selection, pinkish red coloured apple. 24 kg per plant/ The av. yield in the 20th yr. small nuts of 4 g Av. wt.

Vengurla-4. Hybrid, apple of red colour, Bigger nuts as 8 g Av. wt.

VR-1. Selection from Tamil Nadu. Late flowering. Av. yield/tree 7.4 kg/ tree, Nuts are medium 5 g.

VR-2. Selection from Tamil Nadu, late flowering. Av. yield/tree 6 kg., nut 5 g.

Ullal-1. Selection from Kerala 19 kg yield/tree. Nut wt. 7 g.

Ullal-2. Yield 18 kg/tree, late flowering Dec.-Feb., Av. wt. of nut 6 g., Selection from A.P.

BLA-39-4. Selection from seedling progeny of tree no. 39 of Andhra Prasesh, Av. yield 20 kg in the 15th yr. Flowering Nov.- Dec., Nuts 6.2 g.

K 22-1/ Selection from Karnataka. Flowering Dec.-Jan. nut. wt. 6.20 kg., Kernel 1.60

NDR-2-1. Dec.-Jan. Flowering, Nuts 7-25 g., Kernel wt. 2 g.

Cultivars recommended for different parts of India.

Andhra Pradesh. BPP_4, BPP_6, VR_{12}.

Goa. V-1, V-4, VR-12.

Karnataka. BLA-39-4, VRI-1, VRI-2, ULLAL-1, ULLAL-2.

Karela. BLA-39-4, K-22/1, NDR-2-1.

Maharashtra. V-1, V-4, VRI-2.

Orissa. WBDC-V, T_{40}, BPP-2.

Tamil Nadu. BRI-1, VRI-2.

Harvesting and Marketing

Flowering in cashew takes place in stages and therefore the fruits are also ready in stages, so that the process of harvest may continue for a period extending over two months. It is best to collect the fruits that fall to the ground each day. Yield differs widely (10-75 kg per plant). One tree produces on an average 30-35 kg of apples. The apple contains about 65% juice with vitamin C, sugar and minerals of about 5 lakh tons of apples produced every year only a small quantity is put to economic use in the production of liquor and vinegar. A well raised plantation should yield 2.5 tons of nuts per ha. On an average 4.6 tons of raw nuts provide one tone of kernels.

Nuts are imported to India from Tanzania and Mozambique for shelling by hand. The shells are usually roasted before cracking. Oil-bath process ensures uniform and quick roasting and the recovery of cashew-shell oil increases. After shelling the testas are removed and the nuts are vacuum packed for export.

Due to propagation through seed the trees exhibit numerous variations in yield, shape, colour and size of "apple" etc. Three categories of cashew are found according to the colour of the "apple" *viz.* red, yellow and a mixture of red and yellow.

Insect Pests and Diseases

The root and stem borers (Plocaederus ferrigineus and Batocera rofomaculata). These may cause the death of the tree. Plugging of holes with cotton soaked in coaltar and kerosene oil mix (7 : 2) or 0.2% carboryl and removal of the dead and diseased twigs are recommended.

Green leaf-eating caterpillars and dark brown thrips. These damage leaves and flowers. These can be effectively controlled with a spray of endosulfan (0.05%).

Die-back (Corticum salminicolor). Shoots dry up from the tip downwards. Spread of the disease is prevented by spray of 1% Bordeaux mixture twice, in May-June and October.

CHAPTER

36

HAZEL-NUT

(*Corylus avellana* L.)

Betulaceae

Origin and Distribution

The hazel-nut, called filbert, cobnut, barcelona nut (after a place in Spain) European hazel etc. is also known as *findak* or *bindak* (Hindi) thangi or urni (Punjab) and *bhotia badam* (Kumaon). Filberts are long and shaped like human finger nails while the cobnuts are roundish and sometimes flattened. The cultivar Turkish hazel (*Corylus colurna*) which is different from European hazel (*C. avellana*) also grows in the colder regions of India. Hazel nut, a cool temperate climate crop, requiring moderate rain may be found growing in India at altitudes of 1,800 to 3,300 meters. The trees show a variation in height from 3 to 14 meters, smaller ones are known as bush and the tall ones trees.

Hazel-nut has a high calorific values with carbohydrate, fat, protein and phosphorus as 17.7, 60.90, 12.7 and 0.35 per cent respectively. The kernel yields 51% golden yellow oil. Hazel nut is grown in India, Afghanistan, U.S.A., Italy, Spain, Turkey and France.

Ecology. Cobnuts and improved selection *i.e.*, filberts do well in cool temperate regions on almost all moderately fertile soils. Light sandy soils prove better. Rich soils do not suit, as well. Good drainage with ample moisture is a must for success.

Propagation

Seed propagation is possible by sowing the seeds in spring after stratification for 3 months at 4.5°C. Hazel-nut displays suckering tendency which may be used for propagation. Vegetative methods of propagation seem to be more sound and profitable than seedling raising in hazel-nut for obvious reasons. Cutting, layering, graftage etc. are the chief methods of propagation. 25 cm long cuttings of two year old wood, made in the dormant season, tied in bundles are stored in sand for callusing before planting in well prepared beds with two-third of the length below the soil for rooting. Layering succeeds better than cutting. Graftage has better possibilities of improving the crop. shield budding, cleft budding, tongue grafting and crown grafting are generally practised in India. About 480 plants per hectare may be planted at a distance of 3.5 meters both ways. Some cultivars need a distance of about 6 meters between the plants, all round.

Cultural Practices

Plantations of two or more cultivars in the same orchrd serve as a safeguard against self-stertility in the hazel-nut. Cross incompatibility is not known in filbersts and therefore any two or more cultivars may be grown side by side with success. Besides developing an open centre, a tree of medium height and luxuriant vegetative canopy in the early stages of the plant needs high humus and liberal supplies of both phosphorus and nitrogen, prefers non-acidic soils for good performance. Bearing trees must not be pruned until the male flowers have shed pollen. Suckers and surplus as well as older laterals should be removed but not the thinner ones. Pruning should be done only lightly and experience alone gives the insight depending on the particular condition of the plant and its environment.

Flowering starts in March or April, the male flowers appear in the shape of catkins and the rather insignificant female flowers with their red stigma protruding separately on the same plant *i.e.*, the plant is monoecious. Frost is more inimical to the male than the female flowers.

Harvesting and Uses

It is only after 7 years that there is a possibility of getting a crop of nuts from the plants. With the age of the trees yield also increases. Maximum fruiting takes place at the age of about 15 years. As soon as the husk enclosing the nut assumes a brown hue it should be harvested, dried in the sun or air, and the husk covering removed. The yield may be 4-5 kg/plant in full bearing. The nut is considered to be a tonic and is often available in the drug shops. The oil is used for the manufacture of high class soaps. Oil expressed from the nuts works as a tonic, aphrodisiac, stomachic and may also be used against cough. If kept mixed with common salt the bright red colour of the nuts may be retained. Roasting in oven at 200°C for 7 minutes is considered best. Insect pests and disease are no serious problem for hazel-nut plants in India.

Cultivars

Barcelona (France, Spain, U.S.A.), Trebizond (Turkey) and Kentish cob (England) are some of the popular cultivars. Besides these, cultivars like Red filberts, Kentish filbert, Atlas, Cosford Cob, Hemples, Person's Prolific, Tonollo, Du Chilly, Nottingham, Nook Scak. Daviana, Brik Nut, Du Province, White Aveline, Knight's large, Purple leaved are well recognised.

CHAPTER

PECAN NUT

(*Carya illinoensis* Koch)

Juglandaceae

Origin and Distribution

A tree of South Eastern United States and Maxico grown in Uttaranchal, Himachal Pradesh and the Punjab. Pecan nut or Hickora nut. Brittor or Caraya peacan can be grown on elevations of 900 to 1,800 meters above sea level in the Himalayan regions. Heavier rainfall (beyond 100 cm/yr.), severe frost in spring and excessive heat in summers is unwelcome to pecan plants. The climatic requirements resemble that of walnut. Pecan nut, a native of U.S.A. is a favourite of the Americans as a rich source of fat (71.4%), protein (12.5%), water (3.59%), phosphoric acid (.03%) and lime (0.12%) contained in its kernel. Grown at elevations of 750 to 1,500 m in Himalayan base with 75-100 cm means annual precipitation.

Deep, fertile and well drained soils rich in organic matter support pecan trees better when slightly acidic (pH 5.56).

Propagation

Seeds need treatment before germination. Water-soaking for 24 hours followed by storage in humid atmosphere at a temperature of 1-3.5°C for 4 months before sowing. The seeds become non-viable in dry atmosphere hence must not be devoid of the moist media and kept cool. Pecan nut may be propagated vegetatively through bark grafting, slot bark grafting, spur budding, patch budding, plate budding and ring or annular in early spring on two year old pecan seedling rootstock. Seed grown plants have their own disadvantages hence are discouraged.

Planting

Pecans may be planted in November or December in 1 × 1 × 1 meter plts, the distance between the plants being 9 meters. This plant is very suitable for roadside planting as well as decorating the dwellings.

Irrigation

The plants need a steady supply of water for their proper growth and regular bearing. Mulching the soil with leaves or any other matter which may help in greater retention of moisture by the soils is helpful. One irrigation every fortnight is considered as minimum during the summers, specially for the young trees.

Manuring

The trees need a guard against deficiency of micronutrients specially of zinc causing rosette. Should such a situation arise foliar application of zinc sulphate at the rate of 1-2 kg per 450 litres of water can be used to correct the malady. Generally application of three quarters of a quintal of FYM and judicious supply of fertilizer per tree every November or December is profitable. The dose of the complete fertilizer may begin with 1/2 kg for young plants and raised to 8 kg depending upon the growth and bearing of the tree.

Pruning

The Purning should aim at developing well shaped strong trees. This may be done by training and pruning which are more meaningful in first few years of the plant's life. It is advisable to have 1.5 metres of the stem in the immediate vicinity of ground clear of any branch. Thereafter the branches should be spaced at a distance of 30-35 cm from one another so that the plant has clear stem below and equidistant branches above.

Harvesting

A tree may, on an average, yield 10 kg of nuts. As the nuts are exposed, by the splitting of the outer covering or husk with maturity in September-October, these are picked.

Culitvars

Cultivars set a limit to the yield potentialities and a proper selection of cultivars for a locality is always paying.

Curtis. Resistant to scab, good quality. Nuts small but with a high percentage of kernel.

Desirable. As the name indicates it is one of the good quality desirable cultivars similar to Stuart with resistance to scab.

Mohan. Large-size nuts with less percentage of kernel.

Moneymaker. Good yield, early crop of medium-size nuts. Not much susceptible to scab.

Moore. Small nut, good yielder but susceptible to scab.

Schley. Though susceptible to pecan weevil and scab, this thin shelled cultivar has nuts of moderate size and excellent quality.

Stuart. An early maturing, regular bearing cultivar, having medium thick shells adapted to a wide variety of soils. It is somewhat resistant though not immune to scab. The nuts are of good size.

Success. Except irregular bearing and high susceptibility to scab, a good cultivar.

In addition, the following are some of the cultivars grown in different parts of the World. Van Deman, Burkeh, Taxas, Prolific, Halbert, Major, Niblack, Indiana, Busseron, Green river, Posey, Warrick.

Insect Pest and Diseases

Borers. The borers damaging stem and leaf may be destoryed by putting in a few drops of carbon bisulphide into the holes and plugging these.

Pecan weevil. These damage the kernels and should be controlled by spraying with Deltamethrin (0.025%) and 15 days thereafter spray of Fenthion (0.05%).

Pecan scab. The planting of the nearly immune varieties *viz.* Stuart, Desirable, Curtis, Brooks are recommended. In the alternate the cultivars less susceptible to scab *i.e.*, Money maker, Curtis may be found useful. This disease may be controlled by the removal of the diseased portions of branches to ensure better aeration and light the spraying with Bordeaux mixture as often as required. The susceptible cultivars must not be grown where scab is a problem.

Rosette. This is a physiological disorder caused by deficiency of zinc. The leaves bunched near the tips of branches are rusty narrowed and twisted and the growth of the branches is completely halted. In severe cases the shoots also show a die back from the tips.

Control rosette formation by appropriate application of zinc sulphate to the soil at the rate of 300 to 500 gms per tree a year to begin with. With advance in age of the plant the quantity of $ZnSO_4$ needed to control rosette formation increases. $ZnSO_4$ should be applied mixed with slaked limed in the proportion 1 : 0.5 : : $ZnSO_4$: Slaked lime.

CHAPTER

38

WALNUT

(*Jugulans regia* L.)

Juglandaceae

Origin and Distribution

Walnut commonly knwon as *"akhrot"* in India is indigenous to north western Himalayas and usually grown at altitudes of 1,200-2,000 meters above sea level. Besides the delicious kernels it provides valuable timber used for furniture, carving and the butts of guns. Walnut eaten three times a week reduces blood cholesterol and the risk of heart attack. Oil extracted from the kernels is used for various medicinal purposes. Though systematic planting is uncommon, Kashmir, the chief centre of Walnut growing in India has scattered trees covering more than 3,000 hectares producing about 21,000 quintals of nuts in a year. Walnut is also grown in France, Italy, California, Rumania and China. In India, it is grown in Kashmir, Kangra, in Himachal Pradesh, and Kumaon Hills in Uttaranchal.

Soil and Preparation

Walnut responds best to well drained clay with a fair lime content or deep silt loam containing abundance of organic matter with a depth of at least 3 meters. It may be grown on soils of marginal and submarginal fertility of even unreclaimed lands. Walnuts when planted in sandy soils need an addition of mulch or well rotten organic manure to about 2 cm depth and 2 meters round the plant. Soil with fluctuating water levels or hard pan in the root zone (3 meters) or alkalinity must be avoided. Soils with low moisture content during the hot summer months make the plant liable to sunburn.

Climate

Walnut flourishes in areas that have a rainfall of over 75 cm and are free from frost in spring as well as extreme heat in summer. Trees must be planted in frost-free place preferably of fairly high grounds. Flowers in the pre-bloom period cannot stand temperature below freezing which ruins flowers badly whereas above 38°C (shade) produces black nuts specially when the humidity is low.

Propagation

Seed propagation is more common but vegetative propagation is more desirable to obtain plants of known quality. *J. hindsii* seedlings are widely used as rootstock. Walnut is occasionally planted on orchard scale in India. It is generally grown on grass lands, terraces, walls and odd corners of holdings. Bush trees are planted and not the tall ones since the latter take much time before coming to the fruit bearing stage. Propagation by seed should be discouraged and vegetative methods popularised. Where seed plants must be raised, or in the case of cultivar *Placentia* which breeds true to type, walnuts should be collected from vigorous and high yielding trees producing big sized nuts of bright brown colour and good cracking quality possessing kernel of bright yellow colour, good taste and flavour. These are stored in a cool dry place or are stratified after harvest till the following December. If the soil is ready for sowing, the nuts are sown immediately 7-10 cm deep in well prepared beds, keeping 30 cm distance between seeds as well as rows. The seeds germinate in March, the root system develops well in the first year and the tops in the second when seedlings may be transplanted in autumn at a distance of 12-13 meters. The trees trained to a single stem upto a height of about 1 meter give good performance. Walnuts may be propagated by whip grafting, inlay bark grafting and patch budding in early spring. Crown grafting and patch budding have succeeded with walnut in Kashmir, Rootstock of *Juglans regia, J. nigra, J. Californica, J. Hindsii, J. rupestris* etc. have yielded encouraging results elsewhere although no systematic work on the suitability of these rootstocks for commercial cultivars has been done in India. Pruning of diseased twigs and scafold branches is also desirable to keep the plants in healthy bearing condition.

Irrigation

Casual irrigation during dry periods encourages better growth and early bearing besides reducing fruit drop at the time of fruit set. It also encourages better filling of the nuts. Preharvest irrigation helps in the dropping of clean nuts greatly facilitating the harvesting operations.

Manuring

Annual application of nitrogenous and phosphatic additives to the walnut trees improves the yield and checks alternate bearing. The nutritional requirement of the plant depends on the age of the plant and soil fertility conditions. In India walnut is rarely manured and possibly its extensive root system is responsible for fair yields.

Pollination

Walnut, a monoecious (like the pecan and filbert) and self-fertile plant presents interesting features. The male flowers or catkins develop on old branches while the female flowers on the tips of the new shoots of the same plant. Pollen is distributed by wind and pollination difficulties may be temporarily obviated by brining catkins of the neighbouring trees close to the female flowers.

Harvesting and Marketing

The trees start bearing at the age of 8-10 years and may continue to bear for about one hundred years. The yielding capacity of trees varies with age, vigour and cultivar. Good yielders may give about 1.75 quintals of dry nut although the average yield taking the off years into consideration is about 40 to 80 kg per tree in one season.

As the hulls split, nuts fall to the ground and are collected. These may also be made to drop mechanically. Grading of the nuts should be done for size and quality to get more returns. Nuts may be bleached to make them attractive. The nuts may be stored in gunny bags and kept in rooms free from humidity but having good aeration. The bark is used for cleaning the teeth and gums by women folk.

Cultivars

Eureka, Placentia, Wilson wonder, Prolific, Franquette and El monter are the well known cultivars of U.S.A., though in India distinct cultivars are little known. Placentia, a thing shelled popular cultivar bears fruits and breeds true to type from seed and may be raised in Chini area of Himachal Pradesh. The walnut trees of Kashmir have been put into 75 groups depending on the external characteristics of the nut. Some of the trees possessing papery shell white colour, good taste and pleasant flavour are grown especially in Chini (Himachal Pradesh), Kulu (Punjab) and Kumaon Hills (Uttaranchal). The desirable standard cultivars needs be extended and the undesirable ones systematically weeded out to raise the production of walnuts.

Insect Pest and Disease

Walnut weevil (Alcides porrectirostris Marshall). The adult weevils feed on petioles, twigs, flowers and young fruits whereas the grubs feed inside the fruits and reduce the kernels into black mass. Attacked fruits drop prematurely with grubs inside. Collect and destory the fallen fruits.

Diseases

Die back (Glomerella cingulata). Irregular dark spots are seen on leaves which involve the entire leaf causing defoliation. Spray copper oxychloride @ 3 g/litre of water before rainy season to control the disease. Repeat the spray in the case of severe infestation.

□□□

GLOSSARY

ABSCISSION—The shedding of a leaf, flower or fruit as a result of the formation of layer of loosely adhering cells at its base that break apart readily.

ABSCISSION LAYER—A layer of thin-walled cells formed at the base of the organ.

ABSORPTION—The process of taking-up water and holding it in pores or open spaces as a sponge.

ACCLIMATIZATION—Adjustment in tolerance shown by a species in the course of several generations in a changed environment ; adapting to a new set of conditions.

ACID—A condition in which acid elements viz., nitrogen, sulphur, or chlorine over-balance the basic elements. Associated with greater concentration of hydrogen ions than of hydroxyl ions.

ACRE—A unit of land measure equal to 43,560 sq ft ; 4,840 sq. yds ; 4,047 sq. mts., or 0.4 hectare.

ACUMINATE—Drawn out into a long point ; tapering, pointed.

ADSORPTION—The adhesion of molecules of gases or dissolved substance to the surfaces of solid particles. It is distinguished from absorption, which is not a surface phenomenon.

ADVENTITIOUS—Out of place, arising from the secondary meristem. Arising of roots at unexpected places.

AEROBIC—Active in the presence of free oxygen.

AEROBIOSIS—Existing in the presence of oxygen, as in atmosphere for example.

AEROPHYTE—A plant growing attached to an aerial portion of another plant (EPIPHYTE).

AEROSOL—Type of insecticide, antibiotic or germicide formulation, which is dispersed as very fine droplets on application, forming smoke if particles are solid, and fog, if liquid.

AFTER-RIPENING—Completion of maturation process within seeds to enable germination. The physiological changes taking place in the dormant seed to permit germination.

AGGREGATE FRUIT—The product of the fusion of several ovaries in the same flower as in strawberry and raspberry. It is different from a collective fruit (mulberry and pineapple) formed from a complete inflorescence.

AIR LAYER—The shoot of a plant on which roots are made to develop by means of some stimulation or manipulation.

AITIONOMIC—Ability to develop parthenocarpic fruits only in response to some stimulus external to the ovary.

AKENE—Dry, one seeded, small, indehiscent fruit of one carpel, seed-like in appearance, as the so-called "seeds" of the strawberry. ACHENE.

AKHROT—Walnut (*Juglans regia* L.) Family : *Juglandaceae.*

ALLELE—Allelomorph, one of any pair of alternative hereditary characters ; gene which can occupy the some locus as another gene in a particular chromosome.

ALPHONSO—The mango *par excellence* grown in Ratnagiri (Maharashtra) and Balsar (Gujarat).

ALMOND—*Badam* (H.), *Prunus amygdalus* Batsch. Family Rosaceae. Delicious nut, tree.

ALTERNATE BEARING—*Biennial bearing* ; Bearing heavy and lean crops in alternative years viz., mango. Good cultural practices can change 'off' years into 'on' years.

AMBIA BAHAR—The second blossom in some fruit trees appearing in December-January.

AMBRI KASHMIRI—Apple cultivar indigenous to the Kashmir valley. Good cropper, regular bearer ; fruits : medium, oblong, elliptical and tapering with white flesh and yellowish green skin.

AMRA—*Hog plum (Mangifera pinnata* L.*f.*). Family : Anacardiaceae.

ANAB-E-SHAHI—(also called Malta). Grape cultivar mostly grown in Hydrabad; sweet, heavy cropper.

ANABOLISM—Process of building of body tissues from simple raw materials supplied by the products of digestion of food.

ANAEROBIC—Active in the absence of free oxygen.

ANALOGUES—Organs of different plants or animals with like function but of unlike origin.

ANAR—*Pomegranate (Punica granatum* L.). Family : Punicaceae.

ANHYDROUS—Dry, without water.

ANNONA SQUAMOSA—Custard apple or sugarapple (*Sherifa, Sitafal*). Bullock's heart, *A. reticulata.* Family : Annonaceae.

ANNUAL—Completion of life cycle within one year.

ANNUAL RINGS—Growth rings visible in cross-section of woody plants stems.

ANTAGONISM—Of salts, a mutual counteraction of their influence on cell permeability.

ANTHER—The pollen producing organ of flower.

ANTHESIS—The opening of flower bud, the duration of life of a flower from the opening of the bud to the setting of fruit. Full blossom.

ANTHOCYANIN—Pigments in plant sap giving scarlet, purple or blue colour to leaves, stems and flowers in plants.

ANTIBIOTICS—Substances which interfere with the life cycle of the microbes in the body without causing any bad effect on the animal body, viz., sulpha drugs, penicillin, auromycine, streptomycine.

ANTI-OXIDANT—Material used to prevent oxidation or discolouration food.

APICAL—The point or tip e.g., of a stem, shoot etc.

APOMIXIS—Asexual production of seedlings in the flowers without fusion of gametes. The seedlings are similar in character to the parent plant.

APPLE—Fruit of species of genus *Malus* of family Rosaceae. *Malus sylvestris* Mill.

APRICOT—*Khobani, Zardalu. Armeniaca vulgaris* Lamk. *Prunus armeniaca* L. Family Rosaceae.

ARID—Potential water losses by evaporation and transpiration more than precipitation.

ARIL—The fleshy mass surrounding each seed as in litchi.

ARMS—The branches of more than one year of age. Fruiting spurs, fruiting canes, renewal spurs and replacement spurs are present on the arms.

ARTOCARPUS ALTILIS—*A. incisa* L., (*Breadfruit*) ; *Katahal* (H.) *A. hetrophyllus* Syn. *A. integrifolia* (Jackfruit) Family Moraceae.

ASCORBIC ACID—Vitamin C ; hexauronic acid ; $C_6H_6O_8$; water soluble vitamin found in fruits and vegetables. In good quantity in citrus fruits and Aonla.

ASEXUAL PROPAGATION—Vegetative propagation or vegetative reproduction of plants by methods other than the seed. There is no meiosis and fusion of gametes.

ASPHYXIANT—Gas for killing pests through respiratory difficulties.

ASSIMILATION—Conversion of substances taken in from the outside into living tissue of plants or animals.

AUTOGAMY—When a flower is fertilized by its own pollen.

AUTONOMIC—As referred to parthenocarpy, the ability to set fruit without the stimulus resulting from pollination.

AUTOTROPHIC—Capable of utilizing carbon dioxide and of obtaining energy for the reduction of carbon dioxide and other life processes from the oxidation of inorganic elements or from light.

AUXIN—A substance synthesized by the plant influencing growth at some point other than the point of its synthesis, comparable to hormone in animals ; $C_{18}H_{32}O_5$.

AVOCADO—(Butter fruit) *Persea americana* Mill. ; *Persea gratissima, P. drymifolia* Cham & Schelect.

AXILLARY—In the angle formed by the leaf with stem.

BACTERICIDE—A substance capable of destroying bacteria.

BAGHU GOSHA—Also called *Citron Des Carmes;* pear cultivar grown on large scale in Kashmir.

BANANA (Kela)—*Musa paradisiaca* L. (*M. sapientum* L.). Family Musaceae. 20% of the area under fruits in India is under banana.

BANGALORA—Syn. *Totaparii, Neelam* and *Romani* cultivars of mango. Heavy cropper, excellent keeping quality and plenty of fibreless hard pulp. Poor in taste. Grown in Andhra Pradesh and elsewhere in South India.

BANGALORE PURPLE—Also called *Bangalore blue*, cultivar of grape grown in Karnataka State.

BARAMASI—Group of cultivars of mango whose flowering and cropping is spread throughout the year.

BARAMASIA LEMON—Cultivars of lemon popular in Western U.P. and Punjab, medium sized rounded fruit, pulp juicy with good flavour.

BARAMASI CHIKU—Popular chiku (*Achras zapota* L.) variety of West Bengal.

BARREN—Unproductive, unable to support vegetation, incapable of producing offspring or seed.

BASRAI DWARF—Banana cultivar. Syn. *Vamankeli, Mauritius* of South India and *Kabuli* of W. Bengal and Orissa with large and curved fruits, skin dull yellow or greenish yellow, soft and sweet flesh. Bunch-weight 30 kg on an average with about 130 fruits. Good for dry areas.

BEAUTY—Medium sized, deep-red plum (*Preunus domestica* L.) cultivar with cling stone, prolific and regular bearer with delicious early fruits.

BEAUTY OF BATH—Cultivars of *Lithi* (*Litchi chinensis* Sonn.) grown in Bihar.

BER—Jujube (Chinese date). *Zizyphus mauritiana* Lamk. Family Rhammaceae.

BERRY—Small pulpy fruit with seeds immersed as in grape or raspberry. Pericarp fleshy or pulpy (banana, tomato). The dry seeds of certain plants as the coffee berry, which is really the seed contained in the fleshy, true berry.

BIENNIAL—Life span of more than one year but not more than two years.

BIENNIAL BEARING—Producing fruit in alternate years.

BIOSYNTHESIS—The building up or synthesis of complex compound from simple substances in a living body.

BIOTIC—Concerning life.

BISEXUAL—Hermaphrodite. Presence of both male and female structures.

BLACK ALKALI—Alkaline soil with sodium carbonate-organic matter mix, black or brown incrustation on the surface following evaporation.

BLACK CURRANT – Umtao (*Anlidesma ghesaembilla* Gaertn. ; *A. paniculata* Roxb.) Trees with edible fruits found in tropical Himalayas and North-East India.

BLACK END – A disorder of certain fruits, especially pears, in which the apical fleshy tissue dry out, turn black and become hard and woody. Usually associated with drought in some form.

BLACK HEART – A diseased condition of the woody tissues in which the inner layers become darkened in colour, impregnated with certain substances, lose their ability to serve as conducting and storage tissue. Apparently caused by low winter temperature.

BLANCHING – Heat treatment given to fruits and vegetables before processing. Stem cooking followed by a cold dip of 10-12 seconds. Pre-cooking.

BLEACHING – Any process which lightens the colours. In plants, by destruction of chlorophyll or anthocyanin pigments.

BLENHEIM ORANGE – Cultivar of apple, large, roundish : skin yellow with red streaks. Good cropper. Flesh : yellow, sweet crisp and juicy ; self unfruitful.

BORDEAUX MIXTURE – Copper fungicide containing copper-sulphate solution, lime and water in different proportions (4 : 4 : 50 ; 5 : 5 : 50, etc.) prepared fresh before use.

BONSAI – Japanese art of growing miniature trees and shrubs by extreme dwarfing.

BOUND WATER – Water held by the cell against freezing.

BRAMBLE – Cane bush with spines. Fruit is a berry (*e.g.* raspberry, blackberry).

BRANCH – A subsidiary stem arising from the main stem or another branch.

BREBA – One of the crops of the pistillate fig tree, the first to mature in the spring.

BUD – The portion of a plant capable of developing into leaves, shoots, flowers etc. It is meristematic in nature and is protected by scale leaves.

BUD-BREAK – Resting buds resume growth.

BUDDING – A special type of grafting using a single bud as a scion. The bud is introduced under the bark of the stock and allowed to grow.

BUD-MUTATION – *Bud sport* – A somatic mutation or variation, arising in bud, producing an abnormal branch.

BUDSTICK – A shoot or twig used as a source of buds for budding.

BULB – A plant structure (onion) having concentric layers of fleshy leaves attached to a disc-shaped stem plate at the base, functions as storage for food and for reproduction.

BULBLETS – Small bulbs produced around the base of the mother bulb.

BULLOCK'S HEART – Common custard apple *Annona squamosa, A. reticulata* Family – Annonaceae.

CALLUS—Protective thin walled parenchyma cell covering which forms over a cut or wounded surface in plants ; it develops from cambium or other exposed meristem.

CALYX—The sepals of a flower as a group.

CAMBIUM—Thin mucilaginous, cellular layer between the wood of a tree and its inner bark which is converted into xylem on the inner side and the phloem on the outer side. Generally applied to the lateral meristem which produces secondary xylem and phloem in dicots, leading to secondary growth.

CANES—The externally woody, internally pithy stems of brambles and vines. Young leafy growth of current season (a shoot) after maturity.

CANDY—Fruit or vegetable product, a preserve which has been heavily impregnated with sugar.

CANNING—A method of preserving food in a fresh condition by sealing in airtight containers.

CAPE GOOSEBERRY—*Peruvian cherry ; Rasbhari, Makoya* (H.), *Physalis peruviana* L. (*P. edulis* Sims) of family Solanaceae. Fruit, a berry, acidic, pleasant taste, propagated by seed, vegetative propagation also possible.

CAPILLARY WATER—Water held in soil between soil-particle. It is available to plants.

CAPRIFIG—The wild or "male" fig, the uncultivated form.

CARBOHYDRATE—Energy containing product of plants consisting of carbon, hydrogen and oxygen.

CARBON BISULPHIDE—Used as fumigant for grains in storage, destructive to mice and redents.

CARISSA—(*Caraunda* or *Karonda*). *Carissa carandas* L. family ; Apocynaceae.

CARYOPSIS—Superior, one celled, one seeded, indehiscent fruit with a thin dry membranous pericarp, inseparably united with seed as in barley, wheat etc. Achene with pericarp and testa united. Gramineae.

CASHEW—*Cashew nut, Kaju* (H.), *Anacardium occidentale* L. Family : Anacardiaceae.

CATABOLISM—*Katabolism*. The breaking down phase of metabolism.

CAUSAL AGENT—The organism or condition directly responsible for causing a disease.

CELL—Structural and functional unit surrounded by a cell wall as in plants.

CELL WALL—The membranous covering of a cell secreted by the cytoplasm in growing plants. Consists largely of cellulose but may contain silica as in some algae, and chitin in fungi.

CENTRAL LEADER—System of training fruit tree in which it is trained to form a trunk extending from the soil to the top of the tree.

CHAMAGAMY—Pollination only after the opening of the flower.

CHANCE SEEDLING – A fruit tree which grows from seed scattered by wind, bird etc. ; sometimes bears fruits.

CHAUBATIA PASTE – Fungicide containing equal quantities of red lead powder, copper carbonate and lanoline ; developed in Chaubatia (U.P.), India.

CHELATES – Metal-containing organic compounds specially useful in supplying minerals to plants.

CHEMOTROPSIM – A bending or turning in response to chemical stimulus.

CHICORY – *Kasni, Cichorium Intybus* L. Family : Compositae. Roots used for blending and mixing with coffee, to the extent of 40% in French coffee.

CHIKU – *Sapodilla* or *Sapota (chicle gum) Achras sapota* L. Family : Sapotaceae.

CHIP BUDDING – A single bud with a large piece of wood attached to it is removed from the scion wood and fitted to similar cut in the stock and tied. Common in grapes.

CHLOROPHYLL – Green colouring matter in leaves.

CHLOROPICRIN – Tear gas.

CHLOROSIS – A diseased condition showing yellowing of the plant between the veins on upper side of leaves due to loss of chlorophyll causing its deficiency.

CITTRANGE – A hybrid (*Poncirus trifoliata* × *Citrus sinensis*) bearing very acid fruits. A good stock for citrus. Family : Rutaceae.

CITRON – *Baranimbu, Bijaura* or *galgal. Citrus medica* L. *the parent species of several cultivars. Citrus tuberosa* Mill. ; *C. cedrata* Rag ; *C. aurantium* var. *medica.* Wight and Arn. ; *C. medica* var. *medica* proper Hook. f.. Family. Rutaceae, Thick fruit rind.

CITRUS – Rutaceae family. Third in importance among Indian fruits : *Malta, Santra,* lime, lemon, grape fruit are some of the important citrus fruits. Evergreen fruits with thick rind or peel encasing sections of juicy flesh. *C. Medica* L. Citron is the parent species of several cultivars, *e.g., Limonum,* the lemon, var. *acida,* the lime, var. *limitta* the sweet lime. *C. aurantium* L. is the orange with its cultivars *Bergamia,* the Bergamot orange (perfume obtained). *Bigaradia* or *amra* the bitter orange, used in marmalade, *C. decumana* the shaddock or pummelo with its cultivar the grape fruit. *C. sinensis,* the Malta or Portugal oranges. *C. santara* or kumquat etc. *C. nobilis* Lour. is the true mandarin orange.

CLADOPHYLL – A stem-like leaf.

CLEFT GRAFTING – Wedge grafting. Scion is in the form of a wedge common in apple, par, plum etc.

CLEISTOGAMY – Close fertilization within the unopened flower due to self-pollination *e.g. Amphicarpaca, Laminum, Oxalis, Viola,* especially the last.

CLIMACTERIC PEAK – The maximum point of respiration rate of mature fruit.

CLONE–This is produced from one sexually produced individual (or from a mutant) and maintained by asexual propagation only. It is an aggregate of individual organisms.

COLLECTIVE FRUIT–Fruit formed from a complete inflorescence or from several flowers as of mulberry pineapple, *Ficus, Morus, Platanus*.

COLLOID–A state of a substance where the units are very large molecules or molecule complexes. Colloids diffuse slowly or not at all through plant or animal membrane, cell membrane or cell wall. Colloids do not form true solution.

COLLOIDAL–Gelatinous with cementing or adhesive action as in clayey/soils or humus. The colloidal state of matter as a solid is amorphous *e.g.*, starch, gelatin and gums.

COMPANION CROPS–Two non-competitive crops grown in the same area at the same time. One is of short-term and the other of long duration.

COMPATIBLE–Varieties that will set fruit when cross-pollinated or make a successful graft union when intergrafted.

COMPATIBILITY–(*i*) Of sex cells, the ability to unite and form a fertilized egg that can grow to maturity. (*ii*) Congeniality as determined by the degree of success of the union between stock and scion.

COMPOST–A well rotten organic matter with C : N ratio 10 : 1 or near about; valued for its nutritive value.

CONTACT HERBICIDE–A herbicide that kills a plant primarily by contact with plant tissue rather than by internal absorption.

CORDON–A system of training grape vine where the trunk is not developed straight but is trained in a horizontal direction along the lower wire of trellis at a height of 60 cm. Short fruiting spurs of 2-4 buds each are kept at pruning.

CORKY CORE–A disorder in which the tissue around the core turns brown, and becomes more or less corky, usually accompanied by the development of a bitter taste.

CORM–Structure similar the bulbs in appearance. These are enlarged modified stems in contrast to compacted fleshy leaves (bulbs).

CORMELS–Small corms produced around the base of the mother corm.

COROLLA–The petals of a flower as a group.

COULURE–The failure of blossoms to set, resulting in a premature drop. *Cf.* millerandage, a condition in grapes where the ovary persists but the seeds remain small or do not attain usual size ; Conditions being the same as in culture.

CRAZY TOP–A disorder of citrus trees associated with an alkaline soil reaction and resulting in a peculiar type of growth.

CRENATE–Possessing short, rounded teeth, or scalloped edges.

CRINKLE—A disorder of apples in which the surface of the fruit becomes roughened. Supposed to be a form of drought injury.

CROP ROTATION—Growing of several different crops in a definite sequence, avoids continuous cropping with the same crop, benefits the soil and the farmer through recuperation of nitrogen.

CROSS-POLLINATION—Transfer of the pollen from the anther of one plant to stigma of another plant not in the same clone.

CROWN—The perennial top stem structure of a perennial herbaceous plant.

CUDDAPAH ALMOND—*Buchanania lanzan* Spreng. *Buchanania latifolia* Roxb. *Chiraunji* (H). Family : Anacardiaceae.

CURRENT—Plants and fruits of the family Saxifragaceae, European gooseberry, black current, red current, Dried seedless resins, cultivars of *vitis vinifera*, used in confectionary.

CUTICLE—Layer of cutin or wax on the outer surface of leaves and fruits.

CUTTAGE—Method of propagation from severed parts of the parent plant prior to rooting.

CUTTING—Any part detatched from the plant which is used to form new roots and shoots and develop into a plant.

CYTOPLASM—The whole of the protoplasm of a cell excluding the nucleus.

CYTOKININ—Group of plant growth regulator governing nucleic acid and protein metabolism, cell division and organ initiation capable of delaying senescence.

DAKH—A cultivar with compact bunch, thick skin, subacid pulp, purple round grapes, prolific bearer but of inferior quality.

DARJEELING ORANGE—Sikkim orange similar to Assam (Sylhet) orange.

DATE—Delicious and nutritious fruit of the date palm. *Phoenix dactylifera* L. It is cured by sun or artificially.

DATE-PALM—*Khajur* (H). *Phoenix dactylifera* L. Family : Palmaceae.

DATE-PLUM—*Amlok* (H) *Diospyros lotus* L. Family : Ebenaceae giving edible fruits. Seeds sedative. Fruits contain tannic acid, invert sugar and malic acid. Stock for Japanese Persimmon.

DAY-NEUTRAL PLANT—Plants in which flowering is not influenced by day-length.

DEBLOSSOMING—Removing blossoms from a plant.

DECIDUOUS—Having no living leaves on the plant during dormant season.

DEFICIENCY DISEASE—Hidden hunger, pathological condition due to lack of nutrition, physiological disease.

DEFOLIATE—To strip off leaves.

DEFOLIATION—Loss of leaves.

DESICCATION—Drying out.

DESSERT FRUIT–Any fruit which may be served after meals.

DEOXYRIBONUCLEIC ACID (DNA)–A molecule constituted by repeating sub-units of ribose, phosphate and nitrogenous bases (adenine, guanine, cytosine and thymine).

DICHOGAMY–The female and male parts mature at different times.

DICOT–(Dicotyledons), Plant with two cotyledons or seed leaves in its seed *e.g.*, mango, apple etc.

DIFFERENTIATION–The process of changes in composition, structure, and functions of cells and tissues during growth.

DICHOGAMY–The absence of simultaneous maturation of stamens and stigmas in a flower so that cross fertilization is ensured. Stamen ripe before stigma, protoandry ; stigma ripe before stamen, protogyny.

DICLINY–Male and female organs separate and in different flowers.

DIELDRIN–Insecticide chemically related to aldrin with longer residual toxicity. Acts as stomach and contact poison. Posionous to mammals.

DIGESTION–Breakdown of foods into simple constituents by enzymes.

DIMORPHISM–Presenting two forms as long and short growths or permanent and deciduous branches.

DIOECIOUS–Unisexual, the male and female flowers in different individuals.

DIPLOID-(2N)–Double set of chromosomes. Germ cells have only one set (haploid, N) somatic cells have two sets *i.e.* these are diploid.

DISACCHARIDES–A compound sugar yielding two simple sugars on hydrolysis, $C_{12}H_{22}O_{11}$; a carbohydrate *viz.* sucrose, lactose, maltose.

DISBUDDING–Removal of flower or shoot buds.

DORMANCY–A resting condition in an organism caused by external factors is known as quiescence. The development ceases until more favourable conditions appear. When due to internal factors it is popularly called "Rest".

DORMANT–Applied to buds when these are not actively growing and to plants when they are not in leaf. Also called as resting, inactive, quiescent, stage of development in an organism.

DORMANT PERIOD–Time during which no growth occurs because of unfavourable environmental or internal causes.

DOUBLE WORKING–Grafted twice. A scion is grafted to an intermediate stock which, in turn, is grafted to rootstock. It is primarily done as a measure to derive the advantage of the graft when that between scion and rootstock is not possible due to incompatibility between the two.

DROUGHT–A period of dryness with soil moisture deficiency within the zone of the plant root coupled with low atmospheric humidity.

DRUPE–A stone fruit in which the seed is encased in a hard shell inside fleshy tissue (*e.g.*, peach, plum, cherry).

DWARFED—Restricted plant size without loss of health and vigour.

ECOLOGY—The science that deals with the study of inter-relationships of organisms in nature and to their environment.

EDAPHIC—Influence or relationship of soil or media to plant growth in contrast to atmospheric influences.

EDCT—Ethylene dichloride and carbon tetrachloride mixture in 75 : 25 ratio. Fumigant, available as liquid, volatile at room temperature.

ELLIPTICAL—Narrowish, tapering quality on both the sides ; oval shaped.

EMASCULATION—The artificial removal of the anthers from the flower before they dehisce.

EMBRYO—A miniature plant within a seed formed by the union of male and female gametes (zygote). As the seed germinates it gives rise to a seedling.

EMBRYO CULTURE—Growing seedlings from excised embryos on nutrient media through micropropagation.

EMBRYOGENIC—Pertaining to the development of the embryo.

EMBRYO SAC—The cell in the ovule in which the embryo is formed.

ENDOCARP—The inner layer of the wall of a fructified ovary.

ENDOSPERM—The nutritive material stored within seed, originally deposited within the embryo sac.

ENDRIN—Insecticide, contact and stomach poison.

ENZYME—A complex chemical substance formed by living cells, accelerates specific changes in other substances without undergoing any change in itself.

EPICOTYL—The upper portion of the embryo axis or seedling above the cotyledons and below the first true leaves.

EPINASTY—More growth in upper surface than in the lower surface which causes the leaf to curl downward.

EROSION—Wearing away of soil by wind and moving water.

ESPALIER—Training of fruit tree planted against the wall in which main branches are trained in a plane parallel to the walls in a geometric design.

ETIOLATION—Caused by deficiency of light resulting in lack of chlorophyll, increased stem elongation, reduction of leaf surface besides several physiological disorders.

EVERGREEN—Retain functional leaves throughout the year.

EXCISING—Removing or extracting, as an embryo from a seed or ovule.

EXOCARP—The outer layer of the wall of fructified ovary.

EXOCORTIS—A shelling off of the bark of trees.

EXOGENOUS—Produced outside, opposite of endogenous.

EXOSMOSIS—The passing out through a membrane of a gas or liquid.

EXPLANT—Living tissue taken out from the body and placed for culture in a tube with sterilised media for growth.

EXTENSIVE PRODUCTION—Use of large areas of land with a minimum of labour and capital.

FALSE FRUIT—A fruit in which, besides the ovary, other parts of the flower *viz.* receptacle or complete inflorescence take part as in cashew.

FECUNDATION—The fusion of two gametes to form a new cell.

FECUNDITY—The ability of flowers to produce seeds that will germinate.

FEEDER ROOTS—Fine roots and root branches with a large absorbing area (root hairs).

F_1—The first generation progeny of a cross.

FERTILITY—(1) of flowers, the capacity of producing seeds that will germinate; (2) of soils, the inherent crop producing power.

FERTILIZATION—Fusion of male and female cells following pollination is sexual propagation.

FILAMENT—The stalk supporting the anther in a flower.

FLOWER—Structure for sexual reproduction in plants. Also the attractive colourful bloom for aesthetic purposes.

FOLIAGE—The leaves of a plant.

FOLIAR FEEDING—Applying liquid nutrients to the leaves.

FOOT-CANDLE—The light intensity at a distance of one foot from a source of light of 1 candle power.

FORCING CROPS—Those grown out of season as in green house.

FORKERT BUDDING—Budding, modification of shield budding. Good success in mango.

FREE WATER—Water released by the cell when freezing occurs in intercellular spaces. (*see* bound water)

FRENCHING—A disease characterized by loss of colour of leaves between the veins.

FRUIT. The edible product of a woody or perennial plant which in its development is closely associated with a flower (botanically, a fruit is a ripened ovary).

FRUITFULNESS—The capacity of producing fruit.

FRUITING CANE—The basal part of a mature cane with 8-12 buds left after pruning to produce fruit.

FRUITING HABIT—The location and manner in which fruit is borne on woody plants.

FRUITING SPUR—The spur intended to bear fruits, normally with 2-4 buds.

FRUIT-SETTING—A development of ovary and adjacent tissues following the blossoming period.

FUNGICIDE– That which destroys or kills fungi.

GAMETE–A unisexual cell which must fuse with another gamete to produce a new individual. Sexually active reproductive cells, haploid in nature.

GENE–A group of base parts in the DNA molecule in the chromosome that determines one or more characters hereditary in nature.

GENUS–Singular of genera. A group of closely related species. Similar genera are grouped in a family.

GERMINATION–Resumption of growth of the seed embryo.

GIBBERELLIC ACID–$C_{19}H_{22}O_6$. Tetracyclic dihydroxy lactonic acid, one of the gibberellins. A plant growth regulator, effects elongation of shoots of even a genetic dwarf.

GIRDLING–Destroying an area of barkaround the periphery of a stem on the living plant.

GLUCOSIDES–Nitrogenous compounds that yield sugar and some other substances, usually aromatics, on hydrolysis.

GRADING–Separating or sorting or produce according to size or quality.

GRAFTAGE–Method of inserting buds, twigs, or shoots on other stems or roots for fusion of tissues.

GRAFTING–Placement of a detatched branch's (or scion) cambium in close contact with the cambium of a rooted stem (rootstock) in a manner so as to effect the union of the two forming a new plant.

GRAPEFRUIT–Popular breakfast fruit, rich in vitamin C and thiamin. Family : Rutaceae ; *Citrus paradisi* Macf.

GRAVITATIONAL WATER–Water in excess of a soil's holding capacity, removed by drainage.

GROUND COVER–Plants used in lieu of grass for holding soil and protecting it.

GROWING SEASON–Period between beginning of growth in the Spring and cessation of growth in the fall.

GROWTH REGULATOR–Any substance or physical factor, either external or internal, that influences a growth process ; in a more restricted sense, a substance that influences growth.

GUANINE–A basic nitrogenous compound related to uric acid, one of the purines.

GUMMOSIS–A disorder, particularly of stone fruits and citrus. Fruits in which there are copious exudation or deposits of gum. Symptom of disease or disorder due to mechanical injury, pathogenic fungi, insects etc.

GUM SPOT–A disorder of stone, citrus and certain other fruits in which there are small local deposits of gums in tissues of fruit, shoot or any other organ.

GYNOECIUM–The pistil or pistils of a flower.

HARDENING—Subjecting plants to adverse conditions to hasten tissue maturation for increasing hardiness.

HARDINESS—Quality which causes plant to resist injury from unfavourable temperatures.

HARD PAN—An impervious layer of soil or rock which prevents vertical movement of water.

HARDWOOD CUTTINGS—Cutting from mature hardwood for propagation as in the case of pomegranate, grape, fig, plum, etc.

HEADING BACK—One of the several forms of pruning. Heading back removes only a portion leaving another portion from which new growths can take place. As against this, thinning out removes entirely a shoot, spur cane, branch, etc.

HEARTWOOD—Central cylinder of xylem tissue in a woody stem.

HERBACEOUS—Non-woody. Dies back to the crown during the dormant season.

HERBICIDE—A substance capable of killing weeds.

HERMAPHRODITE—A flower with both stamens and pistils.

HETEROSIS—Increase in vigour, yield, size, growth or function of a hybrid progeny in relation to the parents. Also known as hybrid vigour.

HETEROSTYLY—The presence of styles of two or more forms or two or more lengths.

HETEROTYPIC—Reduction division of a cell.

HETEROZYGOUS—Endowed with "mixed" hereditary factors. Not a pure line. Different genes of a Mendelian pair present in the same cell or organism.

HOMOTYPE—Part or organ, like any other in structure.

HOMOTYPIC—As applied to cell division, involving the usual process of karyokinesis.

HOMOZYGOUS—Purity of type. A pure line.

HORMONE—A substance synthesized by an animal organism that is able to influence growth at some point other than the point of synthesis; comparable to an auxin in plants. Sometimes the term is used in broad sense to include auxins.

HORTICULTURE—Art and science of treating of fruits, vegetables and ornamental plants.

HUMUS—Partially decomposed organic matter.

HYBRID—The first generation progeny from a cross from different varieties, strains, or inbred lines. Off-spring bred of two different species, varieties or classes of plants or animals.

HYDROLYSIS—Chemical splitting by taking up the elements of water.

HYDROPHYTE – Aquatic plant that normally grows in water, Hygrophyte.

HYGROSCOPIC COEFFICIENT – The percentage of soil water retained when in contact with a saturated atmosphere and in the absence of any source of moisture.

IMBIBITION – The process of absorption, usually by a solid *viz.*, the seed imbibes water.

IMPLANT – A grafted portion of a tissue ; infix or insert.

IMPERFECT – In flowers, unisexual.

IMPOTENCE – Inability to produce functional gametes of one sex or the other; sometimes used in a more general sense to denote sterility.

INARCHING – Approach grafting; followed in case of trees namely mango, sapota, guava, aonla, loquat where scion remains a part of tree till its union with stock is complete. The scion is severed off 2-3 months after the union.

IMCOMPATIBLE – Different varieties which do not successfully cross-pollinate or intergraft.

INCOMPATIBILITY – Of sex cells, inability to unite and form a fertilized egg that can grow to maturity.

INDIAN BLACKBERRY – Jambolana. *Jamun* (H.) *Syzygium cumini* (L.). Family : Myrtaceae.

INDIAN BUTTER TREE – *Bassia butyracea* Roxb, seeds yield butter-like substance. Sapotaceae. *B. pollida* yields gutta parcha. Flowers of *B. latifolia* Roxb. (*mahua* in Hindi) are edible.

INDICATOR PLANTS – Plants characteristic of a specific soil and site condition.

INSECTICIDE EMULSION – When active ingredient is not soluble in some common and cheap solvent it is dissolved in small quantity of any suitable solvent and then water is added to required degree of concentration.

INSECTICIDE SOLUTION – Insecticide whose active ingredient is dissolved in some cheap and readily available solvent, *viz.*, water.

INSECTICIDE SOLVENT – Substances that dissolve insecticide so that its spray may be possible. Some of the common solvents are water, acetone, ethanol, benzene, xylene, linseed oil, groundnut oil, cotton seed oil, turpentine, kerosene, petroleum and ether.

INSECTIFUSE – An insect repellent that drives away insects.

IPM – Integrated Pest Management.

INTER-CROPPING – Growing short-term crops between rows of trees in a young orchard for full utilization of land.

INTERFERTILITY – The ability of one cultivar to set fruit and produce seeds that will germinate when pollinized by another variety.

INTERFRUITFULNESS The ability of a cultivar of fruit plant to form seedless or seed-containing mature fruits following pollination by another specific cultivar, also called cross fruitfulness.

INTERMEDIATE STOCK—Stock between scion and another stock.

INTERNODE—The portion of a stem between two nodes.

INTERSEXUALISM—A term referring to the varying degrees of development of the two sex organs in the same plant, relative maleness and femaleness of the plant.

INTERSTERILITY—Inability of one variety when pollinized by another variety to set fruit and produce viable seeds that will germinate.

IN VITRO—Artificial condition *viz.* in test tube or out of the living body.

IN VIVO—Within the living organism.

JUNE DROP—The abscission of partly developed fruits (often occurring in June).

JUVENILES STAGE—Early or vegetative phase of growth characterized by carbohydrate utilization.

KAGZI NIMBU—Sour lime. *Citrus aurantifolia* Swing. Family : Rutaceae.

KHORIAL—Sapota, chiku, *Achras zapota* L. Family : Sapotaceae.

KHOBANI—Apricot (*Khurmani*). *Armeniaca vulgaris* Lamk. *Prunus armemaia* L. Family : Rosaceae.

KNIFFEN—System of training of grape vine named after William Kniffen. The arms of the vine are tied to horizontal wire on the same level above the ground.

LANCEOLATE—Shaped like a lance or spearhead.

LATENT BUD—A bud, usually more than one year old, which may remain dormant indefinitely or may develop under certain conditions.

LATERAL—From the side as of a stem.

LAYERAGE—Method of asexual plant propagation, rooting of root, stem and branches from parts still attached to the mother plant. It is also known as marcotage.

LEACHING—Washing of soluble nutrients down the profile of soil.

LOAM—A soil which is neither sandy nor clayey, well equipped with organic matter. Clay (7-27%), silt (28-50%) and sand (52%).

LOCATION—A climate or geographic area.

LONG DAY—More than 14 continuous hours of day-light.

LOPPING—Pruning or cutting away aerial parts of shrubs or trees.

LUCKNOW 49—A cultivar of large, round, white fleshed excellent quality guava; dwarf, spreading and profuse bearing, compares well with *Safeda*.

LYCOPENE—$C_{40}H_{56}$, a carotenoid responsible for the red colour of tomato fruit.

MACRO ELEMENT—Elements present and also required in relatively large amounts by plants, *viz.*, N, P, K, Ca, H, etc.

MALATHION – 0, 0-dimethyl-S (1, 2 dicarbethoxyethyl) phosphodithioate. An organophosphorus compound. Acaricide and insecticide ; used as contact insecticide or stomach poison. Some fumigant action is also there. Used in control of storage pests, aphids, mango hoppers, young stages of mealy bugs thrips etc.

MAMME – One of the crops of the caprifig or "male" fig, the first to mature in the spring. The fruits of this crop winter over as comparatively large specimens.

MAMMONI – One of the crops of the caprifig or "male" fig which sets in June and matures in late summer.

MANDARIN ORANGE – (Tangarine). *Citrus reticulata* Blanco ; *Santra* (loose jacket yellow oranges). Family : Rutaceae.

MATURATION – Accumulation of solids in tissues. Ripening ; completion of germ-cell development.

MATURITY – Stage of full growth, development or ripeness.

MCPA – 2-methyl + chlorophenoxy acetic acid ; a selective weedicide closely resembling 2, 4-D in weed killing property.

MEIOSIS – The reduction of diploid chromosome number to haploid by two successive nuclear divisions. Genetic seggregation takes place.

MERISTEMATIC – Capable of cell division.

MESOPHYLL – The inner or middle, spongy portion of the leaf.

MESOPHYTE – A plant that grows normally under conditions of moderate humidity, that neither requires nor survives culture in water or extreme drought.

METASYSTOX – (Methyldemeton) 0, 0-dimethyl-0-2-ethylmercaptoethyl thionophosphate. Organophosphorus compound, far less toxic than systox, a systemic insecticide for aphids and mites.

METAXENIA – The supposed direct influence of the pollen of pollens present on the parts of the seed and fruit which lie outside the embryo and endosperm.

METHOXYCHLOR – Dichloromethoxyphenyl trichloroethane. A chlorinated hydrocarbon insecticide.

METHYL BROMIDE – CH_3Br. Very toxic insect fumigant, gas 3.3 times heavier than air used for fumigating stored grain.

MICRO ELEMENT – Trace element. Element essentially needed in small quantity, for the nutrition ; also termed as minor element *viz.*, Cu, Zn, Mn, Mg, B.

MILLERANDAGE – A condition in the grape where the ovary persists but the seeds remain small or do not attain usual size ; produced by conditions similar to those that lead to coulure.

MIST PROPAGATION—Spraying water in mist form to leafy cuttings in the rooting stage to make up for water loss due to transpiration. The atmosphere is made more humid and desiccation is not allowed to occur.

MITOSIS—Nuclear division leading to duplication of chromosomes takes place to divide to yield nuclei identical with original nucleus. It may accompany cultivar division (cytokinesis).

MODIFIED LEADER—System of training fruit trees intermediate between central leader and open centre.

MONOCOT—(Monocotyledons) Plant with one cotyledon in its seed, *e.g.*, coconut, date-palm, banana.

MONOECIOUS—The stamens and pistils in separate flower but on the same plant.

MULCH—Plant residues and other materials placed on the surface of the soil to conserve moisture, control erosion, runoff, check temperature, weed growth, protect plants from temperature effects.

MYCORRHIZA—Symbiotic association of fungus with the roots of higher plants.

NARROW LEAFED—Having leaves of much greater length than width *e.g.*, grasses, cereals, sugarcane.

NEMATICIDE—A substance capable of destroying nematodes or eelworms. CS_2, paradichlorobenzene ($C_6H_4Cl_2$) or PDB, EDB, DBCP, methyl isothiocynate are examples.

NEW WOOD—The current year's growth in plants.

NUCLEIC ACID—Phosphorus containing acids, usually combined with protein present in cell nuclei.

NURSERY—Establishment of growing, handling, or retailing woody and herbaceous plants.

NUTRIENTS—Elements available through soil, air and water which the plant utilizes in growth, development and reproduction.

OBCORDATE—Heart-shaped with a notch at the apex.

OBLATE—Spheroid, flattened at the poles *e.g.* the earth.

OBOVATE—Ovate with the broader end towards the apex.

OFF-SHOOT—A lateral shoot or branch which arises from one of the main stems of a plant.

OFF-TYPE—Any notable deviation from the normal or standard.

OFF-YEAR—The year in which an alternate bearing plant does not produce or produces very small number of fruits.

OSMOSIS—Passage of materials through a membrane from areas of higher concentration to areas of lower concentration.

OVAL—Broadly elliptical ; egg-shaped.

OVARY—The flask-shaped female organ of the flower which contains ovules. The undeveloped fruit.

OVATE—Egg-shaped, the broader end at the base.

OVULE—The undeveloped seed. A body containing a female gamete which develops into a seed ; a small egg or an egg-like structure.

PARTHENOCARPIC—The formation of fruit without fertilization and without seed or with embryo-less seed.

PARTHENOCARPY—The production of fruits without the fertilization.

PARTHENOGENESIS—The development of the unfertilized egg into the usual product of fertilization without previous union of gametes.

PARTS PER MILLION (ppm)—Weight units per million weight units of soil, on oven dry basis ; expresses gm/million gm of soil or kg per million kg of soil, or in the case of the soil or other solutions weight units/million weight units of solution.

PEDICEL—The support of a single flower of an inflorescence.

PEDUNCLE—The support of an inflorescence or a flower stalk.

PERENNATION—A lasting stage, referring particularly to the persistence of fruit longer after its usual season of maturity.

PECTIN—Any of the fruit juice substances which form a colloidal substance and are derived from pectose in the process of ripening. Available from guava, citrus fruit, apple wastes; used in jelly making.

PERENNIAL—In the case of plant, living for more than two years.

PERFECT—Hermaphrodite flowers.

PETALS—The usually showy and attractive structures around the essential organs of the flower. Petals in a group are known as corolla.

PETIOLE—Leaf stem.

PHLOEM—A complex tissue of sieve tubes and parenchyma. Responsible for translocation of food in solution. The inner bark of woody plant.

PHOTOPERIODISM—Response of plants to the daily duration of light.

PHOTOTROPISM—The influence of light on the direction of growth as seen in plants.

pH SCALE—A means of expressing acidity or alkalinity where a reading of 7 is neutral as of distilled water, below 7 it is acidic and beyond 7 it is alkaline. A pH of 6 is 10 times more acid than a pH of 7 and a pH of 8 is 10 times more alkaline than a pH of 7 and so on.

PHYLLOCLADE—A leaf-like or chlorophyll-containing stem that performs the photosynthetic function like ordinary leaves.

PHYLLODY—A disorder in which a shoot or branch develops a much larger number of leaves than the normal.

PINCHING BACK—Pruning the soft succulent tips of a shoot to stimulate lateral growth.

PISTIL—The female part of a flower consisting of ovary, ovules, style and stigma.

PLANT PROPAGATION—The perpetuation or increase in number of plants.

POLLEN—The dust like spores produced in the anther containing the male germ cell.

POLLINATION—Transfer of pollen from anther to stigma.

POLYEMBRYONY—The production of more than one embryo in an ovule.

POLYGAMO-DIOECIOUS—With hermaphrodite and unisexual flowers on different individuals of the same species.

POLYGAMOUS—With hermaphrodite and unisexual flowers.

POLYSACCHARIDE—A carbohydrate which yields a large but indefinite number of simple sugars on hydrolysis ; usually colloids.

POME—An inferior, indehiscent, two or more celled fleshy fruit, such as apple, pear, quince, having many seeds instead of a stone.

PRIMORDIA—Initials or beginnings of structures.

PROCESSING—Preparing for future use as in canning, freezing, preserving and dehydrating.

PRODUCTIVE STAGE—The period of fruitfulness, productivity, or accumulation of carbohydrates.

PROFICHI—One of the crops of the caprifig or 'male' fig, the second to mature in the Spring. The fruits of this crop appear as small buttons in the late fall or early winter.

PROGENY—The young or seedlings of plant. The offspring of an animal or plant.

PROLIFERATION—Abnormal bush-like growths. A rapid and repeated production of new parts, as the formation of leafy parts from flora parts.

PROTANDRY—The dehiscence of pollen before the stigma is receptive.

PROTOGYNOUS, DIURNALLY SYNCHRONOUS DICHOGAMY—The male and female parts mature at different times (dichogamy), it synchronises in such a way that all the open flowers on a tree are female at one time and male at a different time. The synchronisation is diurnal for each tree, flowers are functionally female during one part of the day and male at another. The female part (pistol) matures earlier (protogynous). The flowers function as being unisexual. Example, Avocado.

PROTOGYNY—The pistil being receptive before the anthers have ripe pollen.

PRUNING—The judicious removal of leaves, shoots, twigs, branches, or roots of a plant to increase its usefulness. Method of pruning followed depends on the growing and fruiting habit of the plant.

PSEUDO-HERMAPHRODITE – Functional unisexuality in the presence of apparently well developed stamens and pistils

PUMMELO – *Citrus decumana* (Shaddock, H. *Chakota*) – Family : Rutaceae.

PURINES – A group of nitrogenous organic compounds such as uric acid, xanthine and caffein. The compounds have hydrogen and nitrogen.

PYRIDINE – A nitrogenous base which is the nucleus of many organic compounds, for example, Nicotine.

PYRIMIDINES – A group of basic nitrogenous compounds related to purines, a product of nucleic acid cleavage.

QUALITY – Various factors including flavour, texture, appearance, odour, and food value that influence the produce. Characteristic that helps in differentiating a product regarding its acceptability to the consumer.

QUINCE – *Cydonia vulgaris* Pers, *Bihi* (H.). Family Rosaceae, fruits edible. Grown in Kashmir, the Punjab and Nilgiris. Also used as a root stock for pear.

QUINONE – An oxidation product of benzene.

QUINTAL – A measure of weight ; 100 kilograms.

REJUVENATION – Stimulation of new growth on old plants usually accomplished by pruning, cleft and bridge grafting.

RENEWAL SPUR OR FOUNDATION SPUR – Grape canes or spur near the trunk cut back to two buds to provide new fruiting wood in a favourable season.

REPLACEMENT SPUR – Water sprout suckers arising from the arm directly chosen to produce for the following year. It is useful in shortening the arm which is otherwise lengthened when a fruiting cane is chosen from the base bud of the renewal spur consistently.

REPRODUCTIVE STAGE – Ability to increase in numbers. Development of organs capable of giving rise to new plants.

RESPIRATION – Gaseous exchange by which the plant utilises oxygen and gives off carbon dioxide ; oxidation of food by plants for energy release.

RESPIRATORY COEFFICIENT – The amount of carbon dioxide given off divided by the amount of oxygen utilised in respiration.

REST PERIOD – Period of non-visible growth, controlled by internal factors.

RHIZOME – Horizontal underground stem with roots and sends up shoots at its nodes. Ginger, turmeric, arrow-root, banana, are examples.

RINGING – Removing a narrow strip of bark round the periphery of the stem to prevent downward translocation of food beyond that point.

RING NECK – A disorder as in avocado in which there is a drying out and shrivelling of surface areas on the fruit pedicels.

RIPENESS – State of complete maturation prior to breakdown.

RODENTICIDE—A rodent poison.

ROUGING—Removing and destroying undesired plants, such as those infected with systemic diseases or of inferior quality.

ROOT HAIRS—Epidermal outgrowths from cell wall of the root, specialised for water and nutrient absorption.

ROOTSTOCK—The seedling or plant over which a shoot is grafted *viz.*, almond on peach for almond, *Prunus* on myrobalan for plum. Seedling obtained from seeds of vigorous trees serve as rootstock for guava, mango, jamun etc.

ROOT TUBER—The fleshy root of a herbaceous perennial plant with buds or eyes in the upper regions.

ROSETTE—A condition in which the internodes are much shortened giving the leaves a bunched or clustered appearance.

SAP-WOOD—Outer portion of xylem tissue and the area of upward movement of water and nutrients in a woody stem.

SCAFFOLD BRANCHES—The primary branches arising from the tree trunk.

SCARIFICATION—(*Scarifying*) Injuring or scratching the seed coat by acids, for example, to affect early germination by breaking dormancy. Mostly used in temperate woody scions viz. apple, pear, peach, etc.

SCION—Piece of shoot inserted on another plant in the process of grafting.

SEED—A ripened ovule consisting of a dormant plant in miniature. It bridges the gap between two generations.

SEEDLING—A young plant grown from seed.

SEED TECHNOLOGY—Science dealing with the production, harvesting, processing, testing, packaging, storing and marketing of seeds.

SEED TREATMENT—Application of fungicides to seeds for protection against smut, damping off, and seed-decay organisms.

SELF-POLLINATED—Transfer of pollen from the anther to the stigma of the same plant.

SEPALS—The leaf like structures encasing a flower-bud and later subtending the open flower.

SERRULATE—Finely serrate ; having a series of small notches.

SEXUAL PROPAGATION—Increase in plant numbers by reproductive methods, through seeds.

SHOOT—Current season's growth of stem with leaves and buds.

SHORT-DAY—Less than 12 continuous hours of (daylight) day-length.

SIDE DRESSING—Fertilizer application alongside row crops during growth.

SITE—A selected place of operation in a locality.

SLIP—A herbaceous or soft wood cutting.

SOD CULTURE A method of orchard soil management in which a permanent perennial crop is grown between the trees, mowed once or twice during the growing season and then allowed to remain on the ground. A limited area around the tree is hoed, spaded or tilled. SOD MULCH.

SOUR SAP – A sour or fermented condition of the sap, often associated with a high water table or other root disturbance and usually followed by the death of the tree.

SPUR – Short, woody stem (branch) which is the principal fruiting area of many tree-fruits.

STAMEN – The male part of flower consisting of anther and filament.

STAPLE CROP – A crop commonly produced and having ready market.

STARTER SOLUTION – Fertilizer dissolved in water and applied immediately following transplanting.

STERILIZATION (Soil) – Inactivation by steam or chemicals to prevent disease, insect, or weed problems.

STERILITY – The inability to produce viable seeds.

STIGMA – The sticky upper surface of pistil on which the pollen germinates.

STIPULE – One of the two foliaceous or membraneous processes developed at the base of a leaf-petiole ; sometimes in the form of tendril or spine.

STOCK – The root or stem on which a graft is made.

STOLON (runner) – Horizontal stem near the ground surface which forms plants at nodes.

STOMATE – Pore in the epidermis of a leaf.

STRAIN – A special type selected from a variety.

STRATIFICATION – Subjection of imbibed seeds, to temperature 2 – 10°C in a well aerated media, for some period before germination to complete after-ripening and break the rest period.

STUNTED – Plant restricted in size and vigour owing to unfavourable environment.

STYLE – The "neck-like" support joining the ovary and stigma in a flower.

SUBORDINATION – Serve heading-back of all but one of multiple leaders to restore apical dominance.

SUBTROPICAL – Meso-thermal temperature area, in part tropical.

SUCCESSION CROP – A crop following another crop on the same area in the same season.

SUCCESSIVE PLANTINGS – Varied planting dates of the same crop extending the harvest period.

SUCCULENT Watery, as stems and leaves high in water content. Also a group of plants with thick, fleshy leaves and stems and high in water content.

SUCKER—Plant arising from adventitious buds on the roots or underground parts of some trees like guava, pomegranate, banana and pineapple.

SUMMER FALLOW—Allowing land to lie idle for a season to conserve water for use in the following season.

SUN-SCALD—Injured or killed bark or epidermal tissue due to excess light intensity or other unfavourable conditions.

SUPER COOLING—Cooling below the freezing point without solidification.

SYMBIOSIS (Mutualism)—Two organisms live in close physiological association for mutual benefit (Legume root nodules having *Rhizobia* spp).

SYNCARP—Aggregate fruit with united carpels *viz.* mulberry and pineapple.

SYNGAMY—Fusion of gametes in fertilization.

SYNGENESIS—Formation of embryo partly from the male and party from the female elements.

SYSTOX—(Demeton) ; 0, 0-diethyl S-(2-ethylthio) ethyl phosphorothioate; organo-phosphorus compound. Insecticide used against aphids and mites.

TAP ROOT—A vertical, central and deeply penetrating root with few laterals.

TERMINAL—The distal end of a shoot, twig or branch.

THINNING OUT—Pruning of entire stems or branches to give remaining stems or branches more space and more reserve food for improving growth.

TISSUE TESTS—Determination of plant food needs by chemical analyses of leaves or stems.

TOP WORKING—Converting an undesirable plant into a suitable (*e.g.* pollination or commercial purposes) cultivar by shield budding or cleft, crown or tongue grafting through cutting back the main scaffold branches and inserting one or more scions of desirable cultivar in each cut end.

TORUS—(1) The receptacle of a flower, part of the axis on which the flower parts are inserted. (2) The thickening in the centre of the membrane in bordered pits.

TRAINING—Shaping or adapting plants to specific forms by pruning, fastening, or bracing.

TRANSITION STAGE—The integration period of juvenile and productive stages of growth.

TRANSLOCATION—Movement of water, minerals, and food within the plant.

TRANSPIRATION—Exhalation of water-vapour from plant body into the atmosphere through the cuticle, stomata and lenticels.

TRANSPLANTING—Moving plants from one container or media to another.

TREE SURGERY—Art of pruning involving removal of large limbs, treating wounds, and bracing weak trunks and crotches.

TRUNK—The permanent framework of the vine or tree, the undivided main stem.

TUBER—A thick, fleshy, underground stem (*e.g.* potato) specialised for food storage and capable of giving rise to new plants.

TWIG—One-year old branch or stem of a woody plant.

VACUOLE—In cells, the cell sap surrounded by protoplasm.

VAPAM—Sodium-N-methyl dithiocarpamate, dihydrate.

VARIETY—A group of strains or single strain of closely related plants of common origin which have similar characteristics and can be differentiated on the basis of structural and functional characters from another groups of plants. Termed as Cultivar also.

VEGETABLE—The edible product of a herbaceous garden plant.

VEGETATIVE—Non-sexual, may be associated by significant increase in size.

VIABILITY—Ability of seeds to germinate.

VIRESCENCE—A disorder in which certain vegetative or floral structures become and remain abnormally vegetative and green.

WATER BERRIES—A disorder of the grape in which the fruits are watery and fail to ripen properly.

WATERCORE—A disorder, especially of apples, in which the fleshy tissues, either around the core or near the surface, becomes watery, hard and glossy.

WATER-LOGGED—Devoid of aeration of the soil due to blocking of the pore spaces by the filling up of water.

WATER SPOT—A disorder of orange fruits characterized by a water soaked appearance.

WATER SPROUTS—Rapid growing shoots that arise from adventitious or latent buds on branches or trunks.

WEED—A plant growing out of place.

WILTING—The drooping and loss of freshness of plant foliage.

WILTING COEFFICIENT—(Wilting Point, Wilting Range). The percentage of moisture held by the soil within the root feeding zone when permanent wilting of plants takes place.

WILTING PERCENTAGE—(Permanent wilting point). The percentage of water in soil in which plants wilt and fail to recover when placed in an atmosphere saturated with water vapour.

WITCHES BROOM—(Malformation). A disorder of the stem of its branch in which there is extreme amount of branching, resulting in a very bushy condition. Sometimes caused by fungus attack.

WITHER TIP—A form of dieback reported to be due to copper deficiency.

XANTHIN—A basic nitrogenous compound related to uric acid, one of the purines. Insoluble part of yellow colouring matter of flowers (soluble part of the yellow colouring matter of flowers is xantheine).

XANTHOPHYLL—$C_{43}H_{56}O_2$. A carotenoid pigment responsible for the pigmentation of skin. Associated with chlorophyll and carotene in plants imparting a golden yellow to ivory colour according to concentration.

XENIA—The direct influence of foreign pollen on the endosperm due to the phenomenon of double fertilization which may allow the male parent to affect the appearance of seed.

XEROPHYTE—A plant that can endure extreme drought where the evaporation stress is high and the water supply low.

XYLEM—A complex tissue of tough, fibrous, lignified elongated cells forming vessels and woody tissue ; part of the vascular bundle responsible for upward movement of water and minerals from roots to the above-ground parts of the plant.

XYLOPOROSIS—A disorder, associated with lack of compatibility between stock and scion, characterised by pores or pits in the wood and corresponding pegs in the bark:

YELLOW PASSION FRUIT—*Passiflora edulis flavicarpa*. Family Passifloraceae.

ZARDALU—See Apricot.

ZINC ARSENATE—A component of insecticidal sprays, $Z_3^-(AsO_2)_2$.

ZINC ARSENITE—Also used as an insecticide $Zn(AsO_2)_2$.

ZINC CHLORIDE—A fungicide and wood preservative, $ZnCl_2$.

ZINC DEFICIENCY—Conditions brought about in fruit trees *viz.* citrus causing mottle leaf, interveinal chlorosis with necrosis, reduced leaf size, reduced fruit production etc.

ZINC OXIDE—A fungicide and seed fumigant.

ZINC PHOSPHIDE—A raticide and insecticide.

ZYGOTE—Fertilized egg, cell arising from the fusion of gametes.

SELECT REFERENCES

Annual Review of Plant Physiology (Vols. 8, 11, 12, 13, 14, 15, 16, 17, 18, 19, 24), Annual Rev. Inc. Palo Alto, California, 1957-73.

Atwal, A.S., *Agricultural Pests of India and South-East Asia,* Kalyani Publishers, New Delhi, 1977.

Bajwa, M.S. Improving Fruit Quality, *Ind. Hort.,* XXI, No. 3.

Beevers, Herry, *Respiratory Metabolism in Plants.* Harper and Row Ltd., 1960.

Bonner, J. and Varner, J.E. *Plant Biochemistry,* Acadmic Press, N.Y., 1965.

Bose, T.K. and Mitra S.K. *Fruits : Tropical and Subtropical,* Naya Prokash, Calcutta, 1990.

Chadha K.L. (ed). Handbook of Horticulture. (2001), ICAR Publication.

Chibrall, A.C., *Protein Metabolism in the Plant,* Yale University Press, London, 1964.

Cheema, G.S., Bhat, S.S. and Naik, K.C., *Commercial Fruits of India,* Macmillan & Co. Ltd., 1954.

Christopher, E.P., *Introductory Horticulture,* McGraw Hill Book Company, INC., N.Y., 1958.

Cohen, Georges, N., *The Regulation of Cell Metabolism,* Hermann, France, 1967.

Crafts, A.S., *Translocation in Plants,* Hold, Rinehert and Winston, N.Y., 1961.

Datta, C., *Plant Physiology*, New Central Book Agency, Calcutta, 1968.

Denisen, Ervin, L., *Principles of Horticulture,* The Macmillan Company, N.Y., 1958.

Devlin, Robert M. and Barker, Allen V., *Photosynthesis,* Affiliated East-West Press Pvt. Ltd., New Delhi, 1971.

De Robertis, E.D.P., Saez, Francisco A and De Robertis, E.M.F., *Cell Biology,* Toppan Company Ltd., Tokyo, 1975.

Edmond, J.B., Senn, T.L. and Andrews, F.s. *Fundamentals of Horticulture,* McGraw Hill Book Company, N.Y. 1964.

Edmond, J.B., Musser, A.M. and Andrews, F.S., *Fundamentals of Horticulture,* McGraw Hill Book Company, Inc., N.Y. 1957

Evans., L.T., *Environmental Control of Plant Growth,* Academic Press, N.Y., 1963.

Gardner, Victor R., *Basic Horticulture.* The Macmillan Company, 1951.

Gardner, V.R., Bradford F.C. and Hoker, H.D., (Jr.). *The Fundamentals of Fruit Production,* McGraw Hill Book Company, N.Y. 1952.

Giese, A.C., *Photophysiology,* Vol. 1, Academic Press, N.Y. 1964.

Gourley. J.H. and Homlett, F.S., *Modern Fruit Production,* Macmillan Company, N.Y., 1961.

Handbook of Horticulture, 2001

Hartmann, Hudson T. and Kester, Dale E., *Plant Propagation,* Prentice-Hall of India Pvt. Ltd., 1976.

Hay, Roy, *Reader's Digest Encyclopedia of Garden Plants and Flowers,* Hodder and Stoughton Ltd., London, 1971.

Hillman, William S., *The Physiology of Flowering,* Hold, Rinehort and Winston, N.Y., 1961.

Indian Horticulture Vols. I-XX.

Janick, Jules, *Horticultural Science,* W.H. Freeman and Company, London, 1963.

Jensen, P. Boysen, *Growth Hormones in Plants.* McGraw Hill Book Company, Inc., N.Y., 1963.

Klein, R.M. (ed.) *Plant Growth Regulation*, Iowa State University Press, America.

Kolesnikov, V., *Fruit Biology,* Mir Publishers, Moscow, 1969.

Kozlowski, Teodore T., *Tree Growth*, The Ronald Press Company, N.Y., 1962.

Kramer, Paul J. and Kozlowski, Theodove T., *Physiology of Trees,* McGraw Hill Book Company, N.Y., 1960.

Lal Girdhari, Siddappa G.S. and Tandon, G.L., *Preservation of Fruits and Vegetatives,* ICAR, New Delhi, 1960.

Leopold, A. Carl. *Auxins and Plant Growth,* University of California Press, Barkeley, 1955.

Leopold A. Carl., *Plant Growth and Development,* McGraw Hill Book Company, N.Y., 1964.

Mahlstede, John P. and Haber, Ernest S., *Plant Propagation,* Wiley and Sons, Inc., N.Y., 1957.

Meyer Bernard S. and Anderson, Donald, B., *Plant Physiology,* D. Van Nostrand Company Inc., N.Y., 1952.

Meinder, M. and Mansfield, T.A., *Physiology of Stomata,* Tata McGraw Hill Publishing Co. Ltd., New Delhi 1968.

Misra, K.K. and Pandey R.M. (ed.). *Phal Utpadan* (1999). G.B. Pant University of Agr. and Tech., Pantnagar Publication.

Muzik, T.J., *Weed Biology and Control.* McGraw Hill Book Company, N.Y., 1970.

Nair, M.G.K. Insects and Mites of Crops in India (1986), ICAR Publication.

Nord, F.F., *Advances in Enzymology,* Vol. XII, Interscience Publishers Inc., N.Y., 1951.

Nord, F.F., *Advances in Enzymology,* Vol. XIII, Interscience Publishers Inc., N.Y., 1952.

Pathak, R.K. Phal Vrikcha Parivardhan, ICAR Publishers (1996).

Pathak, R.R., D. Pandey and Pandey V.S., Propagate temperate fruit through hardwood cuttings, *Inc. Hort.,* Vol. 22, No. 2.

Plant Research & Development. Vol. V., Information for Scientific Cooperation, 1977.

Price, C.A., *Molecular Approaches to Plant Physiology,* McGraw Hill Book Company, N.Y., 1977.

Ram, Sant, Aonla—A Tree You Must Grow, *Int. Agric.* XII, No. 10.

Randhawa, G.S., *Use of Plant Growth Regulators and Gibberellins in Horticulture,* ICAR, New Delhi, 1971.

Rao, Seshagiri D., *A Hand Book of Plant Protection,* S.V. Rangaswamy and Co. Pvt. Ltd., Bangalore, 1972.

Reader's Digest, *Complete Library of Garden,* Vols. I, II and III, Reader's Digest Association, London.

Reddy D. Bap, *Plant Protection in India,* Allied Publishers Pvt. Ltd., Bombay, 1968.

Rudnick, Dorothea, *Cell, Organism and Milieu,* The Ronald Press Company, N.Y., 1959.

Ruhland, W., *Encyclopaedia of Plant Physiology.*

Rosenberg, H.R., *Chemistry and Physiology of the Vitamins,* Interscience Publishers Inc., N.Y., 1945.

Saha L.R., *Handbook of Plant Protection* (1999), Kalyani Publishers, New Delhi.

Salmon, S.C. and Hanson, A.A., *The Principles and Practice of Agricultural Research,* Leonard Hill, London, 1964.

Salisbury, Frank, B., *The Flowering Process,* Pergamon Press, Oxford, 1933.

Salisbury, F.B. and C. Ross, *Plant Physiology*, Prentice Hall of India Pvt. Ltd., 1974.

Sant Ram, *Aonla* (1993) G.B. Pant University of Agri. and Tech. Publication.

Sarabhog, A.K. Test Book of Mycology, ICAR Publication, 2000.

Schery, R.W. *Plants for Man,* George Allen and Unwin Ltd., 1954.

Sharma, Brij Bhushan and Sharma H.C., *Fruit Cultivation in India,* Ministry of Agriculture, Government of India (1993).

Shoemaker, James S. and Teskey, Bengamin, J.E., *Practical Horticulture,* John Wiley and Sons, Inc., N.Y., 1955.

Singh, Amar, *Plant Physiology,* Asia Publishing, Bombay, 1967.

Singh L.B., *The Mango*, Leonard Hill (Books) Ltd.

1	2	3	4	5	6
4.	Apricot : **Khubani** (*Prunus armeniaca*) T.F.	(*i*) Tongue grafting (*ii*) T-budding (*iii*) Chip budding	February June September	Wild apricot and wild peach	T- budding gives better result.
5.	Avocado/Butter fruit (*Persea americana*) St.F.	(*i*) Seeds (*ii*) Shield budding (*iii*) Forkert budding (*iv*) Side grafting (*v*) Whip grafting (*vi*) Cleft grafting (*vi*) Air layering	September-October June-July June-July June-July June-July June-July June-July	Mexican and West Indian race of Avocado	Budding and grafting are most popular
6.	Bael (*Aegle marmelos*) St.F.	(*i*) Seeds (*ii*) Patch budding (*iii*) Suckers (*iv*) Stool layering	April-May June-July Rainy season Rainy season	– – – –	Seedlings used for root stock. Ideal method with a 80%-90% success.
7.	Banana (*Musa sp.*) Tr.F.	(*i*) Suckers (*ii*) Rhizome (*iii*) Corms (*iv*) Shoot-tip (*v*) Bits (*vi*) Tissue culture	May-June May-June May-June May-June May-June	– – – – –	Sword suckers should be used. – – Plants require longer time to bear fruit.

1	2	3	4	5	6
8.	Ber (*Zizyphus maeritiana*) St.F.	(*i*) Seeds (*ii*) I-budding (*iii*) T- budding (*iv*) Patch budding	February-March May- July May-July May-July	— —	Soak seeds for 48 hrs in water or for 6 hrs in conc. H_2SO_4 or in 200 ppm GA to improve germination. I-budding is the most sucessful.
9.	Bread fruit (*Artocarpus incisa*) Tr.F.	(*i*) Seeds (*ii*) Root cuttings (*iii*) Root suckers (*iv*) Air-layering (*v*) Inarching	June-July June-July June-July July-August July-August	— — — — —	—
10.	Cherry (*Prunus avium*) T.F.	(*i*) Seeds (*ii*) Tongue grafting (*iii*) Shield budding (*iv*) Patch budding	May, November-Dec. February-March July July	— Seedlings of Mazzard (*Prunus avium*) and Mohaleb (*P. mahaleb*) **Vegetatively propagated** $F_{12/1}$ and Cdt	— Most success with 90%.
11.	Cashew nut, kazu (*Anacardium occidentale*) Tr.F.	(*i*) Seeds (*ii*) Air layering (*iii*) Mound layering (*iv*) Stooling (*v*) Epicotyl grafting (*vi*) Soft wood grafting	June-July April-October April-October April-October April-October April-October	Seedlings of Cashew — —	No eeaed dormancy. Easiest method, poor field establishment. Absence of tap root is a disadvantage. Mortality in nursery is high. Most successful and commercially viable technique.

1	2	3	4	5	6
12.	Date-Palm : **Khajoor** (*Phoenix dactylifera*) St.F.	(*i*) Seeds (*ii*) Off- shoots Suckers	June-July March-April	— —	Inferior quality fruits. —
13.	Citrus (*a*) Acid lime (*Citrus aurantifolia*) St. F.,	(*i*) Seeds (*ii*) Air layering	September, February-March June-July	— — — — —	Seeds have no domancy sow after harvesting —
	(*b*) **Sweet lime (C. Limettiodes)**	(*i*) Hard wood cuttings (*ii*) Air layering (*iii*) T-budding (*iv*) Sub-terminal leafy cuttings	February, August-September June- July September February-March February-March	Sweet lime, Karana Khatta Jambhiri (Rough lemon) (*C. Jambhiri*)	Shallow rooted and surfaces feeder. 100% success in rooting if dipped in 50 and 100 ppm IBA for 24 hr or 2000 ppm for 10 seconds.
	(*c*) Lemon (*C. limon*)	(*i*) Stem cutting (*ii*) Air layering (*iii*) T- budding	February August- September. June-July September, February-March	— — Karana Khatta (*Citrus Karana*)	Budded plants are precocisus and produce uniform crop.

FPI.

1	2	3	4	5	6
	(*d*) Mandarin : Orange-Santara (*C. reticulata*)	(*i*) Seeds (*ii*) T- budding	February- March February March September-October	– Rough Lemon, Rangpur lime (*C.Limonia*), Karana Khatta, Cleoptra (*C.reshni*) Troyer citrange (Hybrid) (Trifoliate Orange × Sweet Orange)	–
	(*e*) Sweet Orange/Malta. *C. Sinensis*	(*i*) Seeds (*ii*) T-budding (*iii*) Patch budding	February-March February-March Sept.-October	Karana Khatta, Jambhiri, Rangpur lime	–
14.	Chiku (*Manilkara achras*) Tr.F.	(*i*) Seeds (*ii*) Air layering (*iii*) Inarching	June-July June-July July-August	Chiku, rayan/khirani (*Mimusops hexandra*), mahua	– – Commercial method
15.	Custard apple (*Annona squamosa*) Tr.F.	(*i*) Seeds (*ii*) Inarching (*iii*) Side grafting	July-August July-August July-August	Seedings of Ramphal (*A. reticulata*), Sharifa (*A. squamosa*)	Fresh seeds germinate better. Treating the seeds with GA_3 at 500 ppm assists germination.

16.	Fig : Anjeer (*Ficus carica*) St.F.	(*i*) Hard wood cuttings (*ii*) Air layering (*iii*) Side grafting (*iv*) Shield budding (*v*) Patch budding (*vi*) Seeds (*vii*) Tissue culture	Dec.-January June-July June-July May- June May-June June-July June-July	*Ficus glomerata* and *Ficus palmata*	Easy and Commercial method — For top working. For top working. Used only for breeding new cultivars.
17.	Grape (*Vitisvinifera*) St. F.	(*i*) Seeds (*ii*) Dormant hard wood cutting (*iii*) Air layering (*iv*) Bench grafting (*v*) Green wood grafting (*vi*) Tongue grafting (*vii*) Cleft grafting (*viii*) Chip budding	February Dec.-January June-July Dec.- January Dec.-January Dec.-January Dec.- January Dec.-January	St. George (*V. rupestris*) Ganzin No.1, Couderc 1613, Harmony, Dog-Ridge, Salt Creek	Seeds stratification at 0.5 to 4°C for about 12 weeks before sowing for best result. Commercial and easy method — Budding and grafting are used for special purpose.

1	2	3	4	5	6
18.	Guava (*Psidium guajava*) Tr.F.	(*i*) Seeds	February-March	—	Due to 35.6% cross pollination seed is not recommended for commercial propagation.
		(*ii*) Air layering	June-July	Bottle brush and Jamun to avoid the infection of wilt.	100% success after IBA (3000 ppm) application.
		(*iii*) Stooling	June-July		Easiest and cheapest method.
		(*iv*) Patch budding	June- July		Ideal with high percentage of success.
		(*v*) Chip budding	June-July		High percentage of success.
		(*vi*) T budding	June-July		
		(*vii*) Inarching	June-July		—
		(*viii*) Veneer grafting	June-July		
19.	Jack-fruit (*Artocarpus heterophyllus*) Tr.F.	(*i*) Seeds	June-July	Seedlings of Jack fruit.	Common method of propagation
		(*ii*) Air layering	June-July		
		(*iii*) Stooling	June-July		
		(*iv*) Inarching	June- July		
		(*v*) Epicotyl grafting	April-May		
		(*vi*) Forkert budding	May-July		
		(*vii*) Chip budding	May-July		
		(*viii*) Patch budding	May-July		Most successful in Northern India.

1	2	3	4	5	6
20.	Jamun (*Syzygium cumini*) St.F.	(*i*) Seeds (*ii*) Patch budding (*iii*) Forkert	June-July March, July-August (in low rainfed area)	Seedlings of Jamun	Most common.
21.	Karonda (*Carissa carandas*) St.F.	(*i*) Seeds (*ii*) Air layering	June-July July- August	– –	Sow the seeds immediately after extraction. Very successful.
22.	Litchi (*Litchi chinensis*) St.F.	(*i*) Seeds (*ii*) Air-layering (*iii*) Stooling (*iv*) Splice grafting (*v*) Inarching grafting	June-July May- June May-June June-July June-July	Seedlings of Litchi and *Nephelium longana*	Generally to raise seedings. Common and easiest method. – –
23.	Loquat (*Eriobotrya japonica*) St.F.	(*i*) Seeds (*ii*) Air-layering (*iii*) Inarching (*iv*) Shield budding	March June-July June-July February-March	Seedlings of Loquat, apple pear, mespilus and Cydonia oblanga.	– Highly successful – –

1	2	3	4	5	6
24.	Mango (*Mangifera indica*) Tr.F.	(*i*) Seeds	June-August	Seedlings of mango seeds.	Used for raising root stock.
		(*ii*) Cutting	July-August		Cutting from mature shoot by treatment with IBA (5000 ppm).
		(*iii*) Layering	July-August		Applicable under high atmospheric humidity, good for developing clonal root stocks.
		(*iv*) Stooling	June-July		Ringing of shoots and treatment with 5000 ppm IBA.
25.	Mango	(*i*) Inarching (*ii*) Veneer grafting (*iii*) Soft wood grafting (*iv*) Epicotyl/stone grafting (*v*) Shield budding	June-August July-August July-August July-August March-Sept.	—	Most widely practised method over 95% success. Gaining popularity in Northern India. — Grafts are ready to be transplanted in one year.
26.	Mangosteen (*Garcinia mangostana*) Tr.F.	Seeds	June-July	—	Common method of multiplication, seeds are azygotic, produce trees resembling mother plant.

1	2	3	4	5	6
27.	Mulberry (*Morus* spp. St.F.	(*i*) Hard wood cuttings (*ii*) T-budding (*iii*) Ring budding	December-Jan. February-March	Seedings of mulberry	–
28.	Papaya (*Carica papaya*) Tr.F.	Seeds	June-September		Most practical and commercial method, seeds taken should be from isolated condition or sibmating.
29.	Passion fruit (*Passiflora edulis*) Tr.F.	(*i*) Seeds (*ii*) Cuttings (*iii*) Grafting-cleft	March-April June-July July-August	*Passiflora flavicarpa* (golden passion fruit)	Commercial Commercial To avoid attack of fusarium wilt.
30.	Peach : **Aru.** (*Prunus persica*) T.F.	(*i*) Seeds (*ii*) Hard wood cutting (*iii*) Grafting-Tongue or cleft (*iv*) T-budding or Ring budding	November, November, February January-February – hills Nov.-January – plains. May-June – hills April-June and September	Peach, Plum, Apricot, Almond, Peach/almond hybrids viz. GF 556, GF 667, western sand Cherry (P. *bessye*) and Nanking Cherry (*P. tomentosa*)	Seedlings used for root stock. Treat the stem cuttings with IAA 4000 ppm for five seconds.

1	2	3	4	5	6
31.	Pear : **Naspati** (*Pyrus communis*) T.F.	(*i*) Seeds (*ii*) Cuttings (*iii*) Root suckers (*iv*) Stooling (*v*) T-budding (*vi*) Tongue grafting	November- February Nov.-February Oct.-December May- June April-September Dec.-January	Seedlings of pear *Kainth* and *Shiara*; clonally propagated quince A (*Cydonia oblonga*).	For raising root stocks —do— —do— —do— Common method of propagation
32.	Pecan (*Carya illinoensis*) T.F.	(*i*) Seeds (*ii*) Patch budding (*iii*) Ring budding (*iv*) Tongue grafting	November, February May End of June and early-March	Seedlings of pecan viz. Apache, River-side, Elliott, Curtis.	To raise root stocks. Most successful Most successful Most successful
33.	Persimmon : **Kaku** (*Diospyros kaki*) T.F.	(*i*) Seeds (*ii*) T-budding (*iii*) Veneer grafting (*iv*) Cleft grafting (*v*) Tongue grafting	November-Feb. Spring or early Autumn September September September	Seedlings of *D.kaki*, *D. lotus*, *D.virginiana*	Stratification for 60-90 days, used to raise root stocks. 80% success. For larger diameter stock above 80% success. For smaller diameter stock with 60-65% success.

1	2	3	4	5	6
34.	Phalsa : (*Grewia* sp.) St.F.	(*i*) Seeds (*ii*) Hardwood cuttings (*iii*) Air-layering	July-August November-Dec. July-August	– – –	Commercially adopted method, no variability. For success growth regulators (IBA 2500–3000 ppm) should be used.
35.	Pineapple : (*Ananas comosus*) Tr.F.	(*i*) Suckers (*ii*) Slips (*iii*) Crowns	June-July June-July June-July	–	Uniform-sized material of mono-type should be used.
36.	Plum : **Aloocha** (*Prunus salicina*) T.F.	(*i*) Hard wood cuttings (*ii*) Mound layering (*iii*) Tongue grafting (*iv*) Chip-budding (*v*) T-budding	November, Feb. May-June February Mid-Feb. July-August	Wild apricot (Zardalu) seeds used for raising seedling root stock. **Clonal root stock**-Myro balan B(*P.Cerasifera*).	To raise root stocks To raise root stocks Most commonly used and ideal method. Gives good success. Plant growth is poor.
37.	Pomegranate : **Anar** (*Punica granatum*) St.F.	(*i*) Seeds (*ii*) Hard wood cutting (*iii*) Soft wood cutting (*iv*) Air layering (*v*) Mound layering (*vi*) Side grafting (*vii*) Veneer grafting	June-July June-July June- July Rainy season and Nov.-December. Throughout the year July July		– Difficult to root easily Girdling and etiolation are required. Treating with IBA for 30 seconds. Treating with IBA (10,000 ppm) for profuse rooting. For top working For top working

38.	Hazelnut (*Corylus avellana*) T.F.	(*i*) Seeds	Spring season	Seedlings of *Corylus columa (Bhotia badam)*	–
		(*ii*) Tongue grafting	Early- March		Commercially adopted method
		(*iii*) Cleft grafting	Early-March		Commercially adopted method
		(*iv*) Shield budding	Mid-August		
		(*v*) Stool layering	Rainy season		To raise root stock.
39.	Strawberry (*Fragaria anansa*)	(*i*) Seeds	Sept.-October	–	Used for the development of new cultivars.
		(*ii*) Runners	Sept.-October	–	Commercial method
		(*iii*) Crown	Sept.-October	–	Tedious and expensive
40.	Walnut : **Akharot** (*Juglans regia*) T.F.	(*i*) Seeds	November	Seedlings of *J.regia* and *J.hindsii.*	For raising root stock.
		(*ii*) Patch budding	July-August		
		(*iii*) Ring budding	July-August		
		(*iv*) Veneer grafting	July-August		
		(*v*) Cleft grafting	February-March		
		(*vi*) Tongue grafting	February-March		

Abbreviation : T.F. = Temperate fruit, Tr.F. = Tropical fruit, St.F. = Sub-tropical fruit.

APPENDIX II. PESTICIDES AND THEIR USE

Chemical name of Pesticides	Trade Name	Formulation	Qty. l/ml/lit water	Used against
1	2	3	4	5
INSECTICIDES **(Inorganic Compounds)** ***I. Arsenic Compounds*** 1. Lead arsenate	–	WP	2-6 g	Stomach and contact poison. Leaf eating caterpillars
2. Calcium arsenate	–	–	–	Stomach poison; leaf eating insects.
3. White arsenic	–	–	–	Poison baits for rats, grass-hoppers, army worms, ants, cock-roaches.
4. Paris green	–	–	–	Slugs
5. Sodium arsenite				Highly toxic to plants, bait for ants and locusts.
II. Fluorine Compounds 1. Sodium flouride	–	–	–	Cockroaches, cutworms, grasshoppers
III. Other Inorganic Compounds 1. Sulphur		Diluted with inert dust		Mites, thrips etc.
2. Lime Sulphur		Rock lime 500 g + Sulphur 675 g + Water 22.5 lit		Mites scale insects

1	2	3	4	5
3. Zinc Phosphide 4. Barium carbonate 5. Thallium sulphate				Field rats Rat Rats and ants.
Synthetic Organic Compounds *IV. Chlorinated hydrocarbons* 1. Gamma BHC	Lindane, Lintaf	6 G 20 EC	20 kg. (*i*) 3.75 lit/ha with irrigation water. (*ii*) 2 ml/lit	Termites and Stem borers Avoid its use in user soil. Termite Beetles, hairy caterpillars.
2. Endosulfan	Thiodan, Hildan, Endocil, Hexasulfan etc.	35 EC	0.05%	Sucking insects, caterpillars, borers, safer to bees and natural enemies of pest, highly toxic to fish.
V. Organophosphorus Compounds Phosphates (Derivatives of Phosphoric acid)				
1. Monocrotophos	Nuvacron, Monocil	40 EC	0.05%	Chewing, boring and sucking insects.
2. Dichlorovos (DDVP)	Nuvan, Vapona	70 EC	0.15%, 0.1%	Sucking and leaf eating insects, used as fumigant against mosquitoes and other dipterous insects.

1	2	3	4	5
IX. Synthetic Pyrethroides				
1. Cypermethrin	Cypermar Cythrion Cymbush Cyperkil	10 EC 10 EC 25 EC 25 EC	0.01%	Semi-looper and pod-borer
2. Decamethrin	Decis	2.8 EC	0.003%	– do –
X. Acaricide				
1. Dicofol	Kelthane	20 EC	0.05%	Phytophagous mites.
2. Aramite	Aracide, Niagaramite	3-4% dust 15 WP 25 EC	20 – 25 kg/ha 0.05 %	Controls mites.
XI. Nematicide				
1. EDB	Dowfume : W-40, Soilfume 40, Dowfume 85, Soilfume 85 etc.		60 l/ha	
2. DD	Telone, Vidden D, D-D.		250-500 l/ha	Cyst and lesion nematode, Root knot nematode. Root knot nematode.
XII. Rodonticides				
1. Warafarin	Rodafarin, Ratafin	Dust and liquid form	1 Part Warafarin + 19 parts of baits	Anticoagulant in action.

1	2	3	4	5
2. Zinc phosphide	–	85% powder	1 kg smeared bajra or sorghum or cracked wheat + 20 g vegetable oil + 20 g powdered sugar + 20 g zinc phosphide.	–
3. Aluminum phosphide	Celphos, Phostoxin	Tablet or pellet form of 3 g.	1/2 tab of 3 gm or 2 tab. of 0.6 gm	Plug the borrow with mud to check the escaping gas.
XIII. Biological Insecticides/ Agents				
1. Azadirachtin	Nimbecidin, Margocide, Nimark, Nimazol F, Bioneem Nimbasol Neem Gold Rakshak Sukarina	0.03% W/W 0.15% W/W	500 ml -2.5 lt/ha.	Not harmful to mammals. Effective against Mealybug, white fly, fruit borer etc.
2. Neem oil	RD-9, Ropalin-93 EC	93 EC	1.5%-2.0%	Red scale insect of citrus, mealy bugs.
3. Neem seed Kernel extract	NSKE		4%	Mango mealy bugs.
4. Mahua oil	–	–	1%	Aphid.
5. Crude garlic oil			1%	Mango mealy bug.
6. *Glomus mosseae*	Endomycorrhiza	–	–	Suppresses the Lesion nematode (*Pratylenchus pruni*) multiplication

1	2	3	4	5
7. *Verticillium lecanii*	–		$10^6 \times 10$ spores/ml + 0.005% Quinlphos 10^9 spores/ml + 0.1% Teapol	Green scale of citrus, Mango hopper. *Aphis gossypii* of guava.
8. *Cryptolaemus montrouzieri*	–	–	2500-3000 beetles/ha.	**Maconellicoccus hirsutus** of grape, citrus mealy bug (*Planococuss citri*), green shield scale of guava.
9. *Trickogramma embryophagum*	–	–	2000 adults/tree	Fruit moth of apple.
10. *Aphelinus mali*			1000 adults/infested tree	Wooly apple aphid.
12. *Encarsia perniciosi*			2000 adults/infested tree	San Jose Scale of Apple.
13. *Chilocorus bijugas*			20 adults/infested tree	San Jose Scale of Apple.
14. *C. nigrita*			15 adults/infested tree	Red Scale of Citrus.
15. *Rodolia Cardinalis*	–	–	10 beetles/infested tree	Mealy bug (*Icerya purchasi*) of citrus
16. *Liptomastrix dactylopii*			5000 adults per tree	*Planococcus citri.*
17. *Bacillus thuringiensis* var. *Kurstabi* (Bt)	Bio-lap, Dilep	–	0.5 a.i.	Lemon butterfly
18. *Granulosis* Virus (GV)	–	–	8×10^7 virus capsules/ml	Fruit moth of apple.

1	2	3	4	5
FUNGICIDES **A. Sulphur fungicides** *I. Inorganic sulphur Fungicides* 1. Elemental Sulphur	Sulfex	80 WP	2 gm/lit	Powdery mildew diseases.
II. Organo Sulphur Fungicides 1. Zineb	Hexathane, Dithane Z-78	75 WP	2 g/lit	Rust and Anthracnose
2. Mancozeb	Dithane M-45, Indofil-45	75 WP	2 g/lit	Downy mildew, Anthracnose, blight and fruit rot.
3. Ziram	Ziride	80 WP	2 g/lit	Different diseases of fruits.
4. Ziram	Cuman L.	27LC	4 ml/lit	–do–
5. Thiram	Thiram	50 % powder 80 WP	2-3 g kg., 6 g/lit of water	Seed treatment, Soil drenching for soil borne diseases.
6. Farbam	Hexaferb, Ferberk	50 WP	1 g/lit	Wide variety of fungal diseases.
7. Vapam	Vapam	Stock solution with 33% anhydrous salt.	1 g/lit	Soil borne diseases.
B. Copper Fungicides 1. Bordeux mixture	–	–	5:5:50 : $CuSO_4$ – 4.5 kg + Hydrated lime 4.5 kg + water 450 lit.	Blight, downy mildew, fruit rot.
2. Bordeux paste	–	–	$CuSO_4$:450 g + Hydrated lime : 450 g + Water : 4.5 lit	Used for dressing wounds and cut surfaces.

1	2	3	4	5
3. Chaubatia paste	–	–	Copper carbonate : 800 g + Red lead : 800 g + Lanolin or linseed oil : 1 lit	Used on pruned parts o. temperate fruits.
4. Copper oxychloride (COC)	Blitox-50	50 WP	3 g/lit	Foliar diseases of fruits.
C. Heterocyclic Nitrogen Compound				
1. Captan	Captaf	50 WP 75% Powder	0.5%	Drenching of soil
2. Captafol/Difoltan	Difoltan	80 WP		Collar rot, foot rot and Anthracnose of mango.
D. Others				
Boric acid	–	–	3%	Vegetative seeding material treatment.
E. Other Organic Fungicides	–			
1. Dinocap	Karathane	48 EC	0.5 – 1.0 ml/lit	Powdery mildew diseases.
2. Quintozene	PCNB, Brassicol	10%, 20%, 40% a.i. or 75% WP	0.5 – 1.0 ml/lit	Powdery mildew diseases. Soil treatment or bulb dip.
F. Systemic Fungicides				
1. Benomyl	Benlate	50 WP	1 g/lit	Powdery mildew of apple.
2. Carbendazim	Bavistin	50 WP 60 WP	1 g/lit	Ascomycetes Deuteromycetes and various Basidiomycetes fungi; not effective against phycomycetes.

1	2	3	4	5
3. Thiophanate methyl	Topsin 50 WP	50 WP 70 WP	0.5 g/lit	Broad spectrum fungicide.
G. Morpholines				
Tridimorph	Calixin	75 EC	0.5 ml/lit.	Sigatoka disease of banana.
H. Acyalanine Compounds				
Metalaxyl	Ridomil, Apron SD	WP, granules, SD	2.5 g/lit	Effective against oomycetes fungi, seed treatment and downy mildew of vines
I. Antifungal antibiotic				
Aureofungin	Aureofungin Sol.		0.01%	Canker of apple.
J. Antibacterial antibiotics				
Streptomycin	Agrimycin, Phytomycin, Streptocyclin		0.01%	Fire blight of apple. Citrus Canker.
K Antagonists				
1. *Trichoderma viradae*	Ecofit, Fungi nil, Trichoguard, Defence-SF, Tricho-X, Bioguard, Biocon			Root rot of citrus, Wilt of guava
2. *T. harzianum*	Pant Biocontrol Agent-1			Bunchy rot of grape.

1	2	3	4	5
4.	Oxyflurofen Pre.	4.0	Mango	Just after sowing seeds in the nursery and just after transplanting in orchard.
			Banana	Just after planting suckers.
			Grape	Just after planting in nursery.
5.	Paraquat Post emer.	12.0 6.0 3.0	Mango, Banana Grape Litchi Guava	5-6 months after pre-emergent spray. When weeds become 12-15 cm tall. One spray 2 months prior to rainy season and one after 2 months after rainy season. One spray in rainy season and another in winter.
6.	Glyphosate Post em.	6.0	Mango, Banana, Papaya, Grape	6 months after pre-emergent spray 6 months after pre-emergent spray When weeds become 12-15 cm tall.
7	Fluchloralin PPI & Pre-em.	4.0	Papaya	One week after transplanting.
8.	Butachlor Pre-em.	4.0	Papaya	One week after transplanting.
9.	Alachlor Pre, early post	4.0	Papaya	One week after transplanting.
10.	2,4-D Post-em.	2.0	Guava Peach	One spray in established orchard yearly. After harvesing fruits.
11.	Bromacil Pre-em.	2.25 2.25	Pineapple, Sapota.	Soon after planting of suckers. Spray alongwith Diuron (2.25 kg/ha and Bromasil) Diuron (2.25 kg/ha) + Bromacil to be applied in two-year old orchard.
12.	Chloroxuron Pre. em.	2.0	Strawberry	After planting.

PPI = Pre-Plant incorporation, Pre-em. = Pre-emergent. Post-em. = Post-emergent.

APPENDIX IV. LIST OF BANNED INSECTICIDES

1. Aldrin,
2. Benzene hexachloride (BHC),
3. Calcium cyanide,
4. Chlordane,
5. Copper acetoarsenite,
6. Dibromochloropropane (DBCP),
7. Endrin,
8. Ethyl mercury chloride (EMC),
9. Ethyl parathion,
10. Heptachlor,
11. Menazon,
12. Nicotine sulphate,
13. Nitrofen,
14. Paraquat dimethyl sulphate,
15. Pentachloro nitrobenzene (PCNE),
16. Pentachlorophenol (PCP),
17. Phenyl mercury acetate (PMA),
18. Sodium methane arsonate (MSMA),
19. Tetradifon,
20. Texaphene,
21. Methomyl 12.5% L,
22. Methomyl 24% Formulation,
23. Phosphamidon 85% Sl

Source : Handbook of Horticulture (2001).

□□□